Advances in Analytical Strategies to Study Cultural Heritage Samples

Advances in Analytical Strategies to Study Cultural Heritage Samples

Editors

Maria Luisa Astolfi
Maria Pia Sammartino
Emanuele Dell'Aglio

Basel • Beijing • Wuhan • Barcelona • Belgrade • Novi Sad • Cluj • Manchester

Editors
Maria Luisa Astolfi
Sapienza University of Rome
Rome, Italy

Maria Pia Sammartino
Sapienza University of Rome
Rome, Italy

Emanuele Dell'Aglio
Sapienza University of Rome
Rome, Italy

Editorial Office
MDPI
St. Alban-Anlage 66
4052 Basel, Switzerland

This is a reprint of articles from the Special Issue published online in the open access journal *Molecules* (ISSN 1420-3049) (available at: https://www.mdpi.com/journal/molecules/special_issues/Analytical_Heritage).

For citation purposes, cite each article independently as indicated on the article page online and as indicated below:

Lastname, A.A.; Lastname, B.B. Article Title. *Journal Name* **Year**, *Volume Number*, Page Range.

ISBN 978-3-0365-8993-0 (Hbk)
ISBN 978-3-0365-8992-3 (PDF)
doi.org/10.3390/books978-3-0365-8992-3

© 2023 by the authors. Articles in this book are Open Access and distributed under the Creative Commons Attribution (CC BY) license. The book as a whole is distributed by MDPI under the terms and conditions of the Creative Commons Attribution-NonCommercial-NoDerivs (CC BY-NC-ND) license.

Contents

About the Editors vii

Maria Luisa Astolfi
Advances in Analytical Strategies to Study Cultural Heritage Samples
Reprinted from: *Molecules* **2023**, *28*, 6423, doi:10.3390/molecules28176423 1

Marine Cotte, Victor Gonzalez, Frederik Vanmeert, Letizia Monico, Catherine Dejoie, Manfred Burghammer, et al.
The "Historical Materials BAG": A New Facilitated Access to Synchrotron X-ray Diffraction Analyses for Cultural Heritage Materials at the European Synchrotron Radiation Facility
Reprinted from: *Molecules* **2022**, *27*, 1997, doi:10.3390/molecules27061997 7

Elettra Barberis, Marcello Manfredi, Enrico Ferraris, Raffaella Bianucci and Emilio Marengo
Non-Invasive Paleo-Metabolomics and Paleo-Proteomics Analyses Reveal the Complex Funerary Treatment of the Early 18th Dynasty Dignitary NEBIRI (QV30)
Reprinted from: *Molecules* **2022**, *27*, 7208, doi:10.3390/molecules27217208 29

Ilaria Costantini, Julene Aramendia, Nagore Prieto-Taboada, Gorka Arana, Juan Manuel Madariaga and Juan Francisco Ruiz
Study of Micro-Samples from the Open-Air Rock Art Site of Cueva de la Vieja (Alpera, Albacete, Spain) for Assessing the Performance of a Desalination Treatment
Reprinted from: *Molecules* **2023**, *28*, 5854, doi:10.3390/molecules28155854 41

Francesco Cardellicchio, Sabino Aurelio Bufo, Stefania Mirela Mang, Ippolito Camele, Anna Maria Salvi and Laura Scrano
The Bio-Patina on a Hypogeum Wall of the Matera-Sassi Rupestrian Church "San Pietro Barisano" before and after Treatment with Glycoalkaloids
Reprinted from: *Molecules* **2023**, *28*, 330, doi:10.3390/molecules28010330 59

Rita Reale, Giovanni Battista Andreozzi, Maria Pia Sammartino and Anna Maria Salvi
Analytical Investigation of Iron-Based Stains on Carbonate Stones: Rust Formation, Diffusion Mechanisms, and Speciation
Reprinted from: *Molecules* **2023**, *28*, 1582, doi:10.3390/molecules28041582 75

Lidia Kozak, Andrzej Michałowski, Jedrzej Proch, Michal Krueger, Octavian Munteanu and Przemyslaw Niedzielski
Iron Forms Fe(II) and Fe(III) Determination in Pre-Roman Iron Age Archaeological Pottery as a New Tool in Archaeometry
Reprinted from: *Molecules* **2021**, *26*, 5617, doi:10.3390/molecules26185617 91

Claudia Colantonio, Paola Baldassarri, Pasquale Avino, Maria Luisa Astolfi and Giovanni Visco
Visual and Physical Degradation of the Black and White Mosaic of a *Roman Domus* under Palazzo Valentini in Rome: A Preliminary Study
Reprinted from: *Molecules* **2022**, *27*, 7765, doi:10.3390/molecules27227765 103

Maurizio Aceto, Elisa Calà, Federica Gulino, Francesca Gullo, Maria Labate, Angelo Agostino and Marcello Picollo
The Use of UV-Visible Diffuse Reflectance Spectrophotometry for a Fast, Preliminary Authentication of Gemstones
Reprinted from: *Molecules* **2022**, *27*, 4716, doi:10.3390/molecules27154716 121

Ilaria Serafini, Kathryn Raeburn McClure, Alessandro Ciccola, Flaminia Vincenti, Adele Bosi, Greta Peruzzi, et al.
Inside the History of Italian Coloring Industries: An Investigation of ACNA Dyes through a Novel Analytical Protocol for Synthetic Dye Extraction and Characterization
Reprinted from: *Molecules* 2023, 28, 5331, doi:10.3390/molecules28145331 145

Adele Bosi, Greta Peruzzi, Alessandro Ciccola, Ilaria Serafini, Flaminia Vincenti, Camilla Montesano, et al.
New Advances in Dye Analyses: In Situ Gel-Supported Liquid Extraction from Paint Layers and Textiles for SERS and HPLC-MS/MS Identification
Reprinted from: *Molecules* 2023, 28, 5290, doi:10.3390/molecules28145290 167

Alice Dal Fovo, Marina Martínez-Weinbaum, Mohamed Oujja, Marta Castillejo and Raffaella Fontana
Reflectance Spectroscopy as a Novel Tool for Thickness Measurements of Paint Layers
Reprinted from: *Molecules* 2023, 28, 4683, doi:10.3390/molecules28124683 183

Pablo Aguilar-Rodríguez, Sandra Zetina, Adrián Mejía-González and Nuria Esturau-Escofet
Microanalytical Characterization of an Innovative Modern Mural Painting Technique by SEM-EDS, NMR and Micro-ATR-FTIR among Others
Reprinted from: *Molecules* 2023, 28, 564, doi:10.3390/molecules28020564 201

Andrea Macchia, Lisa Maria Schuberthan, Daniela Ferro, Irene Angela Colasanti, Stefania Montorsi, Chiara Biribicchi, et al.
Analytical Investigations of XIX–XX Century Paints: The Study of Two Vehicles from the Museum for Communications of Frankfurt
Reprinted from: *Molecules* 2023, 28, 2197, doi:10.3390/molecules28052197 215

Ilaria Costantini, Kepa Castro, Juan Manuel Madariaga and Gorka Arana
Analytical Techniques Applied to the Study of Industrial Archaeology Heritage: The Case of *Plaiko Zubixe* Footbridge
Reprinted from: *Molecules* 2022, 27, 3609, doi:10.3390/molecules27113609 239

Zeynep Alp, Alessandro Ciccola, Ilaria Serafini, Alessandro Nucara, Paolo Postorino, Alessandra Gentili, et al.
Photons for Photography: A First Diagnostic Approach to Polaroid Emulsion Transfer on Paper in Paolo Gioli's Artworks
Reprinted from: *Molecules* 2022, 27, 7023, doi:10.3390/molecules27207023 255

Margherita Longoni, Carlotta Beccaria, Letizia Bonizzoni and Silvia Bruni
A Silver Monochrome "Concetto spaziale" by Lucio Fontana: A Spectroscopic Non- and Micro-Invasive Investigation of Materials
Reprinted from: *Molecules* 2022, 27, 4442, doi:10.3390/molecules27144442 271

Marina Creydt and Markus Fischer
Artefact Profiling: Panomics Approaches for Understanding the Materiality of Written Artefacts
Reprinted from: *Molecules* 2023, 28, 4872, doi:10.3390/molecules28124872 283

Anna Filopoulou, Sophia Vlachou and Stamatis C. Boyatzis
Fatty Acids and Their Metal Salts: A Review of Their Infrared Spectra in Light of Their Presence in Cultural Heritage
Reprinted from: *Molecules* 2021, 26, 6005, doi:10.3390/molecules26196005 319

Anna Irto, Giuseppe Micalizzi, Clemente Bretti, Valentina Chiaia, Luigi Mondello and Paola Cardiano
Lipids in Archaeological Pottery: A Review on Their Sampling and Extraction Techniques
Reprinted from: *Molecules* 2022, 27, 3451, doi:10.3390/molecules27113451 345

About the Editors

Maria Luisa Astolfi

Maria Luisa Astolfi is Researcher of Analytical Chemistry at Sapienza University of Rome (Italy), where she received her Master's degree in Chemistry and her PhD degree in Chemical Sciences. Her research interests focus on the development of new analytical methods in environmental, analytical, and bioanalytical fields based on spectroscopic techniques. In particular, she develops innovative approaches for the elemental characterization of environmental and biological complex matrices. Dr. Astolfi's contribution to the field includes more than 79 manuscripts, various congress communications, one patent, and one book chapter, and she has served as a guest editor for seven Special Issues with demonstrable scientific impact (based on Scopus, h-index 24, 80 papers, 1422 citations). She is an Editorial Board member of five scientific journals and has peer reviewed several manuscripts.

Maria Pia Sammartino

Maria Pia Sammartino is a senior researcher in Analytical Chemistry with a PhD in Chemical Sciences. She has over 25 years' experience in leading research in the field of Analytical Chemistry in the application of analytical methods, including modeling studies (with multivariate and chemometric techniques) of physical chemistry on the kinetics and thermodynamics of degradation processes, for the characterization of cultural heritage materials and their degradation products, both for conservation and for microclimatic monitoring of the environments in which they are stored. The objective of her research is the development of innovative and improved analytical procedures compared to those regulated or commonly used. She is author and co-author of more than 95 original research articles in peer-reviewed journals with H-index 24 (Scopus). She has supervised several PhD theses and dissertations, and has served as a reviewer for many scientific articles in international scientific journals.

Emanuele Dell'Aglio

In 2017, Emanuele Dell'Aglio graduated cum laude in Science and Technology in the Conservation of Cultural Heritage at Sapienza University of Rome, Italy. He is currently a freelancer and collaborates with research institutes, *Soprintendenze*, and restorers in diagnostic investigations relating to restoration projects. In particular, he deals with the conservation and restoration of archaeological wood. He has participated in national and international projects on the monitoring of degradation phenomena and the development of alternative products for restoration. He is the author of 15 original research articles in peer-reviewed journals.

Editorial

Advances in Analytical Strategies to Study Cultural Heritage Samples

Maria Luisa Astolfi [1,2]

1. Department of Chemistry, Sapienza University of Rome, Piazzale Aldo Moro 5, 00185 Rome, Italy; marialuisa.astolfi@uniroma1.it
2. Research Center for Applied Sciences to the Safeguard of Environment and Cultural Heritage (CIABC), Sapienza University of Rome, Piazzale Aldo Moro 5, 00185 Rome, Italy

Citation: Astolfi, M.L. Advances in Analytical Strategies to Study Cultural Heritage Samples. *Molecules* **2023**, *28*, 6423. https://doi.org/10.3390/molecules28176423

Received: 22 August 2023
Accepted: 25 August 2023
Published: 4 September 2023

Copyright: © 2023 by the author. Licensee MDPI, Basel, Switzerland. This article is an open access article distributed under the terms and conditions of the Creative Commons Attribution (CC BY) license (https://creativecommons.org/licenses/by/4.0/).

The advancements of civilization are based on our ability to pass on the events and knowledge of the past so that the next generations can start from an ever-higher level of expertise. The memory of the past can be kept alive by preserving historical and artistic assets, the cultural heritage of every nation.

In light of the above, this Special Issue (S.I.) aimed to collect studies that describe interesting/relevant problems in analyzing cultural heritage samples and suggest analytical strategies to solve/manage them. A total of 19 manuscripts were published (16 original research articles and 3 reviews). The submitted papers cover different aspects of cultural and archeological heritage protection, conservation, and restoration.

The sample matrix complexity belonging to cultural heritage, including archeological samples, generally requires a multi-analytical approach. The integrated use of different analytical techniques, with the preference for noninvasive or micro-invasive ones, allows an in-depth understanding of the original materials and their degradation processes and helps obtain innovative solutions for the restoration and conservation of artworks.

The paper by Cotte et al. [1] shows the advantages of techniques based on synchrotron radiation and X-ray powder diffraction (XRPD) to study ancient cultural heritage materials. In particular, high-angular-resolution X-ray powder diffraction (HR-XRPD) allows accurate characterization of crystalline materials, while micro X-ray powder diffraction (µXRPD) mapping provides additional information on the 2D or 3D distribution of crystalline phases at the micrometer scale. The authors outline the potential of these instrumental techniques, specific hardware and software developments to facilitate and speed up data acquisition and processing, and show recent applications to pigments, paints, ceramics, and wood.

Recently, enormous advances have been seen in analytical strategies for studying bioarcheological materials, particularly through mass spectrometry. The work of Barberis et al. [2] offers a focused approach to the chemical characterization of human tissues, embalming compounds, and layering organic ceramic materials. The authors applied a rapid and noninvasive functionalized film method to collect various compounds, ranging from macromolecules such as proteins to small molecules such as organic acids, and perform nontargeted analysis of the human remains of an Egyptian 18th dynasty individual named Nebiri.

The synergistic action of air pollution, climatic conditions, and biological contamination can harm the preservation of historical and artistic heritage. In particular, prehistoric artworks, including paintings and engravings, are particularly fragile as they are exposed to different environmental impacts. With the use of nondestructive elemental and molecular spectroscopies, it is possible to choose the most appropriate conservation and restoration strategy for prehistoric artworks and better understand their evolution over time.

In work from Costantini et al. [3], micro-energy dispersive X-ray fluorescence spectroscopy (µ-EDXRF), Raman spectroscopy, and X-ray diffraction (XRD) enabled the characterization of the degradation product, mineral substrate, and pigments in microsamples belonging to the open-air rock art site of Cueva de la Vieja (Alpera, Albacete, Spain). In addition, an extensive phenomenon of biological activity was also found in almost all samples analyzed.

The degradation of stone materials with the formation of unsightly and chromatic stains may be related to the biological activity of specific microorganisms and promoted by environmental factors such as temperature, humidity, and illumination. The work by Cardellicchio et al. [4] investigates the degradation on a hypogeum wall of the "San Pietro Barisano" rupestrian church, located in the Sassi of Matera (Italy), one of the UNESCO World Heritage Sites. The authors present an innovative and ecological approach using the biocidal action of a mixture of natural glycoalkaloids extracted from the unripe fruit of Solanum nigrum. The disappearance of the biological patina was assessed visually and using X-ray photoelectron spectroscopy (XPS), scanning electron microscopy (SEM), and energy dispersive X-ray spectroscopy (EDS). Bio-patina analyses revealed the bio-calcogenicity of some native microorganisms, which could, once the bio-cleaning phase was completed, be isolated and reintroduced on the wall surface to act as consolidants.

Among the various forms of degradation affecting stone artifacts, colored stains that form on surfaces in contact with metals or alloys are of particular concern. In the work of Reale et al. [5], the mechanisms of rust formation and diffusion in carbonates are defined. The different oxidation states of Fe are studied through the combined use of various analytical techniques, such as optical microscopy (O.M.), SEM-EDS, XPS, and Mössbauer spectroscopy. A better understanding of the composition and evolution of iron-based stains may enable more proper care of archeological artifacts and specific treatments for more effective and safe rust removal.

Studies of Fe speciation in archeological ceramics may be critical to gain insights into the various types of clay used and the different techniques and temperatures of clay firing. The study by Kozak et al. [6] reports the results of Fe speciation in ceramics obtained from archeological sites. UV-Vis molecular absorption spectrophotometry was used to determine Fe(II) and Fe(III). At the same time, the elemental composition was determined by inductively coupled plasma optical emission spectrometry (ICP-OES) and EDXRF spectrometry.

Among the most beautiful expressions of ancient art is the Roman mosaic. The preliminary study by Colantonio et al. [7] suggests a multi-analytical approach using SEM-EDS, XRF, nuclear magnetic resonance (NMR), Fourier-transform infrared spectroscopy (FT-IR), and gas chromatography–mass spectrometry (GC-MS) to assess the causes of chemical degradation of a black-and-white Roman mosaic at Palazzo Valentini near the Roman Forum (Italy). The study's results may improve the knowledge of the Roman mosaic and its construction phases and enable the most appropriate strategies for its conservation to be undertaken.

Another topic addressed in this S.I. is the study and evaluation of the authenticity of precious stones belonging to cultural heritage. The work of Aceto et al. [8] suggests the use of UV-visible diffuse reflectance spectrophotometry with fiber optics (FORS) as a rapid and simple method for the preliminary identification of gemstones. It describes applying chemometric pattern recognition methods for processing large spectral datasets.

Two papers, by Serafini et al. [9] and Bosi et al. [10], focus on the study of dyes. In the former, the authors present and successfully apply an ammonia-based extraction protocol and an innovative purification step based on the use of an ion-pair dispersive liquid-liquid microextraction (IP-dLLME) for the purification and preconcentration of synthetic dyes before high-performance liquid chromatography (HPLC-HRMS) analysis. Synthetic dyes represent a relatively new field in cultural heritage analysis and are becoming increasingly important due to the growing demand for conservation interventions in 19th–20th century art productions. The case study of Serafini et al. [9] illustrates the advantages and need for a multi-analytical approach to analyzing unknown artifacts. The combined use of Raman spectroscopy and mass spectrometry can help interpret spectra and determine compounds in synthetically dyed objects [9].

The work from Bosi et al. [10] highlights the importance of developing analytical methods to preserve integrity and obtain increasingly comprehensive information on unique objects with priceless historical or artistic significance. The authors propose a minimally invasive approach for identifying madder, turmeric, and indigo dyes via hydrogel-supported extraction and subsequent surface-enhanced Raman spectroscopy (SERS) analysis on the gel after the addition of colloidal Ag pastes. In situ gel-supported microextraction was suitable for multi-analytical dye identification in wool and tempera by SERS and HPLC-MS/MS, after appropriate re-extraction and purification of the analytes [10].

Noninvasive cross-sectional analysis of pictorial stratigraphy in paintings, essential for adopting the most appropriate cleaning strategies or assessing adhesion and compactness between layers, is a major challenge in cultural heritage science. To date, no technique can noninvasively and uniquely measure the thickness of pictorial layers, as these are generally composed of heterogeneous and optically opaque materials. In the work of Dal Fovo et al. [11], the possibility of obtaining stratigraphic information on the pictorial layers from their reflectance spectra is explored. The use of a multi-analytical approach with the combination of Raman spectroscopy, laser-induced breakdown spectroscopy (LIBS), optical coherence tomography (OCT), FORS, and Vis-NIR multispectral reflectance imaging on single layers of ten pure acrylic colors revealed the existence of a clear correlation between the spectral response of diffuse reflectance and the micrometer thickness of acrylic paint layers.

In the paper by Aguilar-Rodríguez et al. [12], the combination of different analytical techniques (O.M., SEM-EDS, NMR spectroscopy, attenuated total reflectance (ATR) FTIR, micro-ATR-FTIR, and GC/MS) allowed for the characterization of the composition and painting technique of the mural Paisaje Abstracto painted in 1964 by Rafael Coronel with an unknown synthetic medium on a wood panel support. This study shows an innovative painting technique with poly(methyl methacrylate) (pMMA) as a binder and hypothesizes that a higher pMMA/MMA proportion could affect the mechanical properties and preservation of the mural, producing a rigid pictorial layer that fractures easily.

Studying paints and their components, such as binders, dyes, or pigments, may allow us to acquire useful information to differentiate original from non-original materials and date objects and artifacts. In the work of Macchia et al. [13], the combined use of noninvasive in situ techniques, such as portable O.M. and multispectral imaging, and nondestructive laboratory techniques, namely ATR-FTIR and SEM-EDS, allowed us to obtain information on the compositional nature of the paints of two vehicles in the Museum of Communication in Frankfurt (Germany), which were designed for the German postal and telecommunications service. The study was also helpful in identifying authentic materials to be preserved and areas that needed restoration to preserve the historical and testimonial importance of these vehicles.

The study and preservation of objects, infrastructures, and works belonging to the industrial archeological heritage is a relatively new area of research that emerged in the 1970s. Costantini et al. [14] aimed to identify the original color of the Ondarroa footbridge in Spain with a view to its future restoration. Raman spectroscopy and μ-EDXRF were used to characterize the pictorial layers of the footbridge. Colorimetric analyses were also performed to discover the different pigments used. The authors were also able to highlight the atmospheric impact on the preservation of the rotating bridge due to both the effect of marine aerosol and the presence of acidic compounds in the environment from anthropogenic activities.

Contemporary art is affected by a globally influenced, culturally diverse, and technologically advanced world in which it is increasingly easy to communicate and exchange information. Because of this, contemporary artworks are a dynamic combination of materials, methods, concepts, and subjects. The conservation of contemporary artworks has become a new and exciting field of study, which needs continuous research because of the ever-changing and more recent materials. The work of Alp et al. [15] provides preliminary knowledge about the materials of works by contemporary artist Paolo Gioli. It diagnoses Polaroid emulsion transfers via noninvasive analysis with FORS, Raman, and FTIR spectroscopies.

The work of Longoni et al. [16] confirms the importance of scientific investigations in responding to specific conservation and restoration problems of modern and contemporary artworks. The authors, after an initial visual inspection in visible and ultraviolet (U.V.) light, apply noninvasive and micro-invasive techniques (XRF spectroscopy, FTIR, Raman spectroscopy, and SEM-EDS) to investigate the materials in a silver monochromatic painting by Lucio Fontana, preserved in the exhibition hall of the San Fedele Church in Milan (Italy). The analytical data obtained made it possible to identify the composition of the metallic varnish and the underlying dark layer, both from the point of view of the pigments and binders used.

The review by Creydt and Fischer [17] discusses using omics technologies and various sampling techniques for analyzing written artifacts. The authors show that each omics strategy brings different information allowing better interpretation of data and a significant increase in knowledge of the written heritage.

Two reviews concern lipid residues and their degradation products, such as fatty acids and metal soaps, which are of great importance for the study of archeological objects and oil paintings [18,19]. Organic archeological residues can provide important information about past cultures, such as dietary habits, rituals, and medical practices [19]. In particular, Filopolou et al.'s [18] review explores the possibility of using infrared spectroscopy to investigate and distinguish fatty acids and their metal soaps. The authors show the diagnostic use of the spectroscopic characteristics of some typical fatty monoacids, diacids, and their Ca, Na, and Zn salts.

The review by Irto et al. [19] reports on artificial aging studies to elucidate the mechanisms of lipid degradation in archeological contexts and discusses methodologies for sampling and extracting specific lipid biomarkers in ancient ceramics. The authors stress the need to introduce innovative, miniaturized protocols to reduce the use of chemicals and avoid extractions with organic solvents, which are often laborious and environmentally unfriendly.

Acknowledgments: I would like to thank all the authors who have chosen to report their results in this Special Issue and thank the contributions of the Academic Editors, especially Maria Pia Sammartino and Emanuele Dell'Aglio for their invaluable support also as co-guest editors; all the peer reviewers; and the editorial team members.

Conflicts of Interest: The author declares no conflict of interest.

References

1. Cotte, M.; Gonzalez, V.; Vanmeert, F.; Monico, L.; Dejoie, C.; Burghammer, M.; Huder, L.; de Nolf, W.; Fisher, S.; Fazlic, I.; et al. The "Historical Materials BAG": A New Facilitated Access to Synchrotron X-ray Diffraction Analyses for Cultural Heritage Materials at the European Synchrotron Radiation Facility. *Molecules* **2022**, *27*, 1997. [CrossRef] [PubMed]
2. Barberis, E.; Manfredi, M.; Ferraris, E.; Bianucci, R.; Marengo, E. Non-Invasive Paleo-Metabolomics and Paleo-Proteomics Analyses Reveal the Complex Funerary Treatment of the Early 18th Dynasty Dignitary NEBIRI (QV30). *Molecules* **2022**, *27*, 7208. [CrossRef]
3. Costantini, I.; Aramendia, J.; Prieto-Taboada, N.; Arana, G.; Madariaga, J.M.; Ruiz, J.F. Study of Micro-Samples from the Open-Air Rock Art Site of Cueva de la Vieja (Alpera, Albacete, Spain) for Assessing the Performance of a Desalination Treatment. *Molecules* **2023**, *28*, 5854. [CrossRef] [PubMed]
4. Cardellicchio, F.; Bufo, S.A.; Mang, S.M.; Camele, I.; Salvi, A.M.; Scrano, L. The Bio-Patina on a Hypogeum Wall of the Matera-Sassi Rupestrian Church "San Pietro Barisano" before and after Treatment with Glycoalkaloids. *Molecules* **2023**, *28*, 330. [CrossRef] [PubMed]
5. Reale, R.; Andreozzi, G.B.; Sammartino, M.P.; Salvi, A.M. Analytical Investigation of Iron-Based Stains on Carbonate Stones: Rust Formation, Diffusion Mechanisms, and Speciation. *Molecules* **2023**, *28*, 1582. [CrossRef]
6. Kozak, L.; Michałowski, A.; Proch, J.; Krueger, M.; Munteanu, O.; Niedzielski, P. Iron Forms Fe(II) and Fe(III) Determination in Pre-Roman Iron Age Archaeological Pottery as a New Tool in Archaeometry. *Molecules* **2021**, *26*, 5617. [CrossRef]
7. Colantonio, C.; Baldassarri, P.; Avino, P.; Astolfi, M.L.; Visco, G. Visual and Physical Degradation of the Black and White Mosaic of a Roman Domus under Palazzo Valentini in Rome: A Preliminary Study. *Molecules* **2022**, *27*, 7765. [CrossRef]

8. Aceto, M.; Calà, E.; Gulino, F.; Gullo, F.; Labate, M.; Agostino, A.; Picollo, M. The Use of UV-Visible Diffuse Reflectance Spectrophotometry for a Fast, Preliminary Authentication of Gemstones. *Molecules* **2022**, *27*, 4716. [CrossRef]
9. Serafini, I.; McClure, K.R.; Ciccola, A.; Vincenti, F.; Bosi, A.; Peruzzi, G.; Montesano, C.; Sergi, M.; Favero, G.; Curini, R. Inside the History of Italian Coloring Industries: An Investigation of ACNA Dyes through a Novel Analytical Protocol for Synthetic Dye Extraction and Characterization. *Molecules* **2023**, *28*, 5331. [CrossRef] [PubMed]
10. Bosi, A.; Peruzzi, G.; Ciccola, A.; Serafini, I.; Vincenti, F.; Montesano, C.; Postorino, P.; Sergi, M.; Favero, G.; Curini, R. New Advances in Dye Analyses: In Situ Gel-Supported Liquid Extraction from Paint Layers and Textiles for SERS and HPLC-MS/MS Identification. *Molecules* **2023**, *28*, 5290. [CrossRef] [PubMed]
11. Dal Fovo, A.; Martínez-Weinbaum, M.; Oujja, M.; Castillejo, M.; Fontana, R. Reflectance Spectroscopy as a Novel Tool for Thickness Measurements of Paint Layers. *Molecules* **2023**, *28*, 4683. [CrossRef]
12. Aguilar-Rodríguez, P.; Zetina, S.; Mejía-González, A.; Esturau-Escofet, N. Microanalytical Characterization of an Innovative Modern Mural Painting Technique by SEM-EDS, NMR and Micro-ATR-FTIR among Others. *Molecules* **2023**, *28*, 564. [CrossRef] [PubMed]
13. Macchia, A.; Schuberthan, L.M.; Ferro, D.; Colasanti, I.A.; Montorsi, S.; Biribicchi, C.; Barbaccia, F.I.; La Russa, M.F. Analytical Investigations of XIX–XX Century Paints: The Study of Two Vehicles from the Museum for Communications of Frankfurt. *Molecules* **2023**, *28*, 2197. [CrossRef] [PubMed]
14. Costantini, I.; Castro, K.; Madariaga, J.M.; Arana, G. Analytical Techniques Applied to the Study of Industrial Archaeology Heritage: The Case of Plaiko Zubixe Footbridge. *Molecules* **2022**, *27*, 3609. [CrossRef] [PubMed]
15. Alp, Z.; Ciccola, A.; Serafini, I.; Nucara, A.; Postorino, P.; Gentili, A.; Curini, R.; Favero, G. Photons for Photography: A First Diagnostic Approach to Polaroid Emulsion Transfer on Paper in Paolo Gioli's Artworks. *Molecules* **2022**, *27*, 7023. [CrossRef] [PubMed]
16. Longoni, M.; Beccaria, C.; Bonizzoni, L.; Bruni, S. A Silver Monochrome "Concetto spaziale" by Lucio Fontana: A Spectroscopic Non- and Micro-Invasive Investigation of Materials. *Molecules* **2022**, *27*, 4442. [CrossRef] [PubMed]
17. Creydt, M.; Fischer, M. Artefact Profiling: Panomics Approaches for Understanding the Materiality of Written Artefacts. *Molecules* **2023**, *28*, 4872. [CrossRef] [PubMed]
18. Filopoulou, A.; Vlachou, S.; Boyatzis, S.C. Fatty Acids and Their Metal Salts: A Review of Their Infrared Spectra in Light of Their Presence in Cultural Heritage. *Molecules* **2021**, *26*, 6005. [CrossRef] [PubMed]
19. Irto, A.; Micalizzi, G.; Bretti, C.; Chiaia, V.; Mondello, L.; Cardiano, P. Lipids in Archaeological Pottery: A Review on Their Sampling and Extraction Techniques. *Molecules* **2022**, *27*, 3451. [CrossRef] [PubMed]

Disclaimer/Publisher's Note: The statements, opinions and data contained in all publications are solely those of the individual author(s) and contributor(s) and not of MDPI and/or the editor(s). MDPI and/or the editor(s) disclaim responsibility for any injury to people or property resulting from any ideas, methods, instructions or products referred to in the content.

Article

The "Historical Materials BAG": A New Facilitated Access to Synchrotron X-ray Diffraction Analyses for Cultural Heritage Materials at the European Synchrotron Radiation Facility

Marine Cotte [1,2,*], Victor Gonzalez [3,*], Frederik Vanmeert [4,5,*], Letizia Monico [4,6,7,*], Catherine Dejoie [1], Manfred Burghammer [1], Loïc Huder [1], Wout de Nolf [1], Stuart Fisher [1], Ida Fazlic [1,8], Christelle Chauffeton [9,10,11], Gilles Wallez [9,11,12], Núria Jiménez [13], Francesc Albert-Tortosa [13], Nati Salvadó [13], Elena Possenti [14], Chiara Colombo [14], Marta Ghirardello [15], Daniela Comelli [15], Ermanno Avranovich Clerici [4,16], Riccardo Vivani [17], Aldo Romani [6,7], Claudio Costantino [6,7], Koen Janssens [4,8], Yoko Taniguchi [18], Joanne McCarthy [1], Harald Reichert [1] and Jean Susini [1,†]

1 European Synchrotron Radiation Facility, 71 Avenue des Martyrs, 38000 Grenoble, France; catherine.dejoie@esrf.fr (C.D.); burgham@esrf.fr (M.B.); loic.huder@esrf.fr (L.H.); wout.de_nolf@esrf.fr (W.d.N.); stuart.fisher@esrf.fr (S.F.); ida.fazlic@esrf.fr (I.F.); mccarthy@esrf.fr (J.M.); reichert@esrf.fr (H.R.); jean.susini@synchrotron-soleil.fr
2 Laboratoire d'Archéologie Moléculaire et Structurale (LAMS) CNRS UMR 8220, UPMC Univ Paris 06, Sorbonne Université, 5 place Jussieu, 75005 Paris, France
3 Université Paris-Saclay, ENS Paris-Saclay, CNRS, PPSM, 91190 Gif-sur-Yvette, France
4 Antwerp X-ray Imaging and Spectroscopy laboratory (AXIS) Research Group, NANOLab Centre of Excellence, University of Antwerp, Groenenborgerlaan 171, 2020 Antwerp, Belgium; ermanno.avranovichclerici@uantwerpen.be (E.A.C.); koen.janssens@uantwerpen.be (K.J.)
5 Paintings Laboratory, Royal Institute for Cultural Heritage (KIK-IRPA), Jubelpark 1, 1000 Brussels, Belgium
6 CNR-SCITEC, c/o Department of Chemistry, Biology and Biotechnology, University of Perugia, Via Elce di Sotto 8, 06123 Perugia, Italy; aldo.romani@unipg.it (A.R.); claudio.costantino@studenti.unipg.it (C.C.)
7 Centre of Excellence SMAArt and Department of Chemistry, Biology and Biotechnology, University of Perugia, Via Elce di Sotto 8, 06123 Perugia, Italy
8 Rijksmuseum, Conservation and Restoration, P.O. Box 74888, 1070 DN Amsterdam, The Netherlands
9 Chimie ParisTech, PSL University, CNRS, Institut de Recherche de Chimie Paris, 11 rue Pierre et Marie Curie, 75005 Paris, France; c.chauffeton@chimieparistech.psl.eu (C.C.); gilles.wallez@sorbonne-universite.fr (G.W.)
10 Cité de la Céramique Sèvres-Limoges, place de la Manufacture, 92310 Sèvres, France
11 Centre de Recherche et Restauration des Musées de France (C2RMF), Porte des Lions, 14 quai François Mitterrand, 75001 Paris, France
12 UFR 926, Sorbonne Université, 75005 Paris, France
13 Departament d'Enginyeria Química EPSEVG, Universitat Politècnica de Catalunya (UPC)·BarcelonaTech Av. Víctor Balaguer s/n, 08800 Vilanova i la Geltrú, Spain; nuria.jimenez.garcia@upc.edu (N.J.); francesc.albert.tortosa@upc.edu (F.A.-T.); nativitat.salvado@upc.edu (N.S.)
14 Institute of Heritage Science, National Research Council, ISPC-CNR, Via R. Cozzi 53, 20125 Milan, Italy; elena.possenti@cnr.it (E.P.); chiara.colombo@cnr.it (C.C.)
15 Politecnico di Milano, Physics Department, Piazza Leonardo da Vinci 32, 20133 Milano, Italy; marta.ghirardello@polimi.it (M.G.); daniela.comelli@polimi.it (D.C.)
16 Department of Materials Science and Engineering, 3mE, Delft University of Technology, Mekelweg 2, 2628 CD Delft, The Netherlands
17 Pharmaceutical Science Department, University of Perugia, Via del Liceo 1, 06123 Perugia, Italy; riccardo.vivani@unipg.it
18 History and Anthropology, Faculty of Humanities and Social Sciences, University of Tsukuba, 1-1-1 Tennodai, Tsukuba 305-8577, Japan; taniguchi.yoko.fu@u.tsukuba.ac.jp
* Correspondence: cotte@esrf.fr (M.C.); victor.gonzalez@ens-paris-saclay.fr (V.G.); frederik.vanmeert@uantwerpen.be (F.V.); letizia.monico@cnr.it (L.M.)
† Current address: Synchrotron SOLEIL, L'Orme des Merisiers, 91190 Saint-Aubin, France.

Citation: Cotte, M.; Gonzalez, V.; Vanmeert, F.; Monico, L.; Dejoie, C.; Burghammer, M.; Huder, L.; de Nolf, W.; Fisher, S.; Fazlic, I.; et al. The "Historical Materials BAG": A New Facilitated Access to Synchrotron X-ray Diffraction Analyses for Cultural Heritage Materials at the European Synchrotron Radiation Facility. *Molecules* **2022**, *27*, 1997. https://doi.org/10.3390/molecules27061997

Academic Editors: Maria Luisa Astolfi, Maria Pia Sammartino and Emanuele Dell'Aglio

Received: 18 February 2022
Accepted: 17 March 2022
Published: 20 March 2022

Publisher's Note: MDPI stays neutral with regard to jurisdictional claims in published maps and institutional affiliations.

Copyright: © 2022 by the authors. Licensee MDPI, Basel, Switzerland. This article is an open access article distributed under the terms and conditions of the Creative Commons Attribution (CC BY) license (https:// creativecommons.org/licenses/by/ 4.0/).

Abstract: The European Synchrotron Radiation Facility (ESRF) has recently commissioned the new Extremely Brilliant Source (EBS). The gain in brightness as well as the continuous development of beamline instruments boosts the beamline performances, in particular in terms of accelerated data acquisition. This has motivated the development of new access modes as an alternative to standard proposals for access to beamtime, in particular via the "block allocation group" (BAG) mode. Here, we present the recently implemented "historical materials BAG": a community proposal giving to 10 European institutes the opportunity for guaranteed beamtime at two X-ray powder diffraction (XRPD) beamlines—ID13, for 2D high lateral resolution XRPD mapping, and ID22 for high angular resolution XRPD bulk analyses—with a particular focus on applications to cultural heritage. The capabilities offered by these instruments, the specific hardware and software developments to facilitate and speed-up data acquisition and data processing are detailed, and the first results from this new access are illustrated with recent applications to pigments, paintings, ceramics and wood.

Keywords: synchrotron; X-ray diffraction; cultural heritage; beamtime access; paintings; pigments; ceramics; artistic; crystallography; structural analyses

1. Introduction

Synchrotron radiation facilities are increasingly used to study ancient materials from cultural heritage [1–3]. Assets of synchrotron radiation (SR)-based techniques are many, in particular including the following: (i) the beam brightness, which offers small probes (down to tens of nanometers) combined with high beam intensity (>10^{12} photons per seconds) and consequently high acquisition speed and high data quality; (ii) the energy tunability, a key property for absorption spectroscopy techniques, from X-ray to UV-vis and infrared range. Many techniques such as X-ray fluorescence (XRF), X-ray powder diffraction (XRPD), X-ray absorption spectroscopy (XAS), and phase contrast tomography are now commonly used for the study of art history and for the knowledge and conservation of our cultural and natural heritage [2]. More specifically, the possibility to use these methods with mapping capabilities, sub-micrometric resolution and high sensitivity makes them highly suitable for the multi-modal micro-analysis of highly heterogeneous and multi-layered tiny fragments from artifacts [1].

The main assets of XRPD-based techniques, as compared to the other techniques listed above, is not only the possibility to identify crystalline phases within complex mixtures, but also to obtain information regarding their crystallite size, orientation, microstrain and their crystal structure. Several XRPD-based set-ups and beamlines are available and complement each other. They differ in their technical characteristics, namely maximum sample size and weight, sample environment, beam size range, X-ray energy range, X-ray spectral bandwidth, detection modalities, acquisition speed, level of automation, versatility, etc. Notably, two complementary XRPD-based techniques are increasingly used by the cultural heritage community: high angular resolution X-ray powder diffraction (HR-XRPD) and micro X-ray powder diffraction (µXRPD) mapping. The former is key for the accurate characterization of crystalline materials (e.g., identification and quantification, structure refinement and crystallite size determination, description of the crystallographic structure). The latter yields more qualitative information but provides additional insight into the 2D or 3D distribution of crystalline phases at the micrometre scale. When applied to historical materials, these analyses can give clues about both the early life of the materials (creation) and later phases of the life of the materials (degradation, past and present conservation interventions) [4,5]. The same pigment can exist in different compositions, as a consequence of different synthesis procedures (by varying the nature and ratio of starting reagents, pH, etc.) or refinement steps. This can translate into different pigment grades, colors, prices, stability, etc. [4,6]. Identifying precisely such formulations can give insight into the decision of an artist to choose one or another quality of the same pigment or its availability in a specific time period or region. As an example, lead white, a ubiquitous pigment used since antiquity for paintings, is usually composed of a mixture of cerussite ($PbCO_3$) and hydrocerussite ($Pb_3(CO_3)_2(OH)_2$). Different ratios of these compounds and different crystallite sizes can be associated with different post-synthesis treatments of the white powder obtained after lead corrosion following the traditional production of lead white [7]. This information was used to explain the presence of different lead whites in

Leonardo da Vinci's *Virgin and Child with St. Anne* [8]. Determining precisely the pigment composition is also very important for conservation purposes. As an example, the various chrome yellows (PbCr$_{1-x}$S$_x$O$_4$, with $0 \leq x \leq 0.8$) used by Van Gogh in *the Sunflowers* have different yellow-orange hues, related to their different chemical composition and crystalline structure, each with their own photochemical stability [6,9–11]). In general, in addition to crystalline phase identification, determining their distribution within a multi-layer system can reveal the way the artist applied artistic materials (e.g., in a painting process), but it can also highlight materials which were not in the original artwork, for example materials added during later modifications of the artwork (by the artist, by conservators), or materials formed or deposited during degradation processes.

At synchrotron radiation facilities, continuous efforts are dedicated not only to optimize techniques in terms of speed, lateral resolution, sensitivity, but also to implement new techniques. This usually relies on the implementation of new optics, mechanics, electronics, detectors, and less often on the upgrade of the synchrotron source itself. The European Synchrotron Radiation Facility (ESRF, Grenoble, France) has been benefitting from a major upgrade program (EBS—Extremely Brilliant Source), its main component being a revolutionary new electron storage ring concept that increases the brilliance and coherence of the X-ray beams produced by a factor of 100 (brilliance up to some 10^{22} ph/s/0.1%/mm^2/mrad^2). The full exploitation of the ESRF-EBS calls for new paradigms in order to address new long-term sustainability challenges related to the unprecedented X-ray beam properties. By way of example, some experiments which would have typically required several hours per sample can be carried out now in a few minutes. However, the time spent in writing a two-page proposal, completing its technical and scientific review, in planning the experiment, in setting-up the beamline, etc., stays fundamentally the same. It was therefore essential to develop new access routes to ESRF beamlines. In the context of the H2020 European project STREAMLINE, new access modes are in development at the ESRF, inspired by the success of the block allocation group (BAG) system to schedule beamtime used in structural biology for several years (https://www.esrf.fr/CommunityAccess (accessed on 1 February 2022)). This entails grouping together user experiments requiring the same beamline set-up in a single project rather than having many users submitting individual projects for one or two shifts (e.g., 8 h of beamtime). This not only saves time in assessing proposals and setting-up the beamlines, but also maintains users' flexibility in the choice of projects and samples, and fosters collaborations and synergy within user communities. Ten European institutes (Rijksmuseum, Amsterdam, The Netherlands; TU Delft, The Netherlands; CNR-SCITEC, Perugia; Courtauld Institute of Art, London, UK; Politecnico di Milano; Centre de Recherche et de Restauration des Musées de France, Paris, France; Institut de Recherche de Chimie de Paris, Paris, France; Universitat Politècnica de Catalunya, Barcelona, Spain; University of Antwerp, Belgium and the ESRF, Grenoble, France) have proposed such a (Heritage) BAG for structural investigations of historical materials. Within the Historical Materials BAG, different projects are grouped together that all require structural information obtainable by X-ray powder diffraction at the ESRF, either through µXRPD/µXRF mapping at ID13 or HR-XRPD at ID22. Through the Heritage BAG, regular access to ID13 and ID22 (once every six months) is provided for a 2-year period (2021–2023) to the partners, which is renewable upon request and reviewed at the end of the 2-year period.

Below we present the experimental set-ups offered through the BAG, the hardware and software developments implemented in the context of the BAG, and some recent applications. Most of the applications are related to pigments and paintings; however, the instruments can be exploited to analyze any artistic materials, as illustrated with the examples of enamels and wood.

2. High-Lateral Resolution 2D X-ray Diffraction Mapping at ID13

2.1. Beamline Description

ID13 is an ESRF undulator beamline dedicated to high-lateral-resolution diffraction and scattering experiments using focused monochromatic X-ray beams [12]. Two endstations, a micro-branch (beam size ~2 × 2 µm^2) and a nano-branch (beam size down to 100 nm), are operated in time-sharing mode. For the BAG, the microbranch was preferred for different reasons: (i) a larger beam, to better fulfil powder diffraction conditions over single-crystal-like diffraction when the beam size is smaller than the crystallite size; (ii) a large range of sample stage scanning motors, allowing to mount large samples (used for example to analyze centimetric papyrus fragments) or large sample holders (see next section). Samples are mounted vertically, perpendicular to the X-ray beam. The energy of the incident beam is chosen around 13.0 keV, typically in the pre-edge region of Pb, a ubiquitous element in paintings. The energy is usually chosen to slightly excite Pb L$_3$-edge XRF, but without saturating the XRF detector, nor attenuating too much the transmission of the beam through the material, and consequently the XRPD intensity. The beam is focused to ~2 × 2 µm^2 (flux ~2 × 10^{12} ph/s, at I = 128 mA electron beam current) using a compound refractive lens set-up (CRL) mounted in a transfocator. For the study of beam-sensitive samples, the flux can be reduced (typically by a factor of 10) by detuning the gap of the undulator. XRPD maps are obtained by raster-scanning the samples and collecting 2D XRPD patterns, in transmission, with a Dectris EIGER 4 M single photon counting detector that acquires frames with 2070 × 2167 pixels (75 × 75 µm^2 pixel size) at a rate up to 750 Hz. A dwell time of 10 ms is usually sufficient to detect most of the crystalline phases, with a reduced risk of beam damage (see below). XRF spectra are collected simultaneously with XRPD patterns, using a Vortex EM detector and XIA readout electronics. A detailed description of the set-up is shown in Figure S1, Supplementary Materials.

2.2. Sample Preparation and Mounting

For optical and electron microscopic observations, historical materials are regularly prepared as transversal cross-sections. Such preparations are compatible with the ID13 set-up if the sample dimension, composition and density in the direction perpendicular to the cross-section surface allow for sufficient transmission of X-rays at ~13 keV. The resin blocks should be resized to the minimum dimensions in three directions, in order to reduce the X-ray absorption by the resin and the space necessary to mount each sample on the sample holders. Users are strongly encouraged, whenever possible, to prepare thin sections, which offer a much better control of sample thickness and consequently of X-ray transmission. Using thin sections allows having a controlled and homogenous probed voxel over the 2D surface [13]. For this purpose, the ID21 microtome is regularly used to prepare thin sections of ~5–10 µm. In the case of paint mock-ups applied on polycarbonate sheets, sectioning can be performed without any prior embedding of the samples [14]. If such sections are sufficiently large (>~1 mm) or in the case of a slice from a resin-embedded sample, slices can be glued on two edges, keeping the 2D analysis region completely free of any mounting material (Figure S2, right). Alternatively, and more particularly in the case of historical small and precious samples, and/or when the fragments are fragile and prompt to break when sliced, a piece of tape can be deposited on the surface of the cross-section to maintain the paint structure during sectioning. This procedure as well as the preparation of thin sections from existing and precious historical cross-sections are described in detail in the supporting information of [15].

For materials which cannot be sliced with a microtome (e.g., glass, ceramics), a procedure consisting of double-side polishing is recommended (for further details see supporting information of [16]).

Cross-sections and thin sections are then mounted on specific sample holders (Figure S2). This step is crucial to the success of the experiment. Indeed, because of the design of the ID13 microscope, which is equipped downstream with an on-axis optical microscope, changing the sample set requires a long (~30 min) procedure (see details in supporting information). The success of the BAG relies upon the user mounting as many samples as possible on the same sample holder to reduce the set-up time with respect to data acquisition time. To benefit from the higher speed of the horizontally scanning motor, the direction where the sample is most heterogeneous is usually oriented horizontally.

2.3. Data Acquisition

While the ID13 set-up is controlled by BLISS commands, all the steps related to samples (navigation, focus and selection of regions and points of interest (ROIs and POIs), data acquisition) can be performed easily by non-expert users thanks to the graphical user interface (GUI) Daiquiri (cf. Figure S3). It was primarily developed for ID21 [17] and was recently deployed to ID13, and successfully commissioned and used for the BAG project. This GUI is another key element for the success of the BAG. In only a few minutes, any untrained user can perform the following:

(1) Define the sample name (which will automatically define the structure of data saving, with one folder per sample), and possibly add comments about the sample;
(2) Navigate on the sample holder by clicking on plus/minus steps on the sample stage motors or by clicking directly on the video image. A mosaic photograph of the entire sample holder (collection of optical images taken while raster-scanning the sample holder over 2D large regions) permits the user to observe, grab and queue positions of interest for all the samples at once;
(3) Define a ROI over the 2D area to be scanned. A unique number is associated with each ROI, which will be used in data naming;
(4) Select the conditions for each map (pixel size, dwell time, detector(s), low/high flux (LF/HF) to mitigate beam damage, see below), position of the XRF detector (to mitigate detector saturation);
(5) As an alternative to standard XRPD mapping, a single-point acquisition mode has been implemented for the purpose of beam damage studies. This mode allows the selection of POIs and the repeated acquisition of thousands of XRPD patterns at unique positions, to monitor the evolution in the XRPD patterns (peak position, intensity and width) as a function of accumulated dose;
(6) Build a queue of all the above ROIs and POIs scans and organize them along a priority list.
(7) Once the experimental set-up is back to data acquisition mode (see supporting information), the queue can be easily launched and continuously indicates the on-going and remaining scans.
(8) Data produced during each scan is given a unique and automatic identifier and a proper place within the experiment folder.
(9) Daiquiri also offers the possibility to visualize results in real time, such as XRF emission or XRPD intensity over a pre-set range of channels or angles, respectively. These images can be displayed, superimposed on the registered optical light image, providing in real time first diagnostics about the sample composition.

The so-called ICAT tools implemented for the ESRF data policy strategy (https://www.esrf.fr/datapolicy (accessed on 1 February 2022)) offer four panels: (i) the dataset list (with metadata of each sample and tools for downloading data); (ii) the electronic logbook (automatically filled with BLISS/Daiquiri command lines and error messages, but also appended by users); (iii) shipping information for remote experiments; (iv) information on and management of experiment participants.

Last but not least, the remote access mode, Guacamole, implemented at the ESRF as a remedy to the limited access to the facility imposed by COVID-19, makes all these steps easily available to anyone worldwide. This is an important step for the cultural heritage community, giving a chance to non-expert end-users (conservators, art historians, archaeologists, etc.) to join remotely the on-going experiments and use the X-ray beamline as easily as they would use an optical microscope at home.

2.4. Data Processing and Data Analysis

Data are produced as .h5 files, following the NeXuS convention and the ESRF data policy. The 2D XRPD patterns are azimuthally integrated using dedicated Jupyter notebooks, based on the PyFAI software package [18]. The processing notebooks are open-source and freely available at https://gitlab.esrf.fr/loic.huder/juno (accessed on 1 February 2022). The first two notebooks (0 and 1) are used for the calibration of the set-up. The next two (2a and 2b) deal with the azimuthal integration of 2D maps and series of repeated scans, respectively. After having defined the sample and dataset names as well as the integration parameters, any non-expert user can run the notebook through the Jupyter interface. The plots of the average integrated XRPD pattern, and the map (or series) of the integrated XRPD intensity shown in the notebooks offer a primary diagnostic of data quality and processing. Integrated data is saved as .h5 files and can be converted to .edf format by the notebook.

For the analysis of XRPD maps, the XRDUA software offers the most advanced tools for a precise, quantitative Rietveld refinement-based fit of data in .edf format [19]. Alternatively, the PyMca ROI imaging software [20] offers tools for the calculation of map intensities over 2-theta/q regions of interest, for the creation of average XRPD patterns over a selection of pixels, or for performing principal component analyses, non-negative matrix approximation calculations, etc. PyMca also offers the possibility to batch-fit XRF data, as well as the combined analysis of XRPD and XRF data in master/slave panels. Finally, a third Jupyter notebook offers tools for fast fitting of the .h5 files, as a linear combination of a set of reference patterns. The development of machine-learning-based tools for the analysis of μXRPD maps is subject of an ongoing PhD project, to further improve this time-consuming step.

2.5. Assessment of Radiation Damage

The risk of radiation damage must not be neglected, even more so in the context of the higher flux offered by EBS and for the study of hybrid and humid materials such as paint samples. Various actions have been initiated to assess, understand and mitigate radiation damage when analyzing cultural heritage materials [21,22]. Radiation damage manifests itself in at least two effects: (i) the modification of the sample, in particular its optical aspect but also its composition, which consequently leads to erroneous results in future chemical analyses; (ii) the alteration of the measured XRPD data (formation or disappearance of XRPD peaks, peak broadening). Radiation damage can be mitigated and prevented by reducing the X-ray dose hitting the sample, in particular by decreasing the flux and/or the dwell time [22]. In order to define the safety thresholds and establish safe procedures for the analyses of (historical) paint fragments in the context of the BAG, dedicated experiments were carried out on model samples to evaluate the two above points. Since paint samples constitute the majority of the materials under investigation within the BAG, radiation damage studies were carried out on a series of model oil paints prepared with different pigments. As an example, Figure 1 shows selected results obtained on a lead white pigment (mostly hydrocerussite). Some results obtained on chrome yellow pigments are presented as well in the supporting information. To assess and understand the beam damage, optical microscopy images and Fourier-transform infrared (FTIR) spectro-microscopy maps were recorded at the ID21 beamline prior and posterior to a series of ID13 μXRPD maps, acquired at different flux, different dwell time and as single or triple acquisitions (see details in the supporting information). XRPD patterns were also collected repeatedly at POIs, in different flux conditions, to monitor their evolution over time. The results show that with increasing flux and/or dwell time a yellowing of the sample is observed (Figure 1a). Furthermore, modification of the FTIR signal, notably in the C-H and C=O stretching mode domains, is seen, in particular with a decrease in C=O ester and the formation of C=O acid peaks (Figure 1b,c). After repeated XRPD acquisitions at the same point, a shift in position of the peaks, a decrease in the total diffracted intensity and a broadening of the diffraction peaks are observed. These changes are associated with amorphization leading to a loss of crystalline order in the material, thus reducing the intensity of the Bragg diffraction (Figure S4). Additionally, new peaks appear and can be partially attributed to the formation of metallic lead, Pb(0), through the reduction of Pb(II) (Figure 1d). Nevertheless, employing low flux (flux ~10^{11} ph/s), and short dwell time (10 ms), which were the conditions preferred for the analyses of painting fragments, the yellowing of the paint is not noticeable, and no modification is detected in the FTIR nor XRPD data (map 1 in Figure 1a–c, and scan 1 in Figure 1d) guaranteeing the validity of the measurement. The complete set of results of this study will be detailed in a forthcoming publication.

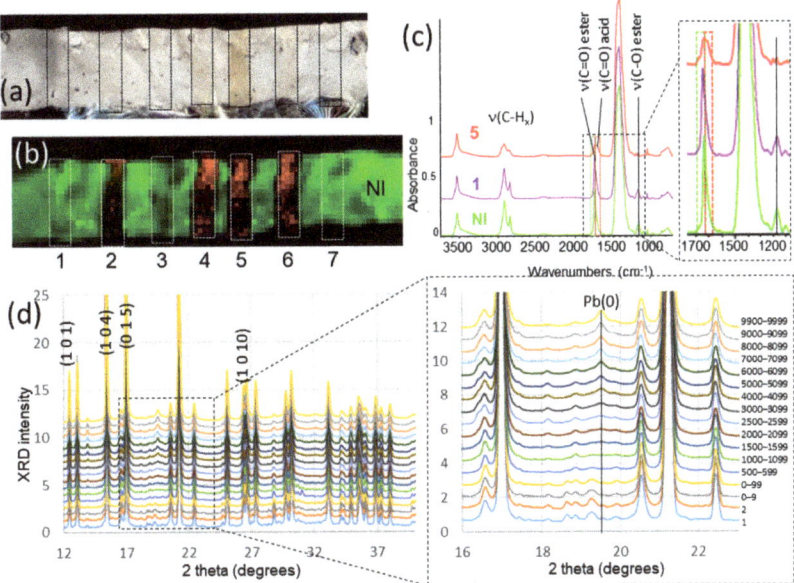

Figure 1. Assessment of radiation damage on a lead-white (mostly hydrocerussite) oil paint mock-up. Seven µXRPD maps were acquired at ID13, with the following conditions (1 or 3 repeats; different dwell times; at low flux (LF) or high flux (HF)): 1: 1 scan, 0.01 s, LF; 2: 1 scan, 0.01 s, HF; 3: 1 scan, 0.03 s, LF; 4: 1 scan, 0.03 s, HF; 5: 1 scan, 0.1 s, HF; 6: 3 scans, 0.01 s for each scan, HF; 7: 3 scans, 0.01 s for each scan, LF. The pixel size was 2.5×2.5 µm^2 to avoid overlap between two consecutive points. The map width was 50 µm and the height sufficient to cover the entire thickness of the paint sample. The maps are represented as rectangles in (**a**,**b**). (**a**) Optical microscopy after µXRPD. (**b**) µFTIR map acquired in transmission mode (beam size 15×15 µm^2, pixel size 10×10 µm^2, 50 cumulated scans per spectrum) after performing the µXRPD maps. The red/green display shows the integrated intensity over the ν(CO) acid range (1683–1724 cm^{-1}) and ν(CO) ester range (1726–1759 cm^{-1}), respectively. These regions are displayed in (**c**) by a red and green rectangle, respectively. (**c**) Average FTIR spectra calculated over map 1, map 5 and a non-irradiated (NI) region. (**d**) XRPD 1D patterns measured with repeated acquisitions of 10 ms at LF; from bottom to top: 1st scan, 2nd scan, average of the first 10 scans, of the first 100 scans, and then average from x to x + 99, for x = 500, 1500, 2000, 2500, 3000, 4000, 5000, 6000, 7000, 8000, 9000, and 9900. The peaks assigned with Miller index (h k l) values are those for which the evolution of the position and full-width-at-half-maximum (FWHM) is reported in Figure S4.

3. High-Angular Resolution X-ray Diffraction at ID22

3.1. Beamline Description

Pioneering studies using HR-XRPD for the study of cultural heritage materials were performed at the former BM16 beamline at the end of the 1990s [23], paving the way for the exploitation of synchrotron radiation in Heritage Science. In the field of cultural heritage, HR-XRPD is used for phase identification and quantification, microstructure characterization and, in some cases, complete structure determination [24,25]. Fast screening of a series of samples can also be carried out, a convenient way to look for example at modern reproductions obtained in the laboratory in controlled atmosphere (e.g., humidity, temperature, for aging/long-term degradation studies).

The ID22 beamline (formerly BM16 and ID31) combines a continuous range of incident energies (from 6 to 80 keV) with high brightness, also enhanced by the new EBS source, thus offering the possibility to carry out high-quality HR-XRPD. A highly monochromatic beam ($\Delta E/E \sim 10^{-4}$, E being the energy of the incident X-ray beam) of about $\sim 1 \times 1$ mm^2 and of low-divergence arrives on the sample, and diffracted photons are measured by scanning the 2θ circle which holds an EIGER2 2M-W CdTe pixel detector positioned behind a set of 13 Si(111) analyzer crystals. Because of the small acceptance of an analyzer crystal, precise 2θ angles of diffraction are defined, yielding very narrow resolution function, with a resulting FWHM of about 0.0025° (2θ) at 35 keV for the 111 reflection of a NIST Si 640c standard. The efficiency of detecting the diffracted radiation can be increased by operating multiple crystals in parallel [26], and thirteen channels, 2° apart from each other, are currently available. The presence of a 2D detector combined with the analyzer crystals offers additional flexibility in terms of data handling and processing [27], improving both peak shape at low diffraction angles and counting statistics at high diffraction angles, resulting in an overall increase in the quality of high-resolution powder diffraction data [28]. In the context of the BAG, standardized operating conditions are provided, with an incident radiation of 35 keV (λ = 0.3542 Å) to reduce absorption by the sample as well as limiting potential radiation damage. One of the requirements for the BAG is to have the samples compatible with the robotic sample changer (see part 3.2) in order to maximize data collection efficiency. However, more diverse setups (energy, sample stage, sample environment, etc.) are accessible through standard proposal calls.

3.2. Sample Preparation and Mounting

Two main types of samples are being analyzed through the BAG at ID22: free powders and artwork micro-fragments, both compatible with the use of borosilicate glass capillaries as sample holders. The choice of the capillary size for the free powders depends on the sample content and corresponding X-ray absorption coefficient, which can be calculated from https://11bm.xray.aps.anl.gov/absorb/absorb.php (accessed on 1 February 2022). In order to maximize the diffraction signal, a μR value (μ: linear absorption coefficient; R: capillary radius) below 1.5 is recommended. Capillaries are then mounted on a magnetic base, to be handled by the robotic sample changer in an automated way (see the robot at former ID31 in action: https://www.youtube.com/watch?v=ACMScnxOYkM (accessed on 1 February 2022). The most precious samples or samples requiring careful positioning are usually mounted manually on the spinner of the ID22 diffractometer.

3.3. Data Acquisition and Data Analysis

Similar to ID13, ID22 operates with the new BLISS control system and uses similar ICAT capabilities. In order to reduce preferred orientation, the samples are usually rotated (nominal speed close to 1000 rpm), and diffraction data are collected through continuous scanning of the 2θ circle at a chosen speed (from 0.5 to 30°/min, 2°/min being the default one) and at the appropriate acquisition time in order to achieve a step size of about 0.0005° (2θ). Several scans per sample can be collected to improve the statistics. The diffraction signals collected over the 13 channels (and over the different scans) are then combined using dedicated beamline software [29], and the final high-resolution powder diffraction pattern is obtained as a three-column ascii file (.xye). A new data processing suite taking full advantage of the EIGER2 detector behind the analyzer crystals is under implementation [28]. Data analysis can be carried out with various software offering crystal phase identification, microstructure analysis through Le Bail or Pawley fits, and phase quantification or structure studies through Rietveld refinement.

3.4. Assessment of Radiation Damage

The effects of radiation damage on the sample are sometimes visual, with a change of color (usually a darkening) under the X-ray beam. As observed during ID13 experiments, radiation damage usually results in diffraction peak shifting and/or broadening, and can be assessed by looking at the evolution of the diffraction signal between two scans. To reduce such effects, the incident beam can be attenuated, higher energies used, and faster scans implemented. If repeated scans are recorded on the same position, after comparison, the scans not showing signs of radiation damage can be merged to increase counting statistics. In the case of powders in capillaries, a fast scan is generally carried out (>15°/min), and repeated several times on fresh zones of the capillary [30].

4. Some Recent Examples of Studies Performed within the BAG Project

4.1. Revisiting the Bamyian Buddhist Paintings to Obtain More Insight into Pigment Syntheses and Early Oil Painting Practices in the Silk Road

As a first illustration of the BAG capabilities, Figure 2 reports an updated analysis of an iconic example, presented in a pioneer publication about the application of SR micro-analytical techniques for the analysis of painting fragments [31]. These fragments were sampled from 6th–9th C. Buddhist wall paintings from Bamyian, Afghanistan, and analyzed at the ESRF in 2006. Thanks to SR-based micro-infrared analyses, the most surprising scientific outcome was the identification of the use of the oil technique in these very early wall paintings; nevertheless, results obtained with µXRPD/µXRF were also very informative. In particular, different forms of lead white were identified based on the variable cerussite to hydrocerussite ratios, revealing the use of different lead white qualities, or the different use (and evolution) of the same pigment. Information about the nature of degradation compounds (e.g., palmierite ($(K,Na)_2Pb(SO_4)_2$), anglesite ($PbSO_4$), moolooite ($CuC_2O_4.nH_2O$) and atacamite ($Cu_2Cl(OH)_3$)) were obtained as well. At the time of these first experiments, the beam size available at the former ESRF ID18F beamline was 15×1 µm^2 (hor.×ver.). The flux and detector technology imposed long acquisitions (5 s) but above all, a long lead time (10 s per pixel). To cover a corpus of ~40 samples, a compromise was made between map size and resolution, and maps were collected as 2 or 3 vertical profiles only, to preserve the resolution in the direction of the paint stratigraphy (vertical axis). Each map, of only ~2×100 pixels^2 (hor.×ver.) necessitated about 50 min. With the present EBS-ID13 instrument, the same samples could be reanalyzed, with larger fields of view (up to 500×500 pixels^2) covering the entire cross-sections, with a square and smaller pixel of 1×1 µm^2, and still within a reasonable amount of time (only 18 min for the 500×500 pixels^2 map). The larger field-of-view offers better statistics on the results. The smaller pixel size allows imaging smaller details. As an example, Figure 2 shows a comparison of data obtained in 2006 and in 2021, on thin sections from the same sample from Foladi cave 4, ca. 7–8th C. AD. The sample is a multi-layer system composed of (from depth to surface): a brown earthen plaster render covered with sizing layers, a white ground layer, a red layer, covered with a thin white layer. The final thick green layer highlights some whitish degraded aspect on its topmost surface. Ancient painters used such superimposition of complementary colors (here orange/white/green) to produce deep greens. XRPD data revealed that the two white layers are composed of different lead whites. The large and high resolution map acquired at ID13 reveals big cerussite particles (up to 50 µm diameter) in the ground white layer, while it appears much more finely ground (and with a higher amount of hydrocerussite) in the white layer above. This confirms a deliberate use of two distinct lead white qualities. Regarding the degradation layer, the ID13 map reveals that palmierite, anglesite, and atacamite form a quasi-continuous layer on top of the green layer. These new results provide a better understanding of the paint technologies used in these very early oil paintings and for their conservation.

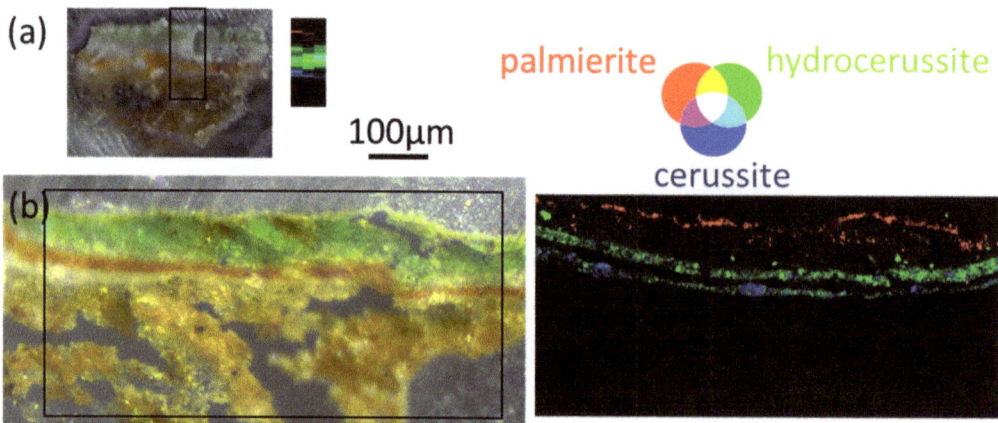

Figure 2. µXRPD maps of two sections of a fragment from a Bamiyan wall painting, acquired (**a**) in 2006 at the ESRF at the former ID18F beamline (map size: 150×60 µm^2, pixel size: 1×20 µm^2), and (**b**) in 2021 at beamline ID13 (map size: 800×370 µm^2, pixel size: 1×1 µm^2). The two acquisitions are displayed with the same scale (see text for technical details).

4.2. Composition and Stability of Pigments Invented during the Industrialization Period (End of 18th- Beginning of 20th C.)

From the end of the 18th C., with the rise of modern chemistry and the discovery of new elements and new minerals, artists and craftsmen had new materials at their disposal, in particular pigments outperforming traditional pigments in tints and hues. This led to major evolutions in artistic practices, in paintings (e.g., impressionism) but also in glasses and ceramics. The more controlled synthesis procedure has an effect not only on the optical properties, but also on the morphology of the pigment, on its chemical composition and crystalline structure. Often, slight variations in the composition translate into a high variation of color, but also sometimes stability. HR-XRPD is therefore essential for distinguishing subtle variations in composition of crystalline artistic materials.

4.2.1. Deepening the Knowledge of Formulations of Cadmium Red Pigments

Cadmium reds are a class of 20th century artists' pigments described by the formula $CdS_{1-x}Se_x$. For their vivid orangish/reddish tone and excellent covering power, many well-known modern and contemporary painters, such as Jackson Pollock, often employed cadmium reds [32–34]. Recently, the extensive study of artificially aged oil paint mock-ups made up of $CdS_{1-x}Se_x$ with different x values, provided first evidence of the tendency of cadmium reds toward photo-degradation and proved that the conversion of $CdS_{1-x}Se_x$ to cadmium sulfates and/or oxalates is influenced by the oil binding medium and moisture and depends on the Se content [33]. Thus, the proper understanding of the overall pigment formulation and the stoichiometry of $CdS_{1-x}Se_x$ is highly relevant in view of the preventive conservation of paintings containing cadmium reds.

$CdS_{1-x}Se_x$ pigments have been largely studied by XRPD techniques [32,33]. In particular, the peak positions of the XRPD pattern are strongly affected by the x value, since the progressive substitution of an S atom with a larger Se one leads to a linear increase in unit cell parameters [35]. Furthermore, the accurate fit of the peak profiles can reveal microstructural features (such as crystallite size, and lattice defects and distortion) and provide quantitative information on both different crystalline phases and the amorphous phase. Thanks to the recent advancements of instrumental techniques and computational methods for data treatment it is often possible to obtain accurate structural and microstructural properties also from conventional X-ray sources, although the use of synchrotron radiation

X-ray beams allows overcoming the intrinsic limitations of laboratory diffractometers, especially in terms of intensity, resolution and peak profile description.

As an example, Figure 3 shows a comparison of a portion of the Rietveld plots of an historical cadmium red pigment powder dated back to ca. 1960–1970 and produced by Kremer (hereafter called powder 442), in which the experimental XRPD pattern is compared with that calculated from a structural model, thus enabling the refinement of the above cited structural parameters (the Rietveld method) [36]. Figure 3a shows the (100), (002), and (101) peaks of $CdS_{1-x}Se_x$, collected with a laboratory diffractometer (PANalytical X'Pert Pro in reflection geometry, X'Celerator detector, Ni filtered CuK_α radiation). These peaks seem to show only two contributions, which come from two $CdS_{1-x}Se_x$ phases (indicated with I and II Roman numerals). The Rietveld refinement procedure successfully converged with reasonably low agreement factors, and the refined x values for the two phases are reported inside the box Figure 3a. Figure 3b shows the same portion collected at the ESRF beamline ID22 (35 keV radiation). With this pattern it was possible to clearly identify five $CdS_{1-x}Se_x$ phases, indicated with I–V Roman numerals, some of them with a very small difference in Se content (refined x values are reported in Figure 3b). Figure 3c shows the detail of the (101) peaks of the ID22 pattern, in which the contribution of the five phases to the global profile has been reported in different colors. The weight % of the five phases plus barite are also reported, as obtained by the Rietveld analysis.

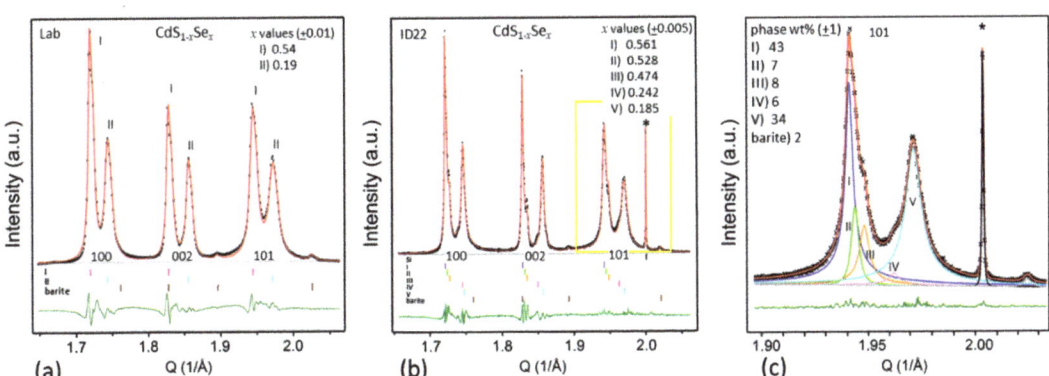

Figure 3. Portions of the Rietveld plots of the historical cadmium red pigment powder 442, showing the (100), (002), and (101) reflections of $CdS_{1-x}Se_x$ phases, collected using (**a**) the laboratory diffractometer and (**b**) the ESRF-ID22 beamline. (**c**) Enlargement of the frame marked in yellow in (**b**) showing, in different colors, the contributions of the five $CdS_{1-x}Se_x$ phases to the (101) reflection. The refined x values (and their standard deviation) for the different phases are reported in box (**a**) and (**b**), while the weight % of the five identified phases, plus barite, are reported in box (**c**), as resulting from the Rietveld analysis [37]. Black cross: experimental data; red line: calculated profile; dark green line: difference curve. Vertical marks indicate the calculated positions of Bragg reflections for each crystalline phase. The peak of silicon, used as internal standard, is marked with an asterisk (*).

Note the fact that the data recorded with the synchrotron radiation X-ray source (Figure 3b) show a negligible instrumental contribution to the peak profile as compared to that recorded with the conventional X-ray source (Figure 3a). To obtain an idea of the instrumental contribution to the total peak broadening, it is sufficient to observe the sharpness of the 111 reflection of the added crystalline silicon as internal standard (marked with an asterisk in Figure 3b,c), as compared to the broadening of the other peaks. Being a highly crystalline sample with ideal crystallite size, the Si peak broadening can be specifically associated with instrumental effects. The virtual absence of instrumental broadening enables a better evaluation of the microstructural characteristics of the sample, which contribute to the diffraction peak shape. So, while the laboratory pattern might be reasonably well refined using an isotropic size and microstrain model, the peak shape of the ID22 pattern allowed the refinement of an anisotropic model, from which the average crystallite size and microstrain along different crystallographic axes can be estimated.

4.2.2. Understanding Paint Degradation in Picasso Cadmium Yellows

The degradation of modern cadmium yellow paints ($CdS/Cd_{1-x}Zn_xS$) has been the object of intensive research [38–43]. The study of micro-fragments of historical paintings and of paint mock-ups established that the degradation process consists of the photo-oxidation of the original form of cadmium sulfide into cadmium sulfate. This process is fostered by environmental conditions [42] and by the presence of chlorine residues [40,41,43]. Research is still ongoing to assess how the degradation rate is influenced by the synthesis method, and in particular by the resulting pigment properties (e.g., particle size) and composition (presence of residues/secondary products) [44]. Further research on historical samples and paint mock-ups is still required to expand our knowledge on other factors influencing the degradation pathways of cadmium yellow paints.

Pablo Picasso's *Femme* (*Époque des "Demoiselles d'Avignon"*, 1907, Fondation Beyeler, Riehen/Basel, Switzerland, Inv. 65.2) was the object of a wide conservation project, concerning the history and pictorial techniques of the painting. During this study, a comparison of the painting with an old slide from the museum archives revealed that the CdS-based paints had retained their original bright yellow color only in some areas, while they had turned brownish in other areas of the painting [45]. As these paints were subjected to the same environmental conditions and hence to the same natural ageing, they represent an important example for understanding the reasons behind the different stability of the various CdS-based paints. To evaluate the differences down to the micrometric level, two paint fragments from the different CdS areas were selected and studied employing SR μXRPD techniques (Figure 4).

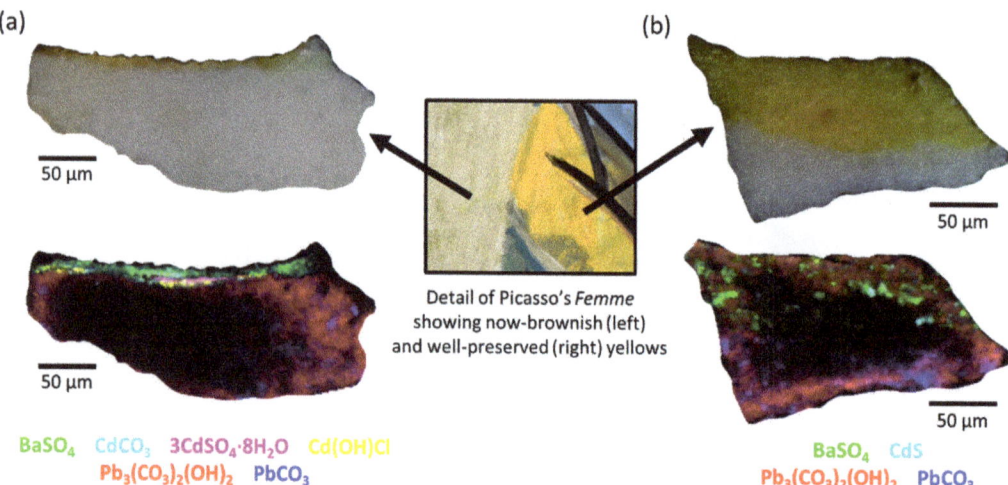

Figure 4. Detail of Picasso's *Femme* containing both now-brownish (left) and vibrant (right) yellows. Visible image and SR-μXRPD distribution maps of crystalline phases identified in (**a**) now-brownish and (**b**) vibrant yellow micro-samples, extracted from representative areas of the painting. Samples were prepared as cross-sections embedded in resin. The colors of the different crystalline compounds identified are indicated in legend. The darker region in the center of the μXRPD maps is due to the poor transparency of the sample to X-rays (measurements acquired in transmission mode).

Analysis conducted at ID13 confirmed that Picasso employed (at least) two different cadmium yellow paints [46]. The well-preserved yellow is a mixture of crystalline CdS (hexagonal and cubic forms) with two different extenders, lead white (hydrocerussite) and barium sulfate. In contrast, the now-brownish yellow is composed of amorphous or poorly crystalline CdS, indeed no diffraction peaks are present, but CdS was detected through µXANES measurements at S K-edge and Cd L_3-edge at ID21. In this case, CdS is mixed only with one extender, barium sulfate. Additional crystalline compounds, such as cadmium hydroxychlorides, sulfates and carbonates, were also identified in the now-brownish layer, which can be associated either to residuals of the pigment synthesis method (hydroxychlorides and carbonates) [40,41,43] or to paint degradation (sulfates) [38,42]. This finding provides the first clear evidence of the different crystallinity of CdS in the two yellow paints employed by Picasso, a difference that can be ascribed to the production methods of the pigment [43,47]. The different crystallinity of CdS in the two paints, along with some residual of the starting reagents (i.e., Cd(OH)Cl), may have influenced the paint stability, leading to a severe degradation in paint layers where the poorly crystalline and highly reactive pigment with high presence of residues of the synthesis was used [48]. This finding paves the way to further tailored research on model paints in the framework of the Heritage BAG to correlate paint degradation with synthesis methods.

4.2.3. Tracking the Origin of the Color of "Thénard's Blue", from the Manufacture Nationale de Sèvres

"Thénard's Blue" ($CoAl_2O_4$) is a spinel pigment, created in the early 19th century for the *Manufacture Nationale de Sèvres*. Its adaptation into a porcelain glaze required nearly 80 years of trial-and-error research. Experiments at ID13 and ID22 were combined to follow and understand, from structural and optical points of view, its complex evolution during the formation of a low-fire porcelain glaze [49]. To this aim, a series of pigments with composition $Co_{1-x}Al_{2+2x/3}O_4$ ($0 \leq x \leq 1$) were prepared by co-precipitation and fired at 850–1400 °C. Then, porcelain glazes were prepared by mixing these pigments (33 w%) with a Pb-rich flux and firing at 880 °C (i) in a crucible to obtain powder samples and (ii) on a porcelain substrate (Figure 5a).

HR-XRPD at ID22 on powder samples was used to determine the crystal structure of the pigments, (i) alone, and (ii) embedded in the glaze after the firing process. Quantitative phase analysis with an internal standard enabled the determination of the solubility of the pigment and the composition of the glass phase, in particular the proportion of metallic oxides (Figure 5e). Complementary UV-visible spectroscopy showed the oxidation state of cobalt and its environment, and how its color was changed once in the glaze.

Besides, cross-sections of the glazed porcelain samples were observed by scanning electron microscopy coupled with an Energy Dispersive Spectrometer (SEM-EDS) analysis, and thin sections were analyzed at ID13 to acquire µXRF and µXRPD maps. SEM revealed the distribution of the pigment particles in the glaze (Figure 5c). EDS and µXRF detected Co, but could not differentiate Co from the glaze matrix and from the pigment due to the small size of the crystallites and the high penetration of electrons and X-rays, respectively. Conversely, the µXRPD maps allowed a selective analysis of the cobalt contained in the pigment, revealing the exact location of the spinel phase $CoAl_2O_4$. Comparing µXRF and µXRPD maps allowed understanding how cobalt dissolves and diffuses inside the glass matrix. As shown in Figure 5d, the Co-XRF signal (in red) is in proportion to $CoAl_2O_4$ XRPD signal (in blue) and is much more intense towards the surface of the glaze, indicating a higher concentration of dissolved cobalt at the glass surface.

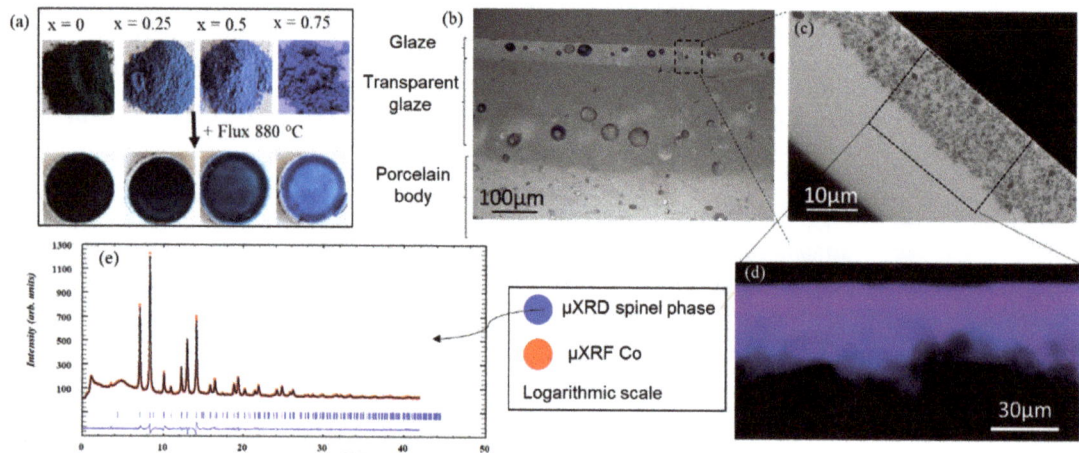

Figure 5. Study of the stability of $Co_{1-x}Al_{2+2x/3}O_4$ spinel pigments in a porcelain glaze. (**a**) Pigments x = 0; 0.25; 0.5 and 0.75 fired at 1000 °C and their respective glazes; (**b**) Optical microscopy image of the cross-section of a painted porcelain; (**c**) SEM image of the glaze layer of glazed sample (x = 0.25 at 1000 °C); (**d**) µXRF and µXRPD map of glazed sample (x = 0.25 at 1000 °C) acquired on ID13, blue = µXRPD signal of the spinel phase, red = µXRF signal of Co (K-edge); the XRPD pattern of $CoAl_2O_4$ in the glaze is then compared with the (**e**) HR-XRPD diagram of the pigment sample, (x = 0.25, at 1000 °C) acquired on ID22 (35 keV).

Determining (i) the location of the cobalt in the glass phase, (ii) the location of the pigments in the glaze, (iii) the evolution of the spinel structure during firing, (iv) the amount of dissolved pigment and (v) the evolution of the color (linked to the oxidation state of the cobalt and its environment), gave us a good understanding of the dissolution and recrystallization mechanisms and will be discussed in a forthcoming paper. Eventually, the long pending issues of coloration and instability met when adapting Thénard's Blue into a porcelain glaze found their explanation by this original combination of µXRF/µXRPD mapping on ID13 and HR-XRPD at ID22.

4.3. Applications to Conservation Studies

SR-techniques are used not only to identify the original materials used by craftsmen and artists and their possible degradation products, but they can also provide information about chemical reactions involved in conservation treatments and characterization of the penetration depth of the conservation products and efficiency of these treatments. As an example, Ca K-edge 2D µXANES has been recently used at beamline ID21, in combination to µXRPD mapping to study the stratigraphic distribution of calcium-based consolidants applied in limestones [50]. Some experiments in the BAG have been similarly dedicated to conservation purposes.

4.3.1. Revealing the Interactions of Inorganic Conservation Treatments with Mg-Containing Frescos

The structural analysis of new crystalline phases formed in painted plasters after inorganic-mineral treatments and the investigation of their distribution within the porous matrixes are in high demand and a challenging task, as high phase selectivity, sensibility to trace phases and micrometric spatial resolution are simultaneously required.

Several inorganic-mineral treatments are available and ammonium oxalate (AmOx, $(NH_4)_2C_2O_4 \cdot H_2O$) is one of the most widely used for the conservation of calcium carbonate stone materials (both natural and artificial) [51–53].

Here, a series of studies by µXRPD at ID13 and µXRF at ID21 (see technical details in Supporting Information) focused on the interactions of AmOx treatment applied to Mg-containing historical frescos with the following aims:

- Identify the new oxalate phases crystallized after the AmOx treatment in the presence of Mg-rich and Ca-rich regions of the fresco;
- Localize the different oxalate phases with respect to each other, as well as to explore their distribution in the different regions of the fresco stratigraphy.

As an example, Figure 6 summarizes some µXRF and µXRPD outcomes collected in correspondence to the fresco painting. The stratigraphy of the fresco (shown in the optical image of Figure 6a as (1) plaster, (2) *intonachino*, (3) ~10 µm portion of *intonachino* with iron-based pigments) is well distinguishable in the µXRF distribution maps of Ca, Mg and Fe (Figure 6b).

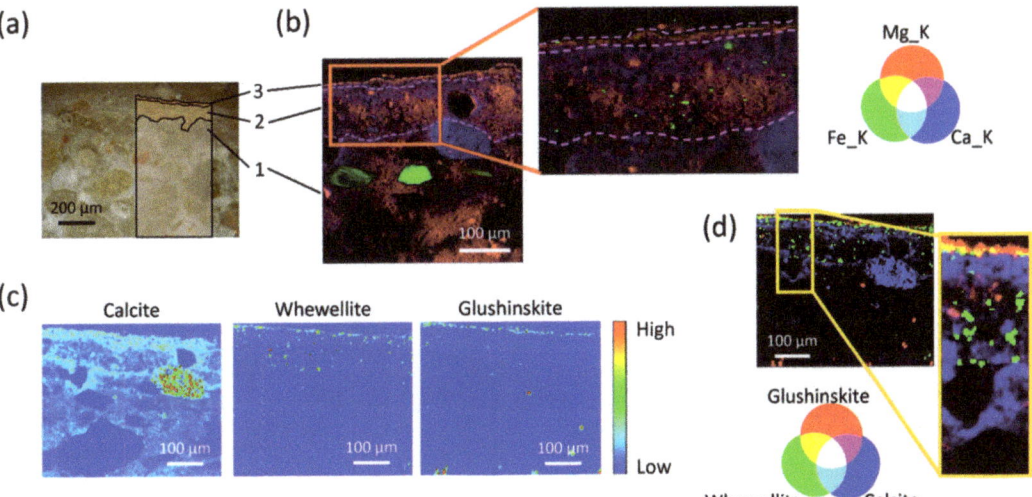

Figure 6. Analysis of the interactions of inorganic-mineral conservation treatments (AmOx) with Mg-containing frescos by a combination of µXRPD and µXRF mapping carried out at ID13 and at ID21, respectively. (**a**) Optical image of the fresco stratigraphy in cross-section: (1) plaster, (2) *intonachino*, (3) ~10 µm external portion of *intonachino* with iron-based pigments. The investigated ROI was about ~500 × 400 µm² at ID13 and ~400 × 400 µm² at ID21. (**b**) RGB correlation of µXRF distribution maps of calcium (Ca K), magnesium (Mg K) and iron (Fe K). The inset highlights the presence of iron-based pigments in the external portion of the *intonachino*; (**c**) µXRPD distribution maps of calcite, whewellite and glushinskite presented in a colourmap spanning from low (blue) to high (red) values of relative intensity; (**d**) RGB correlation of the µXRPD distribution maps of calcite, whewellite and glushinskite.

Two different classes of reaction products are formed after the AmOx treatment: magnesium oxalates (glushinskite) and calcium oxalates (whewellite and weddellite). Magnesium oxalate and calcium oxalates are crystallized in different regions of the sample, with glushinskite formed close to the surface and whewellite localized in the sub-surface portion of the fresco painting (Figure 6c,d). The localization of weddellite has been studied but it is not discussed here. No iron-oxalates have been detected.

The oxalate phases are formed within the *intonachino* as well as in the plaster substrate. The comparison of the µXRF map of calcium and the µXRPD map of calcite shows that: (i) in the portion of *intonachino* with iron-based pigments (layer 3), calcium ions are all ascribed to calcium oxalates phases (no calcite); (ii) below the Fe-containing portion, the calcium ions are ascribed to both calcium oxalates phases and calcite of the matrix (Figure 6c). These findings demonstrate that in the iron-based region of *intonachino* most of the original calcite is converted to calcium oxalate phases. It follows that the oxalate framework restores the microstructural cohesion of the fresco stratigraphy. In addition, the newly formed calcium oxalate phases have a low solubility even to acid environments, providing an advantageous acid resistance for painted carbonate substrates exposed to polluted urban regions. Results of the complete study will be reported in a forthcoming publication. Above all, the significant results obtained thanks to the ESRF-EBS and the high-spatial resolution 2D µXRPD–µXRF mapping at ID13 and ID21 open up new insights in the field of conservation treatments applied to painted plasters and historical painted materials.

4.3.2. Assessing Structural Damage in Wood Vessels

Even though most SR-based X-ray analyses are dedicated to evaluating the composition of inorganic pigments and paintings, the same techniques can also be used to analyze various materials, including (bio)organic materials, such as wood, and to contribute to their preservation. For instance, Sorres X, a 14th century cabotage ship, is a unique example of the few medieval vessels preserved in the Mediterranean [54]. It was unearthed in 1990, desalted, cleaned, treated, as usual, with polyethylene glycol (PEG) to consolidate its wood, and since 2011, it has been kept in the Museu Marítim de Barcelona (Figure 7a). It currently shows some recurrent sulfur and iron-containing efflorescence, which may be related to more severe underlying problems threatening its integrity (Figure 7b–d). Sulfur species (H_2S, HS^- and S^{2-}), common in marine environments, can lead to the formation of more oxidized species (sulfates, etc.), which can cause a significant volume expansion of the wood. In turn, iron species may lead to the formation of iron sulfide and pyrite or act as catalyzers and contribute to the degradation of wood components and PEG polymers into small organic acids (cf. [55,56], and references within).

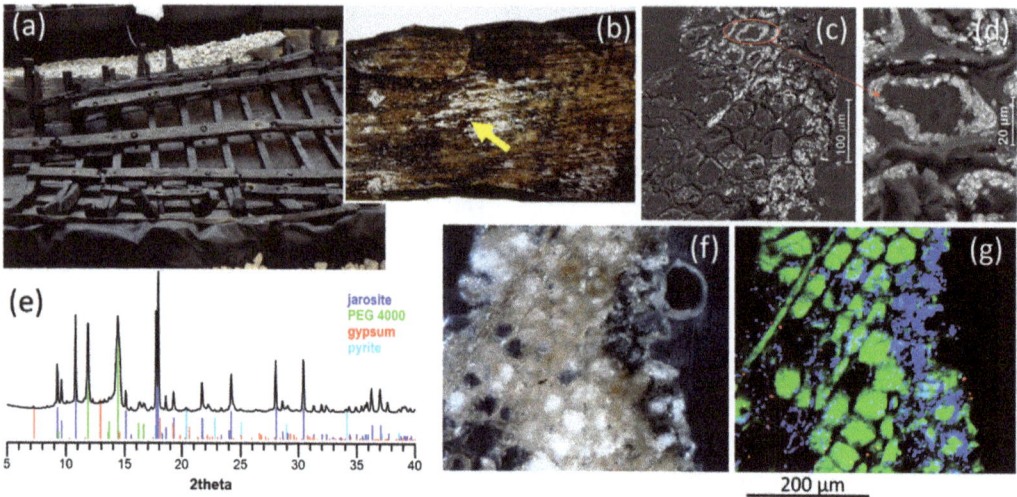

Figure 7. Study of the degradation of a waterlogged medieval timber hull. (**a**) View of the vessel Sorres X (Museu Marítim de Barcelona), which has a total length of 9.5–10 m. (**b**) Detail of efflorescence on pine wood. Back-scattering SEM images of (**c**) a thin cross section of the same pine wood sample and (**d**) detail showing crystal accumulations on the inner side of the cell wall. (**e**) Average diffractogram of a transversal 20 µm-microtomed section of the same pine wood sample, compared to reference patterns and (**f**) corresponding microscopic image and (**g**) False-color µXRPD map, where the inner part of the wood is on the left and the wood surface is on the right of the images. Colors in the µXRPD map represent PEG (green), jarosite (blue) and gypsum (red); 456 × 386 µm² map, obtained with a 2.5 × 2.5 µm² beam, with a step of 1 × 1 µm².

In this context, the objective of the study is to determine the identity, extent and distribution of sulfur, iron and other mineral species within the wood, in order to assess internal wood damage and influencing factors (such as the kind of wood, depth, proximity to old iron bolts). Analyses by μXRF and μXRPD at ID13 allowed us to discriminate among chemical species, based on their distinct diffraction patterns (Figure 7e), and a precise mapping of crystalline compounds with high spatial resolution in large maps from microtomed cross-section slices (Figure 7f), such as the ones in Figure 7g. Sulfur and iron-containing salts with different hydration degrees and oxides are present, together with PEG, in different holm oak, pine and elm wood samples, showing the complexity of redox processes occurring on and within the wood. In this particular sample, gypsum ($CaSO_4 \cdot 2H_2O$, in red) appears only as spots, apparently without a definite distribution, whereas jarosite ($KFe_3(SO_4)_2(OH)_6$, in blue) and lower amounts of pyrite (FeS, see the diffraction pattern) tend to accumulate near the surface (right side of the map) and around collapsed cells. This is consistent with the distribution of iron and sulfur, but also, for example, potassium and calcium, determined by μXRF (results not shown). PEG (in green) has penetrated through the rays (long thin tubular structures—such as the one on the upper left corner of the map—used for the radial conduction of water, minerals and organic substances in the plant) and into the cells, preventing them from collapsing. These rays may not only contribute to PEG diffusion, but also migration of salt and degrading bacteria [57].

Further analysis will help ascertain whether these compounds are ubiquitous and/or homogeneously distributed in the vessel (similar to the Mary Rose ship [56]) or if they tend to accumulate near the surface, as in the Baltic ships Vasa, the Crown and Riksnyckeln [58], all of them extensively studied hulls.

5. Conclusions

As shown in the various examples above, the instruments offered at ID22 and ID13 are highly complementary and useful for the characterization of cultural heritage materials. HR-XRPD is very efficient for the precise and sensitive detection of crystalline phases, their identification, and the characterization of their microstructural and structural properties. As shown in the examples above, the low detection limit allows us to detect minor phases, and the high angular resolution allows us to differentiate phases (pigments) with slightly different structures/stoichiometry. Complementarily, μXRPD imaging provides unique insight into the stratigraphical distribution of these phases at the micrometer scale. Although most of the applications concern paintings, the same techniques can be applied to ceramics, wood, etc.

The Historical Materials BAG started in fall 2021, and two beamtime sessions have already been allocated, two days at ID22 and four days at ID13. Thanks to the optimization of sample mounting and data acquisition, and together with the improved capabilities of the EBS source and the beamlines, 74 samples + 3 references, and a record of 186 samples + 2 references were analyzed by 12 and 15 end-users in these two experiments, respectively (with both on-site and remote control). Noteworthy, these end-users represent almost as many individual scientific projects and institutes, demonstrating clearly the high potential of the BAG system. In total, seven nationalities are represented (six from European countries). Beyond an increased efficiency to collect data, the BAG demonstrated a high impact in informing and training a new user community, in particular PhD students, who represent more than half of the users. Additionally, it provides the possibility for non-expert users to join a collective effort and to obtain support in all steps of the analytical workflow (from beamtime allocation, sample preparation, data collection, to data analysis). This new ad hoc access model is therefore a remarkable step forward to make synchrotron-based XRPD analyses a standard method for the characterization and preventive conservation of our cultural heritage.

Supplementary Materials: The following supporting information can be downloaded at: https://www.mdpi.com/article/10.3390/molecules27061997/s1, Figure S1: The ID13 X-ray microscope at the micro-branch. The X-ray beam is displayed as the solid red line, stopped by the beam stop. The dotted red lines represent XRPD signal. For readability, the XRF detector is not represented. It is mounted orthogonal to the beam, on the left side in the beam axis. (Thanks to L. Lardière, ID13, ESRF). Figure S2: Two examples of sample holders at ID13. Scale bar in each picture represents 1cm. On the left, a 4 mm diameter 8-holes holder with fragments of model paints. Samples are glued on tape. On the right, a 2 mm diameter 30-holes holder with thin sections of model and historical paint samples. Sections are either pure paints (three top rows) or painting fragments embedded in resin (three bottom rows). Sections are glued on both edges with carbon tape, such that the XRPD signal comes from the paint only (no support material, see zoomed picture). Figure S3: The Daiquiri graphical user interface and its main functionalities. The sample shown here is used for radiation damage studies. A series of maps and points were selected for the analyses detailed in Section 2.5, Figure S4: Evolution of peak position (left) and FWHM (right) of the (1 0 1), (1 0 4), (0 1 5) and (1 0 10) XRPD peaks in a lead white (mostly hydrocerussite) oil paint mock-up during repeated acquisitions of 10ms at low flux. Figure S5: Assessment of radiation damage on a chrome yellow ($PbCr_{0.2}S_{0.8}O_4$) oil paint mock-up. Seven μXRPD maps were acquired at ID13, with the following conditions (one or three repeats; different dwell times; at low flux (LF) or high flux (HF)): 1: 1 scan, 0.01 s, LF; 2: 1 scan, 0.01 s, HF; 3: 1 scan, 0.03 s, LF; 4: 1 scan, 0.03 s, HF; 5: 1 scan, 0.1 s, HF; 6: 3 scans, 0.01 s for each scan, HF; 7: 3 scans, 0.01 s for each scan, LF. The pixel size was 2.5×2.5 μm^2 to avoid overlap between two consecutive points. The map width was 50 μm and the height sufficient to cover the entire thickness of the paint sample. The maps are represented as rectangles in (a) and (b). (a) Optical microscopy after μXRPD. (b) μFTIR map acquired in transmission mode (beam size 15×15 μm^2, pixel size 10×10 μm^2, 50 cumulated scan per spectrum) after performing the μXRPD maps. The red/green display shows the integrated intensity over the ν(CO) acid range (1674–1724 cm^{-1}) and ν(CO) ester range (1731–1759 cm^{-1}), respectively. These regions are displayed in (c) by a red and green rectangle, respectively. (c) Average FTIR spectra calculated over map 1, map 5 and a non-irradiated (NI) region. References [59,60] are cited in the Supplementary Materials.

Author Contributions: M.C., V.G., F.V. and L.M coordinated the Historical materials BAG project. C.D., M.B., L.H., W.d.N. and S.F. contributed to the technical implementation of the project. I.F., E.A.C., F.V. and K.J. contributed to the radiation damage study, C.C. (Christelle Chauffeton) and G.W. to the study of Sèvres pigment, N.J., F.A.-T. and N.S. to the study of wood, E.P. and C.C. (Chiara Colombo) to the study of Mg frescos, M.G. and D.C. to the study of Picasso's painting, L.M., R.V., A.R., C.C. (Claudio Costantino) to the study of cadmium red pigments, Y.T. to the study of Bamiyan paintings and J.M., H.R. and J.S. coordinate(d) the implementation of BAGs in general at the ESRF. All authors have read and agreed to the published version of the manuscript.

Funding: The Historical materials BAG has been implemented with support from the European Union's Horizon 2020 research and innovation programme under grant agreement No 870313, Streamline, which also supports APC. L. H. is funded via the PANOSC project (European Union's Horizon 2020 Research and Innovation programme under Grant Agreement No. 823852). I. F. is funded by a grant from the Marie Skłodowska-Curie COFUND Programme "InnovaXN" (contract number 847439 with the European Commission). The ESRF beamtime was granted through the peer-review BAG proposal HG-172 at ID13 and ID22 and in-house beamtime at ID21. The project on wood vessels received financial support from MINECO (Spain), Ref. PID2019-105823RB-I00. The project on cadmium red pigments was financially supported from the Horizon 2020 project IPERION-HS (H2020-INFRAIA-2019-1, GA No. 871034) and the Italian project AMIS (Dipartimenti di Eccellenza 2018–2022, funded by MIUR and Perugia University). V.G. has received funding from the European Union's Horizon 2020 research and innovation programme under the Marie Skłodowska-Curie grant agreement No 945298-ParisRegionFP. E.A.C. and K.J. would like to acknowledge the project Smart*Light funded by the Interreg V Flanders-Netherlands program with financial support from the European Regional Development Fund (ERDF) for funding in assessing the damage caused by X-rays to model paints. K.J. acknowledges FWO (Brussels) for financial support through grants G054719N and I001919N. F.V. and K.J. acknowledge BELSPO (Brussels) for funding of the FEDtWIN mandate *Macro-Imaging*. C.C. (Christelle Chauffeton) is grateful to the Fondation Bettencourt-Schueller and the Cité de la Céramique Sèvres-Limoges, France for founding her work in the frame of her PhD thesis. A part of this study has been funded by the Grant-in-Aid for Scientific Research [18700680],

JSPS Grant-in-Aid for Young Scientists (B) from the Ministry of Education, Culture, Sports, Science and Technology of Japan.

Institutional Review Board Statement: Not applicable.

Informed Consent Statement: Not applicable.

Data Availability Statement: The data presented in this study are available on request from the corresponding author.

Acknowledgments: I.F. and M.C. thank Katrien Keune, Stichting Het Rijksmuseum, Jitte Flapper, Akzo Nobel Coatings International, and Bas De Bruin, University of Amsterdam, for their participation in the 847439 InnovaXN PhD project. The Afghan government and National Research Institute for Cultural Properties, Tokyo (NRICPT), are thanked for providing samples from Bamiyan wall paintings. We acknowledge the collaboration of the Museu Marítim de Barcelona (MMB) for providing wood samples. The authors wish to thank Markus Gross and Fondation Beyeler (Riehen/Basel, Switzerland) for providing the samples from Femme and allowing their study and colleagues from the Getty Conservation Institute, Douglas MacLennan, Catherine Schmidt Patterson, Alan Phenix, Herant Khanjian and Karen Trentelman. The TS Lab & Geoservices snc (Pisa, Italy) is acknowledged for developing an ad hoc effective protocol to prepare polished thin sections of frescos on polycarbonate films suitable for the µXRPD-µXRF investigations at ID13 and ID21.

Conflicts of Interest: The authors declare no conflict of interest.

Sample Availability: Samples of the compounds are not available from the authors.

References

1. Cotte, M.; Genty-Vincent, A.; Janssens, K.; Susini, J. Applications of synchrotron X-ray nano-probes in the field of cultural heritage. *Cr. Phys.* **2018**, *19*, 575–588. [CrossRef]
2. Bertrand, L.; Cotte, M.; Stampanoni, M.; Thoury, M.; Marone, F.; Schöder, S. Development and trends in synchrotron studies of ancient and historical materials. *Phys. Rep.* **2012**, *519*, 51–96. [CrossRef]
3. Janssens, K.; Cotte, M. Using Synchrotron Radiation for Characterization of Cultural Heritage Materials. In *Synchrotron Light Sources and Free-Electron Lasers: Accelerator Physics, Instrumentation and Science Applications*; Jaeschke, E., Khan, S., Schneider, J.R., Hastings, J.B., Eds.; Springer International Publishing: Cham, Switzerland, 2019; pp. 1–27.
4. Gonzalez, V.; Wallez, G.; Calligaro, T.; Cotte, M.; De Nolf, W.; Eveno, M.; Ravaud, E.; Menu, M. Synchrotron-Based High Angle Resolution and High Lateral Resolution X-ray Diffraction: Revealing Lead White Pigment Qualities in Old Masters Paintings. *Anal. Chem.* **2017**, *89*, 13203–13211. [CrossRef] [PubMed]
5. Gonzalez, V.; Cotte, M.; Vanmeert, F.; de Nolf, W.; Janssens, K. X-ray Diffraction Mapping for Cultural Heritage Science: A Review of Experimental Configurations and Applications. *Chem. Eur. J.* **2020**, *26*, 1703–1719. [CrossRef]
6. Monico, L.; Janssens, K.H.; Miliani, C.; Brunetti, B.G.; Vagnini, M.; Vanmeert, F.; Falkenberg, G.; Abakumov, A.M.; Lu, Y.; Tian, H.; et al. The Degradation Process of Lead Chromate in paintings by Vincent van Gogh studied by means of Spectromicroscopic methods. Part III: Synthesis, characterization and detection of different crystal forms of the chrome yellow pigment. *Anal. Chem.* **2013**, *85*, 851–859. [CrossRef]
7. Gonzalez, V.; Calligaro, T.; Wallez, G.; Eveno, M.; Toussaint, K.; Menu, M. Composition and microstructure of the lead white pigment in Masters paintings using HR Synchrotron XRD. *Microchem. J.* **2016**, *125*, 43–49. [CrossRef]
8. Gonzalez, V.; Hageraats, S.; Wallez, G.; Eveno, M.; Ravaud, E.; Réfrégiers, M.; Thoury, M.; Menu, M.; Gourier, D. Microchemical analysis of Leonardo da Vinci's lead white paints reveals knowledge and control over pigment scattering properties. *Sci. Rep.* **2020**, *10*, 21715. [CrossRef]
9. Monico, L.; Janssens, K.H.; Miliani, C.; van der Snickt, G.; Brunetti, B.G.; Cestelli Guidi, M.; Radepont, M.; Cotte, M. The Degradation Process of Lead Chromate in paintings by Vincent van Gogh studied by means of Spectromicroscopic methods. Part IV: Artificial ageing of model samples of co-precipitates of lead chromate and lead sulfate. *Anal. Chem.* **2013**, *85*, 860–867. [CrossRef]
10. Monico, L.; Janssens, K.; Hendriks, E.; Vanmeert, F.; van der Snickt, G.; Cotte, M.; Falkenberg, G.; Brunetti, B.G.; Miliani, C. Evidence for Degradation of the Chrome Yellows in Van Gogh's Sunflowers: A Study Using Noninvasive In Situ Methods and Synchrotron-Radiation-Based X-ray Techniques. *Angew. Chem.* **2015**, *127*, 14129–14133. [CrossRef]
11. Vanmeert, F.; Hendriks, E.; van der Snickt, G.; Monico, L.; Dik, J.; Janssens, K. Chemical Mapping by Macroscopic X-ray Powder Diffraction (MA-XRPD) of Van Gogh's Sunflowers: Identification of Areas with Higher Degradation Risk. *Angew. Chem. Int. Ed.* **2018**, *57*, 7418–7422. [CrossRef] [PubMed]
12. Riekel, C.; Burghammer, M.; Davies, R. Progress in Micro-and Nano-Diffraction at the ESRF ID13 Beamline. In *IOP Conference Series: Materials Science and Engineering*; IOP Publishing: Bristol, UK, 2010; p. 012013.

13. Pouyet, E.; Fayard, B.; Salome, M.; Taniguchi, Y.; Sette, F.; Cotte, M. Thin-sections of painting fragments: Opportunities for combined synchrotron-based micro-spectroscopic techniques. *Herit. Sci.* **2015**, *3*, 1–16. [CrossRef]
14. Pouyet, E.; Lluveras-Tenorio, A.; Nevin, A.; Saviello, D.; Sette, F.; Cotte, M. Preparation of thin-sections of painting fragments: Classical and innovative strategies. *Anal. Chim. Acta* **2014**, *822*, 51–59. [CrossRef] [PubMed]
15. Gonzalez, V.; Gourier, D.; Calligaro, T.; Toussaint, K.; Wallez, G.; Menu, M. Revealing the origin and history of lead-white pigments by their Photoluminescence properties. *Anal. Chem.* **2017**, *89*, 2909–2918. [CrossRef]
16. Wang, T.; Zhu, T.Q.; Feng, Z.Y.; Fayard, B.; Pouyet, E.; Cotte, M.; De Nolf, W.; Salomé, M.; Sciau, P. Synchrotron radiation-based multi-analytical approach for studying underglaze color: The microstructure of Chinese Qinghua blue decors (Ming dynasty). *Anal. Chim. Acta* **2016**, *928*, 20–31. [CrossRef]
17. Fisher, S.; Oscarsson, M.; De Nolf, W.; Cotte, M.; Meyer, J. Daiquiri: A web-based user interface framework for beamline control and data acquisition. *J. Synchrotron Radiat.* **2021**, *28*, 1996–2002. [CrossRef] [PubMed]
18. Kieffer, J.; Valls, V.; Blanc, N.; Hennig, C. New tools for calibrating diffraction setups. *J. Synchrotron Radiat.* **2020**, *27*, 558–566. [CrossRef] [PubMed]
19. De Nolf, W.; Vanmeert, F.; Janssens, K. XRDUA: Crystalline phase distribution maps by two-dimensional scanning and tomographic (micro) X-ray powder diffraction. *J. Appl. Crystallogr.* **2014**, *47*, 1107–1117. [CrossRef]
20. Cotte, M.; Fabris, T.; Agostini, G.; Motta Meira, D.; de Viguerie, L.; Solé, V.A. Watching kinetic studies as chemical maps using open-source software. *Anal. Chem.* **2016**, *88*, 6154–6160. [CrossRef] [PubMed]
21. Monico, L.; Cotte, M.; Vanmeert, F.; Amidani, L.; Janssens, K.; Nuyts, G.; Garrevoet, J.; Falkenberg, G.; Glatzel, P.; Romani, A.; et al. Damages Induced by Synchrotron Radiation-Based X-ray Microanalysis in Chrome Yellow Paints and Related Cr-Compounds: Assessment, Quantification, and Mitigation Strategies. *Anal. Chem.* **2020**, *92*, 14164–14173. [CrossRef] [PubMed]
22. Bertrand, L.; Schöeder, S.; Anglos, D.; Breese, M.B.H.; Janssens, K.; Moini, M.; Simon, A. Mitigation strategies for radiation damage in the analysis of ancient materials. *TrAC Trends Anal. Chem.* **2015**, *66*, 128–145. [CrossRef]
23. Walter, P.; Martinetto, P.; Tsoucaris, G.; Brniaux, R.; Lefebvre, M.A.; Richard, G.; Talabot, J.; Dooryhee, E. Making make-up in Ancient Egypt. *Nature* **1999**, *397*, 483–484. [CrossRef]
24. Dejoie, C.; Autran, P.-O.; Bordet, P.; Fitch, A.N.; Martinetto, P.; Sciau, P.; Tamura, N.; Wright, J. X-ray diffraction and heterogeneous materials: An adaptive crystallography approach. *Cr. Phys.* **2018**, *19*, 553–560. [CrossRef]
25. Cotte, M.; Autran, P.-O.; Berruyer, C.; Dejoie, C.; Susini, J.; Tafforeau, P. Cultural and Natural Heritage at the ESRF: Looking Back and to the Future. *Synchrotron Radiat. News* **2019**, *32*, 34–40. [CrossRef]
26. Hodeau, J.-L.; Bordet, P.; Anne, M.; Prat, A.; Fitch, A.; Dooryhee, E.; Vaughan, G.; Freund, A.K. Nine-Crystal Multianalyzer Stage for High-Resolution Powder Diffraction between 6 keV and 40 keV. In *Crystal and Multilayer Optics*; International Society for Optics and Photonics: Bellingham, WA, USA, 1998; pp. 353–361.
27. Dejoie, C.; Coduri, M.; Petitdemange, S.; Giacobbe, C.; Covacci, E.; Grimaldi, O.; Autran, P.-O.; Mogodi, M.W.; Šišak Jung, D.; Fitch, A.N. Combining a nine-crystal multi-analyser stage with a two-dimensional detector for high-resolution powder X-ray diffraction. *J. Appl. Crystallogr.* **2018**, *51*, 1721–1733. [CrossRef]
28. Fitch, A.; Dejoie, C. Combining a multi-analyzer stage with a two-dimensional detector for high-resolution powder X-ray diffraction: Correcting the angular scale. *J. Appl. Crystallogr.* **2021**, *54*, 1088–1099. [CrossRef] [PubMed]
29. Wright, J.P.; Vaughan, G.B.; Fitch, A.N. Merging data from a multi-detector continuous scanning powder diffraction system. *Comm. Crystallogr. Comput.* **2003**, *1*, 92.
30. Cockcroft, J.K.; Fitch, A. Experimental Setups. In *Powder Diffraction: Theory and Practice*; Dinnebier, R.E., Billinge, S.J.L., Eds.; Royal society of chemistry: London, UK, 2008; pp. 20–57.
31. Cotte, M.; Susini, J.; Solé, V.A.; Taniguchi, Y.; Chillida, J.; Checroun, E.; Walter, P. Applications of synchrotron-based micro-imaging techniques to the chemical analysis of ancient paintings. *J. Anal. At. Spectrom.* **2008**, *23*, 820–828. [CrossRef]
32. Grazia, C.; Rosi, F.; Gabrieli, F.; Romani, A.; Paolantoni, M.; Vivani, R.; Brunetti, B.G.; Colomban, P.; Miliani, C. UV-Vis-NIR and microRaman spectroscopies for investigating the composition of ternary CdS1− xSex solid solutions employed as artists' pigments. *Microchem. J.* **2016**, *125*, 279–289. [CrossRef]
33. Monico, L.; Rosi, F.; Vivani, R.; Cartechini, L.; Janssens, K.; Gauquelin, N.; Chezganov, D.; Verbeeck, J.; Cotte, M.; d'Acapito, F.; et al. Deeper insights into the photoluminescence properties and (photo)chemical reactivity of cadmium red (CdS1-xSex) paints in renowned 20th century paintings by state-of-the-art investigations at multiple length scales. *Eur. Phys. J. Plus* **2022**, *137*, 311. [CrossRef]
34. Huckle, W.; Swigert, G.; Wiberley, S.E. Cadmium pigments. Structure and composition. *Ind. Eng. Chem. Prod. Res. Dev.* **1966**, *5*, 362–366. [CrossRef]
35. Al-Bassam, A.; Al-Juffali, A.; Al-Dhafiri, A. Structure and lattice parameters of cadmium sulphide selenide (CdSxSe1− x) mixed crystals. *J. Cryst. Growth* **1994**, *135*, 476–480. [CrossRef]
36. Young, R. (Ed.) *The Rietveld Method*; Oxford University Press: Oxford, UK, 1993.
37. Larson, A.C.; von Dreele, R. *Generalized Crystal Structure Analysis System*; Los Alamos National Laboratory: Los Alamos, NM, USA, 2004.
38. Van der Snickt, G.; Dik, J.; Cotte, M.; Janssens, K.; Jaroszewicz, J.; De Nolf, W.; Groenewegen, J.; van der Loeff, L. Characterization of a degraded cadmium yellow (CdS) pigment in an oil painting by means of synchrotron radiation based X-ray techniques. *Anal. Chem.* **2009**, *81*, 2600–2610. [CrossRef] [PubMed]

39. Van der Snickt, G.; Janssens, K.; Dik, J.; de Nolf, W.; Vanmeert, F.; Jaroszewicz, J.; Cotte, M.; Falkenberg, G.; van der Loeff, L. Combined use of Synchrotron Radiation Based Micro-X-ray Fluorescence, Micro-X-ray Diffraction, Micro-X-ray Absorption Near-Edge, and Micro-Fourier Transform Infrared Spectroscopies for Revealing an Alternative Degradation Pathway of the Pigment Cadmium Yellow in a Painting by Van Gogh. *Anal. Chem.* **2012**, *84*, 10221–10228. [PubMed]
40. Mass, J.L.; Opila, R.; Buckley, B.; Cotte, M.; Church, J.; Mehta, A. The photodegradation of cadmium yellow paints in Henri Matisse's Le Bonheur de vivre (1905–1906). *Appl. Phys. A* **2013**, *111*, 59–68. [CrossRef]
41. Pouyet, E.; Cotte, M.; Fayard, B.; Salomé, M.; Meirer, F.; Mehta, A.; Uffelman, E.S.; Hull, A.; Vanmeert, F.; Kieffer, J.; et al. 2D X-ray and FTIR micro-analysis of the degradation of cadmium yellow pigment in paintings of Henri Matisse. *Appl. Phys. A* **2015**, *121*, 967–980. [CrossRef]
42. Monico, L.; Chieli, A.; De Meyer, S.; Cotte, M.; de Nolf, W.; Falkenberg, G.; Janssens, K.; Romani, A.; Miliani, C. Role of the Relative Humidity and the Cd/Zn Stoichiometry in the Photooxidation Process of Cadmium Yellows (CdS/Cd$_{1-x}$ZnS) in Oil Paintings. *Chem. Eur. J.* **2018**, *24*, 11584–11593. [CrossRef]
43. Monico, L.; Cartechini, L.; Rosi, F.; Chieli, A.; Grazia, C.; De Meyer, S.; Nuyts, G.; Vanmeert, F.; Janssens, K.; Cotte, M.; et al. Probing the chemistry of CdS paints in The Scream by in situ noninvasive spectroscopies and synchrotron radiation x-ray techniques. *Sci. Adv.* **2020**, *6*, eaay3514. [CrossRef]
44. Ghirardello, M.; Otero, V.; Comelli, D.; Toniolo, L.; Dellasega, D.; Nessi, L.; Cantoni, M.; Valentini, G.; Nevin, A.; Melo, M.J. An investigation into the synthesis of cadmium sulfide pigments for a better understanding of their reactivity in artworks. *Dye. Pigment.* **2021**, *186*, 108998. [CrossRef]
45. Comelli, D.; MacLennan, D.; Ghirardello, M.; Phenix, A.; Schmidt Patterson, C.; Khanjian, H.; Gross, M.; Valentini, G.; Trentelman, K.; Nevin, A. Degradation of cadmium yellow paint: New evidence from photoluminescence studies of trap states in Picasso's Femme (Époque des "Demoiselles d'Avignon"). *Anal. Chem.* **2019**, *91*, 3421–3428. [CrossRef]
46. Ghirardello, M.; Gonzalez, V.; Monico, L.; Nevin, A.; MacLennan, D.; Schmidt Patterson, C.; Burghammer, M.; Réfrégiers, M.; Comelli, D.; Cotte, M. *Application of Synchrotron Radiation-Based Micro-Analysis on Cadmium Yellows in Pablo Picasso's Femme (Époque des "Demoiselles d'Avignon")*; Politecnico di Milano: Milano, Italy, 2022; (to be submitted).
47. Leone, B.; Burnstock, A.; Jones, C.; Hallebeek, P.; Boon, J.; Keune, K. The Deterioration of Cadmium Sulphide Yellow Artists' Pigments. In Proceedings of the ICOM-CC 14th Triennial Meeting (James & James, 2005), The Hague, The Netherlands, 12–16 September 2005; pp. 803–813.
48. Levin, B.D.; Finnefrock, A.C.; Hull, A.M.; Thomas, M.G.; Nguyen, K.X.; Holtz, M.E.; Plahter, U.; Grimstad, I.; Mass, J.L.; Muller, D.A. Revealing the nanoparticle composition of Edvard Munch's The Scream, and implications for paint alteration in iconic early 20th century artworks. *arXiv* **2019**, arXiv:1909.01933.
49. Chauffeton, C. Etude et Prospection Physico-Chimique d'un pigment historique de la Manufacture Nationale de Sèvres: Le Bleu Thénard. Ph.D Thesis, Paris Sciences et Lettres (ComUE), Paris, France, 2021.
50. Monico, L.; Cartechini, L.; Rosi, F.; de Nolf, W.; Cotte, M.; Vivani, R.; Maurich, C.; Miliani, C. Synchrotron radiation Ca K-edge 2D-XANES spectroscopy for studying the stratigraphic distribution of calcium-based consolidants applied in limestones. *Sci. Rep.* **2020**, *10*, 14337. [CrossRef] [PubMed]
51. Matteini, M. Inorganic treatments for the consolidation and protection of stone artefacts. *Conserv. Sci. Cult. Herit.* **2008**, *8*, 13–27.
52. Possenti, E.; Colombo, C.; Realini, M.; Song, C.L.; Kazarian, S.G. Time-Resolved ATR–FTIR Spectroscopy and Macro ATR–FTIR Spectroscopic Imaging of Inorganic Treatments for Stone Conservation. *Anal. Chem.* **2021**, *93*, 14635–14642. [CrossRef] [PubMed]
53. Calore, N.; Botteon, A.; Colombo, C.; Comunian, A.; Possenti, E.; Realini, M.; Sali, D.; Conti, C. High Resolution ATR µ-FTIR to map the diffusion of conservation treatments applied to painted plasters. *Vib. Spectrosc.* **2018**, *98*, 105–110. [CrossRef]
54. Pujol i Hamelink, M. Medieval shipbuilding in Catalonia, Spain (13th–15th centuries): One principle, different processes. *Int. J. Naut. Archaeol.* **2016**, *45*, 283–295. [CrossRef]
55. Norbakhsh, S.; Bjurhager, I.; Almkvist, G. Impact of iron (II) and oxygen on degradation of oak–modeling of the Vasa wood. *Holzforschung* **2014**, *68*, 649–655. [CrossRef]
56. Aluri, E.R.; Reynaud, C.; Bardas, H.; Piva, E.; Cibin, G.; Mosselmans, J.F.W.; Chadwick, A.V.; Schofield, E.J. The Formation of Chemical Degraders during the Conservation of a Wooden Tudor Shipwreck. *ChemPlusChem* **2020**, *85*, 1632–1638. [CrossRef]
57. Fors, Y.; Nilsson, T.; Risberg, E.D.; Sandström, M.; Torssander, P. Sulfur accumulation in pinewood (Pinus sylvestris) induced by bacteria in a simulated seabed environment: Implications for marine archaeological wood and fossil fuels. *Int. Biodeterior. Biodegrad.* **2008**, *62*, 336–347. [CrossRef]
58. Fors, Y.; Grudd, H.; Rindby, A.; Jalilehvand, F.; Sandström, M.; Cato, I.; Bornmalm, L. Sulfur and iron accumulation in three marine-archaeological shipwrecks in the Baltic Sea: The Ghost, the Crown and the Sword. *Sci. Rep.* **2014**, *4*, 4222. [CrossRef] [PubMed]
59. Gettens, R.J.; Kühn, H.; Chase, W.T. 3. Lead white. *Stud. Conserv.* **1967**, *12*, 125–139.
60. Cotte, M.; Pouyet, E.; Salome, M.; Rivard, C.; De Nolf, W.; Castillo-Michel, H.; Fabris, T.; Monico, L.; Janssens, K.; Wang, T.; et al. The ID21 X-ray and infrared microscopy beamline at the ESRF: Status and recent applications to artistic materials. *J. Anal. At. Spectrom.* **2017**, *32*, 477–493. [CrossRef]

Article

Non-Invasive Paleo-Metabolomics and Paleo-Proteomics Analyses Reveal the Complex Funerary Treatment of the Early 18th Dynasty Dignitary NEBIRI (QV30)

Elettra Barberis [1,2], Marcello Manfredi [1,2,*], Enrico Ferraris [3], Raffaella Bianucci [4,5,†] and Emilio Marengo [2,6,†]

1. Department of Translational Medicine, Università del Piemonte Orientale, 28100 Novara, Italy
2. CAAD, Center for Translational Research and Autoimmune and Allergic Diseases, Università del Piemonte Orientale, 28100 Novara, Italy
3. Fondazione Museo delle Antichità Egizie di Torino, 10100 Torino, Italy
4. Dipartimento di Culture e Società, Università di Palermo, 90121 Palermo, Italy
5. The Ronin Institute, Montclair, NJ 07042, USA
6. Department of Sciences and Technological Innovation, Università del Piemonte Orientale, 15121 Alessandria, Italy
* Correspondence: marcello.manfredi@uniupo.it
† These authors contributed equally to this work.

Citation: Barberis, E.; Manfredi, M.; Ferraris, E.; Bianucci, R.; Marengo, E. Non-Invasive Paleo-Metabolomics and Paleo-Proteomics Analyses Reveal the Complex Funerary Treatment of the Early 18th Dynasty Dignitary NEBIRI (QV30). *Molecules* **2022**, *27*, 7208. https://doi.org/10.3390/molecules27217208

Academic Editors: Maria Luisa Astolfi, Maria Pia Sammartino and Emanuele Dell'Aglio

Received: 4 October 2022
Accepted: 23 October 2022
Published: 25 October 2022

Publisher's Note: MDPI stays neutral with regard to jurisdictional claims in published maps and institutional affiliations.

Copyright: © 2022 by the authors. Licensee MDPI, Basel, Switzerland. This article is an open access article distributed under the terms and conditions of the Creative Commons Attribution (CC BY) license (https://creativecommons.org/licenses/by/4.0/).

Abstract: Biochemical investigations were carried out on the embalmed head of Nebiri (Museo Egizio, Turin; S-5109)—an 18th Dynasty Ancient Egyptian dignitary—and on the canopic jar containing his lungs (Museo Egizio, Turin; S. 5111/02) with the aim of characterizing the organ's (lung) specific paleo-proteins and of identifying the compounds used in his embalming "recipe". The application of a functionalized film method allowed us to perform a non-invasive sampling. Paleo-proteomics confirmed the presence of lung tissue-specific proteins (organ specific) as well as the presence of proteins linked to severe inflammation. Paleoproteomics and paleometabolomics further allowed the identification of the main components of Nebiri's embalming recipe: animal fats and glue, balms, essential oils, aromatic plants, heated Pistacia, and coniferous resins. Both the use of Pistacia and coniferous resins in an early 18th Dynasty individual confirm Nebiri's high social status. The technique applied offers a targeted approach to the chemical characterization of human tissues, embalming compounds, and organic materials layering in pottery. The ability of the functionalized film method to harvest all types of compounds, from macromolecules (i.e., proteins) to small molecules (i.e., organic acids) opens a new path in the study of ancient material culture; furthermore, it allows to perform untargeted analysis, which is necessary when no a priori information is available.

Keywords: non-invasive chemical analyses; paleo-proteomics; paleo-metabolomics; New Kingdom; Nebiri

1. Introduction

Over the past two decades, the application of analytical chemistry to bio-archeological materials has made enormous progress. More specifically, through the application of mass spectrometry, minimum amounts of ancient molecules extracted from micro-samples have been identified and quantified [1,2]. Recently, new methods for the non-invasive analysis of cultural materials were developed [3–9]. Functionalized resins were successfully employed to investigate the compounds used in ancient paintings and frescoes [10,11]; commercially available skin sampling strips were applied to archeological materials to identify paleo-proteins [9]; and other techniques were developed for the identification of in situ hydrogel extraction of proteinaceous binders [5,6] or to reveal the taxonomic identification of ancient archeological materials [8].

Naturally preserved and embalmed bodies from different archeological contexts represent a powerful source of information; over the last decades, investigations of Ancient Egyptian mummies and associated funerary equipment have been extensively performed.

For example, Habicht et al. carried out a multidisciplinary investigation on the supposed remains of Queen Nefertari, the royal spouse of pharaoh Ramses II [12]. Similarly, Bianucci et al. performed a multidisciplinary investigation on the embalmed corpses of the Royal Architect Kha and his wife Meryt [13], while Jones et al. identified the embalming recipe and the evolution of early funerary treatments in a prehistoric Egyptian mummy [14]. Proteomics analysis of two 4200-years-old embalmed mummies dated to the First Intermediate Period provided molecular insight into their health conditions, suggesting evidence of acute inflammation and severe immune response [15]. More recently, the application of untargeted metabolomics for chemical characterization of canopic jars' content and mummy samples from Ancient Egypt led to the identification of thousands of ancient molecules [16,17]; however, no characterization of specific embalming recipes per individual or per organ sampled was obtained [15,16].

The present research focuses on the application of a non-invasive functionalized film method [4] to the analysis of the human remains of an 18th Dynasty Egyptian individual named Nebiri.

2. Results and Discussion

2.1. Paleo-Proteomic Investigation of Nebiri's Remains

Mass spectrometry was used to sequence ancient protein residues in a dedicated laboratory for the analysis of ancient materials. We applied shotgun proteomics using LC-MS/MS for the analysis of proteins harvested from the head and the lung of Nebiri (Figure 1). The proteins were extracted from the external table of the right parietal bone (in an area where the original textiles were lacking), the scalp, and the lungs using functionalized films; they were then digested prior to the mass spectrometry analysis. Each analyzed sample was preceded and followed by at least one blank injection in order to assess peptide carryover. All the consumables used were new to avoid environmental contaminations. Common protein contaminants were included in the database searches and removed whether identified.

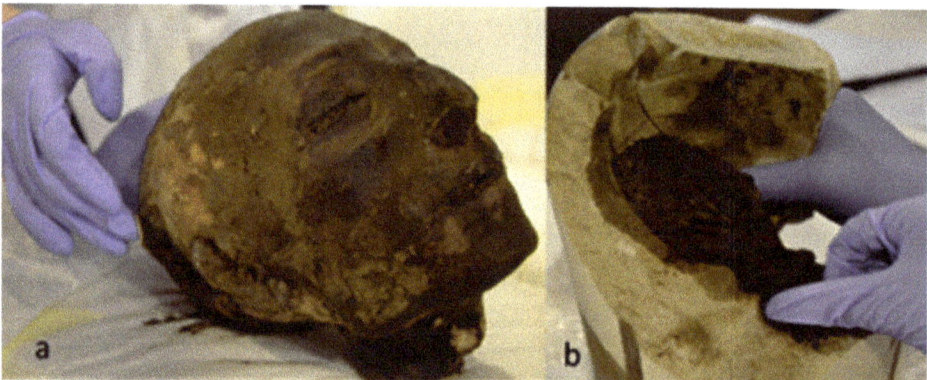

Figure 1. Pictures of the head of Nebiri (**a**) and of the canopic jar containing the lungs (**b**) taken during the non-invasive sampling procedure.

2.1.1. Paleo-Proteins Extracted from the External Surface of the Right Parietal Bone

Several collagens and keratins were identified with the right parietal bone. While human collagen alpha-1(I) chain (CO1A1_HUMAN), collagen alpha-2(I) chain (CO1A2_HUMAN), and collagen alpha-1(III) chain (CO3A1_HUMAN) were the most abundant proteins, two collagens from fowl, namely collagen alpha-1(I) chain (CO1A1_CHICK) and collagen alpha-1(IX) chain (CO9A1_CHICK), with unique and specific peptides, were detected (Table 1). Collagen alpha-2(I) chain (CO1A2_ONCMY) with two unique peptides originated from a fish species was also identified.

Table 1. Proteins identified in the head and in the lung of Nebiri. Sample, proteins, accession name, score, and number of peptides identified for each protein are reported.

Sample	Proteins	Accession Name	Score	N. of Peptides
External table of the right parietal bone	Collagen alpha-1(I) chain	CO1A1_HUMAN	1436	23
	Collagen alpha-2(I) chain	CO1A2_HUMAN	1127	22
	Collagen alpha-1(I) chain	CO1A1_CHICK	764	12
	Collagen alpha-2(I) chain	CO1A2_ONCMY	66	3
	Keratin, type I cytoskeletal 9	K1C9_HUMAN	464	5
	Keratin, type I cytoskeletal 10	K1C10_HUMAN	236	5
	Keratin, type II cytoskeletal 1	K2C1_HUMAN	225	5
	Keratin, type II cytoskeletal 2 epidermal	K22E_HUMAN	40	3
	Keratin, type II cytoskeletal 5	K2C5_HUMAN	39	3
	Collagen alpha-1(IX) chain	CO9A1_CHICK	207	3
	Collagen alpha-1(III) chain	CO3A1_HUMAN	82	2
Lung tissue (most abundant proteins)	Serum albumin	ALBU_HUMAN	942	12
	Protein S100-A9	S10A9_HUMAN	640	6
	Ig alpha-1 chain C region	IGHA1_HUMAN	469	5
	Alpha-1-antitrypsin	A1AT_HUMAN	376	7
	Cathepsin G	CATG_HUMAN	367	5
	Hemoglobin subunit beta	HBB_HUMAN	350	3
	Histone H2A type 1	H2A1_HUMAN	340	3
	Isoform H14 of Myeloperoxidase	PERM_HUMAN	288	7
	Alpha-1-antichymotrypsin	AACT_HUMAN	236	6
	Collagen alpha-1(III) chain	CO3A1_HUMAN	210	5
	Ig gamma-1 chain C region	IGHG1_HUMAN	209	5
	Hemoglobin subunit alpha	HBA_HUMAN	190	3
	Neutrophil defensin 1	DEF1_HUMAN	177	3
	Lysozyme C	LYSC_HUMAN	166	2
	Histone H4	H4_HUMAN	158	2
	Histone H2B type F-S	H2BFS_HUMAN	137	4
	Peroxiredoxin-2	PRDX2_HUMAN	125	2
	Actin, cytoplasmic 1	ACTB_HUMAN	120	2
	Tubulin beta-2B chain	TBB2B_HUMAN	118	4
	Fibrinogen beta chain	FIBB_HUMAN	110	4
	Neutrophil elastase	ELNE_HUMAN	104	2
	Isoform Gamma-A of Fibrinogen gamma chain	FIBG_HUMAN	100	3
	Band 3 anion transport protein	B3AT_HUMAN	77	2
	Isoform 2 of Haptoglobin	HPT_HUMAN	77	2
	Isoform 2 of Complement C4-A	CO4A_HUMAN	76	3
	Isoform 2 of Heat shock protein HSP 90-alpha	HS90A_HUMAN	76	2
	Myeloblastin	PRTN3_HUMAN	75	2

Table 1. Cont.

Sample	Proteins	Accession Name	Score	N. of Peptides
Scalp	Collagen alpha-1(I) chain	CO1A1_HUMAN	1145	12
	Collagen alpha-2(I) chain	CO1A2_HUMAN	1010	13
	Collagen alpha-2(I) chain	CO1A2_CHICK	134	3
	Collagen alpha-1(III) chain	CO3A1_HUMAN	400	9
	Keratin, type I cytoskeletal 9	K1C9_HUMAN	327	5
	Keratin, type I cytoskeletal 10	K1C10_HUMAN	103	3

As for the presence of collagens from fowl, the proteins CO1A1_CHICK and CO9A1_CHICK from *Gallus gallus* were identified. Both proteins were characterized by ancient modifications, indicating that the collagen was a component of the original material and did not come from contamination with more recent conservation treatments. While it is reported that animal glues were widely used as binders and adhesives, especially in Egyptian cartonnage, here the first scientific evidence of a proteinaceous material—an animal glue from fowl—directly extracted from the surface of a human ancient bone is provided.

Interestingly, domestic fowls were raised during the 18th Dynasty as reported in the literature [18]. In addition, the presence of fish collagen suggests that fish glue adhesives were also used. Although Ancient Egyptian records do not describe their preparation process in detail, it is known that these adhesives were used in the embalming procedures. Fishing was a highly diffused practice in Ancient Egypt, and fish glue would have been made by melting fish/fish scraps over a fire and then applied with the help of a brush/or a spatula [19].

The fish collagen identified has been classified as belonging to the species *Oncorhynchus mykiss* (Rainbow trout) and not to the Nile Tilapia, as one would have expected; this result is mostly likely due to the fact that the full protein database of Tilapia spp. is not yet available and only a few proteins have been characterized thus far. Although the use of fish oil was reported in a previous study on Egyptian mummies [13], the authors did not find specific markers attributed to this kind of animal oil.

2.1.2. Lung Proteins

The proteomic analysis of lung samples revealed the presence of 60 unique human proteins (Table 1); among them are several specific biomarkers of lung tissue. Lung is undoubtedly a major "immunological organ" since it contains a considerable amount of lymphoid tissue. We identified neutrophil defensins (DEF1_HUMAN), which are antimicrobial peptides present in large amounts in the neutrophil [20]; hemoglobins (HBA_HUMAN and HBB_HUMAN), which are involved in oxygen transport from the lung to the various peripheral tissues; the neutrophil serine proteases cathepsin G (CATG_HUMAN) and neutrophil elastase (ELNE_HUMAN), which are involved in immune-regulatory processes and exert antibacterial activity against various pathogens [21]; and haptoglobin (HPT_HUMAN), which is known to be associated with the host-defense response to infection and inflammation and is expressed at a high level in lung cells [22]. The identified proteins were also subjected to gene ontology classification based on biological processes with the Cytoscape software and the ClueGO plug-in. The analysis showed several lung biological processes such as immune response, defense response to bacteria, and oxygen transport (Figure 2). No proteins of animal origin were detected.

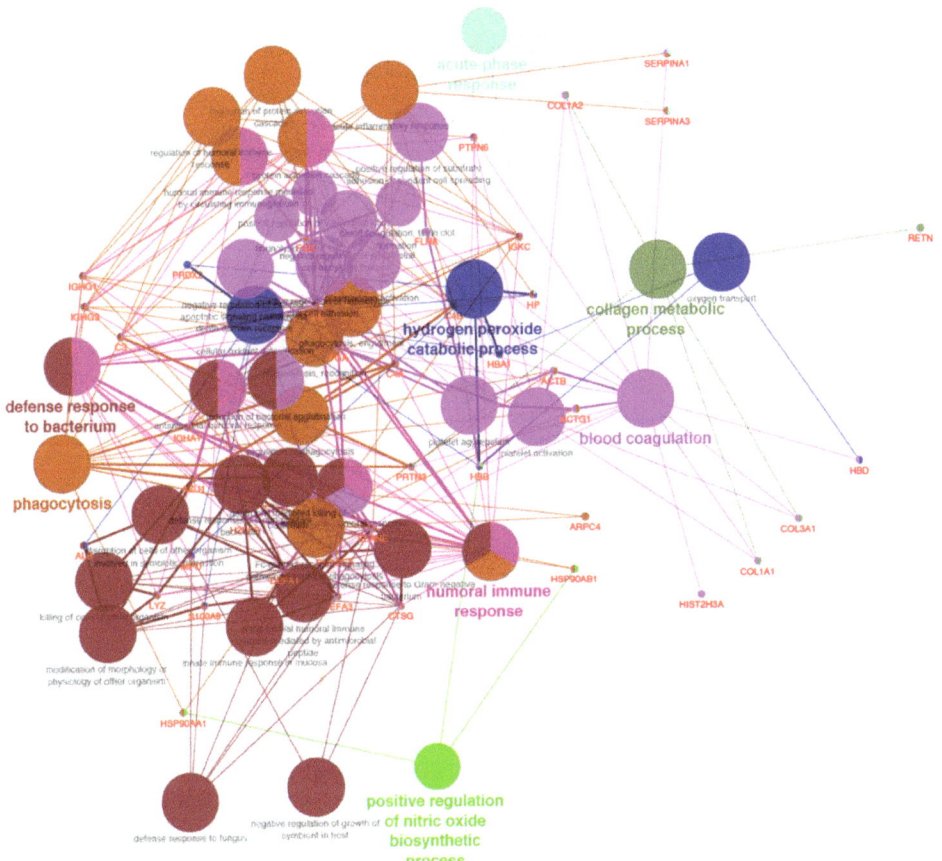

Figure 2. Cytoscape-based ClueGo pathway analysis and visualization. Enriched pathways were obtained from the Kyoto Encyclopaedia of Genes and Genome (KEGG) database. The figure reports the functionally grouped networks of identified proteins. Terms are linked based on κ-score (≥ 0.4); edge thickness indicates the association strength while node size corresponds to the statistical significance for each term. Biological processes are also reported.

2.1.3. Proteins from the Scalp

Proteins extracted from the scalp were mainly human collagens (CO1A1_HUMAN, CO1A2_HUMAN and CO3A1_HUMAN) and human keratins (K1C9_HUMAN and K1C10_HUMAN). Besides these proteins, an animal collagen, namely collagen alpha-2(I) chain (CO1A2_CHICK), was identified. All these proteins bear modifications specific for ancient materials; in addition, the presence of collagen from fowl confirmed the use of an original treatment with an ancient animal glue.

2.1.4. Ancient Protein Damage

Although proteins survive longer than DNA, they still decay naturally over time. The identification of diagenetic protein modifications was used to distinguish paleo-proteins from modern ones. Hydroxylation of proline, which is one of the main modifications of collagen, was identified in all analyzed samples. Deamidation and aminoadipic acid from lysine are more specifically related to degradation. Deamidation is usually associated with protein biological aging; it plays an important role in protein degradation and has been correlated with the time of aging [23]. Aminoadipic acid from lysine is the most important age-dependent form of oxidative damage. The presence of 2-aminoadipic acid in the ancient samples can be associated with the decomposition that occurred immediately after Nebiri's death [24]. As reported in Table 2, all samples taken from the parietal bone, lung, and head skin were characterized by the presence of deamidation and aminoadipic acid from lysine. These results confirmed the ancient origin of the samples analyzed.

Table 2. Ancient modifications identified in proteins extracted from Nebiri's remains. The name of the protein, type of modification, and number of modified amino acids in each sample for each protein are reported.

Protein	Modification	Scalp	Parietal Bone	Lung
Collagen alpha-1(I) chain	Deamidated (NQ)	8	11	4
Collagen alpha-1(III) chain	Deamidated (NQ)	3	2	2
Collagen alpha-2(I) chain	Deamidated (NQ)	3	4	1
	Lys-> AminoadipicAcid (K)	-	1	-
Collagen alpha-5(VI) chain	Deamidated (NQ)	-	1	-
Collagen alpha-3(VI) chain	Deamidated (NQ)	-	-	1
Collagen alpha-1(IX) chain	Deamidated (NQ)	1	-	-
Collagen alpha-1(V) chain	Deamidated (NQ)	-	-	1
	Lys-> AminoadipicAcid (K)	-	-	1
Collagen alpha-1(XXVIII) chain	Deamidated (NQ)	-	1	-
Collagen alpha-1(X) chain	Lys-> AminoadipicAcid (K)	-	1	-

2.2. Identification of Small Molecules with Untargeted Metabolomics Analysis

Paleo-metabolomic analysis was performed to investigate small molecules extracted from Nebiri's lungs, from the textiles, which originally wrapped the lung, and from the inner surface of the lung's canopic jar; this was done in order to identify the chemical composition of the embalming "recipe" used to treat the body of the ancient Egyptian dignitary. The paleo-molecules were extracted using functionalized films and then analyzed using mono-dimensional and comprehensive GCxGC-MS. The complete list of identified molecules across the samples are reported in Supplementary Table S2.

2.2.1. Textiles Wrapping the Lungs

The main components characterized in the lung's textile samples were fatty acids, which originally derive from triglycerides. Their distribution indicates that the major compounds are myristic, palmitic, and stearic acids together with few amounts of margaric, oleic, arachidic, azelaic, pimelic, and sebacic acids. The presence of cholesterol and of odd-chain-length fatty acids combined with the presence of squalene, an isoprenoid hydrocarbon abundant in skin lipids, clearly indicates the presence of animal fats [25,26]. Furthermore, the presence of oleic acid as the only unsaturated fatty acid and the absence of nonadecylic acid indicates that the origin of animal fatty acids may be from non-ruminant animals (porcine or fowl) [27]. All these samples were collected from the external side of the bandages, a condition which highly minimizes the possibility of contamination from the lung.

Moreover, the lipid profile points to the use of a plant oil or a mixture of plant oils. In fact, the presence of azelaic, pimelic, and sebacic acids, which form by the degradative oxidation of originally unsaturated fatty acids, suggests the presence of a vegetable oil [28]. The presence of plant oil impregnating the funerary textiles was already observed as the main balm component in previous studies on pharaonic embalming agents [14,27–29].

Aromatic acids characteristic of plant products, vanillic acid, benzoic acid, 4-hydroxybenzoic acid, 4-methoxyphenol, and 4-methylbenzoic acid, were also present in the textiles. These compounds, together with vegetable tannins identified in the samples (pyrogallol, hydroquinone, and myo-inositol), confirm the use of plant extracts although it is impossible to determine the vegetal genus/species. The presence of sucrose and D-mannopyranose suggests the use of a plant gum or sugar as a component of the balm; again, it is not possible to establish the original source, i.e., Acacia.

Coniferous resin was identified in the lung's bandages samples. The mass spectra signal of 15-Hydroxy-7-oxodehydroabietic acid, typical of aged conifer resin, was present along with the terpenoid borneol; however, since borneol can derive from many different plant extracts/essential oils or resins, it cannot be considered a marker of specific botanical species. Beeswax markers and proteinaceous materials, such as animal glue, egg, or milk, were not present in the analyzed samples.

2.2.2. The Jar Containing the Lung

The analysis of the samples taken from the inner surface of the jar containing the lung revealed a slightly different chemical composition. The main fatty acids identified were palmitic and stearic acids, with few amounts of myristic acids and traces of margaric, azelaic, and pimelic acids. The palmitic/stearic ratio of two and the presence of azelaic and pimelic acids indicate the use of a plant oil together with animal fats, as confirmed by the identification in traces of the odd-chain fatty acids pentadecanoic and margaric acids.

The inner surface of the jar was also covered by traces of vegetable tannins such as pyrogallol, hydroquinone, and 4-methylcatechol, thus confirming the use of plant extracts; the conifer resin biomarker 15-Hydroxy-7-oxodehydroabietic acid was also identified.

2.2.3. Lung

The chemical composition of the lung samples were more complex compared with those taken from the textiles and the jar samples. The hydrolysis and oxidation of lung triglycerides and glycolipids generated a different distribution of monocarboxylic and dicarboxylic acids. Palmitic and stearic acids are the main components, followed by myristic, azelaic, and pimelic acids, while margaric and arachidonic are present only in traces.

The presence of odd-chain-length fatty acids and of three cholesterol degradation products, namely 3,7-bis[(trimethylsilyl) oxy]cholest-5-ene, (3ß,7ß)-, 7-ketocholesterol, and cholesta-3,5-dien-7-one confirmed that the analyzed sample is human pulmonary tissue, as already shown by previous histological investigations [30]. The metabolomics analysis also identified other metabolites, especially sugars, formed during the hydrolysis of glycolipids.

The oxidized 15-Hydroxy-7-oxodehydroabietic acid characteristic of the coniferous resin was identified together with D-Pinitol and Resorcinol. Vegetable tannins such as Scyllo-Inositol, o-Toluic acid, Pyroglutamic acid, 3-Phenyllactic acid, 2-furoic acid, 4'-Hydroxyacetophenone, and Myo-Inositol confirmed the use of plant extracts.

Although the most specific triterpenoid of masticadienonate series (moronic, masticadienoic, isomasticadienoic, and oleanonic acids) have not been detected in the monodimensional GC-MS analysis, in the bidimensional GCxGC analysis, we found other biomarker compounds, generally used to identify an antique Pistacia resin (Supplementary Table S1) [31]: penta and tetra-cyclic triterpenes from resins of Pistacia species as oleanane-type molecules, in particular β-Amyrin, with most abundant fragment ion at m/z 218 and Olean-18-en-3-ol, o-TMS with main fragment ions at m/z 189 and 204; and a triterpenoid with dammarane skeleton, Dammaran-3-one, 20,24-epoxy-25-hydroxy with characteristic base peak at m/z 143 [32]. In this case, as well-described by Daifas et al. [33], we confirmed that Pistacia resins were used for embalming treatments, probably for their antibacterial, antifungal, and antiseptic properties [34] or they might have been employed in the preparation of kyphi ointment, as well as used for its religious significance [35]. Although various authors reported the presence of two species of Pistacia, *P. lentiscus* and *P. atlantica* [36], we were not able to discriminate the different species within the sample, possibly due to the very low abundance of characteristic markers. However, the presence of Pistacia and other triterpenoid compounds lead us to confirm that the embalming "recipe" used for

Nebiri was expensive and typically reserved for the royalty or for extremely wealthy nobles and notables [37].

Finally, vegetable tannins and phenols with antiseptic activities, derived from plants as catechol and guaiacol, have been found. The presence of guaiacol in ancient recipes might be linked to the use of wood smoke/cedar wood in the embalming process [38]. The GCxGC-MS analysis also allowed the identification of juniperol (Cupressaceae family), used as an embalming substance in mummies and canopes [16] (Table 3).

Table 3. Nebiri's embalming "recipe": list of main class of compounds associated with the samples and relative assignment.

Class of Compound	Sample	Recipe Products
Linear monocarboxylic saturated fatty acids; Dicarboxylic acids; Hydroxycarboxylic acids; Monounsaturated fatty acids	Lung, canopic jar, head	Plant oils and relative oxidation products
Diterpenoids	Lung, canopic jar, head	Pinaceae resins
Aromatic acids	Lung, canopic jar, head	Vegetable balms
Monosaccharides	Lung, canopic jar, head	Human tissue or gums
Triterpenoids	Lung	Pistacia resin
Tannins	Lung, canopic jar, head	Cedar oil/wood smoke
Collagen proteins	Scalp	Animal glue (fish and fowl)

3. Materials and Methods

3.1. Nebiri's Lung and Head

Nebiri was an ancient Egyptian dignitary who lived 3500 years ago under the reign of Thutmose III (1479–1424 BCE; 18th Dynasty). His tomb (QV30) was discovered in the Valley of the Queens, in 1904, by the first director of the Egyptian Museum of Turin, Ernesto Schiaparelli (1856–1928). Unfortunately, his tomb was plundered in antiquity by grave robbers and his mummy was deliberately destroyed. Only his head (S.5109) and the canopic chest (S. 5110; s. 5111/02; S-5112; S-5113) were preserved. Previous studies showed that the man described as "Chief of Stables" died of acute decompensation of chronic heart failure when he was 45–60 years old [39].

In the present study, non-invasive sampling was carried out on Nebiri's head and on the canopic jar containing his lung (S. 5111/02). Access to the content of this specific canopic jar was granted by Museo Egizio of Turin (Italy) since the apical portion of the vase was already broken and portions of the lungs and linen textiles were easily accessible. Paleo-proteins, small molecules, and lipids were identified without damaging the mummy's head and his funerary vase whose long-term preservation is required by the current legislation on Cultural Heritage (Figure 1a,b). Therefore, non-invasive sampling was performed on Nebiri's scalp, on the canopic jar containing the lung, on the lung tissue, and on the textiles originally used to wrap both the head and organ. The functionalized film was placed in contact with the surface of each sample for 5 min, and then removed leaving the original ancient materials unmodified.

3.2. Small Molecules Extraction and Derivatization

Small molecules were extracted from the surface of the samples using three different functionalized films: (i) a mix-bed cation/anion exchange film, (ii) a C4 resin film, and (iii) a C8 resin film. Briefly, the films are based on ethyl-vinyl acetate (EVA) as a binder of ground AG 501 mix-bed cation/anion exchange (from Biorad, Hercules, CA, USA), of C4 resin and of C8 resin (both from Sigma), respectively. The mixture of melted EVA and resins were extruded in the form of a thin film in laboratory a week before the use. Prior to their use on the object, the functionalized films were humidified with ultrapure water and then the water was discharged. The films were positioned with extreme caution on the surface of the samples for 10 min. The metabolite extracts were then eluted from the film with 1 mL of ethanol for 30 min. Then, the strips were removed and the metabolites were subjected to derivatization. The derivatization protocol was performed by adding 20 µL of methoxamine hydrochloride in pyridine (20 mg/mL) and 30 µL of BSTFA. Samples were incubated at 80 °C for 20 min after every addition, and then centrifuged for 15 min (14,500× g at RT). Nitrogen steam was used to gently dry the samples before the gas-chromatography analysis.

3.3. Proteins Extraction and Digestion

The proteins were extracted from the surface of the samples using a film functionalized with a mix-bed cation/anion exchange and C8 resins, as previously described by Barberis et al. [3,11]. The functionalized film was humidified with ultrapure water and then the water was discharged. The film was positioned with extreme caution on the surface of the sample for 10 min. The protein extracts were then eluted from the film with 500 µL of 1.0 M ammonium acetate in a tube for 30 min. Then, the strip was removed and the proteins were first denatured with TFE at 60 °C, reduced with 200.0 mM DTT, alkylated with 200.0 mM IAM, and finally digested with trypsin overnight. The peptide digested were desalted on the Discovery® DSC-18 solid phase extraction (SPE) 96-well Plate 25 mg/well (Sigma-Aldrich Inc., St. Louis, MO, USA).

3.4. GC-MS and GCXGC-MS Analyses

Gas chromatography–time of flight mass spectrometry (GC-TOF/MS) was performed using an Agilent 7890B GC (Agilent Technologies, USA) and Pegasus (BT) TOF-MS system (Leco Corporation, USA) equipped with an Rxi-5 ms column (30 m × 0.25 mm × 0.25 µm, RESTEK, USA), stationary phase 5% diphenyl-95% dimethyl polysiloxane. High-purity helium (99.999%) was used as the carrier gas at a flow rate of 1.00 mL/min^{-1}. Samples were injected in splitless mode at 280 °C. The chromatographic conditions were as follows: initial temperature 40 °C, 5 min isothermal, 8 °C/min up to 300 °C, 20 min isothermal. The MS parameters were as follows: electron impact ionization source temperature (EI, 70 eV) was set at 250 °C; scan range 40/630 *m/z*, with an extraction frequency of 30 kHz. The chromatograms were acquired in TIC (total ion current) mode. Mass spectral assignment was perfomed by matching with NIST MS Search 2.2. Libraries, implemented with the MoNa Fiehns Libraries. For the 2D analysis, a LECO Pegasus BT 4D GCXGC/TOFMS instrument (Leco Corp., St. Josef, MI, USA) equipped with a LECO dual stage quad jet thermal modulator was used. The GC part of the instrument was an Agilent 7890 gas chromatograph (Agilent Technologies, Palo Alto, CA, USA), equipped with a split/splitless injector. The first-dimension column was at 30 m Rxi-5 ms capillary column (Restek Corp., Bellefonte, PA, USA) with an internal diameter of 0.25 mm and a stationary phase film thickness of 0.25 µm, and the second-dimension chromatographic column was a 2 m Rxi-17Sil MS (Restek Corp., Bellefonte, PA, USA) with a diameter of 0.25 mm and a film thickness of 0.25 µm. The carrier gas (helium) was used with a flow rate of 1.4 mL/min. The secondary column was maintained at +5 °C relative to the GC oven temperature of the first column. Additionally, the MS method was the same as the mono-dimensional analysis, while the extraction frequency was 32 kHz, the acquisition rates was 200 spectra/s, and the modulation period was maintained at 4 s for the entire run. The modulator temperature offset was set at +15 °C relative to the secondary oven temperature, while the transfer line was set at 280 °C [40,41].

3.5. LC-MS Analysis and Data Processing

The extracted proteins were analyzed with a micro-LC Eksigent Technologies system (Eksigent, Dublin, OH, USA) that included a micro LC200 Eksigent pump with flow module

5–50 µL, interfaced with a 5600+ TripleTOF system (AB Sciex, Vaughan, ON, Canada) equipped with DuoSpray Ion Source and CDS (Calibrant Delivery System). The stationary phase was a Halo C18 column (0.5 × 100 mm, 2.7 µm; Eksigent Technologies Dublin, USA). The mass spectrometry data were searched using Mascot (Mascot v. 2.4, Matrix Science Inc., Boston, USA) and Protein Pilot; the digestion enzyme was trypsin, with 1 missed cleavage. The instrument was set to ESI-QUAD-TOF and the following modifications were specified for the search: carbamidomethyl (C) as fixed modification, Acetyl (K), Deamidated (NQ), Gln->pyro-Glu (N-term Q), Glu->pyro-Glu (N-term E), Hydroxylation (KP), Lys-> AminoadipicAcid (K), Oxidation (M), Oxidation (P), Trp->Kynurenin (W) as variable modification and hydroxylation of prolines and lysines when collagen was present [25]. A search tolerance of 0.1 Da was specified for the peptide mass tolerance, and 50 ppm for the MS/MS tolerance. The charges of the peptides to search for were set to 2+, 3+, and 4+, and the search was set on monoisotopic mass. The databases employed were Swissprot human reviewed (version 11032016, containing 42,179 sequence entries), cRAP (proteins commonly found in proteomics experiments that are present either by accident or through unavoidable contamination), and Metazoa (version 10082018, containing 103,419 sequences). Only peptides with individual ion scores > 20 were considered for identification purposes. Only proteins presenting two or more unique peptides, after screening for possible contaminants, were considered positively identified.

4. Conclusions

Previous studies pointed to the fact that Nebiri was a high social status individual. A virtopsy performed on the undamaged head of Nebiri showed an extremely careful cosmetic treatment. A similar treatment was observed only in Yuya and Tjuiu, the parents of Queen Tye, the royal spouse of King Amenhotep III (1388–1348 BCE). Nebiri's high status could also be inferred by the title "Chief of Stables"; at the beginning of the 18th dynasty, "ownership of horses and their stables were frequently reserved for high-ranking officials and those related to royalty" [30,38]. The results of the paleo-proteomics and paleo-metabolomics investigation confirm all previous findings. Nebiri's embalming "recipe" was composed of a mixture of animal fat and glue, balms, essential oils, aromatic plants, heated Pistacia, and coniferous resins. Pistacia and coniferous resins, non-native imported resins from the north-eastern Mediterranean, were considered luxury goods and only available for royal and high elite consumption [30]. Although quantitative information related to the single components of the recipe is very hard to obtain on these ancient and complex materials, our results suggest the dominance of plant oil and animal glue in funerary textiles, which are the major "balm" component, while aromatic plant extracts and gums were added to the mixture in minor amounts. Finally, conifer resins and Pistacia were identified only in trace amounts and can be considered the minor components of the embalming recipe.

From a chemical point of view, the method we used in this research offers an appropriate and non-invasive approach for the identification and characterization of human, vegetable, and animal paleo-proteins and paleo-metabolites in ancient remains. The ability of the film to harvest all types of compounds, from macromolecules such as proteins, to small molecules like organic acids, is fundamental to study ancient remains as well as fragile objects; furthermore, it allows performing an untargeted analysis, which is necessary when no a priori information is available. Lastly, the method is very rapid, thus allowing high sample throughput; the adsorbed molecules can be analyzed with any kind of analytical instrument in dedicated settings.

Supplementary Materials: The following supporting information can be downloaded at: https://www.mdpi.com/article/10.3390/molecules27217208/s1, Table S1: Complete list of identified molecules. Table S2: Heated Pistacia markers obtained from the lung sample with the untargeted analysis by GCxGC-MS; the name of the molecules, formula, retention time in first and second dimension and similarity > 700 (index of identification reliability) are given.

Author Contributions: Conceptualization, E.B., M.M. and R.B; methodology, E.B. and M.M.; formal analysis, E.B. and M.M. writing—review and editing, E.B., M.M., E.F., R.B. and E.M. All authors have read and agreed to the published version of the manuscript.

Funding: This research received no external funding.

Institutional Review Board Statement: Not applicable.

Informed Consent Statement: Not applicable.

Data Availability Statement: Not applicable.

Acknowledgments: The authors thank Sara Aicardi and Gianluca Greco from the Fondazione Museo delle Antichità Egizie di Torino.

Conflicts of Interest: The authors declare no conflict of interest.

References

1. Robotti, E.; Bearman, G.; France, F.; Barberis, E.; Shor, P.; Marengo, E. Direct Analysis in Real Time Mass Spectrometry for the Nondestructive Investigation of Conservation Treatments of Cultural Heritage. *J. Anal. Methods Chem.* **2016**, *2016*, 6853591.
2. Poulin, J.; Kearney, M.; Veall, M.-A. Direct Inlet Py-GC-MS analysis of cultural heritage materials. *J. Anal. Appl. Pyrolysis* **2022**, *164*, 105506. [CrossRef]
3. Manfredi, M.; Barberis, E.; Gosetti, F.; Conte, E.; Gatti, G.; Mattu, C.; Robotti, E.; Koman, I.; Zilberstein, S.; Korman, I.; et al. Method for Noninvasive Analysis of Proteins and Small Molecules from Ancient Objects. *Anal. Chem.* **2017**, *89*, 3310–3317. [CrossRef] [PubMed]
4. Manfredi, M.; Barberis, E.; Rava, A.; Poli, T.; Chiantore, O.; Marengo, E. An analytical approach for the non-invasive selection of consolidants in rubber artworks. *Anal. Bioanal. Chem.* **2016**, *408*, 5711–5722. [CrossRef]
5. Calvano, C.D.; Rigante, E.C.L.; Cataldi, T.R.I.; Sabbatini, L. In Situ Hydrogel Extraction with Dual-Enzyme Digestion of Proteinaceous Binders: The Key for Reliable Mass Spectrometry Investigations of Artworks. *Anal. Chem.* **2020**, *92*, 10257–10261. [CrossRef] [PubMed]
6. Ntasi, G.; Kirby, D.P.; Stanzione, I.; Carpentieri, A.; Somma, P.; Cicatiello, P.; Marino, G.; Giardina, P.; Birolo, L. A versatile and user-friendly approach for the analysis of proteins in ancient and historical objects. *J. Proteom.* **2021**, *231*, 104039. [CrossRef] [PubMed]
7. Cicatiello, P.; Ntasi, G.; Rossi, M.; Marino, G.; Giardina, P.; Birolo, L. Minimally Invasive and Portable Method for the Identification of Proteins in Ancient Paintings. *Anal. Chem.* **2018**, *90*, 10128–10133. [CrossRef]
8. Fiddyment, S.; Holsinger, B.; Ruzzier, C.; Collins, M. Animal origin of 13th-century uterine vellum revealed using noninvasive peptide fingerprinting. *Proc. Natl. Acad. Sci. USA* **2015**, *112*, 15066–15071. [CrossRef]
9. Multari, D.H.; Ravishankar, P.; Sullivan, G.J.; Power, R.K.; Lord, C.; Fraser, J.A.; Haynes, P.A. Development of a novel minimally invasive sampling and analysis technique using skin sampling tape strips for bioarchaeological proteomics. *J. Archeol. Sci.* **2022**, *139*, 105548. [CrossRef]
10. Barberis, E.; Marcello, M.; Marengo, E.; Zilberstein, G.; Zilberstein, S.; Kossolapov, A.; Righetti, G. Leonardo's Donna Nuda unveiled. *J. Proteom.* **2019**, *207*, 103450. [CrossRef] [PubMed]
11. Barberis, E.; Baiocco, S.; Conte, E.; Gosetti, F.; Rava, A.; Zilberstein, G.; Righetti, P.G.; Marengo, E.; Manfredi, M. Towards the non-invasive proteomic analysis of cultural heritage objects. *Microchem. J.* **2018**, *139*, 450–457. [CrossRef]
12. Habicht, M.E.; Bianucci, R.; Buckley, S.A.; Fletcher, J.; Bouwman, A.S.; Öhrström, L.M.; Seiler, R.; Galassi, F.M.; Hajdas, I.; Vassilika, E.; et al. Queen Nefertari, the Royal Spouse of Pharaoh Ramses II: A Multidisciplinary Investigation of the Mummified Remains Found in Her Tomb (QV66). *PLoS ONE* **2016**, *11*, e0166571.
13. Bianucci, R.; Habicht, M.E.; Buckley, S.A.; Fletcher, J.; Seiler, R.; Öhrström, L.M.; Vassilika, E.; Böni, T.; Rühli, F.J. Shedding new light on the 18th Dynasty mummies of the Royal Architect Kha and his spouse Merit. *PLoS ONE* **2015**, *10*, e0131916. [CrossRef] [PubMed]
14. Jones, J.; Higham, T.F.G.; Chivall, D.; Bianucci, R.; Kay, G.l.; Pallen, M.J.; Oldfield, R.; Ugliano, F.; Buckley, S.A. A prehistoric Egyptian mummy: Evidence for an 'embalming recipe' and the evolution of early formative funerary treatments. *J. Archaeol. Sci.* **2018**, *100*, 191–200. [CrossRef]
15. Jones, J.; Mirzaei, M.; Ravishankar, P.; Xavier, D.; Lim, D.S.; Shin, D.H.; Bianucci, R.; Haynes, P.A. Identification of proteins from 4200-year-old skin and muscle tissue biopsies from ancient Egyptian mummies of the first intermediate period shows evidence of acute inflammation and severe immune response. *Philos. Trans. A Math. Phys. Eng. Sci.* **2016**, *374*, 20150373. [CrossRef] [PubMed]
16. Brockbals, L.; Habicht, M.; Hajdas, I.; Galassi, F.M.; Rühli, F.J.; Kraemer, T. Untargeted metabolomics-like screening approach for chemical characterization and differentiation of canopic jar and mummy samples from Ancient Egypt using GC-high resolution MS. *Analyst* **2018**, *143*, 4503–4512. [CrossRef]
17. Lebedev, A.T.; Polyakova, O.V.; Artaev, V.B.; Mednikova, M.B.; Anokhina, E.A. Comprehensive two-dimensional gas chromatography-highresolution mass spectrometry with complementary ionizationmethods in the study of 5000-year-old mummy. *Rapid Commun. Mass Spectrom.* **2021**, *35*, e9058. [CrossRef]

18. Coltherd, J.B. The Domestic Fowl in Ancient Egypt. *IBIS* **1966**, *108*, 217–223. [CrossRef]
19. Darrow, F.L. *The Story of an Ancient Art, from the Earliest Adhesives to Vegetable Glue*, 1st ed.; Perkins Glue Company: Lansdale, PA, USA; South Bend, IN, USA, 1930.
20. Aarbiou, J.; Ertmann, M.; van Wetering, S.; van Noort, P.; Rook, D.; Rabe, K.F.; Litvinov, S.V.; van Krieken, J.H.; de Boer, W.I.; Hiemstra, P.S. Human neutrophil defensins induce lung epithelial cell proliferation in vitro. *J. Leukoc. Biol.* **2002**, *72*, 167–174. [CrossRef]
21. Steinwede, K.; Maus, R.; Bohling, J.; Voedisch, S.; Braun, A.; Ochs, M.; Schmiedl, A.; Länger, F.; Gauthier, F.; Roes, J.; et al. Cathepsin G and neutrophil elastase contribute to lung-protective immunity against mycobacterial infections in mice. *J. Immunol.* **2012**, *188*, 4476–4487. [CrossRef]
22. Yang, F.; Haile, D.J.; Coalson, J.J.; Ghio, A.J. Haptoglobin in lung defence. *Redox Rep.* **2001**, *6*, 372–374. [CrossRef] [PubMed]
23. Leo, G.; Bonaduce, I.; Andreotti, A.; Marino, G.; Pucci, P.; Colombini, M.P.; Birolo, L. Deamidation at asparagine and glutamine as a major modification upon deterioration/aging of proteinaceous binders in mural paintings. *Anal. Chem.* **2011**, *83*, 2056–2064. [CrossRef] [PubMed]
24. Cappellini, E.; Jensen, L.J.; Szklarczyk, D.; Ginolhac, A.; da Fonseca, R.A.; Stafford, T.W.; Holen, S.R.; Collins, M.J.; Orlando, L.; Willerslev, E.; et al. Proteomic analysis of a pleistocene mammoth femur reveals more than one hundred ancient bone proteins. *J. Proteome Res.* **2012**, *11*, 917–926. [CrossRef] [PubMed]
25. Buckley, S.A.; Andrew, W.; Stott, A.W.; Evershed, R.P. Studies of organic residues from ancient Egyptian mummies using high temperature-gas chromatography-mass spectrometry and sequential thermal desorption-gas chromatography-mass spectrometry and pyrolysis-gas chromatography-mass spectrometry. *Analyst* **1999**, *124*, 443–452. [CrossRef] [PubMed]
26. Buckley, S.A.; Evershed, R.P. Organic chemistry of embalming agents in Pharaonic and Graeco-Roman mummies. *Nature* **2001**, *413*, 837–841. [CrossRef]
27. Charrié-Duhaut, A.; Burger, P.; Maurer, J.; Connan, J.; Albrecht, P. Molecular and isotopic archaeology: Top grade tools to investigate organic archaeological materials. *Comptes Rendus Chim.* **2009**, *12*, 1140–1153. [CrossRef]
28. Colombini, M.P.; Modugno, F.; Ribechini, E. Caratterizzazione di resine vegetali in reperti archeologici. *Sci. Tecnol.* **2000**, *84*, 1–6.
29. Jones, J.; Higham, T.F.G.; Oldfield, R.; O'Connor, T.P.; Buckley, S.A. Evidence for Prehistoric Origins of Egyptian Mummification in Late Neolithic Burials. *PLoS ONE* **2014**, *9*, e103608. [CrossRef]
30. Loynes, R.D.; Charlier, P.; Froesch, P.; Houlton, T.M.R.; Lallo, R.; Di Vella, G.; Bianucci, R. Virtopsy shows a high status funerary treatment in an early 18th Dynasty non-royal individual. *Forensic Sci. Med. Pathol.* **2017**, *13*, 302–311. [CrossRef]
31. Charrié-Duhaut, A.; Connan, J.; Rouquette, N.; Adam, P.; Barbotin, C.; de Rozière, M.-F.; Tchapla, A.; Albrecht, P. The canopic jars of Rameses II: Real use revealed by molecular study of organic residues. *J. Archaeol. Sci.* **2007**, *34*, 957–967. [CrossRef]
32. Assimopoulou, A.N.; Papageorgiou, V.P. GC-MS analysis of penta- and tetra-cyclic triterpenes from resins of *Pistacia* species. Part. I. *Pistacia lentiscus* var. Chia. *Biomed. Chromatogr.* **2005**, *19*, 285–311. [CrossRef] [PubMed]
33. Daifas, D.P.; Smith, J.P.; Blanchfield, B.; Sanders, G.; Austin, J.W.; Koukoutsis, J. Effects of mastic resin and its essential oil on the growth of proteolytic *Clostridium botulinum*. *Int. J. Food Microbiol.* **2004**, *94*, 313–322. [CrossRef] [PubMed]
34. Nicholson, T.M.; Gradl, M.; Welte, B.; Metzger, M.; Pusch, C.M.; Albert, K. Enlightening the past: Analytical proof for the use of *Pistacia exudates* in ancient Egyptian embalming resins. *J. Sep. Sci.* **2011**, *34*, 3364–3371. [CrossRef] [PubMed]
35. Baumann, B.B. The Botanical Aspects of Ancient Egyptian Embalming and Burial. *Econ. Bot.* **1960**, *14*, 84–104. [CrossRef]
36. Mills, J.S.; White, R. The identity of the resins from the late bronze age shipwreck at Uluburun (Kaş). *Archeometry* **1989**, *31*, 37–44. [CrossRef]
37. Abdel-Maksouda, G.; El-Aminb, A.-R. A review on the materials used during the mummification processes in Ancient Egypt. *Mediterr. Archaeol. Archaeom.* **2011**, *11*, 129–150.
38. Hawass, Z.A.; Saleem, S.N. *Scanning the Pharaohs: CT Imaging of the New Kingdom Royal Mummies*; AUC Press: Cairo, Egypt, 2016.
39. Bianucci, R.; Loynes, R.; Sutherland, M.L.; Lallo, R.; Kay, G.; Froesch, P.; Pallen, M.; Charlier, P.; Nerlich, A. Forensic analysis reveals acute decompensation of chronic heart failure in a 3500-year-old Egyptian dignitary. *J. Forensic Sci.* **2016**, *61*, 1374–1381. [CrossRef]
40. Barberis, E.; Amede, E.; Tavecchia, M.; Marengo, E.; Cittone, M.G.; Rizzi, E.; Pedrinelli, A.R.; Tonello, S.; Minisini, R.; Pirisi, M.; et al. Understanding protection from SARS-CoV-2 using metabolomics. *Sci. Rep.* **2021**, *11*, 13796. [CrossRef]
41. Barberis, E.; Joseph, S.; Amede, E.; Clavenna, M.G.; La Vecchia, M.; Sculco, M.; Aspesi, A.; Occhipinti, P.; Robotti, E.; Boldorini, R.; et al. A new method for investigating microbiota-produced small molecules in adenomatous polyps. *Anal. Chim. Acta* **2021**, *1179*, 338841. [CrossRef]

Article

Study of Micro-Samples from the Open-Air Rock Art Site of Cueva de la Vieja (Alpera, Albacete, Spain) for Assessing the Performance of a Desalination Treatment

Ilaria Costantini [1], Julene Aramendia [1], Nagore Prieto-Taboada [1], Gorka Arana [1], Juan Manuel Madariaga [1,*] and Juan Francisco Ruiz [2]

[1] Department of Analytical Chemistry, Faculty of Science and Technology, University of the Basque Country UPV/EHU, P.O. Box 644, 48080 Bilbao, Spain; ilaria.costantini@ehu.eus (I.C.); julene.aramendia@ehu.eus (J.A.); gorka.arana@ehu.eus (G.A.)

[2] Department of History, Area of Prehistory, Faculty of Education Sciences and Humanities, University of Castilla-La Mancha (UCLM), Avda. de los Alfares 42, 16002 Cuenca, Spain; juanfrancisco.ruiz@uclm.es

* Correspondence: juanmanuel.madariaga@ehu.eus

Abstract: In this work, some micro-samples belonging to the open-air rock art site of *Cueva de la Vieja* (Alpera, Albacete, Spain) were analysed. These samples were collected after and before a desalination treatment was carried out, with the aim of removing a whitish layer of concretion that affected the painted panel. The diagnostic study was performed to study the conservation state of the panel, and to then confirm the effectiveness of the treatment. Micro energy dispersive X-ray fluorescence spectrometry, Raman spectroscopy, and X-ray diffraction were employed for the characterization of the degradation product as well as that of the mineral substrate and pigments. The micro-samples analysis demonstrated that the painted layer was settled on a dolomitic limestone with silicon aggregates and aluminosilicates as well as iron oxides. The whitish crust was composed by sulfate compounds such as gypsum ($CaSO_4 \cdot 2H_2O$) with a minor amount of epsomite ($MgSO_4 \cdot 7H_2O$). An extensive phenomenon of biological activity has been demonstrated since then in almost all of the samples that have been analysed, and the presence of calcium oxalates monohydrate ($CaC_2O_4 \cdot H_2O$) and dehydrate ($CaC_2O_4 \cdot 2H_2O$) were found. The presence of both calcium oxalates probably favoured the conservation of the pictographs. In addition, some carotenoids pigments, scytonemin ($C_{36}H_{20}N_2O_4$), and astaxanthin ($C_{40}H_{52}O_4$) were characterized both by Raman spectroscopy and by X-ray diffraction. Hematite was found as a pigment voluntarily used for the painting of the panels used in a mixture with hydroxyapatite and amorphous carbon. The results of the analyses of the samples taken after the cleaning treatment confirmed a substantial decrease in sulphate formation on the panel surface.

Keywords: Levantine rock art; μ-Raman spectroscopy; μ-EDXRF; XRD; sulfates; biodeterioration

Citation: Costantini, I.; Aramendia, J.; Prieto-Taboada, N.; Arana, G.; Madariaga, J.M.; Ruiz, J.F. Study of Micro-Samples from the Open-Air Rock Art Site of Cueva de la Vieja (Alpera, Albacete, Spain) for Assessing the Performance of a Desalination Treatment. *Molecules* **2023**, *28*, 5854. https://doi.org/10.3390/molecules28155854

Academic Editors: Maria Luisa Astolfi, Maria Pia Sammartino and Emanuele Dell'Aglio

Received: 30 June 2023
Revised: 28 July 2023
Accepted: 1 August 2023
Published: 3 August 2023

Copyright: © 2023 by the authors. Licensee MDPI, Basel, Switzerland. This article is an open access article distributed under the terms and conditions of the Creative Commons Attribution (CC BY) license (https:// creativecommons.org/licenses/by/ 4.0/).

1. Introduction

Several examples of Rupestrian art, which include paintings and engravings, are still in a surprisingly preserved state. The first evidence of rock art known to this day has recently been dated with a minimum age of 45,500 years in Leang Tedongnge (Sulawesi Island, Indonesia), and this dating was based on uranium-series isotope analysis, which was conducted on two small coralloid speleothems overlying the red painting. Prior to this discovery, the first representation created was at least 43,900 years ago from an image from Leang Bulu' Sipong 4 in the limestone karsts of Maros-Pangkep, (South Sulawesi, Indonesia) [1].

Thus, the main discoveries in terms of cave paintings occurred in South Africa [2], Argentina [3], Peru [4], Southeast Asia [5,6], Australia [7], etc., while in Europe, the most important ones were found in France and Spain, and they belonged to the transition period between the Paleolithic and the Neolithic.

Normally, this kind of artistic expression was carried out in closed spaces, such as caves or rock shelters—in other words, in cavities dug out by atmospheric agents, where populations have traditionally found shelter. These spaces suffer constant environmental impacts, making prehistoric artwork particularly fragile. This is why such spaces need constant monitoring in order to preserve them for many years to come. In this way, our research group developed a long-term monitoring methodology to better understand the conservation dynamics of rock art and its evolution over time, and this was based on the use of non-destructive elementary and molecular spectroscopies [8].

The multi-analytical approach has been widely employed in the last twenty years for the diagnostic study of the conservation state of caves and rock shelters where prehistoric art was undertaken [9]. The approach has proven to be fundamental for the study of the composition of raw materials as well as the study of painting technologies [10,11]. The palette of pigments employed was quite reduced, and was essentially composed of mineral based-pigment obtained from the natural resources of the surrounding areas [12]. Although the pigments could be used pure in most cases, a bi-colour pictograph could still be obtained, as in the case of the mixture of hematite and paracoquimbite ($Fe_2(SO_4)_3 \cdot 9H_2O$), and this was discovered for the first time in the Abrigo Remacha rock shelter (Villaseca, Segovia, Spain) [13].

In addition, since these materials are exposed to the open air, several forms of alteration, in the form of discolorations, crusts, and patinas, which are mainly due to the impact of weathering, were characterized thank to the use of portable and laboratory analysis. As reported by Hernanz et al., a crust composed of whewellite, gypsum, calcite, clay, dolomite, α-quartz, anatase, and hematite was detected in several rock art sites in the Iberian Peninsula. Wind-blown dust and surface water runoff may also have contributed to the formation of these layers [14]. In this sense, gypsum and clayish minerals were characterized as the main components of an ochre-coloured accretion covering several parts of the third painted panel of the Hoz de Vicente rock shelter (Minglanilla, Cuenca, Spain), and were responsible for the flaking process that was observed in some areas of the painting panel [15].

In the open-air site of the engraved rock art of the Burrup Peninsula (Western Australia), high concentrations of acidic and nitrate-rich pollution, from nearby industrial complexes, provoked the colour change of the pictographs. The degradation phenomenon was due to the dissolution of manganese oxide (MnO_2) and iron compounds, such as magnetite minerals triggered by acidic rain. This alteration induced the peeling of the rock varnish layer and produced hematite minerals, illite $(K, H_3O)(Al, Mg, Fe)_2 (Si, Al)_4O_{10}[(OH)_2,(H_2O)]$, and kaolinite $[Al_2Si_2O_5(OH)_4]$ [7].

The study of Pozo-Antonio et al. [16] also reported the process of colour change of rock art on a granitic outcrop at the Mougas site of Galicia (Spain). Here, the colour change phenomenon occurs on yellow and red nodes on the surface of the rock art. High temperatures provoked by wildfires cause mineral transformations (of goethite into hematite), and this increases the susceptibility of the rock to the weathering processes. Although most of the degradation phenomena are caused by atmospheric agents, degradation processes caused by anthropic factors have been discussed by Hernanz et al. [17], who detected the presence of electric welding splashes from the erection of protective iron fences around the rock art panels at the site.

In addition, gypsum and other salts, such as jarosite and bassanite, as well as biofilms, were identified in tafoni, and were generated by the weathering of sandstones (Cerro Colorado, Argentina). These secondary products were the result of impact weathering revealing hydroclastic and haloclastic processes. These activities formed active granular disintegration, and flaking and chipping affected the preservation of some painted panels [18]. Another extensive non-invasive study of Argentinean rock shelter paintings was carried out by Rousaki et al. [19]. In this study, gypsum and calcite were commonly found to be responsible for severe degradation in the form of crusts on pigmented as well as non-pigmented areas together with calcium oxalate film. Most recently, the work of

Ilmi et al. [20] demonstrated that the discoloration of Leang Tedongnge (Sulawesi Island, Indonesia) rock paintings was caused by the presence of a grey/yellowish crust composed mainly of gypsum. This was caused by the reaction between the calcium ions dissolved in karst water infiltrations and the sulphate ions of minerals that are deposited on the rock surface.

The presence of biological patinas in the form of calcium oxalate monohydrate (whewellite, $CaC_2O_4 \cdot H_2O$) and dihydrate (weddellite [$CaC_2O_4 \cdot (2 + x)H_2O, x \leq 0.5$]) have been frequently identified in rock shelters. They are generated by the interaction of oxalic acid, a metabolic product of microorganisms, with carbonaceous materials of the substrate [17]. Oxalate layers have also been used for attempting an indirect dating of post-Palaeolithic open-air paintings [21,22]. Although the presence of these compounds was mainly linked to the activity of microorganisms, previous studies on Ethiopian prehistoric rock painting have shown that they can also be the result of the degradation of organic matter, such as binder, that was employed in order to spread the pigments on the substrate [23]. On the other hand, the investigation of Hedges et al. [24] demonstrated the use of pigment containing calcium oxalate derived from local cacti and calcium carbonate that was probably derived from local plant ash. However, these hypotheses have not been entirely refuted, and the origin of oxalate compounds on rock paintings is a current topic that is still being investigated. On the other hand, oxalates occur naturally and are classified as organic minerals and oxalic acid in mineral deposits or in plants, fungi, and lichens, or in the form of deposits in animal tissues that are generated by diagenesis and biomineralization processes [25,26].

In the investigation of the Lower Pecos region in south-western Texas (USA), the colour change of the rock art was caused by the formation of whewellite-rich rock crust with gypsum and clay [27]. In addition, the study demonstrated a paleoclimate change from dry to wet conditions of this area, since the biopatina revealed similarities between whewellite microstructures and the desert lichen Aspicilia calcarea [28].

Therefore, due to the many factors that can degrade these extraordinary works of art, the study of both original and secondary compounds is essential in order to plan an appropriate restoration and conservation strategy for them. The current work was focused on the study of the conservation state of *Cueva de la Vieja*, which is located near Alpera (Albacete, Spain). Indeed, the main evidence of rock art in the Iberian Peninsula comes from the eastern part, and extends along the entire Mediterranean coast. The conservation site has been a UNESCO World Heritage Site since 1998 because the Levantine style used represents a unique artistic expression in the European context. Levantine art is mainly composed of paintings realized in semi-open spaces, such as rocky shelters, and it is characterized by a figurative art that is dominated by scenes of daily life and social activities, such as individual or group hunting scenes, dance, rituals, etc. Levantine art was composed mainly of paintings, and it was realized with red pigments, ochre, and oxides of manganese and iron. However, it has also been documented that there was a use of charcoal of organic origin (wood charcoal and burnt bone) as black pigments, and, moreover, a use of white earths (α-quartz, anatase, muscovite, and illite) and calcined bones (apatite) as white colours [17,29–31]. Numerous scientific studies have been carried out in the last 20 years with the aim of studying original compounds and degradation in order indicate the best conservation strategy [29,32–34].

This study is contextualised in an intervention aimed at recovering the visibility of the pictographs realized in the rock shelter, which include a diagnostic phase, the subsequent cleaning and consolidation interventions by specialized restorers, and then a final phase that will produce new digital tracing. Some samples were taken with the aim of studying the composition of the support and the white layer which covered the paintings, and which did not allow their original appearance to be appreciated from a chromatic point of view. This first approach permitted us to indicate the most adequate cleaning intervention that consisted of desalination of the panel. At the conclusion of the previous works, other samples were taken after the treatment in order to verify its effectiveness. In this work, the elemental and the molecular analyses of all of the micro-samples were performed by means of micro Raman spectroscopy, micro X-ray fluorescence, and X-ray diffraction in a laboratory.

2. Result and Discussion

2.1. Sample Analysis Prior to Cleaning Treatment

Micro X-ray fluorescence analyses were performed to define the elemental composition of the samples. For this purpose, several elemental images were acquired (Figure 1) on both sides of the sample CV01. Although the whitish layer was not appreciable by the naked eye on this sample, according to the restorers, the sulphur was homogeneously distributed throughout the exposed surface of the sample if it was compared to the inner face, where only some areas showed the presence of the element. On the other hand, calcium maps showed the presence of this element distributed throughout the piece. The presence of magnesium stood out in the CV01 sample, and it was more evident on the inner face.

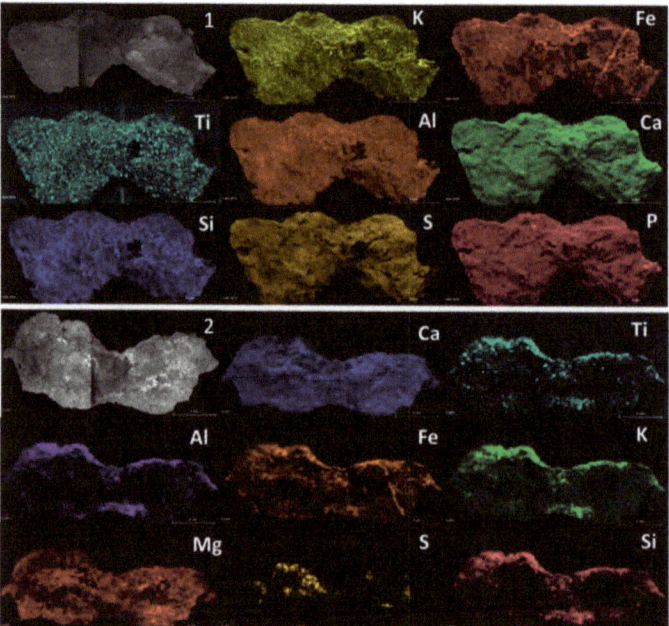

Figure 1. Micro-EDXRF maps of the exterior (1) and inner (2) layer, respectively, of sample CV01.

In all of the micro-samples that were analysed, the presence of sulfur was less perceptible at the elemental level on the inner face, while calcium and silicon were the major elements. The main composition of the samples coming from the rocky support was based on calcium, silicon, magnesium, aluminium, iron, and potassium, with titanium varying slightly in its relative presence. On the other hand, in some samples, aggregates of titanium, zinc, chrome, manganese, copper, and chlorine were evident, which corresponded to the composition of the support. In some samples, sulfur was observed on the internal face, too, mainly in the cracking zone of the samples (see Figure S1 in Supporting Information). This suggested that the formation of sulfur compounds from the exterior part could be responsible for the phenomenon of exfoliation of the support described by the restorers.

Regarding the only micro-sample that showed traces of visible red pigmentation (Figure 2), the elemental map of iron coincided with the red area, suggesting the use of iron oxide for the painting. On the other hand, the elemental maps of this sample did not show a homogeneous layer of sulphur in the surface, as indicated by the rock-bearing analysis, and only some S hotspots were visible.

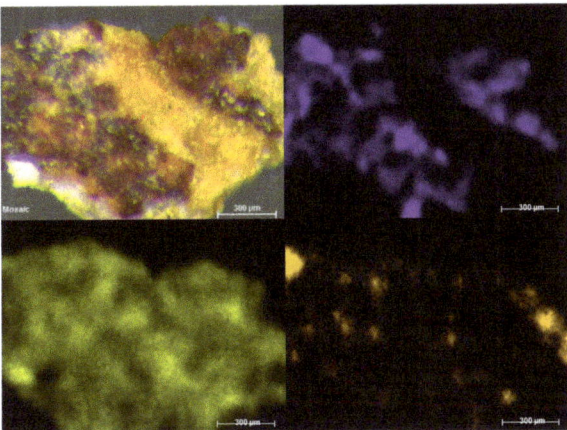

Figure 2. Micro-EDXRF maps of the sample with trace of red pigment.

By means of Raman spectroscopy, in the majority of the points analysed on the external face of all samples, gypsum (CaSO$_4$·2H$_2$O, Raman bands: 180, 414, 492, 620, 670, 1008 and 1135 cm^{-1}, Figure 3a) and calcium carbonate (CaCO$_3$, Raman bands: 154, 282, 712 and 1986 cm^{-1}) were largely detected. Even calcium magnesium carbonate dolomite (CaMg(CO$_3$)$_2$, Raman bands: 178, 300, 724 and 1098 cm^{-1}, Figure 3c) was identified in many analysis points, and was recognized as one of the main compounds of the support. Between the original compounds of the rock, Raman spectra of anatase (TiO$_2$, Raman bands 142, 395, 514, 638 cm^{-1}, Figure S2d), rutile (TiO$_2$, Raman bands: 142, 242, 446, 612 cm^{-1}, Figure S2c), and quartz (SiO$_2$, Raman bands 204, 264, 354, 465, 807 cm^{-1}, Figure S2a) were also recorded. In red and orange grains, Raman analysis showed the presence of hematite iron oxides (Fe$_2$O$_3$, Raman bands: 224, 245, 294, 402, 500, 612, and 1315 cm^{-1}, Figure S2b) and goethite (α-FeOOH, Raman bands: 204, 246, 302, 389, 480 and 550 cm^{-1}), which could be responsible for the orange colour of the stone. Goethite and hematite were identified in individual grains on the specimen and, therefore, their presence does not appear to be due to the presence of pigmentation, and they are present as components of the rock.

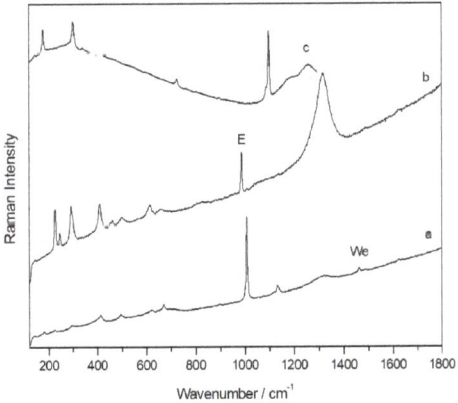

Figure 3. Raman spectra of gypsum with trace of weddellite (We) (**a**), hematite with epsomite (E) (**b**), and dolomite (**c**).

In addition, a band at 985 cm^{-1} was detected in several spectra together with other compounds—mainly gypsum, calcite, or hematite (Figure 3b). This signal could belong to the magnesium sulfate heptahydrate so-called epsomite (MgSO$_4$·7H$_2$O) present on the surface of the sample as a degradation product of dolomite. Moreover, on the internal face, in most of the analysed points, the Raman spectrum of dolomite and calcium sulfate dihydrate or gypsum were observed. In several cases, the same Raman spectrum shows the coexistence of more than one compound at the same point of the analysis. In fact, several spectra have evidenced the presence of a mixture of dolomite, gypsum, and epsomite, which could clearly indicate sulfation of the original material.

This process could be favoured by the infiltration of sulfate-rich water coming from the back of the painted panel, which carries sulphates from the stone, and which accumulates them on the surface after evaporating in the open air. The formation of sulphates, especially with various hydration molecules, such as epsomite, increases the porous pressure within the rocky substrate during the hydration phase. Thus, sulfur was dissolved from rainwater, mobilized, and precipitated during the crystallisation process due to rising temperatures.

In addition, according to the archaeologists, the repeated humidification of the panel has been proven. This practice has been common to all open-air rock art sites since their discovery a couple of decades ago, and the purpose is to improve the visualization of the pictographs to the visitors. Considering that the white formation was located in the middle of the panel, where the pictographs were made, the anthropic factor could possibly be the reason for the formation of the sulphate layer. However, we cannot know exactly where the water used for this practice comes from (probably from the spring adjacent to rock shelter). It is not possible to know the composition of the water used when this practice was carried out. However, the water in the province of Albacete is characterised by a high hardness and the presence of sulphates, so much so that just this year, an osmosis plant for the treatment of drinking water was established to reduce the presence of salts and improve its quality.

The possibility of S being mobilised from the top of the rock shelter, as shown in other studies [8], up to the painted wall by a runoff process seems to be ruled out, as a percolation process from the top of the panel was not evident. On the other hand, the sulfation of the support does not appear to be due to the presence of atmospheric contaminants given that the area where the *Cueva de la Vieja* is located is not highly affected by urban traffic or industrial contamination.

With regard to the degradation by microorganisms, this was detected in both the outer and the inner parts of many of the samples. For example, in the inner face of the CV02 sample, in addition to hematite, calcite, and dolomite, the presence of a very well defined spectra of carotenoid pigments stood out. Specifically, Raman spectra of the carotenoid pigment astaxanthin (C$_{40}$H$_{52}$O$_4$, Figure 4a), the most oxidized species among the carotenoid pigments and synthesized by species such as lichens, were recorded at several points. This identification has been made possible by the main Raman bands at 1001, 1154, and 1508 cm^{-1} and by its overtones (2150, 2298, 2509, and 2654 cm^{-1}), which allowed it to be distinguished from other carotenoid pigments, such as carotene and zeaxanthin. In addition to astaxanthin, Raman spectra of scytonemin (C$_{36}$H$_{20}$N$_2$O$_4$), a pigment generally synthesized by cyanobacteria, were recorded (Figure 4b). The highest intensity Raman bands that allowed its identification were at 1170, 1382, 1554, 1600, 1632, and 1715 cm^{-1}.

As observed under the microscope, the outer face of the sample CuVi03 was characterized by a homogeneous white colour. In this layer, mainly gypsum and dolomite spectra were recorded. Moreover, as observed in the image obtained with the microscope (Figure S3), the traces of hematite were clearly visible with Raman analyses. This could suggest the presence of pigmentation, as it did not appear as loose grains as it did in the previous samples, which suggests an original composition. Therefore, this would indicate the loss of polychrome. At any rate, Raman bands belonging to the calcium oxalates whewellite (192, 204, 220, 248, 895, 1461, 1488 cm^{-1}) and weddelite (138, 1475 cm^{-1}) were also detected in the same area. The presence of calcium oxalates could also have favoured the preservation of pigmentation in that area.

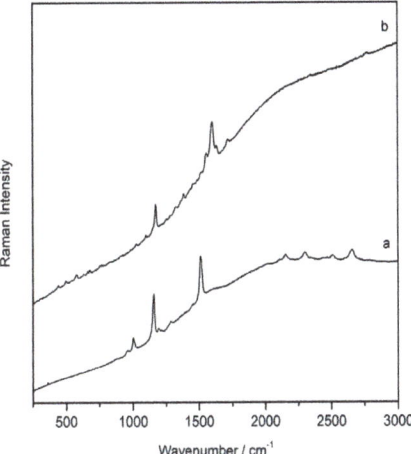

Figure 4. Raman spectra of astaxantin (**a**) and scytonemin (**b**).

Even in the unique sample taken from a painted area, the presence of hematite was recognized by means of Raman spectroscopy as pigment that was voluntarily used. In all of the recorded spectra, the presence of calcium oxalate was also identified along with the broad bands between 1120 and 1650 cm^{-1} that belong to the aluminosilicate compounds of the substrate. Calcium oxalates were even detected in the internal face of the sample. According to previous investigations of the pictorial layer of a painted rock shelter composed mainly of hematite, it was located between layers of calcium oxalate [35], and both were caused by microorganism activity. The presence of this layer and one superior to the pictographs would be factors that have allowed the conservation of prehistoric paintings to this day. Although the presence of calcium oxalate in the internal part of the sample with red pigmentation was also identified in our samples, its small size did not allow a more in-depth study by means of a cross-sectional study of the sample.

In addition, in mixture with the iron oxide, even a single weak band located at 962 cm^{-1}, typical of the hydroxyapatite ($Ca_{10}(PO_4)_6(OH)_2$) (Figure 5a), was visible in the spectra, and this suggested that the red pigment hematite was probably mixed with calcined bones. In the same spectra, moreover, broad bands belonging to amorphous carbon were also evident (Figure 5b). Thus, it is plausible that a black pigment was voluntarily added to hematite to obtain a darker colour.

Finally, molecular analysis of the crystalline part of the samples was performed with X-ray diffraction analysis. As with the other techniques, the analysis was performed on both the external and internal sides. In all samples, sample quartz, dolomite, and calcite high in magnesium were identified as original compounds, which was in agreement with the Raman spectroscopy analysis. Gypsum and compounds related to microbiological activity, such as hydrated and dihydrated calcium oxalate (whewellite and weddelite), were identified as the degradation compounds (Figure S4). Significant differences were observed between the internal and the external part. Indeed, gypsum and oxalates were detected only on the outside of the samples.

Finally, in the case of the sample containing red pigment, this technique could only identify dolomite, hydrated calcium oxalate, and gypsum as impurities. Although Raman spectroscopy identified hematite as the pigment voluntarily used, X-ray diffraction did not reveal its presence. As XRD is only sensitive to crystalline compounds, this could indicate a low crystallization of hematite in this sample. In addition, in this sample, small amounts of sulphates were also identified by all of the analytical techniques, and it is therefore plausible that the whitish sulphate layer affects the integrity of the substrate more than the areas where the pictographs are present. Undoubtedly, the presence of oxalates has guaranteed the conservation of the paintings over the years, and it is for this reason that they seem to be better preserved than the support itself. In any case, the sulfation of the panel had to be treated with the aim of bringing the pictographs back to light. In addition, the exfoliation of the support could cause the loss of the painted areas in the long term.

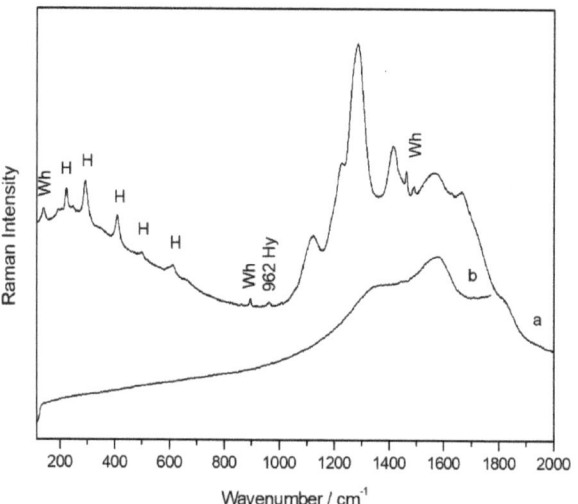

Figure 5. Raman spectra of hematite (H) with the signal of aluminosilicates, whewellite (Wh), and hydroxyapatite (Hy) (**a**), and black carbon (**b**).

From this diagnostic study, it was decided that a desalination treatment by restorers should be carried out on the surface of the panel with the aim of removing the whitish layer that covered the pictographs. According to the report of the restorers, paper dressings impregnated with low mineralization water were used, and were applied directly to the sulfate crust. After a few minutes, they were removed, and then the surface was cleaned with brushes, removing the remains. The operation was repeated a couple of times, and it was completed using only distilled water.

2.2. Samples Analysis after the Cleaning Treatment

After completing the cleaning treatment, it was considered crucial to check its effectiveness and to verify the reduction of salts. The attention of archaeologists was also drawn to a series of points on the panel where an insoluble greyish crust remained after cleaning. These were areas arranged around natural holes in the support, through which there was, perhaps, a certain periodic emanation of water in the passage. In collaboration with the restorer who was in charge of cleaning the panel, samples were collected so that the nature of these grey crusts could be identified.

At first view, there was a considerable reduction in the surface sulphates by the naked eye. Thus, the objective of the analysis of these samples was the characterization of the rock surface after cleaning, verifying the removal of the sulfates and identifying other substances remaining in the rock in order to raise hypotheses about their link to the conservation processes in *Cueva de la Vieja*.

Although micro-EDXRF analyses still identified the presence of sulfur in the samples taken after the treatment, it was heterogeneously present in the surface of the samples. As this is a microanalysis technique, we were not surprised to detect the presence of sulphur after the cleaning treatment. However, the semi-quantitative data from this technique, based on the intensity of the emission lines, indicated a decrease in the percentage of sulphur in the samples collected after desalination. As can be seen from the micro XRF map (Figure 6), the sulfur was still present on the surface of the micro-sample due to the irregular surface, while the distribution of the calcium was homogeneous. Therefore, the treatment did not act in hard-to-reach areas.

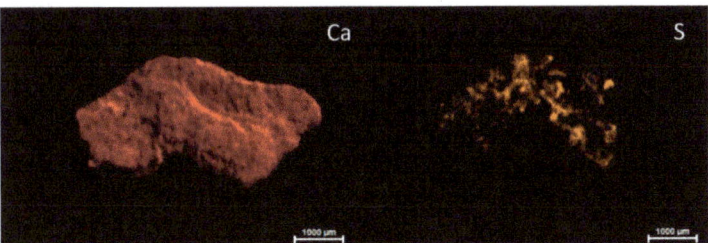

Figure 6. Micro-EDXRF maps of calcium and sulfur distribution after desalination treatment.

However, from an elemental point of view, the analyses carried out on the samples after the cleaning treatment highlighted the presence of other elements whose emission bands were either weak or absent in the previously analysed samples. The presence of strontium associated with calcium was noticed in the samples after cleaning, and the bands of iron, zinc, copper, phosphorus, and manganese appeared much more evident. The emission lines of these elements belonging to the rocky support were partially hidden by the high presence of sulfur in the white layer in some of the samples.

In the three samples that were collected in grey crusts, which had not been solubilised during the cleaning treatment, high percentages of calcium were identified together with silicon to a lesser extent. Raman spectroscopy detected the presence of very sharp Raman bands of calcium carbonate that were always together with the calcium oxalate peaks (Figure S5c). Under the light of the microscope, the samples looked like flat, grey-coloured flakes. No Raman spectra of gypsum were recorded in these samples. On the other hand, in one sample a strong band at 1067 cm^{-1} was visible (Figure S5b). According to the literature, this band, together with the others of low intensity at 722 cm^{-1}, belonged to the sodium nitrate nitratine (NaNO$_3$). This compound was not identified by XRD analysis, probably because its presence was less than the detection limit of this technique. On the other hand, XRD analyses clearly identified the presence of calcium carbonate, silicon dioxide, and hydrated calcium oxalate in all three of the samples. Despite being a soluble compound, some traces of nitrate remained even after cleaning. This compound was only identified in one sample. Considering the fact that these samples were taken from an area where water runs off, the formation of nitrates must be related to traces of organic matter carried away by rainwater. In all samples, the most abundant compounds identified were calcite and calcium oxalate hydrate, which demonstrates that the treatment significantly removed the presence of sulphates without affecting the oxalate film that could contribute to the preservation of the pictographs.

On the other hand, both Raman and XRD analyses demonstrated a very low presence of gypsum in the remaining samples taken from the rock substrate. In these samples, in addition to calcium carbonate, the compounds belonging to the rocky support, dolomite, quartz, and hematite were recognized.

In order to obtain further confirmation of the effectiveness of the cleaning treatment, a statistical analysis of the data obtained by micro EDXRF was carried out.

To extract the maximum information contained in the spectra data, the use of multivariate statistical methods was commonly applied for the data treatment. The most used chemometric methods in the field of cultural heritage is the principal component analysis (PCA) applied to observe clusters of samples based on a specific variable [36]. Principal component analysis was largely employed in many investigations on several materials belonging to cultural heritage [37,38]. This statistic approach allowed the simplification of data interpretation, especially when dealing with large measurement datasets. In the literature, it was used for the identification of execution techniques, classification [39], precedence studies [40], characterization [41], correlation of secondary products and environmental agents [42,43], multispectral imaging treatment [44], colorimetric analysis [45], long-term monitoring [8], etc.

In the field of rock paintings, PCA was employed for the treatment of a data matrix of the LIBS spectra to verify the presence of clusters related to the depth profile analysis of fragments of prehistoric rock wall paintings found at two Brazilian sites. In this way, the clusters were attributed to distinct layers, and the stratigraphy of the samples was characterized [46]. In the study of Linderholm et al. [47], PCA and PLS-DA procedures were applied for near infrared spectroscopy data recorded at a Swedish Stone Age rock painting site (Flatruet, Härjedalen). Chemometric analyses of the EDXRF data was even conducted for the study of the La Peña de Candamo Cave (Asturias, Spain), which was permitted in order to distinguish different painting techniques and the most degraded areas [48].

In the field of cleaning treatment, the use of PCA is quite a new approach to evaluate its effectiveness. In the literature, in relation to cleaning procedures, PCA has mainly been applied for the treatment of colorimetric data in order to monitor natural weathering and cleaning effects on outdoor Bronze monuments patinas [49], on mosaic tiles before and after cleaning [45], or in indoor sustainable cleaning methodology for the sculpted stone of the Duomo of Milan [50].

In our study, performing Principal Component Analysis (PCA) using EDXRF spectra allowed the samples to be considerably separated considering the treatment they underwent (Figure 7). Concretely, the PC2 vs. PC3 plot made it easier to see (i.e., more explicitly) which differences appeared after applying the treatment on the samples. In Figure 7a, it can be observed how the samples before the treatment are grouped mostly on the positive part of PC2, whereas the samples analysed after the treatment are mostly grouped on the negative part of the PC2. It is clear that PC2 (Figure 7b) is the component that better explains the differences induced by the application of the conservation treatment in the samples. By analysing this PC, it can be concluded that sulfur is the main element removed during the treatment (from the elements detected by means of micro EDXRF). Actually, the samples located on the positive side of the PC2 are characterized by a high presence of sulfur, and those located on the negative side are characterized by a near absence of sulfur and a high presence of iron. This fact helps us to understand the clustering of the samples in the PC2 vs. PC3 plot, as the samples collected before the treatment are located in the more sulfur rich area and the samples collected after applying the conservation treatment appear in the region of the plot dominated by high Fe presence and low sulfur. Positive Al, Si, and Fe K bands instead characterize PC3 (Figure 7c). These elements are the main components of the rocky support, and the dispersion of the samples along this PC, especially after the treatment, is explained by the heterogeneity of natural rock. Samples before the treatment seemed to have a sulfur coating that masked rocky composition, preventing EDXRF from detecting this natural heterogeneity.

To better understand the element variations and their relationship with the studied samples, semi-quantitative XRF analyses were performed on the samples' surface. Then, these data were studied by chemometric methods by performing scores and loading bi-plots. First of all, an outlier study was performed through the study of Hotelling T2 vs. Q residuals. After removing outliers, scores and the loading PCA bi-plot (Figure 8) showed, once again, the previously mentioned separation between the non-treated and treated samples. It is wort mentioning, in fact, that both the PCA look very similar regarding the distribution of the samples.

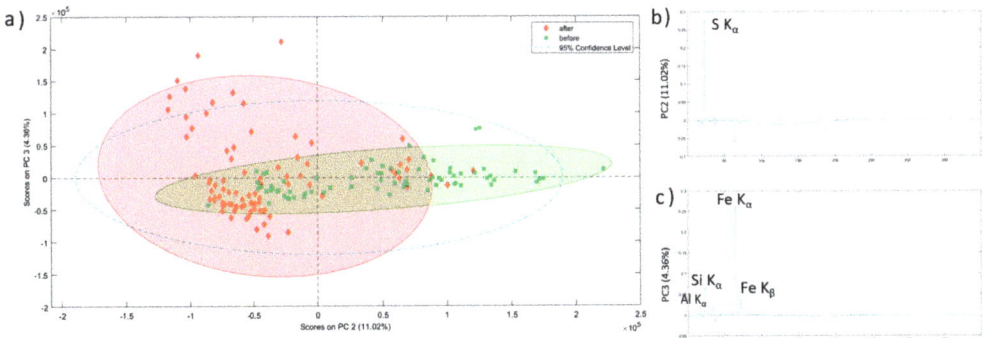

Figure 7. (**a**) PCA score plot of XRF spectra obtained from the sample analysis before and after the conservation treatment. (**b**) PC2 loading plot. (**c**) PC3 loading plot.

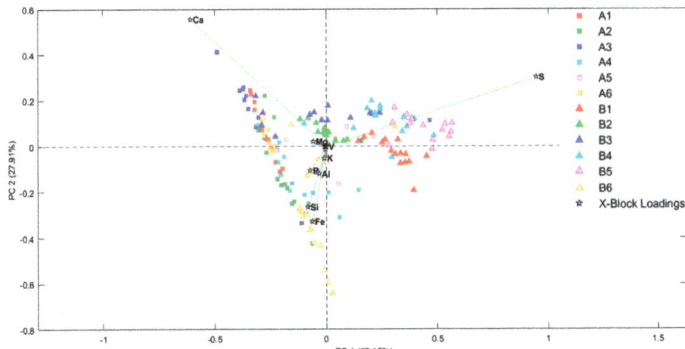

Figure 8. PCA bi-plot of EDXRF spectra obtained from the sample analysis before and after the conservation treatment.

On the other hand, the bi-plot perfectly shows the relationship between the samples and their elemental composition. On the one hand, all of the non-treated samples are highly correlated with sulfur, as observed in the XRF spectra study. However, it must be highlighted that the B6 sample is located near the treated samples, even if it is a non-treated one. The reason for this is that this sample comes from red pictographs and, therefore, has a red coat on its surface. This red coat is composed mainly by iron shifting this sample to the area where iron dominates.

In addition, in this bi-plot, it is appreciated that the treated samples are distributed depending on the Ca and Fe content. This can aid understanding of the different hues observed in the rocky support going from whitish to reddish.

These plots also show some samples or some replicas of the same sample that instead of being non-treated are located nearer the treated samples and vice versa. This fact is again explained by the heterogeneity of a natural system such as a rock-shelter. Considering that EDXRF analysis was performed collimating the X-ray beam to 25 microns, any analyses are subjected to microscopic variation. In this way, it is logical to think that the surface of the rock walls was not perfectly treated, and that minor points remained with a high concentration of sulfur. Likewise, in non-treated samples, some microscopic points with low S presence were located. These cases are just anecdotal, though, compared with the majority of the analysed points from all of the samples (15 points for each sample were recorded).

3. Materials and Methods

3.1. Materials

The rock shelter *Cueva de la Vieja* is located to the east of the province of Albacete, 5 km away from Alpera town, and it was discovered in 1910. It was included in the UNESCO World Heritage list in 1998, as it is one of the most significant sites of the Rock Art of the Mediterranean Basin of the Iberian Peninsula. The rock shelter is a relatively deep-painted panel in full sunlight, where preserved paintings with both a large variety and a large number of figures (human figures, archers, female representations, deer, goats, bulls, horses) are distributed in a panel of 10 m length with the size of the figures ranging from 40 cm to 5 cm. All of these figures painted in red-ochre belong to the Levantine style. Even a group of abstract and geometric motifs are present that correspond to the Schematic style. The most ancient figures in the Levantine style are dated in the Epipaleolithic, which is in the transition from the Paleolithic to the Neolithic period (8000 and 6000 B.C.), while the most modern ones, in the Schematic style, belong to a period between 6000 and 3000 B.C.

The pictographs were barely visible before the cleaning treatment as a whitish layer of concretion covered most of them (Figure S6). The paintings located at both ends of the panel presented a lower incidence of this alteration and, in general, they could be observed much better. On the other hand, alterations such as runoff areas, plaques, small scales, fissures, alveolization, dissolution, and remains of animal activity were also noted. Diagnosis of these alterations and of the processes that generate them was key for verifying whether the panel would withstand cleaning of the whitish crust that makes it difficult to see the cave paintings.

Six samples were analysed before the cleaning treatment. Five of these belonged to the support, and the other one has traces of pigmentation. They were collected in degraded areas where a flaking of the support was clearly visible. After the cleaning, six other samples were collected again (Figure 9). Three of these last samples were collected from an insoluble greyish crust that remained after the cleaning procedure. These areas were located around natural holes in the support through which water could flow periodically.

Figure 9. 3D reconstruction de la *Cueva de la Vieja* after de cleaning treatment where the sampling points (M1–6) are marked.

3.2. Methods

3.2.1. Micro-Energy Dispersive X-ray Fluorescence Spectroscopy (Micro-EDXRF)

The elemental maps were acquired using a M4 TORNADO EDXRF spectrometer (Bruker Nano GmbH, Berlin, Germany). The analyses were performed under vacuum (20 mbar) in order to improve the identification of the lighter elements. The lateral resolution used for spectral acquisitions was 20 µm. The maps were obtained using M-QUANT software. To obtain the quantitative maps, the assignment of the elements and the deconvolution of the spectral information were carried out. The maps were obtained by considering the K-alpha line of each element.

3.2.2. Micro-Raman Spectroscopy

The micro-samples were analysed for their molecular characterization using a confocal Renishaw InVia Raman spectrometer (Renishaw plc, Wootton-under-Edge, UK) coupled to a Leica DMLM microscope (Bradford, UK). The spectra were acquired with the Leica 50× N Plan (0.75 NA) lens with a 2 µm spatial resolution. The minimum theoretical spot diameter, using the 532 nm laser, was for the Leica 50×, 0.9 µm, while using the 785 nm laser, it was 1.7 µm. Additionally, for visualization and focusing, another Leica 5× N Plan (0.12 NA) and a 20× N Plan EPI (0.40 NA) lens were used. For focusing on and searching for points of interest, the microscope implements a motorized stage (XYZ). The power applied was set at the source at a maximum of 50 mW, while on the sample, it was always less than 20 mW in order to avoid possible thermodecomposition of the samples. Normally, 10–300 scans, each lasting 1–20 s, were accumulated to achieve a suitable signal-to-noise ratio at an operating spectral resolution of 1 cm^{-1}.

3.2.3. X-ray Diffraction

The mineralogical composition was characterized by means of the Analytical Xpert PRO X-ray Diffractometer (XRD, PANalytical, Almelo, The Netherlands). The XRD system is equipped with a copper tube, a vertical goniometer (Bragg–Brentano geometry), a programmable divergence slit, a secondary graphite monochromator, and a Pixcel detector. The measurement conditions were set at 40 kV, 40 mA and a scan ranging between 5 and 70° 2theta. Diffractogram interpretation was performed using Win-PLOTR software by comparison with the PDF-4 standards database.

3.2.4. Chemometrics

EDXRF data was analysed using chemometric methods. Two kind of analysis were performed with different aims. On the one hand, PCA analysis was performed using all the EDXRF spectra. After an outlier study, data were mean centered. On the other hand, a semi-quantitative elemental estimation was calculated from the EDXRF spectra. With this dataset and, once again, after an outlier study, the data were mean centered. By this means, a bi-plot study was performed relating scores and loadings. In this case the entire XRF spectrum was employed, while a further statistical treatment was carried out by considering the semi-quantitative data obtained with the aim of finding similarities at elemental levels between all of the samples. For PCA, we used the PLS-Toolbox v.7.0.2 (Eigenvector Research, Wenatchee, WA, USA.) implemented in MATLAB 2010 software (The MathWorks, Natick, MA, USA).

4. Conclusions

According to the results obtained on the micro-samples, it seems evident that the nature of the analysed stone was dolomitic limestone, with aggregates of silica or aluminosilicates as well as the presence of iron oxides, which were mainly hematite. In addition, it has been possible to identify aggregates of anatase, titanium, copper, zinc and manganese. In any case, a much-differentiated composition between the samples was not found. The only sample that presented red pigmentation demonstrated that the pictographs were conducted mainly with hematite mixed with amorphous carbon and calcined bones.

Regarding the external and internal analysis, the significant difference for the micro-samples taken before the cleaning was the presence of a whitish formation in the form of a crust on the external part. Raman and XRD analyses demonstrated that it was mainly composed of gypsum with a minor amount of epsomite. The presence of both sulfates was explained as reaction products of the original material (dolomitic limestone) as well as by the observed relative presence of both degradation compounds. Thus, it appears that a dissolution phenomenon of the original carbonate was taking place, and the following reprecipitation was as sulfates. This reactivity of the material favoured the loss of the consolidant and, therefore, its weakening. Anthropogenic factors and weathering processes contributed to the formation of sulphates on the surface of the painted panel. According to our knowledge of the history of the rock shelter, the practice of humidifiying the panel could be considered as the principal reason for the formation of sulfates, and might even be due to the distribution of the white layer that was exactly on the central area where the pictographs were made.

In addition, it was possible to observe the presence of compounds in almost all of the analysed samples that showed the presence of biological activity by the identification of calcium oxalates monohydrate and dihydrate as well as pigments such as scytonemin or astaxanthin. This indicates that microorganisms, which may also favour the dissolution of the original material and/or the acidification of the support, were affecting the material. On the other hand, it is known from previous studies that the presence of calcium oxalates is not harmful to paints, and that it would preserve their integrity over the centuries. Therefore, evaluating the dangers of the activity of microorganisms for the conservation of paintings can be a complex task.

To the naked eye, this whitish layer was removed thanks to the desalting treatment carried out by the restorers. Since microanalytical techniques have been applied to the study, they still detected the presence of sulfur after the cleaning, although in much smaller quantities. In fact, analyses performed on the samples collected after the desalination treatment showed the presence of calcium oxalates, predominantly whewellite, and showed that calcium carbonate was the main compound. This showed that the treatment removed most of the sulphate compounds without affecting the oxalate film, which is probably one of the main factors that have enabled the pictographs to be preserved.

Statistical treatment of the XRF data using principal component analysis (PCA) confirmed the effectiveness of the treatment, and it highlighted the elemental composition of each sample before and after the treatment. An increased presence of iron, manganese, zinc, copper, and potassium belonging to the rocky substrate is actually evident in the treated samples, which reveals the original nature of the panel.

Long-term monitoring will be carried out in the future by means of a portable instrument in order to check the conservation status of the panel months after the desalination treatment and to verify the influence of weathering processes on the whitish formation.

Supplementary Materials: The following supporting information can be downloaded at: https://www.mdpi.com/article/10.3390/molecules28155854/s1, Figure S1: Distribution of sulfur in the interior face of the sample CuVi04; Figure S2: Raman spectra of quartz (a), hematite (b) with calcite, rutile with gypsum (c) and anatase (d); Figure S3: Microscopic image of traces of hematite; Figure S4: Diffractogram corresponding to the sample CuVi01 and assignment of compounds gupsum, whewellite, quartz, dolomite and weddellite; Figure S5: Raman spectra recorded in the samples taken from grey formation show the presence of calcium oxalate whewellite and silicate compounds (a), calcium carbonate (C) with nitratine (N) (b) and with whewellite (c). Figure S6: Painted panel from *Cueva de Vieja* where the sampling points are marked. The whitish layer is visible on the centre where the pictographs were made.

Author Contributions: Writing-original draft preparation, I.C., formal analysis, I.C. and N.P.-T., writing—review and editing: J.A., supervision, G.A., J.F.R. and J.M.M. project administration, J.F.R., funding acquisition, G.A. and J.M.M. All authors have read and agreed to the published version of the manuscript.

Funding: This research was funded by the DEMORA project (Grant No. PID2020-113391GB-I00), funded by the Spanish Agency for Research (through the Spanish Ministry of Science and Innovation, MICINN, and the European Regional Development Fund, FEDER).

Institutional Review Board Statement: Not applicable.

Informed Consent Statement: Not applicable.

Data Availability Statement: All data generated or analyzed during this study are included in this published article.

Acknowledgments: Technical and manpower support provided by SGIker (UPV/EHU, MICINN, GV/EJ, ERDF, and ESF) is gratefully acknowledged in the XRD analysis. The authors would like to thank the restorers Eudald Guillamet and Laura Ballester. I. Costantini gratefully acknowledges the UPV/EHU for her postdoctoral contract.

Conflicts of Interest: The authors declare no conflict of interest.

Sample Availability: Samples analyzed are available from the authors.

References

1. Brumm, A.; Oktaviana, A.A.; Burhan, B.; Hakim, B.; Lebe, R.; Zhao, J.; Sulistyarto, P.H.; Ririmasse, M.; Adhityatama, S.; Sumantri, I.; et al. Oldest Cave Art Found in Sulawesi. *Sci. Adv.* **2021**, *7*, eabd4648. [CrossRef] [PubMed]
2. Robbins, L.H.; Murphy, M.L.; Brook, G.A.; Ivester, A.H.; Campbell, A.C.; Klein, R.G.; Milo, R.G.; Stewart, K.M.; Downey, W.S.; Stevens, N.J. Archaeology, Palaeoenvironment, and Chronology of the Tsodilo Hills White Paintings Rock Shelter, Northwest Kalahari Desert, Botswana. *J. Archaeol. Sci.* **2000**, *27*, 1085–1113. [CrossRef]
3. Sampietro-Vattuone, M.M.; Peña-Monné, J.L. Application of 2D/3D Models and Alteration Mapping for Detecting Deterioration Processes in Rock Art Heritage (Cerro Colorado, Argentina): A Methodological Proposal. *J. Cult. Herit.* **2021**, *51*, 157–165. [CrossRef]
4. Morillas, H.; Maguregui, M.; Bastante, J.; Huallparimachi, G.; Marcaida, I.; García-Florentino, C.; Astete, F.; Madariaga, J.M. Characterization of the Inkaterra Rock Shelter Paintings Exposed to Tropical Climate (Machupicchu, Peru). *Microchem. J.* **2018**, *137*, 422–428. [CrossRef]
5. Tan, N.H. Rock Art Research in Southeast Asia: A Synthesis. *Arts* **2014**, *3*, 73–104. [CrossRef]
6. Chazine, J.-M.; Setiawan, P. *Discovery of a New Rock Art in East Borneo: New Data for Reflexion*; Collogue UNESCO: Paris, France, 2008.
7. Black, J.L.; MacLeod, I.D.; Smith, B.W. Theoretical Effects of Industrial Emissions on Colour Change at Rock Art Sites on Burrup Peninsula, Western Australia. *J. Archaeol. Sci. Rep.* **2017**, *12*, 457–462. [CrossRef]
8. Aramendia, J.; de Vallejuelo, S.F.-O.; Maguregui, M.; Martinez-Arkarazo, I.; Giakoumaki, A.; Martí, A.P.; Madariaga, J.M.; Ruiz, J.F. Long-Term In Situ Non-Invasive Spectroscopic Monitoring of Weathering Processes in Open-Air Prehistoric Rock Art Sites. *Anal. Bioanal. Chem.* **2020**, *412*, 8155–8166. [CrossRef]
9. Ravindran, T.R.; Arora, A.K.; Singh, M.; Ota, S.B. On-and Off-Site Raman Study of Rock-Shelter Paintings at World-Heritage Site of Bhimbetka. *J. Raman Spectrosc.* **2013**, *44*, 108–113. [CrossRef]
10. Mazel, V.; Richardin, P.; Touboul, D.; Brunelle, A.; Richard, C.; Laval, E.; Walter, P.; Laprévote, O. Animal Urine as Painting Materials in African Rock Art Revealed by Cluster ToF-SIMS Mass Spectrometry Imaging. *J. Mass Spectrom.* **2010**, *45*, 944–950. [CrossRef]
11. Gomes, H.; Collado Giraldo, H.; Martins, A.; Nash, G.; Rosina, P.; Vaccaro, C.; Volpe, L. Pigment in Western Iberian Schematic Rock Art: An Analytical Approach. *Mediterr. Archaeol. Archaeom. Int. Sci. J.* **2015**, *15*, 163–175. [CrossRef]
12. Pitarch, À.; Francisco Ruiz, J.; de Vallejuelo, S.F.-O.; Hernanz, A.; Maguregui, M.; Manuel Madariaga, J. In Situ Characterization by Raman and X-ray Fluorescence Spectroscopy of Post-Paleolithic Blackish Pictographs Exposed to the Open Air in Los Chaparros shelter (Albalate del Arzobispo, Teruel, Spain). *Anal. Methods* **2014**, *6*, 6641–6650. [CrossRef]
13. Iriarte, M.; Hernanz, A.; Ruiz-López, J.F.; Martín, S. μ-Raman Spectroscopy of Prehistoric Paintings from the Abrigo Remacha Rock Shelter (Villaseca, Segovia, Spain). *J. Raman Spectrosc.* **2013**, *44*, 1557–1562. [CrossRef]
14. Hernanz, A.; Ruiz-López, J.F.; Madariaga, J.M.; Gavrilenko, E.; Maguregui, M.; Fdez-Ortiz de Vallejuelo, S.; Martínez-Arkarazo, I.; Alloza-Izquierdo, R.; Baldellou-Martínez, V.; Viñas-Vallverdú, R.; et al. Spectroscopic Characterisation of Crusts Interstratified with Prehistoric Paintings Preserved in Open-Air Rock Art Shelters. *J. Raman Spectrosc.* **2014**, *45*, 1236–1243. [CrossRef]
15. Hernanz, A.; Ruiz-López, J.F.; Gavira-Vallejo, J.M.; Martin, S.; Gavrilenko, E. Raman Microscopy of Prehistoric Rock Paintings from the Hoz de Vicente, Minglanilla, Cuenca, Spain. *J. Raman Spectrosc.* **2010**, *41*, 1394–1399. [CrossRef]
16. Pozo-Antonio, J.S.; Rivas, T.; Carrera, F.; García, L. Deterioration Processes Affecting Prehistoric Rock Art Engravings in Granite in NW Spain. *Earth Surf. Process. Landf.* **2018**, *43*, 2435–2448. [CrossRef]
17. Hernanz, A.; Gavira-Vallejo, J.M.; Ruiz-López, J.F.; Edwards, H.G.M. A Comprehensive Micro-Raman Spectroscopic Study of Prehistoric Rock Paintings from the Sierra de las Cuerdas, Cuenca, Spain. *J. Raman Spectrosc.* **2008**, *39*, 972–984. [CrossRef]
18. Peña-Monné, J.L.; Sampietro-Vattuone, M.M.; Báez, W.A.; García-Giménez, R.; Stábile, F.M.; Martínez Stagnaro, S.Y.; Tissera, L.E. Sandstone Weathering Processes in the Painted Rock Shelters of Cerro Colorado (Córdoba, Argentina). *Geoarchaeology* **2022**, *37*, 332–349. [CrossRef]
19. Rousaki, A.; Vargas, E.; Vázquez, C.; Aldazábal, V.; Bellelli, C.; Carballido Calatayud, M.; Hajduk, A.; Palacios, O.; Moens, L.; Vandenabeele, P. On-Field Raman Spectroscopy of Patagonian Prehistoric Rock Art: Pigments, Alteration Products and Substrata. *TrAC Trends Anal. Chem.* **2018**, *105*, 338–351. [CrossRef]
20. Ilmi, M.M.; Maryanti, E.; Nurdini, N.; Lebe, R.; Oktaviana, A.A.; Burhan, B.; Perston, Y.L.; Setiawan, P.; Ismunandar; Kadja, G.T.M. Uncovering the Chemistry of Color Change in Rock Art in Leang Tedongnge (Pangkep Regency, South Sulawesi, Indonesia). *J. Archaeol. Sci. Rep.* **2023**, *48*, 103871. [CrossRef]

21. Ruiz López, J.F. El Abrigo de los Oculados (Henarejos, Cuenca). In *Actas del Congreso de Arte Rupestre Esquemático en la Península Ibérica: Comarca de Los Vélez, 5–7 May 2004*; García, J.M., Ed.; Dialnet: Barcelona, Spain, 2006; pp. 375–388, ISBN 84-611-2821-4.
22. Ruiz, J.F.; Hernanz, A.; Armitage, R.A.; Rowe, M.W.; Viñas, R.; Gavira-Vallejo, J.M.; Rubio, A. Calcium Oxalate AMS 14C Dating and Chronology of Post-Palaeolithic Rock Paintings in the Iberian Peninsula. Two dates from Abrigo de los Oculados (Henarejos, Cuenca, Spain). *J. Archaeol. Sci.* **2012**, *39*, 2655–2667. [CrossRef]
23. Lofrumento, C.; Ricci, M.; Bachechi, L.; De Feo, D.; Castellucci, E.M. The First Spectroscopic Analysis of ETHIOPIAN Prehistoric Rock Painting. *J. Raman Spectrosc.* **2012**, *43*, 809–816. [CrossRef]
24. Hedges, R.E.M.; Ramsey, C.B.; Klinken, G.J.V.; Pettitt, P.B.; Nielsen-Marsh, C.; Etchegoyen, A.; Niello, J.O.F.; Boschin, M.T.; Llamazares, A.M. Methodological Issues in the 14C Dating of Rock Paintings. *Radiocarbon* **1997**, *40*, 35–44. [CrossRef]
25. Baran, E.J. Review: Natural Oxalates and Their Analogous Synthetic Complexes. *J. Coord. Chem.* **2014**, *67*, 3734–3768. [CrossRef]
26. Sonke, A.; Trembath-Reichert, E. Expanding the Taxonomic and Environmental Extent of an Underexplored Carbon Metabolism—Oxalotrophy. *Front. Microbiol.* **2023**, *14*, 1161937. [CrossRef]
27. Russ, J.; Kaluarachchi, W.D.; Drummond, L.; Edwards, H.G.M. The Nature of a Whewellite-Rich Rock Crust Associated with Pictographs in Southwestern Texas. *Stud. Conserv.* **1999**, *44*, 91–103. [CrossRef]
28. Russ, J.; Palma, R.L.; Loyd, D.H.; Boutton, T.W.; Coy, M.A. Origin of the Whewellite-Rich Rock Crust in the Lower Pecos Region of Southwest Texas and Its Significance to Paleoclimate Reconstructions. *Quat. Res.* **1996**, *46*, 27–36. [CrossRef]
29. López-Montalvo, E.; Villaverde, V.; Roldán, C.; Murcia, S.; Badal, E. An Approximation to the Study of Black Pigments in Cova Remigia (Castellón, Spain). Technical and Cultural Assessments of the Use of Carbon-Based Black Pigments in Spanish Levantine Rock Art. *J. Archaeol. Sci.* **2014**, *52*, 535–545. [CrossRef]
30. Domingo, I.; Chieli, A. Characterizing the Pigments and Paints of Prehistoric Artists. *Archaeol. Anthropol. Sci.* **2021**, *13*, 196. [CrossRef]
31. Beltrán, A. El arte rupestre levantino, cronología y significación. *Caesaraugusta* **1968**, *31*, 7–43.
32. Domingo Sanz, I.; Vendrell, M.; Chieli, A. A Critical Assessment of the Potential and Limitations of Physicochemical Analysis to Advance Knowledge on Levantine Rock Art. *Quat. Int.* **2021**, *572*, 24–40. [CrossRef]
33. Roldán, C.; Murcia-Mascarós, S.; Ferrero, J.; Villaverde, V.; López, E.; Domingo, I.; Martínez, R.; Guillem, P.M. Application of Field Portable EDXRF Spectrometry to Analysis of Pigments of Levantine Rock Art. *X-ray Spectrom.* **2010**, *39*, 243–250. [CrossRef]
34. Mas, M.; Jorge, A.; Gavilán, B.; Solís, M.; Parra, E.; Pérez, P.-P. Minateda Rock Shelters (Albacete) and Post-Palaeolithic Art of the Mediterranean Basin in Spain: Pigments, Surfaces and Patinas. *J. Archaeol. Sci.* **2013**, *40*, 4635–4647. [CrossRef]
35. Iriarte, M.; Hernanz, A.; Gavira-Vallejo, J.M.; de Buruaga, A.S.; Martín, S. Micro-Raman Spectroscopy of Rock Paintings from the Galb Budarga and Tuama Budarga Rock Shelters, Western Sahara. *Microchem. J.* **2018**, *137*, 250–257. [CrossRef]
36. Manuel Madariaga, J. Analytical Chemistry in the Field of Cultural Heritage. *Anal. Methods* **2015**, *7*, 4848–4876. [CrossRef]
37. Musumarra, G.; Fichera, M. Chemometrics and Cultural Heritage. *Chemom. Intell. Lab. Syst.* **1998**, *44*, 363–372. [CrossRef]
38. Visco, G.; Avino, P. Employ of Multivariate Analysis and Chemometrics in Cultural Heritage and Environment Fields. *Environ. Sci. Pollut. Res.* **2017**, *24*, 13863–13865. [CrossRef]
39. Festa, G.; Scatigno, C.; Armetta, F.; Saladino, M.L.; Ciaramitaro, V.; Nardo, V.M.; Ponterio, R.C. Chemometric Tools to Point Out Benchmarks and Chromophores in Pigments through Spectroscopic Data Analyses. *Molecules* **2022**, *27*, 163. [CrossRef]
40. Piña-Torres, C.; Lucero-Gómez, P.; Nieto, S.; Vázquez, A.; Bucio, L.; Belio, I.; Vega, R.; Mathe, C.; Vieillescazes, C. An Analytical Strategy Based on Fourier Transform Infrared Spectroscopy, Principal Component Analysis and Linear Discriminant Analysis to Suggest the Botanical Origin of Resins from Bursera. Application to Archaeological Aztec Samples. *J. Cult. Herit.* **2018**, *33*, 48–59. [CrossRef]
41. Donais, M.K.; Douglass, L.; Ramundt, W.H.; Bizzarri, C.; George, D.B. Handheld Laser-Induced Breakdown Spectroscopy for Field Archaeology: Characterization of Roman Wall Mortars and Etruscan Ceramics. *Appl. Spectrosc. Pract.* **2023**, *1*, 27551857231175850. [CrossRef]
42. Moropoulou, A.; Polikreti, K. Principal Component Analysis in Monument Conservation: Three Application Examples. *J. Cult. Herit.* **2009**, *10*, 73–81. [CrossRef]
43. Brai, M.; Casaletto, M.P.; Gennaro, G.; Marrale, M.; Schillaci, T.; Tranchina, L. Degradation of Stone Materials in the Archaeological Context of the Greek–Roman Theatre in Taormina (Sicily, Italy). *Appl. Phys. A* **2010**, *100*, 945–951. [CrossRef]
44. Colantonio, C.; Pelosi, C.; D'Alessandro, L.; Sottile, S.; Calabrò, G.; Melis, M. Hypercolorimetric Multispectral Imaging System for Cultural Heritage Diagnostics: An Innovative Study for Copper Painting Examination. *Eur. Phys. J. Plus* **2018**, *133*, 526. [CrossRef]
45. Alberghina, M.F.; Barraco, R.; Basile, S.; Brai, M.; Pellegrino, L.; Prestileo, F.; Schiavone, S.; Tranchina, L. Mosaic Floors of Roman Villa del Casale: Principal Component Analysis on Spectrophotometric and Colorimetric Data. *J. Cult. Herit.* **2014**, *15*, 92–97. [CrossRef]
46. Borba, F.S.L.; Cortez, J.; Asfora, V.K.; Pasquini, C.; Pimentel, M.F.; Pessis, A.-M.; Khoury, H.J. Multivariate Treatment of LIBS Data of Prehistoric Paintings. *J. Braz. Chem. Soc.* **2012**, *23*, 958–965. [CrossRef]
47. Linderholm, J.; Geladi, P.; Sciuto, C. Field-Based near Infrared Spectroscopy for Analysis of Scandinavian Stone Age Rock Paintings. *J. Near Infrared Spectrosc.* **2015**, *23*, 227–236. [CrossRef]
48. Olivares, M.; Castro, K.; Corchón, M.S.; Gárate, D.; Murelaga, X.; Sarmiento, A.; Etxebarria, N. Non-Invasive Portable Instrumentation to Study Palaeolithic Rock Paintings: The Case of La Peña Cave in San Roman de Candamo (Asturias, Spain). *J. Archaeol. Sci.* **2013**, *40*, 1354–1360. [CrossRef]

49. Luciano, G.; Leardi, R.; Letardi, P. Principal Component Analysis of Colour Measurements of Patinas and Coating Systems for Outdoor Bronze Monuments. *J. Cult. Herit.* **2009**, *10*, 331–337. [CrossRef]
50. Gulotta, D.; Saviello, D.; Gherardi, F.; Toniolo, L.; Anzani, M.; Rabbolini, A.; Goidanich, S. Setup of a Sustainable Indoor Cleaning Methodology for the Sculpted Stone Surfaces of the Duomo of Milan. *Herit. Sci.* **2014**, *2*, 6. [CrossRef]

Disclaimer/Publisher's Note: The statements, opinions and data contained in all publications are solely those of the individual author(s) and contributor(s) and not of MDPI and/or the editor(s). MDPI and/or the editor(s) disclaim responsibility for any injury to people or property resulting from any ideas, methods, instructions or products referred to in the content.

Article

The Bio-Patina on a Hypogeum Wall of the Matera-Sassi Rupestrian Church "San Pietro Barisano" before and after Treatment with Glycoalkaloids

Francesco Cardellicchio [1], Sabino Aurelio Bufo [2,3,*], Stefania Mirela Mang [4], Ippolito Camele [4], Anna Maria Salvi [2] and Laura Scrano [5]

1. Institute of Methodologies for Environmental Analysis, Italian Research Council (CNR), Contrada Loya, 85050 Tito, Italy
2. Department of Sciences, University of Basilicata, Via dell'Ateneo Lucano 10, 85100 Potenza, Italy
3. Department of Geography, Environmental Management and Energy Studies, University of Johannesburg, Johannesburg 2092, South Africa
4. School of Agriculture, Forestry and Environment, University of Basilicata, 85100 Potenza, Italy
5. Department of European Cultures, University of Basilicata, 75100 Matera, Italy
* Correspondence: sabino.bufo@unibas.it

Abstract: The investigation focused on the deterioration of the walls in the hypogeum of "San Pietro Barisano" rupestrian church, located in the Matera-Sassi (Southern Italy), one of the UNESCO World Heritage sites. The study evaluated the biocide activity of a mixture of natural glycoalkaloids (GAs) extracted from the unripe fruit of *Solanum nigrum* and applied to clean a hypogeum wall surface in the church affected by bio-patinas. The analyzed bio-patina, collected before treatment and, at pre-established times, after treatment, showed changes in chemical composition detected by XPS, accompanied by visible discoloration and biological activity variation. The biocidal action of the glycoalkaloids mixture, directly employed on the wall surface, was effective after about four weeks for most bio-patina colonizers but not for the fungal species that can migrate and survive in the porosities of the calcarenite. Consequently, the cleaning procedure requires the integration of fungicidal actions, combined with the consolidation of the surfaces, to obtain complete bioremediation and avoid subsequent biological recolonization. SEM images and associated microanalysis of pretreated bio-patina have revealed the biocalcogenity of some autochthonous microorganisms, thus preluding to their eventual isolation and reintroduction on the wall surface to act as consolidants once the bio-cleaning phase has been completed.

Keywords: glycoalkaloids; bio-cleaning; XPS; SEM/EDS; cultural heritage; San Pietro Barisano church; Matera-Sassi

1. Introduction

In recent decades, historical and artistic heritage has undergone greater degradation than in the past due to the synergistic action of atmospheric pollution, climate change, and increased biological contamination. The conservation of this heritage also requires the development of innovative, effective, and, at the same time, non-expensive protection strategies. In particular, the biodeterioration of stone materials is related to the biological activity of colonizing microorganisms favored by environmental factors such as temperature, humidity, and lighting. The degradation process begins with surface alterations that produce unsightly and chromatic stains. After the first phase of settlement of primary microorganisms, biodegradation can considerably influence the integrity of the materials with negative consequences on their conservation [1–4]. Photoautotrophic microorganisms are the primary colonizers that can find suitable conditions to secrete exopolysaccharides, alginates, and other compounds useful for binding to the rock. The constitution of this biofilm then favors the settlement of secondary colonizations and heterotrophic microorganisms [5–8]. In recent years, studying biodegradation phenomena in sites of cultural interest has also been one of the activities carried out within the project "Smart Cities, Communities and Social Innovation, SCN_00520", funded by the Italian Ministry of Research and University. The Sassi of Matera (Matera, Basilicata Region, Southern Italy) and the rupestrian churches carved into the limestone have represented important study sites for the extensive degradation phenomena of the surfaces also due to biological colonization [9,10]. In particular, attention was paid to the church of San Pietro Barisano (Figure 1), an example of a typical architectural structure of the Matera-Sassi system. The church also has a hypogeum on the lower level without natural light and enough air exchange. The climatic conditions of the hypogeum, dedicated in the past to the putrefaction of corpses before the final burial in the ossuary (*Putridarium*), were similar but extreme to those of the upper church with a relative humidity of the surfaces above 90% and an ambient temperature between 11 and 17 °C [9]. In this environment, the microclimatic conditions provide an ideal "niche" for developing photosynthetic organisms capable of exploiting the spectral emission of artificial lighting. Consequently, suitable conditions are created for the development of photoautotrophic organisms such as cyanobacteria, green filamentous algae, diatoms, and mosses responsible for forming biofilms on surfaces. At the same time, the products of photosynthesis provide nourishment for the settlement of heterotrophic species such as fungi and bacteria, the development of which produces an increase in biological activity on the colonized surface [10,11]. Biodegradation due to the development of lichens, bacteria, and fungi was also found in places of the upper church due to the high relative humidity and water infiltrations produced by the high porosity of the calcarenite. In Figure 2, for example, particular black crusts can be seen on the altar of St. Joseph [12,13]. For the elimination of biological patinas on the surfaces of historical-artistic artifacts, up to now, both inorganic compounds (such as sodium hypochlorite, hydrogen peroxide, and active chlorine) and organic compounds (such as formaldehyde esters, methyl-phosphates, chloramines, etc.) have been used [14]. These substances can be harmful to both humans and the environment and can also cause damage to architectural materials [15]. The present research work, therefore, was aimed at finding innovative and ecological approaches based on the use of biocides derived from natural compounds, which can interfere, at the molecular level, with the microbial communication system called "quorum sensing" and inhibit the initial phase of microbial biofilm formation [16,17]. Among the natural biocides, glycoalkaloids deserve attention. They are secondary metabolites produced by plants of the *Solanaceae* family, which use them as chemical defenses against pathogens such as fungi, bacteria, and viruses. A series of investigations on glycoalkaloids isolated from different species of plants of the genus *Solanum* has demonstrated antifungal activity [18–20] and antimicrobial action [21] of these compounds. Some glycoalkaloids have revealed antiviral activity. Glycoalkaloids such as solamargine and solasonine can also inactivate various forms of herpes [22]. The solasodine extracted from the flowers of *Solanum dulcamara* also inhibits the growth of *Escherichia coli* and *Staphylococcus aureus*. In this work, the glycoalkaloids' crude extract (containing approximately 510 µM solamargine and 460 µM solasonine as main components) obtained from the berries of *Solanum nigrum* was tested in the hypogeum of San Pietro Barisano for bio-cleaning purposes. The experiment was carried out on a portion of the wall with an extensive biological patina. Bacterial and fungi species on the internal walls were identified with the final aim of choosing the best intervention strategy [13,17,23–25]. The disappearance of the biological patina was evaluated not only by a visual investigation but also by surface analytical techniques such as XPS photoelectron spectroscopy [26,27], scanning electron microscopy (SEM), and energy dispersive X-ray spectroscopy (EDS).

Figure 1. External facade (**a**) and hypogeum (**b**) of the San Pietro Barisano church.

Figure 2. Black crusts on the altar of St. Joseph in the San Pietro Barisano church.

2. Results and Discussion

2.1. XPS Analysis

On the horizontal wall of the hypogeum, chosen as the study area, a sample of surface patina was taken and subjected to XPS analysis before treatment with the glycoalkaloid mixture. The analyses of this sample were compared with those of the samples taken after treatment. Figure 3 shows the XPS wide spectra and the peaks of interest related to the samples taken pre- and post-application of the glycoalkaloid mixture for four weeks.

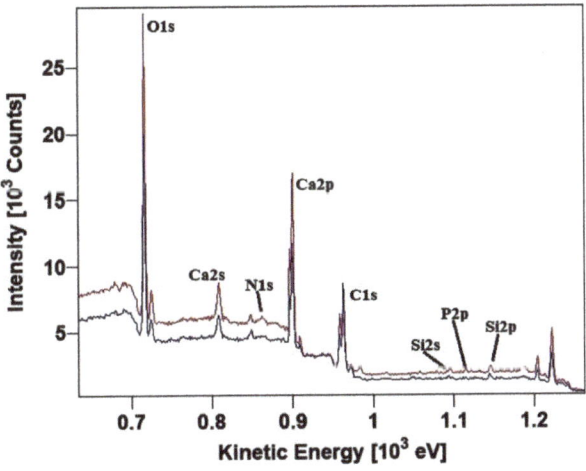

Figure 3. Wide spectra of the samples of the horizontal wall of the hypogeum: pre-(blue) and post-(red) application of glycoalkaloids extract.

The most representative elements in the labeled peaks were acquired at high resolution to proceed with the semi-quantitative analysis via curve-fitting and to derive the chemical composition of the patina before and after treatment with glycoalkaloids. The C1s region is the first to be analyzed (Figure 4) as it contains the carbonate peak, set at 290.0 eV from previous analysis of calcarenite samples and used as an internal reference to correct the shift on binding energies (BEs, eV) due to surface charging [28]. In addition, it is the region most contributive in discerning the organic components associated with the bio-patina's biological activity and their intensity variation after the treatment with glycoalkaloids (Tables 1 and 2).

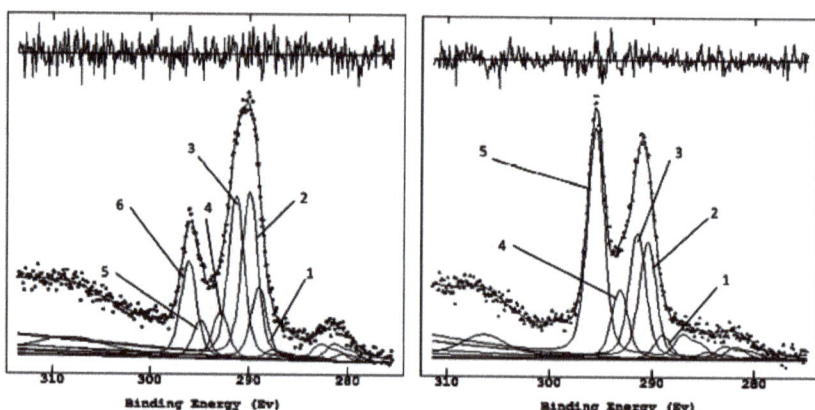

Figure 4. Curve fitting of the C1s region pre-(**left**) and post-(**right**)application of glycoalkaloids.

Table 1. Corrected BE, normalized area, and assignments for C1s regions: pre-treatment sample.

Region	Peak	B.E. Corrected	Assignments	Normalized Area
C1s	1	283.0	C-C (polycyclics, graphite, carbides)	312.5
	2	283.8	C-C (polycyclics, graphite, carbides)	3157.3
	3	285.2	C-C, C-N	6596.0
	4	286.7	C-OH, φ-OH, C=N	6429.4
	5	288.8	$C_2O_4^{2-}$	1877.3
	6	290.0	CO_3^{2-}	5311.7

Table 2. Corrected BE, normalized area and assignments for the C1s regions: post-treatment sample.

Peak	B.E. Corrected	Assignments	Normalized Area
1	283.4	C-C (graphite, carbides, polycyclics)	883.0
2	285.0	C-C	4293.0
3	286.1	C-N, C-O, C-O-C	4267.4
4	287.6	C=O, O-C-O	2351.7
5	290.0	CO_3^{2-}	7345.0

Comparing the C1s carbon regions before and after the treatment with the biocide allows us to detect the decrease of the organic components related to the biological activity [29]. Indeed, it is possible to observe the disappearance of the peak due to calcium oxalate and the increase of the signal associated with the carbonate. These changes confirm the gradual removal of the biological patina, which initially covered the wall and highlighted the calcarenite component, following the biocidal action of the glycoalkaloids spread on the surface and left to act for four weeks. Figure 5 illustrates the Ca2p region relating to the pre- and post-treatment, and Table 3 reports the assignments of the various peaks. Noteworthy, the Ca2p region remains unchanged in shape, within the limits of energy resolution with achromatic sources.

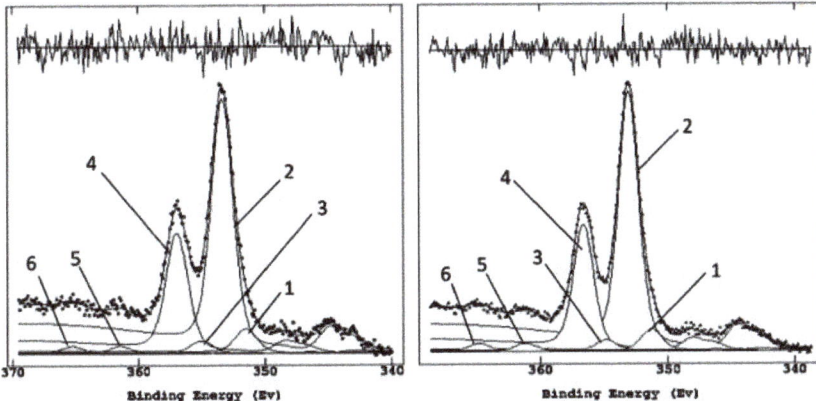

Figure 5. Curve fitting of Ca2p regions pre (**left**) and post (**right**) glycoalkaloids application.

Table 3. Peak assignment, corrected BE, and normalized area for the Ca2p region.

Region	Peak	Pre-Treatment		Post-Treatment		Assignments
		B.E. Corrected	Normalized Area	B.E. Corrected	Normalized Area	
Ca2p	1	345.4	386.9	345.8	614.8	Ca
	2	347.3	3891.9	347.5	7106.2	$CaCO_3$/Ca-alginate
	5	355.4	80.2	355.6	143.3	Shakeup
	6	359.2	80.2	359.3	143.3	Shakeup

Comparing the peaks of pre- and post-treatment and the related normalized areas, calcium level increases due to the rise in the percentage of carbonate. The growth of the calcium carbonate signal demonstrates the cleaning effectiveness; however, the XPS analysis results show the incomplete removal of the surface patina's organic components, which may also include calcium alginate often found in biofilms of the colonizing organisms [30]. Alginate is a component of the cell walls of algae. In this work, the presence of microalgae (diatoms) was confirmed on the hypogeum walls by the SEM/EDS investigation illustrated below.

The O1s regions centered at 533.0 and 532.2 eV, respectively, before and after treatment, include a set of contributions due to the detected oxygenated compounds (C-O, O-H, SiOx bonds, phosphates, sulfates, metal oxides, etc.), some of which are so closely spaced in energy that they cannot be resolved by curve-fitting. However, the total area of O1s made it possible to evaluate the goodness of the assignments through the stoichiometric mass balance.

Table 4 shows the assignments of the peaks relating to minority components identified in the other regions analyzed. The regions refer to elements attributable to silicate and bioclastic compounds of the calcarenite and the metabolites present in the bio-patina, together with the carbonaceous components that make up the biofilm. Finally, after partial and overall mass balance cross-checking, the results were summarized in the pie charts in Figure 6, where the percentage distributions of the various chemical groups determined in the samples before and after treatment with the glycoalkaloid mixture are shown. There was a decrease in the organic carbon bound to the biological fraction present on the wall and a simultaneous increase in the carbonate fraction (from 21.6% to 33.8%) and silicate components typical of calcarenite rock (assuming as reference the chemical composition of the highly porous Gravina's calcarenite, mainly composed of calcium carbonate, as calcite, phyllosilicates, and small percentages of quartz) [31]. The monitored changes confirm the gradual removal of some components linked to the biological activity of the patina covering the wall due to the biocidal action of the glycoalkaloids spread on the surface and left to act for four weeks.

Table 4. Assignment related to Si2p, P2p, and Na1s peaks.

Region	B.E. Corrected (pre-)	B.E. Corrected (post-)	Normalized Area (pre-)	Normalized Area (post-)	Assignments
Si(2p)	102.7	102.8	608.0	839.5	SiC, SiO_2
P(2p)	132.7	133.2	58.5	345.1	$CaPO_4$
Na(1s)	1072.2	1072.4	387.6	613.7	NaH_2PO_4, Na_2HPO_4

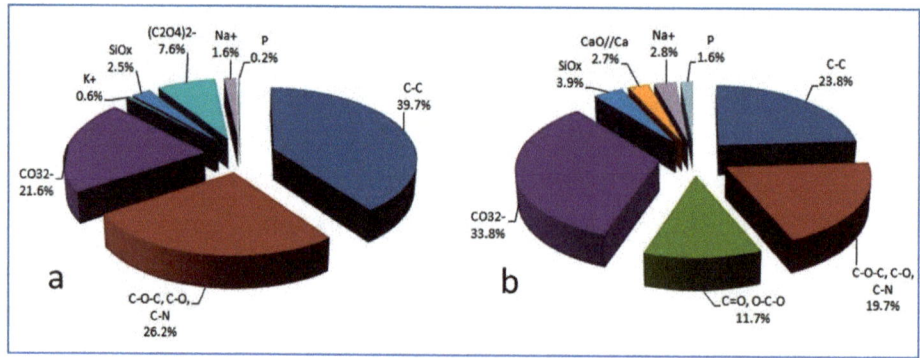

Figure 6. Percentage distribution of the chemical groups determined in the patina samples before (**a**) and after (**b**) treatment with glycoalkaloids.

Samples were taken from the wall in the treated area, even 1 and 2 weeks after application, to evaluate the biocide action according to the time of application. For brevity, below are only reported results for the C1s region, highlighting the most significant functional groups emphasized by the curve-fitting procedure. The comparisons of the C1s regions (Figure 7) are exciting and show that after a week of application, the peak 5 relative to the oxalate was no longer detectable. Moreover, the carbonate signal (peak 6) increased after the second week while the carbonaceous components were reduced.

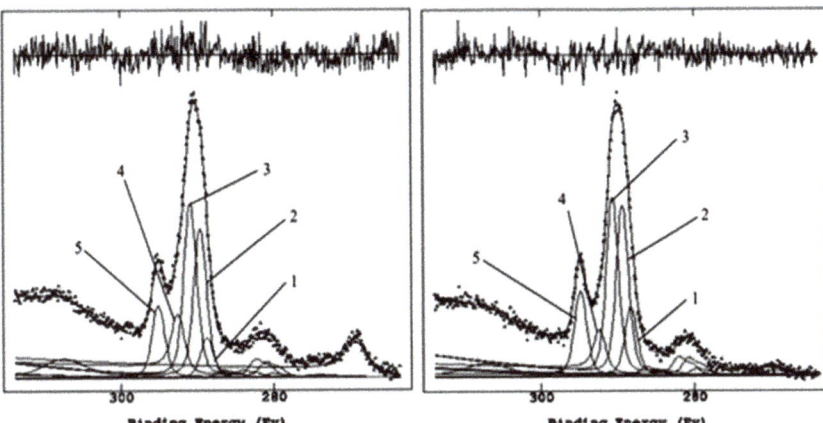

Figure 7. Curve fitting of C1s regions post-1 week (**left**), post-2 weeks (**right**).

It is essential to underline the disappearance of the peak centered at about 289 eV, attributable to oxalate, partially superimposed on the reference peak of carbonate at 290.0 eV but easily resolved by curve-fitting after the treatment.

However, the normalized areas Ca^{2+}/CO_3^{2-} ratio was never unitary, as theoretically expected for the stoichiometric calcium carbonate. The discordance from unity was decreased with the treatment time, and after four weeks, the Ca^{2+}/CO_3^{2-} ratio was nearly 0.90, reaching the lower limit of the uncertainty associated with XPS analysis, i.e., ±10% [28]. Thus, while the increase of the carbonate signal demonstrates the effectiveness of the cleaning action, the non-correct carbonate stoichiometry implies the bonding of calcium ions with other components of the surface patina, which can also include calcium alginate produced in the biofilm by the colonizing microorganisms, anchored on carbonate stones [5,6].

The carboxylic $(COO^-)_2$ interchain crosslinking with Ca^{2+} and the participation of alcohol and ether groups of the alginate chain in the coordination sphere are reported in the literature [32] together with the correspondent XPS assignments. These correspondences ($Ca2p_{3/2}$ 347.7 and C1s 290.5 eV, practically superimposed to $CaCO_3$ positions in both regions) were experimentally confirmed with the XPS acquisition of an alginate-based hydrogel, produced in our laboratory after a prolonged interaction with a Carrara marble surface [33].

Figure 8 shows the color variation of the surficial patina during the four weeks of treatment. During this period, the bio-patina reduced and simultaneously underwent chromatic variations until the initial green color almost wholly disappeared. This confirms that the producing-patina microorganisms were eliminated, demonstrating the effectiveness of the treatment after four weeks, as also shown by the experimental data obtained with the XPS analysis.

Figure 8. Color appearance of the bio-patina in the hypogeum of San Pietro Barisano during the treatment period with glycoalkaloids: (**A**) before the treatment; (**B**) post-1 week; (**C**) post-2 weeks; (**D**) post-4 weeks.

2.2. Biological Analysis

From the bio-patina of the San Pietro Barisano rupestrian church hypogeum, several bacterial and fungal species were isolated and further identified by their morphological key features and molecularly recognized by PCR amplification and sequencing of the two Internal Transcribed Spacers and 5.8 S gene (ITS1-5.8S-ITS2) from the nuclear ribosomal RNA (for fungi) and 16S gene (for bacteria), as shown in Table 5 and Figure 9. All fungal and bacterial nucleotide sequences were deposited in the GenBank (NCBI) under the following accession numbers: *Botryotrichum atrogriseum* (OP888461-OP888463); *Penicillium chrysogenum* (OP890579-OP890581); *Talaromyces pinophilus* (OP894917-OP894919) and *Cladosporium herbarum* (OP890587-OP890590). The sequences of bacteria also received their accession numbers: *Staphylococcus warneri* (CP003668.1); *Brevibacillus* (SRR17297603); *Bacillus cereus* (CP020803.1); *Bacillus mycoides* (GCA_000832605); *Bacillus firmus* (SAMEA4076706).

Table 5. Bacterial and fungal species identified in the bio-patina of the hypogeum.

Bacteria	Fungi
Staphylococcus warneri	*Botryotrichum atrogriseum*
Brevibacillus spp.	*Penicillium chrysogenum*
Bacillus cereus	*Talaromyces pinophilus*
Bacillus mycoides	*Cladosporium herbarum*
Bacillus firmus	

The complex biological colony is typical of humid and poorly lit environments such as hypogea, caves, and catacombs [11,34,35]. In these environments, the organisms find the conditions suitable for their growth: humidity, organic material produced by primary or decomposing organisms, artificial lighting, etc. Among the bacterial species identified, some are toxic species, such as *B. cereus*, a pathogenic bacterium that produces toxins responsible for food poisoning, and *B. mycoides*, capable of causing diseases in some organisms. Among the fungal species, *P. chrysogenum*, a fungus belonging to the *Aspergillaceae* family, is known to be a pathogen that forms blue or grey-green mold. Many fungi identified, such as *T. pinophilus*, have high enzymatic activities capable of degrading various substrates. Species of the genus *Talaromyces* produce enzymes beneficial for degrading biomass and secondary metabolites. However, these enzymes are still poorly characterized due to the lack of complete genetic information. To verify the biocidal action on the fungi found on the hypogeal walls, the glycoalkaloids extracted from *S. nigrum* were tested on *B. atrogriseum* and *T. pinophilus* at the following concentrations: 100, 75, 50, and 25%. The in vitro results showed low biocidal activity for these fungal species due to a restored activity of the fungal colony four weeks after the treatment with glycoalkaloids [12]. Despite the antibacterial activity of glycoalkaloids demonstrated by previous findings [23,36], the studied extract is similarly in vivo characterized by low effectiveness towards fungal species. Under these conditions, the treatment could even increase fungal colonies that can use dead bacterial cells as a carbon source.

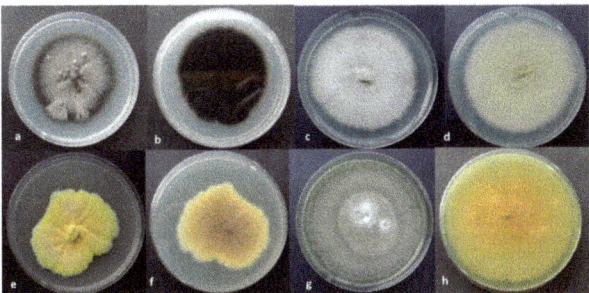

Figure 9. Fungal species isolated from hypogeum walls of the San Pietro Barisano rupestrian church on PDA media (**a,b** = *C. herbarium*; **c,d** = *B. atrogriseum*; **e,f** = *T. pinophilus*; **g,h** = *P. chrysogenum*).

Fungal survival could also be favored by a possible migration inside the stone's pores, given the extensive alveolarization, making biocidal action difficult.

The role of various microorganisms in biofilms is still a complex matter; e.g., it is notable the joint function of compounds such as calcium oxalate, which, although produced by the symbiosis of fungal species and oxalotrophic bacteria, can only be metabolized by the bacteria with the formation of calcium carbonate [37].

It is also worth emphasizing the lower efficacy of glycoalkaloids supported on gels [12]. However, gels, or sometimes cellulose pulp, have proved very useful within the Smart Cities project to work more precisely and avoid the deposition of other contaminants on the surface [31,38–41]. In the case of the hypogeum of San Pietro Barisano church, the experimentation with sodium alginate-based gel did not give satisfactory results due to degradation and liquefaction phenomena of the gel caused by the particular microclimatic conditions [12].

2.3. SEM/EDS Results

Electron microscopy investigations were conducted only on fragments already detached from the hypogeum walls under examination, thus examined before their treatment with glycoalkaloids. The SEM image of Figure 10 highlights the presence of microalgae on the hypogeum walls, in the specific case of diatoms, identified by the presence in the relative EDS spectrum of silicon which is a component of the shell of these organisms. The calcium carbonate crystals on the surface of the hypogeum walls do not have a regular rhombic shape and, as shown in Figure 11 along the same fragment, their arrangement is either linked to the chemical physical equilibrium that is constantly established with the high humidity environments (central image and EDS) and to the activity of the microorganisms present in the patina as evidenced by the calcium carbonate deposits on the fungal hyphae (left image), produced by the fungi themselves.

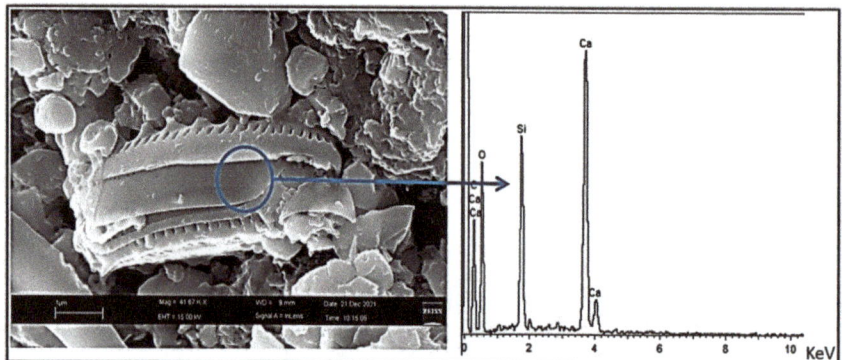

Figure 10. SEM image and associated EDS spectrum of microalgae (diatoms) on the wall.

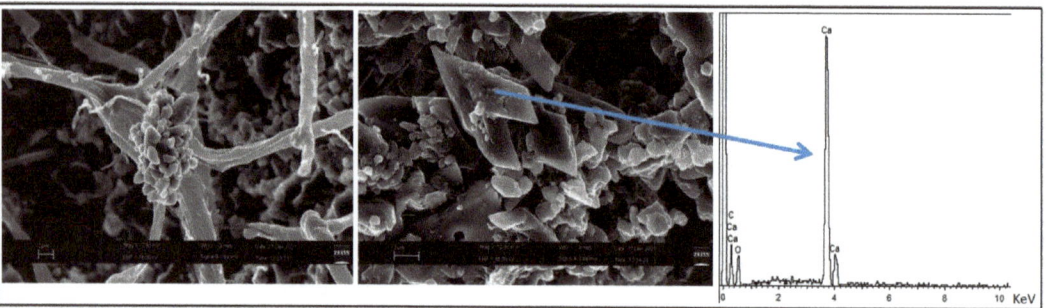

Figure 11. SEM images of calcium carbonate crystals produced by fungal hyphae (**left**) and of calcite (**center**) taken from another area of a fragment free of colored patinas, with-related EDS spectrum (**right**).

3. Materials and Methods

3.1. Sampling

Due to its microclimatic conditions and its walls with extensive biological colonization, the hypogeum of San Pietro Barisano proved to be an ideal site for testing the biocidal efficacy of glycoalkaloids.

Figure 12 illustrates the portion of the wall chosen for the various tests.

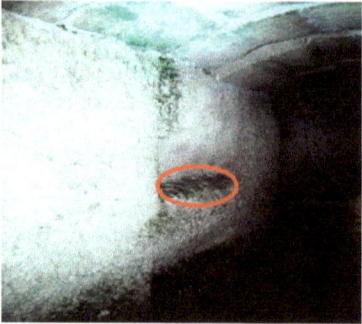

Figure 12. Study area with bio-patina in the hypogeum of San Pietro Barisano church. During the cleaning test, the microclimate conditions of the wall surface portion sampled were locally ranging between 89–92% humidity and 11–16 °C temperature.

A non–invasive sampling was performed according to the standard procedures foreseen by the Normal Recommendations—1/88 [42]. The samples were obtained by gently scraping the hypogeum walls before and after biocide treatment. With the aid of a spatula, samples were collected by removing the bio-patina without affecting the underlying calcareous substrate. The samples were homogenized and stored for XPS and microbiological analyses to determine the chemical composition of the bio-patina and the biological composition.

3.2. Glycoalkaloids Extraction

The glycoalkaloids used in this study were those extracted from unripe berries of *Solanum nigrum*, a spontaneous plant belonging to the *Solanaceae* family, using the method proposed by Cataldi et al. [43]. The freeze-dried berries were placed in an aqueous 1% acetic acid solution. To facilitate the contact between the plant tissue and the extraction solvent, the suspension was stirred for about 2 h and then centrifuged at $3000 \times g$ for 20 min at 4 °C. The pellet was suspended in 1% acetic acid, mixed, centrifuged, and filtered through a 0.22 μm nylon filter (Whatman, Maidstone). The extracts, characterized by liquid chromatography-mass spectrometry (LC-ESI-MS), contained similar amounts of the two main glycoalkaloids, solamargine and solasonine, and other less abundant components [17,23,36].

3.3. Surface Treatment of the Hypogeum Wall

A total of 500 mg of the extracted, lyophilized, and pulverized glycoalkaloids were added to 10 mL of water [17]. The solution was spread directly on the surface of the hypogeum wall, selected for the cleaning treatment, using a special soft brush. Since the hypogeum is an environment with high humidity, it was not necessary to cover the portion of the wall treated to prevent the applied solution from drying out. The surface treatment time was set at four weeks. To monitor the biocidal action of the glycoalkaloids over time, samples were also taken after 1 and 2 weeks. The changes in the bio-patina were followed based on the visual appearance and the results of the XPS analysis.

3.4. X-ray Photoelectron Spectroscopy (XPS)

The samples for XPS analysis were homogenized in an agate mortar and then pressed onto a double-sided copper tape, fixed adequately on a steel sample holder, to be safely introduced in the analysis chamber of the spectrometer. XPS spectra were acquired with a SPECS Phoibos 100-MCD5 spectrometer operating at 10 kV and 10 mA, in medium area (diameter = 2 mm) mode, using a MgKα (1253.6 eV) and AlKα (1486.6 eV) radiations. The pressure in the analysis chamber was less than 10^{-9} mbar during acquisition. Wide spectra were acquired in FAT (Fixed Analyzer Transmission) or FRR (Fixed Retarding Ratio) modes. High-resolution spectra were acquired in the FAT mode, with a constant pass energy of 9 eV and channel widths of 0.1 eV. They were "curve-fitted" using the Googly program, which allows the evaluation of intrinsic and extrinsic features of XPS spectra [44,45]. Peak areas and positions (Binding Energies, BE) as derived by the "curve-fitting" procedure were, respectively, normalized using proper sensitivity factors and referenced to the C1s carbonate carbon set at 290.0 eV [28]. The assignments of the corresponding chemical groups and the relative percentage compositions were derived from the analysis of standard compounds acquired in the laboratory and from the NIST X-ray Photoelectron Spectroscopy Database [46].

3.5. Biological Analysis

To characterize the biological colonization, investigations were conducted at the School of Agricultural, Forestry, Food and Environmental Sciences (SAFE) and Department of European and Mediterranean Cultures (DiCEM) laboratories of the University of Basilicata, Italy.

Samples of powdered bio-patinas, stored in sterile Eppendorf tubes, were suspended in a culture medium to isolate the bacterial strains and proceeded to the subsequent extraction and characterization of the DNA. For the description of bacterial species, the analyses were carried out according to the procedures indicated by Scrano et al. [17]. Bacterial colonies were isolated and characterized based on their different morphologies. The isolated strains were purified, and their DNA was used for molecular identification. Synthetic oligonucleotide primers fD1 (AGAGTTTGATCCTGGCTCAG) and rD1 (AAGGAGGT-GATCCAGCC) were used to amplify the 16S rDNA. As previously reported, PCR mixture and amplification conditions were performed [23,47]. The PCR products were sequenced using Genetic Analyzer 3130xl (Applied Biosystems), and the GenBank and EMBL database performed the DNA similarity using the BLAST suite.

For fungal isolation, samples were collected in sterile vials containing 1 mL of bidistilled water, stored in the refrigerator at 4 °C, and analyzed within 24 h. A 100 µL of fungal suspension was directly placed on Petri dishes containing "Potato Dextrose Agar" (PDA) added with kanamycin (1 mg/L) and streptomycin (1 mg/L). The plates were incubated in an incubator at 24 °C ± 1 °C in the dark for seven days. The pure fungal cultures (PFC) were used for further morphological and molecular investigations and preliminarily identified using a microscope (Axioscope, Zeiss-Germany) based on their macroscopic and microscopic characteristics. The identification of the fungal species was also carried out according to the technique described by Mang et al. [13].

Briefly, PFC mycelium was scraped from the colony's surface and finely homogenized using liquid nitrogen to isolate the genomic DNA (gDNA). The last was extracted from approximately 100 mg of each sample with the NucleoSpin Plant II ™ kit (Macherey-Nagel, Germany) following the manufacturer's instructions. The gDNA of each PFC was then subjected to Polymerase Chain Reaction (PCR) using ITS5 and ITS4 or Bt2a and Bt2b primers which amplified a fragment of the Internal Transcribed Spacer (ITS) and the beta-tubulin gene (TUB-2), respectively. PCR products were detected by 1.2% (w/v) agarose gel electrophoresis, pre-stained with SYBR Safe DNA Gel Stain (Invitrogen Inc., Carlsbad, CA, USA), and photographed with a camera under the transilluminator model Euro-Clone (Pero, Milan, Italy). The amplicons were sequenced by Sanger Dideoxy technology using the same primers as for PCR. All fungal nucleotide sequences obtained in this study were deposited in the Genbank (NCBI). They were also compared with those existing there using the Basic Local Alignment Search Tool (BLASTn) program to identify the fungal isolates at genus and species level [47]. For correct identification of the identified taxa, uncultivated fungal species and those of dubious labels found in the NCBI GenBank were excluded.

3.6. SEM/EDS Analysis

Scanning electron microscopy analyses were carried out on fragments detached from the hypogeum walls at the CNR—IMAA (Institute of Methodologies for Environmental Analysis of the National Research Council, Tito Scalo, Potenza) using a Scanning Electronic Microscope with Field Emission Source (FESEM), Zeiss Supra 40 model, equipped with Oxford INCA. Energy 350 Energy Dispersion Microanalysis System (EDS) and X-act S.D.D. (Silicon Drift Detector) LN2-free detector. This tool allows for obtaining high-resolution (nanometric) images using an "in-lens" detector of secondary electrons. Since these are non-conductive stone samples, a metallization operation with graphite was carried out to obtain a high-quality image. Each sample analyzed different areas at different magnifications, using an acceleration voltage of 15 kV and an aperture of 60 µm. Using SEM, images were obtained both from the secondary electrons, which give information on the surface morphology and from the retro-scattered electrons through which it is possible to recognize differences in the sample's elemental composition. With the microanalysis system, it was also possible to identify these elements from their X-rays emission detected by EDS (energy dispersion spectrometry).

4. Conclusions

The biocidal activity of glycoalkaloids extract (mainly solasonine and solamargine), already demonstrated against some bacterial species, was verified by treating a degraded surface of the hypogeum of the San Pietro Barisano church. The glycoalkaloid solution was left in direct contact with the bio-patina for 4 weeks.

The cleaning effectiveness was visually checked with the color change of the surface. Biological and electron microscopy analyses supplied information on the living components and the characteristics of the surfaces, which are made up of porous calcarenite and offer settlement sites for the various organisms. XPS photoelectron spectroscopy has proved to be an interesting analytical technique for evaluating organic and inorganic components present on the surface and studying the changes induced by the biocide treatment operated by using glycoalkaloids. The visual results show that the biocide acts on the bio-patina by reducing its color and thickness; from the XPS analysis, it was possible to demonstrate the reduction of organic compounds attributable to biological activity.

The results highlighted the removal of the bacterial load but not the effect on the fungal colonies.

Under these conditions, the treatment could increase fungal colonies at the expense of bacteria and use dead bacterial cells as a carbon source. Fungal survival is also favored by a possible migration inside the stone's pores, given the extensive alveolarization present, which makes the biocidal action of the gel difficult.

One of the immediate research perspectives is optimizing the cleaning process by combining the antibacterial actions of glycoalkaloids with those of other natural extracts with antifungal properties possibly supported on a gel, also of natural origin. In this case, the particular microclimatic and humidity conditions present in the hypogeum of the church of San Pietro Barisano are factors to be considered.

The biocidal actions could be integrated with the bio-consolidation of the treated surface to confer resistance to the underlying support against new attacks by biodeteriogens. As reported in the literature and highlighted here by the electron microscopy analyses, some microorganisms can produce compounds chemically and structurally compatible with the carbonate stones by decreasing their porosity. The development of innovative bioremediation techniques based on native colonizing microorganisms capable of inducing carbonate precipitation and structural consolidation represented one of the macro-objectives of the SCN_0520 project.

Author Contributions: Conceptualization, L.S., S.A.B. and A.M.S.; methodology, L.S., A.M.S., S.M.M. and I.C.; validation, L.S., A.M.S., I.C. and S.A.B.; formal analysis, F.C. and S.M.M.; investigation, F.C. and S.M.M.; resources, L.S., S.A.B. and A.M.S.; data curation, F.C., S.M.M., I.C. and A.M.S.; writing—original draft preparation, F.C. and S.M.M.; writing—review and editing, S.A.B., S.M.M. and I.C.; supervision, S.A.B.; project administration, L.S. and A.M.S.; funding acquisition, S.A.B. All authors have read and agreed to the published version of the manuscript.

Funding: This research was funded by the European Union under the funding program PON-MiUR "Smart Cities and Communities and Social Innovation (DD n. 391/Ric. of 5 July 2012)". Project "Product and process innovation for sustainable and planned maintenance, conservation and restoration of cultural heritage"; grant number SCN_00520.

Institutional Review Board Statement: Not applicable.

Informed Consent Statement: Not applicable.

Data Availability Statement: Not applicable.

Acknowledgments: We are grateful to Fausto Langerame for the XPS acquisitions and support to the laboratory activities and to the graduating students Stefania Nigro and Severina Berlingieri for their valuable laboratory support. Thanks to Antonio Lettino, Luca Medici, and Pietro Pasquale Ragone for their collaboration in electron microscopy investigations at the Medical and Environmental Geology Laboratory of the IMAA-CNR institute in Tito Scalo—Potenza, Italy.

Conflicts of Interest: The authors declare no conflict of interest. The funders had no role in the design of the study; in the collection, analyses, or interpretation of data; in the writing of the manuscript; or in the decision to publish the results.

Sample Availability: Samples of the compounds are not available from the authors.

References

1. Kosznik-Kwasnicka, K.; Golec, P.; Jaroszewicz, V.; Lubomska, D.; Piechowicz, L. Into the Unknown: Microbial Communities in Caves, Their Role, and Potential Use. *Microorganisms* **2022**, *10*, 222. [CrossRef] [PubMed]
2. Ortega-Morales, B.O.; Gaylar, C.C. Bioconservation of Historic Stone Buildings—An Updated Review. *Appl. Sci.* **2021**, *11*, 5695. [CrossRef]
3. Soffritti, I.; D'Accolti, M.; Lanzoni, L.; Volta, A.; Bisi, M.; Mazzacane, S.; Caselli, E. The Potential Use of Microorganisms as Restorative Agents: An Update. *Sustainability* **2019**, *11*, 3853. [CrossRef]
4. Warscheida, T.; Braamsb, J. Biodeterioration of stone: A review. *Int. Biodeterior. Biodegrad.* **2000**, *46*, 343–368. [CrossRef]
5. O'Toole, G.; Kaplan, H.B.; Kolter, R. Biofilm formation as microbial development. *Annu. Rev. Microbiol.* **2000**, *54*, 49–79. [CrossRef] [PubMed]
6. Wang, L.-L.; Wang, L.-F.; Ren, X.-M.; Ye, X.-D.; Li, W.-W.; Yuan, S.-J.; Sun, M.; Sheng, G.-P.; Yu, H.-Q.; Wang, X.-K. pH dependence of structure and surface properties of microbial EPS. *Environ. Sci. Technol.* **2012**, *46*, 737–744. [CrossRef] [PubMed]
7. Saiz-Jimenez, C. Biodeterioration vs biodegradation: The role of microorganisms in the removal of pollutants deposited on historic buildings. *Int. Biodeterior. Biodegrad.* **1997**, *40*, 225–232. [CrossRef]
8. Lepinay, C.; Mihajlovski, A.; Touron, S.; Seyer, D.; Bousta, F.; Di Martino, P. Bacterial diversity associated with saline efflorescences damaging the walls of a French decorated prehistoric cave registered as a World Cultural Heritage Site. *Int. Biodeterior. Biodegrad.* **2018**, *130*, 55–64. [CrossRef]
9. Gabriele, F.; Bruno, L.; Casieri, C.; Ranaldi, R.; Rugnini, L.; Spreti, N. Application and monitoring of oxidative alginate–biocide hydrogels for two case studies in "The Sassi and the Park of the Rupestrian Churches of Matera". *Coatings* **2022**, *12*, 462. [CrossRef]
10. Gizzi, F.T.; Sileo, M.; Biscione, M.; Danese, M.; de Buergo, M.A. The conservation state of the Sassi of Matera site (Southern Italy) and its correlation with the environmental conditions analyzed through spatial analysis techniques. *J. Cult. Herit.* **2016**, *17*, 61–74. [CrossRef]
11. Albertano, P.; Moscone, D.; Palleschi, G.; Hermosin, B.; Saiz-Jimenez, C.; Sanchez-Moral, S.; Hernandez-Marine, M.; Urzi, C.; Groth, I.; Schroeckh, V.; et al. Cyanobacteria attack rocks (CATS): Control and preventive strategies to avoid damage caused by cyanobacteria and associated microorganisms in Roman hypogean monuments. In *Molecular Biology and Cultural Heritage*; Routledge: Oxfordshire, UK, 2003; pp. 151–162. [CrossRef]
12. Cardellicchio, F. Monitoring and Conservation of Stone Cultural Heritage: Diagnostic Investigations Using Surface Techniques and Innovative Methodologies for Sustainable Maintenance Interventions. Ph.D. Thesis, University of Basilicata, Potenza, Italy, 2022. Not yet published. 233p.
13. Mang, S.; Scrano, L.; Camele, I. Preliminary studies on fungal contamination of two rupestrian churches from Matera (Southern Italy). *Sustainability* **2020**, *12*, 6988. [CrossRef]
14. Blazquez, A.B.; Lorenzo, J.; Flores, M.; Gómez-Alarcón, G. Evaluation of the effect of some biocides against organisms isolated from historic monuments. *Aerobiologia* **2000**, *16*, 423–428. [CrossRef]
15. Balliana, E.; Ricci, G.; Pesce, C.; Zendri, E. Assessing the value of green conservation for cultural heritage: Positive and critical aspects of already available methodologies. *Int. J. Conserv. Sci.* **2016**, *7*, 185–202. Available online: https://www.researchgate.net/publication/303788227 (accessed on 4 November 2021).
16. Ansari, M.I.; Schiwon, K.; Malik, A.; Grohman, E. Biofilm formation by environmental bacteria. environmental protection strategies for sustainable development. In *Environmental Protection Strategies for Sustainable Development*; Springer: Berlin/Heidelberg, Germany, 2012; Chapter 11; pp. 341–367. [CrossRef]
17. Scrano, L.; Laviano, R.; Salzano, G.; Santacroce, M.; De Franchi, S.A.; Baranek, J.; Bufo, S.A. Natural biocides and bio-calcite: Innovative tools for cultural heritage. In Proceedings of the International Conference Florence Heri-tech: The Future of Heritage Science and Technologies, Florence, Italy, 14–16 October 2020; 949, p. 012096. Available online: https://iopscience.iop.org/article/10.1088/1757-899X/949/1/012096 (accessed on 1 November 2022).
18. Fewell, A.M.; Roddick, J.G. Interactive antifungal activity of the glycoalkaloids α-solanine and chaconine. *Phytochemistry* **1993**, *33*, 323–328. [CrossRef]
19. Fewell, A.M.; Roddick, J.G.; Weissenberg, M. Interactions between the glycoalkaloids solasonine and solamargine in relation to inhibition of fungal growth. *Phytochemistry* **1994**, *37*, 1007–1011. [CrossRef]
20. Shamim, S.; Ahmed, S.W.; Azhar, I. Antifungal activity of allium, aloe, and Solanum species. *Pharm. Biol.* **2004**, *42*, 491–498. [CrossRef]
21. Rani, P.; Khullar, N. Antimicrobial evaluation of some medicinal plants for their anti-enteric potential against multi-drug resistant *Salmonella typhi*. *Phytother Res.* **2003**, *18*, 670–673. [CrossRef]
22. Milner, S.E.; Brunton, N.P.; Jones, P.W.; O'Brien, N.M.; Collins, S.G.; Maguire, A.R. Bioactivities of glycoalkaloids and their aglycones from Solanum species. *J. Agric. Food Chem.* **2011**, *59*, 3454–3484. [CrossRef]
23. Sasso, S.; Scrano, L.; Ventrella, E.; Bonomo, M.G.; Crescenzi, A.; Salzano, G.; Bufo, S.A. Natural biocides to prevent the microbial growth on cultural heritage. *Built. Herit.* **2013**, *1*, 1035–1042. [CrossRef]

24. Caneva, G.; Bartoli, F.; Imperi, F.; Visca, P. Changes in biodeterioration patterns of mural paintings: Multi-temporal mapping for a preventive conservation strategy in the Crypt of the Original Sin (Matera, Italy). *J. Cult. Herit.* **2019**, *40*, 59–68. [CrossRef]
25. Alfano, G.; Lustrato, G.; Belli, C.; Zanardini, E.; Cappitelli, F.; Mello, E.; Sorlini, C.; Ranalli, G. The bioremoval of nitrate and sulfate alterations on artistic stonework: The case-study of Matera Cathedral after six years from the treatment. *Int. Biodeterior. Biodegrad.* **2011**, *65*, 1004–1011. [CrossRef]
26. Scrano, L.; Fraddosio-Boccone, L.; Langerame, F.; Laviano, R.; Adamski, Z.; Bufo, S.A. Application of different methods of surface analysis for the early diagnosis of art-stone (calcarenite) deterioration. *Karaelmas Sci. Eng. J.* **2011**, *1*, 1–14. Available online: https://www.researchgate.net/publication/233862726 (accessed on 4 November 2021).
27. Alessandrini, G.; Toniolo, L.; Cariati, F.; Daminelli, G.; Polesello, S.; Pozzi, A.; Salvi, A.M. A black paint on the facade of a renaissance building in Bergamo, Italy. *Stud. Conserv.* **1996**, *41*, 193–204. [CrossRef]
28. Briggs, D.; Grant, J.T. *Surface Analysis by Auger and X-ray Photoelectron Spectroscopy*; I.M. Publications: Chichester, UK; Surface Spectra Limited: Manchester, UK, 2003; pp. 345–375.
29. Kjærvik, M.; Ramstedt, M.; Schwibbert, K.; Dietrich, P.M.; Unger, W.E.S. Comparative Study of NAP-XPS and Cryo-XPS for the Investigation of Surface Chemistry of the Bacterial Cell-Envelope. *Front. Chem.* **2021**, *9*, 666161. [CrossRef]
30. Boyd, A.; Chakrabarty, A.M. *Pseudomonas aeruginosa* biofilms: Role of the alginate exopolysaccharide. *J. Ind. Microbiol.* **1995**, *15*, 162–168. [CrossRef]
31. Bonomo, A.E.; Amodio, A.M.; Prosser, G.; Sileo, M.; Rizzo, G. Evaluation of soft limestone degradation in the Sassi UNESCO site (Matera, Southern Italy): Loss of material measurement and classification. *J. Cult. Herit.* **2020**, *42*, 191–201. [CrossRef]
32. Chen, J.P.; Hong, L.; Wu, S.; Wang, L. Elucidation of interactions between metal ions and Ca alginate-based ion-exchange resin by spectroscopic analysis and modeling simulation. *Langmuir* **2002**, *18*, 9413–9421. [CrossRef]
33. Campanella, L.; Dell'Aglio, E.; Reale, R.; Cardellicchio, F.; Salvi, A.M.; Casieri, C.; Cerichelli, G.; Gabriele, F.; Spreti, N.; Bernardo, G.; et al. Culture Economy: Innovative strategies to sustainable restoration of artistic heritage. Part I—Development of natural gels for cleaning the stone materials of cultural heritage from iron stains and biodeteriogenic microorganisms. In Proceedings of the Conference: Diagnosis for the Conservation and Valorization of Cultural Heritage, Naples, Italy, 9–10 December 2021; Cervino Ed.. pp. 313–324, ISBN 978 88 95609 61 4.
34. Urzì, C.; De Leo, F.; Krakova, L.; Pangallo, D.; Bruno, L. Effects of biocide treatments on the biofilm community in Domitilla's catacombs in Rome. *Sci. Total Environ.* **2016**, *572*, 252–262. [CrossRef]
35. Bruno, L.; Rugnini, L.; Spizzichino, V.; Caneve, L.; Canini, A.; Ellwood, N.T.W. Biodeterioration of Roman hypogea: The case study of the Catacombs of S.S. Marcellino and Pietro (Rome, Italy). *Ann. Microbiol.* **2019**, *69*, 1023–1032. [CrossRef]
36. Sasso, S.; Miller, A.Z.; Rogerio-Candelera, M.A.; Cubero, B.; Coutinho, M.L.; Scrano, L.; Bufo, S.A. Potential of natural biocides for biocontrolling phototrophic colonization on limestone. *Int. Biodeter. Biodegr.* **2016**, *107*, 102–110. [CrossRef]
37. Martin, G.; Guggiari, M.; Bravo, D.; Zopfi, J.; Cailleau, G.; Aragno, M.; Job, D.; Verrecchia, E.; Junier, P. Fungi, bacteria and soil pH: The oxalate-carbonate pathway as a model for metabolic interaction. *Environ. Microbiol.* **2012**, *14*, 2960–2970. [CrossRef]
38. Casieri, C.; Gabriele, F.; Spreti, N.; Cardellicchio, F.; Scrano, L.; Salvi, A.M. Novel hydrogels for the selective removal of bio-contaminants from stone artworks. A case study: Rupestrian Church Madonna dei Derelitti in Matera. In Proceedings of the XI AIAr National Congress, Naples, Italy, 28–30 July 2021.
39. Gabriele, F.; Tortora, M.; Bruno, L.; Casieri, C.; Chiarini, M.; Germani, R.; Spreti, N. Alginate-biocide hydrogel for the removal of biofilms from calcareous stone artworks. *J. Cult. Herit.* **2021**, *49*, 106–114. [CrossRef]
40. Campanella, L.; Cardellicchio, F.; Dell'Aglio, E.; Reale, R.; Salvi, A.M. A green approach to clean iron stains from marble surfaces. *Herit. Sci.* **2022**, *10*, 79. [CrossRef]
41. Bernardo, G.; Guida, A.; Porcari, V.; Campanella, L.; Dell'Aglio, E.; Reale, R.; Cardellicchio, F.; Salvi, A.M.; Casieri, C.; Cerichelli, G.; et al. Culture Economy: Innovative strategies to sustainable restoration of artistic heritage. Part II—New materials and diagnostic techniques to prevent and control calcarenite degradation. In Proceedings of the Conference: Diagnosis for the Conservation and Valorization of Cultural Heritage, Naples, Italy, 9–10 December 2021; Cervino Ed. pp. 325–334, ISBN 978 88 95609 61.
42. Normal Recommendations 1/88: Macroscopic Alterations of Stone Materials: Lexica, CNR, ICR, 1990, Rome. English Translation by Pusuluri, Pullarao. 2017. Available online: https://www.researchgate.net/publication/320467086_Normal_188_Translated_in_English (accessed on 4 November 2021).
43. Cataldi, T.R.I.; Lelario, F.; Bufo, S.A. Analysis of tomato glycoalkaloids by liquid chromatography coupled with electrospray ionization tandem mass spectrometry. *Rapid Commun. Mass Spectrom.* **2005**, *19*, 3103–3110. [CrossRef] [PubMed]
44. Castle, J.E.; Chapman-Kpodo, H.; Proctor, A.; Salvi, A.M. Curve-fitting in XPS using extrinsic and intrinsic background structure. *J. Electron Spectrosc. Relat. Phenom.* **2000**, *106*, 65–80. [CrossRef]
45. Castle, J.E.; Salvi, A.M. Chemical state information from the near-peak region of the X-ray photoelectron background. *J. Electron Spectrosc. Relat. Phenom* **2001**, *114–116*, 1103–1113. [CrossRef]

46. NIST: National Institute of Standards and Technology. *X-ray Photoelectron Spectroscopy Database, Version 4.1*; NIST: Gaithersburg, MD, USA, 2012. [CrossRef]
47. Sterflinger, K.; Little, B.; Pinar, G.; Pinzari, F.; de los Rios, A.; Gu, J.-D. Future directions and challenges in biodeterioration research on historic materials and cultural properties. *Int. Biodeterior. Biodegrad.* **2018**, *129*, 10–12. [CrossRef]

Disclaimer/Publisher's Note: The statements, opinions and data contained in all publications are solely those of the individual author(s) and contributor(s) and not of MDPI and/or the editor(s). MDPI and/or the editor(s) disclaim responsibility for any injury to people or property resulting from any ideas, methods, instructions or products referred to in the content.

Article

Analytical Investigation of Iron-Based Stains on Carbonate Stones: Rust Formation, Diffusion Mechanisms, and Speciation

Rita Reale [1,*], Giovanni Battista Andreozzi [2,3], Maria Pia Sammartino [1] and Anna Maria Salvi [4,*]

1. Chemistry Department, University of Rome 'La Sapienza' Piazzale Aldo Moro 5, 00185 Rome, Italy
2. Earth Sciences Department, University of Rome 'La Sapienza' Piazzale Aldo Moro 5, 00185 Rome, Italy
3. CNR-IGAG c/o Department of Earth Sciences, Sapienza University of Rome, 00185 Rome, Italy
4. Science Department, University of Basilicata, Viale dell'Ateneo Lucano 10, 85100 Potenza, Italy
* Correspondence: ritareale65@gmail.com (R.R.); anna.salvi@unibas.it (A.M.S.)

Abstract: In cultural heritage, unaesthetic stains on carbonate stones due to their close contacts with metals are of concern for the preservation of sculptures, monumental facades and archeological finds of various origin and antiquities. Rust stains made up of various oxidized iron compounds are the most frequent forms of alteration. The presence of ferric iron on rust-stained marble surfaces was confirmed in previous studies and oriented the choice of the best cleaning method (based on complexing agents specific for ferric ions). However, the composition of rust stains may vary along their extension. As the corrosion of the metallic iron proceeds, if the oxygen levels in the surroundings are low and there are no conditions to favor the oxidation, ferrous ions can also diffuse within the carbonate structure and form a variety of intermediate compounds. In this study, the iron stains on archeological marbles were compared with those artificially produced on Carrara marbles and Travertine samples. The use of integrated techniques (optical and scanning electron microscopy as well as Mössbauer and XPS spectroscopy) with complementary analytical depths, has provided the overall information. Rust formation and diffusion mechanisms in carbonates were revealed together with the evolution of iron speciation and identification of phases such as ferrihydrite, goethite, maghemite, nanomagnetite, and hematite.

Keywords: iron stains; mechanism of rust formation; iron speciation; Mössbauer spectroscopy; XPS; SEM/EDS; OM

Citation: Reale, R.; Andreozzi, G.B.; Sammartino, M.P.; Salvi, A.M. Analytical Investigation of Iron-Based Stains on Carbonate Stones: Rust Formation, Diffusion Mechanisms, and Speciation. *Molecules* **2023**, *28*, 1582. https://doi.org/10.3390/molecules28041582

Academic Editor: Petr Bednar

Received: 29 December 2022
Revised: 26 January 2023
Accepted: 31 January 2023
Published: 7 February 2023

Copyright: © 2023 by the authors. Licensee MDPI, Basel, Switzerland. This article is an open access article distributed under the terms and conditions of the Creative Commons Attribution (CC BY) license (https://creativecommons.org/licenses/by/4.0/).

1. Introduction

The use of carbonate stones in cultural heritage has always interested both architecture and statuary. Unfortunately, stone artifacts (especially those exposed in outdoor environments), despite their resistance, show various forms of degradation, due to chemical, physical or biological interactions recurring over time [1,2]. Among these, the colored stains formed on stone surfaces when in contact with metals or alloy, such as iron-based stains on carbonate stones or bronze statues on stone pedestals [3], are of considerable impact, not only aesthetic. In particular, iron-based stains imply the diffusion of corrosion products and related risk of cracking [4] and can originate from two distinct sources. The first source is the oxidation of the iron coming in contact with the stone from the water supply, mounting nails or screws, decorative elements, gratings, etc. A secondary source is the oxidation of the iron minerals present in the stone, such as pyrite or marcasite or other iron compounds. The first source of stain is the most common, as metallic iron, when exposed to the natural environment (e.g., acid rain, humidity, and temperature variations), is subject to corrosion and produces a colloidal phase of ferrous or ferric hydroxides, which propagate both on the stone surface and inside, through the boundary grains. Subsequently, the dehydration of these oxyhydroxides leads to the formation of different hydrated oxides, which are less soluble and more difficult to remove [5]. Indeed, the choice of the correct cleaning treatment is crucial for the need to combine the best stain removal, based on the solubility of the specific iron phases, with the restoration of the historical material "without any structural damage", in compliance with the theory of cultural heritage conservation [6]. Consequently, to realize an effective cleaning procedure, a better knowledge of the stain compositions is required.

Although the pure phases of iron oxides are well known, as are their topotactic transformations [7], the rust composition varies with the different environments that have determined its growth. Given that rust is mainly found to be composed of nanocrystalline ferrihydrite ($Fe^{3+}_{10}O_{14}(OH)_2$), lepidocrocite (γ-FeOOH), goethite (α-FeOOH), and/or hematite (α-Fe_2O_3) [8,9], the formation mechanisms of these phases within carbonate stones are largely unknown and their interactions with carbonate matrix may be of critical relevance. In fact, iron stains not only disturb the aspect of stones but can also cause chemical-physical damage, as the oxidation of metallic iron to give oxidic phases leads to an increase in volume of the layer surface, due to oxygen penetration in the crystalline lattice [10]. In the case of iron oxidation inside the stone artifacts (bars, rods, or pins), the released corrosion product is higher in volume than the original metal [11], and the ratio between the volume of expansive corrosion product and the volume of iron consumed in the corrosion process is called the "rust expansion coefficient" [12]. The resulting increase in volume associated with the formation of corrosion products gradually induces tensile stresses within the surrounding carbonate matrix and may strongly damage the stone artifacts.

In previous work [13–15] using surface techniques, we easily identified iron (III) compounds on the surface of stained marbles and carbonate stones. We thus proposed as suitable chelators for their removal the natural compounds glutathione and deferiprone dispersed into alginate gel and two proteins of the transferrin family (lactotransferrin, Ltf and ovotransferrin, Ovt) supported by cellulose pulp. The results, very satisfyingly, showed the efficacy of the selected chelating agents, all known for their affinity for ferric ions in vivo, antimicrobial activity and selectivity, ensuring safety for the carbonate structure and for the operators. Each chelant has proved the best cleaning action when dispersed in the most suitable support to be spread on the surfaces to be treated and removed after the given contact time. Overall, the proposed green methodologies are suitable to remove iron stains from carbonate stones as testified by visual inspection, optical microscopy, and XPS, comparing the Fe/Ca ratio before and after cleaning. Also, it was seen that iron (III) removal also always implies the removal of surface contaminants associated with rust deposits, eventually bioactive [16]. The cleaning efficacy of the removal is strongly dependent on (a) chelant properties and bioactivity, (b) volume of water retained by the support, which determines the number of iron complexes to be therein dissolved, and (c) type and entity of the rust deposit and its lateral and in-depth extension. In practice, subsequent applications were foreseen to remove the rust patches completely, leaving the layers underneath to be exposed outside following the first application. Clearly, within the context of cultural heritage, particularly for archeological artifacts, different iron oxidation states can be encountered at the subsurface and boundary interfaces; therefore, highly differentiated treatments may be required.

In this work, we investigated iron stains of (a) historical archeological white marble and (b) artificial iron-stained surfaces of Carrara marble and Travertine, suitably sized for laboratory tests, using complementary analytical techniques, such as optical microscopy (OM), scanning electron microscopy (SEM-EDS), X-ray photoelectron spectroscopy (XPS), and Mössbauer spectroscopy. To better understand the development of rust on the carbonate matrices from these comparisons, the laboratory approach (see Section 4) was simply based on the seasonal monitoring of the corrosion products produced on Carrara marbles and Travertine samples by metallic iron "placed in contact", so as to recall the degradation phenomena affecting ancient marbles over time under the recurring action of atmospheric agents in outdoor and indoor conditions.

The main outcome was the definition of the different states of iron oxidation in rust stains, the identification of the corresponding corrosion products, and insight on their evolution over time.

2. Results

2.1. Optical Microscopy (OM) and Electron Scanning Microscopy Coupled with X-ray Microanalysis (SEM/EDS)

Optical microscopy allowed the choice of the most significant historical iron-stained samples to be used for further analytical investigations. These samples (NS1, 19th century A.D., and NS3 and NS8, 4th century A.D., Figure 1a, b, and c, respectively) show the characteristic pigmentation of iron compounds, due to the oxidation of pins or brackets, penetrated within the carbonate substrates visible in cross sections in reflected light (Figure 1d–f, respectively). Fractures, as in Figure 1a, can be attributed to the insertion of the pin and/or to the volume increase associated with metal oxidation. In all samples, the area surrounding the iron element shows a compact and brown coloration that penetrates the grain boundaries inside the carbonate matrix, drawing reddish brown and yellow–orange bands (Figure 1d–f) similar to cyclic precipitations of iron oxides, known as Liesegang rings [17,18].

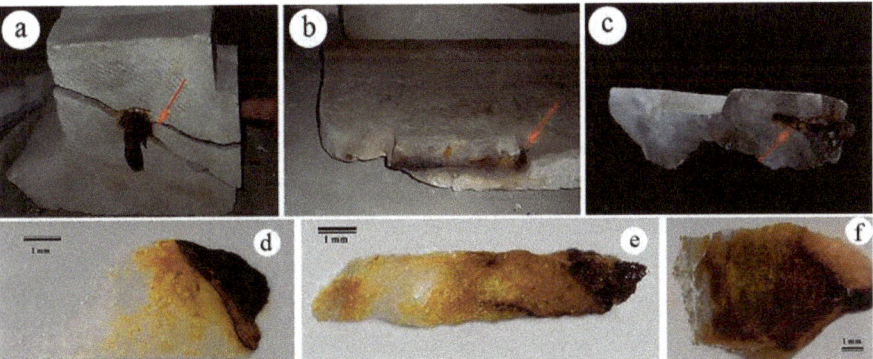

Figure 1. Sampling points, evidenced by red arrows, and images of cross sections acquired in reflected light (25×), respectively, for samples NS1 (**a,d**), NS3 (**b,e**) and NS8 (**c,f**).

OM (with analysis in polarized light) showed important differences in the petrography of the three historical marble samples (Figure 2).

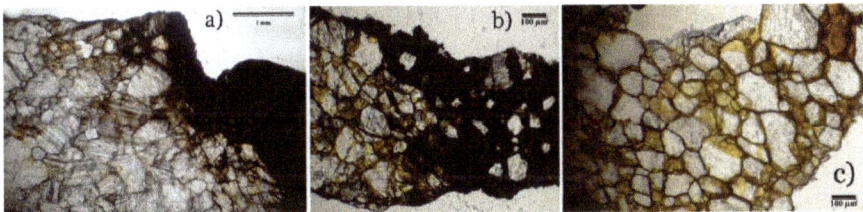

Figure 2. Thin NS1 (**a**), NS3 (**b**), and NS8 (**c**) sections that show the infiltration of iron phases inside the carbonate matrix. OM polarized light analysis, 25×.

NS1, a white calcitic marble, shows an isotropic microstructure, with lobed intergranular sutured contacts and widespread triple junctions, with grain size between 0.55 to 1.03 mm (Figure 2a). Xenoblasts of calcite are visible, with evident rhombohedral and poor lamellar polysynthetic geminations, rare and very small opaque minerals.

NS3, a calcitic marble, shows an isotropic microfabric with irregular shapes and grain size between 0.2 and 0.6 mm, with sutured/lobated grain boundaries (Figure 2b).

NS8 shows a microfabric with size of 0.15–0.35 mm, with intergranular contacts, mainly interlobates, although occasionally pseudolinear, and with an abundance of triple junctions (Figure 2c). In the same sample, in addition to the calcite, few opaque mineral phases are visible. The texture is mainly isotropic; however, there are parallel bands with a thickness of about 1 mm, while the average size of the crystals is smaller (0.06 and 0.15 mm).

In all these samples, the precipitation of iron oxides takes place mainly along the fracture surface, the cleavage planes, and at the grain boundary cavitations, with a higher concentration where the grains are very small (Figure 2). As expected, when the distance from the metal element increases, the concentration of iron oxide hydroxides progressively decreases.

The artificially stained samples (see staining procedure in Section 4), unlike the historical samples, show in the cross section (observed in reflected light) a superficial growth of iron oxides a few micrometers in thickness in both the Carrara marble and Travertine. The diffusion within the calcite matrix reaches about 1 mm maximum in Carrara marble, with a decreasing color intensity, while for Travertine, diffusion filled mainly the voids (Figure 3).

Figure 3. Artificially stained samples. Cross sections images acquired in reflected light (10×) of Carrara marble (**a**) and Travertine (**b**) stained outdoors for two seasons (spring and summer).

More information about the iron compounds and their distribution was obtained by a detailed observation of thin cross sections by SEM/EDS in a portion of the sample where the carbonate matrix is less compact, more porous, and carious, with uneven distribution of iron compounds. The line scan on the NS1 sample shows the distribution profile of Ca and Fe (Figure 4): on the opposite sides, there are the highest concentrations of Fe (left, corresponding to the external part) and Ca (right). The concentration of Fe falls below half at about 250 µm, Fe remains the most abundant elements up to about 800 µm, then there is an area in which Ca and Fe alternately reach similar concentration, and finally Ca becomes prevalent at a depth of about 1.7 mm.

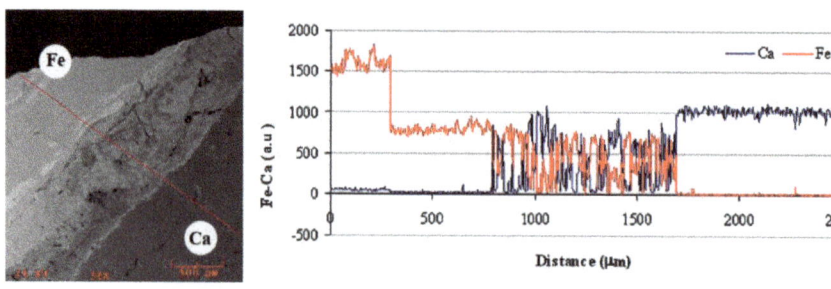

Figure 4. SEM image (**left**) and elemental line mapping (**right**) of iron and calcium obtained for the NS1 samples.

The observation of the thin NS3 section shows how the calcite was substituted by iron-rich phases (see lighter zones in Figure 5a). The elemental analysis of point 1 (Table 1) reveals low iron concentration (0.3 wt%). The carbonate matrix is still dominant, but there is a noticeable penetration of the iron along the cleavage planes [19]. In point 2, a significant precipitation of Fe (Fe 55.6 wt%, Ca 4.3 wt%) increases the intergranular pore spaces surrounding the residual carbonate grains.

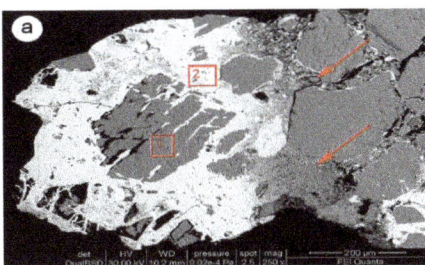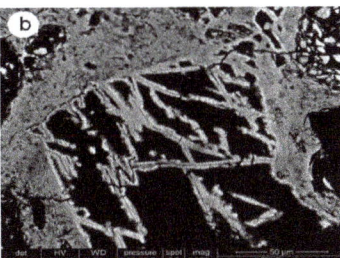

Figure 5. SEM image of sample NS3 showing: (**a**) the replacement of calcite (gray color) by iron rich phases (white color); (**b**) morphological presence of colloid microporous goethite and light gray crystals of lamellar hematite. Red arrows highlight the diffusion front of iron. Points 1 and 2 were analyzed by EDS (see Table 1).

Table 1. EDS element analysis (wt%).

Spectrum	O	Na	Mg	Al	Si	P	S	Cl	K	Ca	Fe
1	43.3	0.1	0.7	0.2	0.3	0.1	0.1	0.1	0.0	54.8	0.3
2	33.0	0.7	0.7	0.4	4.3	0.6	0.2	0.1	0.2	4.3	55.6

In Figure 5a, the red arrows highlight areas with a very porous appearance that represent the contact between the "healthy" portion of carbonate matrix and the diffusion fronts of iron, which irregularly penetrates the calcite body along the intergranular contacts and replaces it. Small fragments of calcite, residual, or precipitate from dissolution process are visible. In Figure 5b, many voids appear partially filled with different iron phases [20] that may be morphologically identified as microporous colloform goethite (dark gray) and lamellar hematite crystals (light gray) [21]. Confirmation of all the natural samples by XRD was impossible because of the small quantity of iron phases in the powder samples, which were below the diffractometer detection limit (about 1 percent of a mineral's content).

2.2. XPS Analysis

A systematic characterization of the artificial stains produced on marble samples was undertaken by collecting the samples gently scraped as powders and homogenized for XPS analysis, with the aim of determining the chemical states of iron and other elements composing the rust deposited as a function of marble type, surface and inner corrosion, and environmental conditions. A well-established curve-fitting procedure was used [22,23].

XPS spectra were processed and resolved in their peak components, thus allowing chemical state assignments (using corrected positions in binding energy, BE eV) and relative percentage composition (by the normalized peak area ratios). From the large number of experiments on Carrara marbles and Travertine (whose comparative characterization is still in progress), in this work we report the curve-fitted Fe 2p region for the stained Carrara samples (Figure 6) with long exposure outdoors (spring/summer) and all the detailed regions acquired (Table 2).

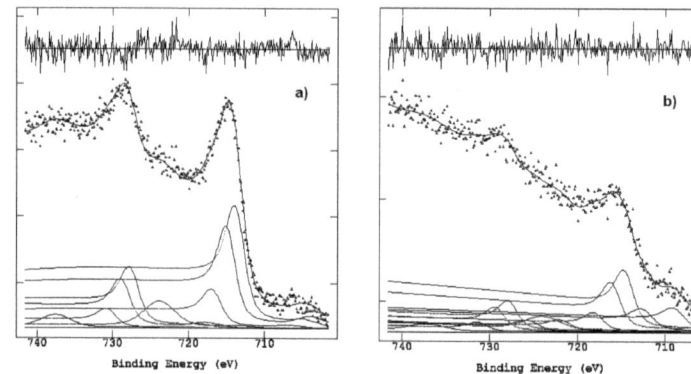

Figure 6. Spectra of the Fe region obtained for the stained Carrara samples with long outdoor exposure: (a) external growth layer; (b) interface zone.

Table 2. Curve-fitting results of the XPS regions of the Carrara marble samples with long outdoor exposure. (a): C1s, O1s, Ca 2p (with superimposed Mg KL_1L_2 signal) and Mg 1s regions; (b): Fe2p regions.

(a)				
Element, Orbital Chemical State	Carrara Surface		Carrara Interface	
	Corrected BE (eV)	Normalized Area	Corrected BE (eV)	Normalized Area
C_{1s} → lower BE carbons *	283.5	122.8	283.3	247.1
C-C (IS)	285.0	1267.8	285.0	1887.4
C-O/C-O-C	286.0	946.9	286.5	328.9
O-C-O/C=O/COO	288.7	341.5	288.6	517.7
CO_3^{2-}/COOR	290.5	723.1	290.1	3075.1
* Typical of carbides, polycyclic compounds etc.				
Ca $2p_{3/2}$ → Ca	---	---	344.8	130.1
nnaCaO/	345.9	44.8	346.7	692.3
$CaCO_3$	347.9	441.7	347.9	2437.6
Ca $2p_{1/2}$ → Ca	---	---	348.3	65.0
CaO	349.5	22.4	350.2	346.1
$CaCO_3$	351.5	220.9	351.4	1194.4
Mg KL_1L_2 *	---	---	352.8	---
SU_1 $CaCO_3$	---	---	355.9	100.1
SU_2 $CaCO_3$	---	---	359.7	121.4
* Auger signal				
O1s-peak 1 (metal oxides)	530.4	2422.3	530.0	1275.8
O1s-peak 2 (calcite matrix)	532.3	4337.8	532.1	11610.0
Mg_{1s} → MgO	1304.8	51.8	1305.4	504.8
(b)				
$Fe2p_{3/2}$				
Fe I → Metallic			704.6	30.0
FeO ↔ Fe(II) compounds	---	---	708.2	28.7
Fe_2O_3 main peak	710.7	387.1	710.2	67.0
Fe_2O_3 MS(I)	711.9	290.3	711.5	53.6
Fe_2O_3 MS(II)	713.8	130.6	713.5	21.4
Fe_2O_3 SU	720.6	185.6	719.6	43.5
$Fe2p_{1/2}$				
Fe I → Metallic	---	---	717.6	15.0
FeO ↔ Fe(II) compounds	---	---	721.2	14.4
Fe_2O_3 main peak	724.1	193.5	723.5	26.1
Fe_2O_3 MS(I)	725.3	145.1	724.8	19.6
Fe_2O_3 MS(II)	727.2	65.3	726.6	9.1
Fe_2O_3 SU	734.2	92.8	732.6	16.0

The curve-fitting results with the proposed identification of the corrosion products formed on the external surface and at the carbonate interface of the iron-stained Carrara marbles were considered the most suitable to be compared with those found in archeological samples by other techniques, in particular Mössbauer spectroscopy (vide infra).

Both Table 2 and Figure 6a indicate the presence of only Fe^{3+} ions in the external growth layer, in agreement with literature data and XPS results previously reported [15], while Fe^{2+} compounds and other Fe-containing phases in reduced chemical states (more difficult to identify) are co-present only in the interface zone, Figure 6b (vide infra).

In more detail, for the external layer, the Fe2p curve-fitting parameters matched exactly those of the standard α-Fe_2O_3 acquired in the XPS laboratory [15], represented by four peaks for each 2p3/2 and 2p1/2 doublet component, comprising one main peak, two split multiplets (MS), and one shake-up (SU) satellite (Figure 6a). In contrast, for the sample collected at the interface, in addition to the Fe^{3+} doublet still represented by the multiplet and shake up satellites, two extra doublets, with their main 2p3/2 components at significantly lower BE (708.2 eV and 704.7 eV), were required to properly fit the whole Fe2p envelope (Figure 6b). To also reduce computational complexity, both components were tentatively represented by only one doublet (2p3/2 and 2p1/2 separated by = 13 eV) with a large FWHM (full width, half maximum) to match the experimental line, without the need to split them into multiplet and shake-up satellites.

Considering in [24] the superimposed spectra of Fe^{3+}, Fe^{2+} well-defined oxides and Fe metallic, also reported in [15], and their relative BE positions well in agreement with literature data [25–27] and the XPS online database [28], the two extra peaks of lowest BE found at the interface are not straightforwardly identifiable, and further consideration comprising all the other detailed regions, elaborated by curve-fitting (Table 2), is required.

In support, we recall the XPS studies of stratified systems, intercalated with organic and inorganic compounds, of worn surfaces, oxide nanoparticles, coupled inorganic nanoparticles–nanocarbon for ultrafast batteries, encapsulated iron carbide catalysts, etc., just to mention some of the relevant publications [29–35].

Experimental results highlight the local information provided by XPS for the element under investigation, dependent on chemical bonds, aggregation sizes, and surrounding environments. Of relevant importance are the chemical shifts reported for intercalated Fe^{3+} and Fe^{2+} ions found at 709.5 and 707.8 eV, respectively, confined Fe^{2+} carbides at 708.2 eV, and a variety of signals ranging from Fe^{3+} to metallic Fe, with the lowest BE feature of Fe 2p3/2 due to strong metallic aggregates at 705 eV, using an internal standard similar to ours based on aliphatic/aromatic C-C set at 284.8 eV.

Based on a complete elaboration of the considerable data set collected during a triennial doctoral research project [36], in light of the above insights, important information can be gleaned from the data reported in Table 2 for each elemental peak, chemical state(s), and relative intensity (peak area). The main outcome, based on peak area mass balance, taking into account the stoichiometric coefficients for the given chemical states, is as follows.

Iron oxide spread on surfaces in the form of more or less hydrated Fe_2O_3 does not affect the calcite structure, leaving the correct $CaCO_3$ stoichiometry.

In agreement with previous work [13–15], iron stains are often associated with organic contaminants that enter the rust composition. Of the contaminants deposited on surfaces, the aliphatic carbons can be recognized by curve-fitting the C1s region and taken as the internal standard, set at 285.0 eV, to correct all the binding energies.

At the interface depths, iron chemical states could be the result of its intercalation within the carbonate structure and confined association. The peaks at the lowest BE side of Figure 6b can thus be assigned to a variety of iron compounds ranging from metallic and organometallic aggregation, through iron carbide and Fe^{2+} oxides prone to further oxidation once exposed to air [28–35].

Concomitantly, comparing the surface and interface Ca:CO_3 stoichiometry, a fairly consistent imbalance was observed at the interface with excess calcium, represented as CaO in Table 2, significantly shifted 0.6 eV to higher BE than surface CaO.

2.3. Mössbauer Analysis

Mössbauer spectra were collected by measuring at room temperature samples made of powder scraped from the inner and outer rust layers of both archeological and artificially stained marble samples.

The samples with long outdoor exposure for one, two, three, and four seasons show very similar Mössbauer spectra dominated by the presence of a strong paramagnetic doublet, with an isomer shift of about 0.36 mm s^{-1} and a quadrupole splitting of about 0.60 mm s^{-1} (Figure 7).

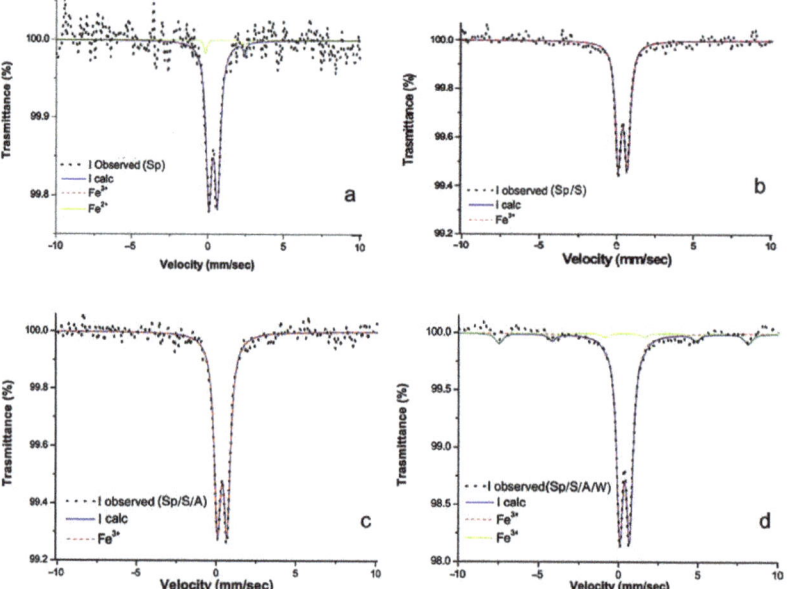

Figure 7. Mössbauer spectra of marble samples artificially stained outdoors during: (**a**) one season (spring, Sp); (**b**) two seasons (spring and summer, Sp/S); (**c**) three seasons (spring, summer, autumn, Sp/S/A); and (**d**) four seasons (spring, summer, autumn, winter, Sp/S/A/W). Observed absorption intensity (I obs, black dots) and calculated total intensity (I calc, continuous blue line) are plotted together with calculated absorption components (dashed red line for the Fe^{3+} central doublet, continuous green line for accessory components).

According to the literature [37], this may be interpreted as due to an irresolvable mixture of the Fe^{3+} oxide hydroxides lepidocrocite, ferrihydrite, nanogoethite, and nanohematite, with the last two exhibiting superparamagnetic behavior because of an average size <10 nm (Table 3). A doublet corresponding to Fe^{2+} was tentatively added to the spectrum of the samples exposed outdoors for one season to verify if any signature of the possible presence of reduced Fe was detectable, as indicated by XPS analysis at the interface zone together with even more reduced Fe as minority components. Unfortunately, the resolution of the spectrum was so low that the presence of Fe^{2+} could be neither confirmed nor excluded (Figure 7a). A weak additional sextet, with IS = 0.33 mm s^{-1} and HF 49 T, was clearly observed only in the spectrum of the samples exposed for four seasons (Figure 7d). A likely attribution of the sextet is to ferrimagnetic maghemite or, less probably, to ferromagnetic nanohematite with average particle size >10 nm, both phases pointing to an evolution of the rusty deposits with time.

Table 3. ^{57}Fe Mössbauer parameters relative to the studied archeological and artificially stained marble samples.

Sample	IS (mm/s)	QS (mm/s)	HF (T)	A (%)	Attribution
Sp	0.34	0.57	–	100	Nanogoethite/lepidocrocite/ferrihydrite/nanohematite
Sp/S	0.37	0.59	–	100	Nanogoethite/lepidocrocite/ferrihydrite/nanohematite
Sp/S/A	0.36	0.61	–	100	Nanogoethite/lepidocrocite/ferrihydrite/nanohematite
Sp/S/A/W	0.37	0.61	–	89	Nanogoethite/lepidocrocite/ferrihydrite/nanohematite
	0.33	–	49	11	Maghemite/hematite
AS25	0.35	0.60	–	14	Nanogoethite/lepidocrocite/ferrihydrite/nanohematite
	0.30	–	48	54	Maghemite/hematite
	0.34	–	50	32	Maghemite/hematite
NS1	0.37	–	36	17	Goethite
	0.41	–	31	5	Nanomagnetite
	0.59.	–	20	24	Nanomagnetite
	0.31	0.75	–	21	Ferrihydrite
	0.76	1.18	–	33	Fe$^{2.5+}$ (IVCT)
NS3	0.27	–	49	37	Maghemite/hematite
	0.35	–	33	5	Nanomagnetite
	0.67	–	20	6	Nanomagnetite
	0.37	0.55	–	27	Lepidocrocite
	0.67	1.15	–	25	Fe$^{2.5+}$ (IVCT)
NS8	0.35	–	50	10	Maghemite/hematite
	0.35	–	36	15	Goethite
	0.35	–	28	22	Nanohematite
	0.35	0.67	–	18	Ferrihydrite
	0.67	1.44	–	35	Fe$^{2.5+}$ (IVCT)

Note: Room temperature measurements. IS = isomer shift relative to α-Fe foil, QS = quadrupole splitting, HF = magnetic hyperfine field, A (%) = percentage of total absorption area. Errors are estimated at about ±0.02 mm/s for IS and QS, about ±0.7 T for HF, and no less than ±3% for both doublet and sextet areas.

By merging the spectra obtained from the different exposures (Figure 8a), it appears evident that they are all dominated by the central quadrupole doublet, likely due to the presence of Fe^{3+} oxyhydroxides, either amorphous or low crystalline (with average size less than 10 nm). By increasing the outdoor exposure, such poorly crystalline, nanometer-sized phases of Fe^{3+} oxyhydroxides reasonably underwent aggregation processes towards relatively larger crystalline Fe^{3+} oxides, such as maghemite and/or hematite. A similar multiple signature was also observed for the sample exposed indoors, AS 25, stained for three months in a highly humid environment (Figure 8b). In this case also, the absorption bands were fitted with a central doublet (IS = 0.35 mm s^{-1}, QS = 0.60 mm s^{-1}), interpreted as lepidocrocite/ferrihydrite/nanogoethite/nanohematite, and two sextets (IS = 0.30 mm s^{-1}, HF 48 T, and IS = 0.34 mm s^{-1}, HF 50 T), interpreted as maghemite/hematite (Table 3). Other contributions due to Fe oxide/oxyhydroxide nanoparticles (emerging as relaxed sextets) are possible, but they could not be fitted with confidence at room temperature. Results obtained from sample AS25 demonstrate that Fe aggregation processes and stain formation were more rapid for the sample exposed indoors than for those exposed outdoors, even for a year. Similar results showing the stronger dependence of stain formation on humidity with respect to time were obtained in the literature [38].

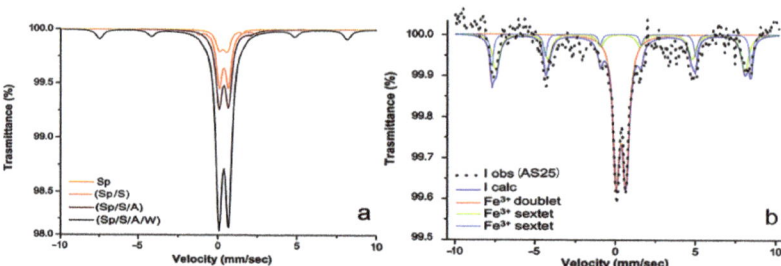

Figure 8. Mössbauer spectra of: (**a**) marble samples stained outdoors after exposure of one-to-four seasons (see Figure 7 for details); (**b**) sample AS25 stained for three months in a highly humid environment. Observed absorption intensity (I obs, black dots) and calculated total intensity (I calc, continuous blue line) are plotted together with calculated absorption components (continuous red line for the Fe^{3+} central doublet, green and cyan for Fe^{3+} sextets).

Absorption spectra of the historical samples NS1 and NS8 are quite similar and show both quadrupole doublets and magnetic sextets (Figure 9). In the NS1 spectrum, the two quadrupole doublets (IS = 0.31 mm s^{-1}, QS = 0.75 mm s^{-1}, and IS = 0.76 mm s^{-1}, QS = 1.18 mm s^{-1}) were considered as due to the possible presence of Fe^{3+} from low crystalline ferrihydrite and iron at a mixed valence $Fe^{2.5+}$ from intervalence charge transfer (IVCT), respectively. Concerning the three sextets, the first, IS = 0.37 mm s^{-1}, HF = 36 T, was attributed to goethite, while the other two, being very relaxed with IS = 0.41 mm s^{-1}, HF = 31 T and IS = 0.59 mm s^{-1}, HF = 20 T, were tentatively attributed to superparamagnetic nanomagnetite (Figure 9a and Table 3). In NS8, the two quadrupole doublets (IS = 0.35 mm s^{-1}, QS = 0.67 mm s^{-1}, and IS = 0.67 mm s^{-1}, QS = 1.44 mm s^{-1}) were attributed to the same Fe species identified in NS1, while two sextets were considered as due to goethite (IS = 0.35 mm s^{-1}, HF = 36 T) and maghemite (IS = 0.35 mm s^{-1}, HF = 50 T), and the third, being very relaxed with IS = 0.35 mm s^{-1} and HF = 28 T, was tentatively attributed to superparamagnetic nanohematite (Figure 9b). Rearranging the obtained pieces of information, in NS1 and NS8, iron stains are tentatively attributed to the combination of Fe^{2+}–Fe^{3+}-bearing amorphous/crystalline phases, such as magnetite nanoparticles, and their oxidation products, likely Fe^{3+} oxide hydroxides of higher crystallinity and larger size. Notably, the oxidation of NS8 is more advanced than that of NS1, as NS8 contains maghemite and nanohematite in place of nanomagnetite. In the literature, a similarly broad sextet pattern was described by a distribution of hyperfine fields attributed to medium-sized nanoparticles [39].

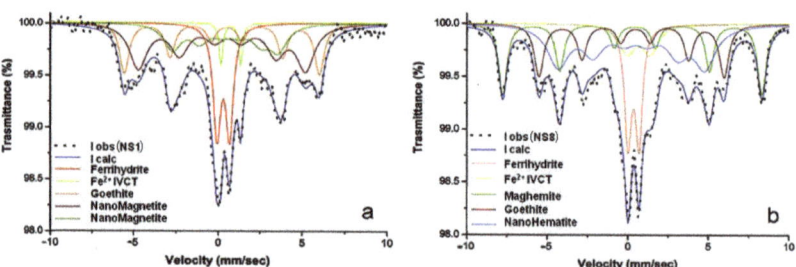

Figure 9. Mössbauer spectra of the historical marble samples: (**a**) NS1; (**b**) NS8. Observed absorption intensity (I obs, black dots) and calculated total intensity (I calc, continuous blue line) are plotted together with calculated absorption components (see text).

The absorption spectrum of the historical sample NS3 sample only appears to be different from the previous ones (Figure 10), though it is made of two quadrupole doublets and three magnetic sextets like them (Table 3). The two quadrupole doublets, one with IS of

0.37 and QS of 0.55 mm s^{-1}, the other with IS of 0.67 and QS of 1.15 mm s^{-1}, are attributed to lepidocrocite and IVCT Fe$^{2.5+}$, respectively. One of the three sextets is well defined, with IS = 0.27 mm s^{-1} and HF = 49 T, and is attributed to maghemite (less probably to relatively large nanohematite), while the other two, being very relaxed with IS = 0.35 mm s^{-1}, HF = 33 T, and IS = 0.67 mm s^{-1}, HF = 20 T, are tentatively attributed to superparamagnetic nanomagnetite (Table 3).

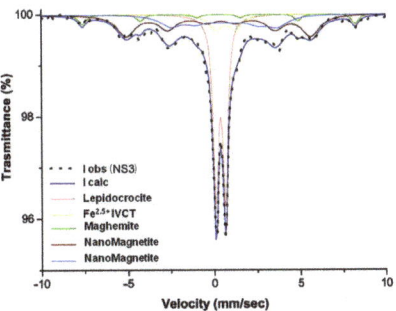

Figure 10. Mössbauer spectra of the historical marble NS3. Observed absorption intensity (I obs, black dots) and calculated total intensity (I calc, continuous blue line) are plotted together with calculated absorption components (see text).

3. Discussion

By comparing studies on real samples differently aged with those on artificial specimens stained at different exposure periods, we were able to make a contribution to the definition of the mechanisms of iron stain formation and its speciation, which will be useful to individuate procedures for better cleaning.

In agreement with the results of Beltran et al. (2016) [40], transport of the iron on carbonate stones mainly occurs as a colloidal solution of oxyhydroxides, which transform with time in more stable forms. In the present investigation, the transformation over time is documented by Mössbauer spectroscopy that shows both in historical and artificial stains the presence of nanometric clusters of Fe that, in the course of time, tend towards larger oxyhydroxides/oxides. Iron diffusion was shown by microscopy techniques to be dependent on structural characteristics and stone porosity. Using archeological samples, it was demonstrated that iron replaces dissolved calcite, moving through the grain boundaries along irregular paths by cycling precipitation of iron oxides. Moreover, XPS allowed discriminating the chemical composition of the iron phases diffused inside the substrate and those grown on the surface. As an example, the two samples in Figure 6 present only carbonaceous contaminants and no others. As reported in Table 2, only magnesium oxide is present, likely as an intrinsic component of the Carrara marble. The results already explained in the XPS section can be compatible with the action of water and oxygen triggering iron corrosion till a potential equilibrium is reached. Corrosion also depends on the adsorption of organic contaminants, the lowering pH of rainwater due at least to the presence of dissolved carbon dioxide, and eventually more acidic pollutants (sulfur and nitrogen oxides) on dissolution and reprecipitation of calcium carbonate and other related factors [41,42].

As expected, the actions of natural weathering, specific environmental pollutants, alternating wetting and drying cycles, and seasonal fluctuations in climatic conditions affect the stones' integrity and the iron corrosion potential, inducing changes in the rust composition and internal diffusion. It is therefore difficult to singly identify the factors responsible for deterioration, as they can act in synergy and/or in temporal sequence with different modalities, eventually including biodegradation processes [43].

We believe, however, that research combining parallel laboratory experiments and advanced analytical techniques could progressively contribute to a more accurate characterization of each case study, the overall outcome providing the right directions for effective maintenance and conservation intervention.

4. Materials and Methods

4.1. Sampling

Several iron-stained samples (Table 4a) of white marble were collected from the archeological remains of ancient Roman baths discovered beneath Palazzo Valentini (Rome, Italy) and the warehouses of ancient marble sculptures of the Vatican Museum, according to UNI 11182 (Description of alteration forms—Terms and definitions) and UNI EN 16085 (Methodology for sampling from materials of cultural property. General rules).

The artificial stains were produced through the oxidation of iron cubes (10 × 10 mm) and nails put in contact with the surface of Travertine (Tivoli quarries) and Carrara marble samples (50 × 50 × 20 mm), both indoors in a highly humid environment and outdoors exposed on a terrace (Table 4b). The outdoor exposition was continued for one year to achieve stains from iron oxides related to one to four seasons.

Table 4. (a) Archeological samples. (b) Artificially stained samples.

(a)			
Sample	Description	Provenience	Date
NS1	Base of bust	Vatican Museum	19th century AD
NS3	Sarcophagus' fragment	Vatican Museum	4th century AD
NS8	Fragment of marble slab	Roman bath Palazzo Valentini (Room 5)	4th century AD

(b)	
Samples	Description
AS25	Travertine stained indoor for three months
Sp	Marble stained outdoor during spring
Sp/S	Marble stained during spring/summer
Sp/S/A	Marble stained during spring/summer/autumn
Sp/S/A/W	Marble stained during spring/summer/autumn/winter

4.2. Analytical Techniques

4.2.1. Optical Microscopy

The preparation of thin and cross sections was performed according to the specifications of the UNI 10922: 2001 Standard (Ref. Recommendations CNR—ICR NorMal 10/82, Natural and artificial stone materials—Preparation of thin sections and cross sections of stone materials). The petrographic investigation was performed by means of optical polarized light microscopy (transmitted and reflected) with a Zeiss Axiolab, equipped with a Nikon D800 camera for digital image capture.

4.2.2. Scanning Electron Microscopy with EDS (Energy-Dispersive Spectroscopy)

Scanning electron microscopy was performed on the samples selected from optical microscopy (NS1, NS3, NS8) using a Jeol JSM 5400 to determine the chemical composition of mineral phases through EDS of backscattering (BSE) images. To produce an electrically conductive surface for SEM, both thin and cross sections were coated using thin-film evaporation of graphite in a vacuum coater, with a thin layer of about 20 nm thickness.

4.2.3. X-ray Photoelectron Spectroscopy (XPS)

XPS spectra were acquired with a SPECS Phoibos 100-MCD5 spectrometer operating at 100 W in FAT (fixed analyzer transmission) and medium area modality (spot diameter ≈2 mm), using Al Kα (1486.6 eV) and Mg Kα (1253.6 eV) achromatic radiation. The pressure in the analysis chamber was higher than 10^{-9} mbar (UHV, ultrahigh vacuum). The use of the double anode (Al/Mg Kα) helps to distinguish XPS signals, varying in kinetic energy with the source employed, from the X-ray-induced Auger signals, dependent on atomic relaxation following photoemission [26]. High-resolution detailed regions (channel width of 0.1 eV) were elaborated by curve-fitting using the home made Googly software [22,23] qualitatively by referring the binding energies to the acquired spectrum α-Fe$_2$O$_3$ (standard Aldrich) for correcting the surface charging and to literature data and quantitatively by normalizing the peaks area with appropriate sensitivity factors [26]. For XPS acquisitions, the surface sampling was performed by carefully scraping the surface of the blank and stained marbles, respectively not in contact and in contact with the iron (nails, cubes). The interface sampling was performed by immediately scraping the inner part of the marble specimens left exposed after having the outer layers removed for the previous Surface analysis. The powders collected were all stored in special inert and sealed plastic containers to avoid external contamination.

Hence, by referring to the acquired spectrum α-Fe$_2$O$_3$ (Sigma Aldrich Hematite 99%, St. Louis, MO, USA) and to literature data mentioned in the relevant paragraphs, the oxidation products derived from curve-fitting for all analyzed samples could be compared within the limits of XPS accuracy ($\pm$10%) [26].

4.2.4. Mössbauer Spectroscopy

The analyzed samples were homogeneous powder gently scraped from the surfaces and interfaces of the stained specimens: three tiles of Carrara and Travertine. Two samples were similarly collected from indoor specimens and from the archeological artifacts, observing the BB.CC. regulations for the sampling modality.

The absorbers were prepared by pressing finely the powder samples mixed with powder acrylic resin (Lucite) to self-supporting disks of about 10 mm diameter. Sample quantity at this stage was strictly dependent on the availability of the examined material, with the drawback of—in some cases—staying at or near the detection limit. The amount used should correspond to about 2 mg of iron oxide, with an absorption density in which the thickness does not affect the Mössbauer results. In our case, these concentrations were very difficult to reach, and to solve this problem long acquisition times (one or two weeks) were used to ensure a good signal/noise ratio in data reading, and in all cases the powders collected at the surface and interface zones were mixed and analyzed as a single sample.

Spectra were collected at 298 K (room temperature, RT) using a ^{57}Co/Rh source and a conventional constant acceleration mode. A multichannel analyzer with 512 channels was used for the data recording at a range of velocity of -10 to $+10$ mm/s. A highly pure sample of α-iron was used to calibrate the speed, and raw data were collected in 512 channels. Spectra were elaborated by Recoil 1.04 [44] software accounting for symmetric Lorentz curves. The χ^2 test was used to individuate the best conditions and uncertainty was obtained by the covariance matrix. A 0.02 mm/s uncertainty resulted for the isomer shift (IS), quadrupole splitting (QS), and magnetic hyperfine field (HF). Uncertainty no lower than $\pm$3% was estimated for the doublet areas.

5. Conclusions

The results obtained concern the mechanisms of iron speciation and rust stain formation over the surface and inner areas of carbonate matrices of either artificially stained marbles exposed over time under different environmental conditions and archeological finds. The analytical methods adopted provided complementary information and reciprocally confirmed the overall outcome. Optical microscopy gave preliminary information and allowed the choice of the most significant samples. SEM/EDS also gave information on the diffusion mechanisms and damage of the pore net. Mössbauer spectroscopy gave information on iron oxidation states, mineral phases possibly formed as a consequence of metallic iron alteration, and insights on their aggregation state. The surface sensitivity of

XPS allowed the chemical states of iron associated with those mineral phases (and possible aggregates) to be determined at a very local nanometer scale.

It was thus possible by means of combined techniques to effectively reveal the effect of oxidation, crystal growth, and hydration/dehydration reactions in the progressive transformation of iron compounds (and rust composition) from metal aggregates into mixed phases containing Fe^{2+}, $Fe^{2.5+}$ and Fe^{3+} ions, such as magnetite followed by oxide hydroxides maghemite, lepidocrocite, goethite, and as the last stage hematite.

The results obtained will have implications for the correct care of archeological artifacts, which—depending on their evolution of iron-based stains—may require different treatments for effective and safe rust removal.

Author Contributions: Conceptualization, R.R. and M.P.S.; methodology, M.P.S., G.B.A. and A.M.S.; software, A.M.S., R.R. and G.B.A.; validation, M.P.S., G.B.A. and A.M.S.; formal analysis, R.R.; investigation, R.R.; resources, M.P.S., G.B.A. and A.M.S.; data curation, R.R.; writing—original draft preparation, R.R.; writing—review and editing, R.R., M.P.S., G.B.A. and A.M.S.; supervision, A.M.S., R.R., G.B.A., M.P.S.; project administration, M.P.S., A.M.S. All authors have read and agreed to the published version of the manuscript.

Funding: This research received no external funding.

Institutional Review Board Statement: Not applicable.

Informed Consent Statement: Not applicable.

Data Availability Statement: Not applicable.

Acknowledgments: The authors wish to thank Luigi Campanella for the research support (SCN_00520 project), Fausto Langerame for XPS acquisitions and technical assistance to the master's student Teresa Lovaglio for spectra elaboration, and Sara Ronca for thin and cross-section analysis.

Conflicts of Interest: The authors declare no conflict of interest.

Sample Availability: Data processing of the analyzed samples are available from the corresponding authors on request.

References

1. Vidorni, G.; Sardella, A.; De Nuntiis, P.; Volpi, F.; Dinoi, A.; Contini, D.; Bonazza, A. Air pollution impact on carbonate building stones in Italian urban sites. *Eur. Phys. J. Plus* **2019**, *134*, 1–10. [CrossRef]
2. appitelli, F.; Cattò, C.; Villa, F. The control of cultural heritage microbial deterioration. *Microorganisms* **2020**, *8*, 1542. [CrossRef]
3. Salvi, A.M.; Langerame, F.; Macchia, A.; Sammartino, M.P.; Tabasso, M.L. XPS characterization of (copperbased) ed stains formed on limestone surfaces of outdoor Roman monuments. *Chem. Cent. J.* **2012**, *6*, S10. [CrossRef]
4. Winkler, E.M. Iron in Minerals and the Formation of Rust in Stone. In *Stone Archit*; Springer: Berlin/Heidelberg, Germany, 1997; pp. 233–240.
5. Macchia, A.; Ruffolo, S.A.; Rivaroli, L.; La Russa, M.F. The treatment of iron-stained marble: Toward a "green" solution. *Int. J. Conserv. Sci.* **2016**, *7*, 323–332.
6. Schwertmann, U. Solubility and dissolution of iron oxides, Iron nutrition and interactions in plants. *Springer Neth.* **1991**, *130*, 1–25.
7. Mohapatra, M.; Anand, S. Synthesis and applications of nano-structured iron oxides/hydroxides—A review. *Int. J. Eng. Sci. Technol.* **2010**, *2*, 127–146. [CrossRef]
8. Dillmann, P.; Mazaudier, F.; Hoerlé, S. Advances in understanding atmospheric corrosion of iron. I. Rust characterisation of ancient ferrous artefacts exposed to indoor atmospheric corrosion. *Corros. Sci.* **2004**, *46*, 1401–1429. [CrossRef]
9. Cudennec, Y.; Lecerf, A. The transformation of ferrihydrite into goethite or hematite, revisited. *J. Solid State Chem.* **2006**, *179*, 716–722. [CrossRef]
10. Bams, V.; Dewaelem, S. Staining of white marble. *Mater. Charact.* **2007**, *58*, 1052–1062. [CrossRef]
11. Anstice, C.; Alonso, C.; Molina, F.J. Cover cracking as a function of bar corrosion: Part I-experimental test. *Mater. Struct.* **1993**, *26*, 453–464.
12. Caré, S.; Nguyen, Q.T.; l'Hostis, V.; Berthaud, Y. Mechanical properties of the rust layer induced by impressed current method in reinforced mortar. *Cem. Concr. Res.* **2008**, *38*, 1079–1091. [CrossRef]
13. Campanella, L.; Dell'Aglio, E.; Reale, R.; Cardellicchio, F.; Salvi, A.M.; Casieri, C.; Cerichelli, G.; Gabriele, F.; Spreti, N.; Bernardo, G.; et al. Development of natural gels for cleaning the stone materials of cultural heritage from iron stains and biodeteriogenic microorganisms. In Proceedings of the XII International Conference Diagnosis, Conservation and Enhancement of the Cultural Heritage, Naples, Italy, 9–10 December 2021.

14. Bernardo, G.; Guida, A.; Porcari, V.; Campanella, L.; Dell'Aglio, E.; Reale, R.; Cardellicchio, F.; Salvi, A.M.; Casieri, C.; Cerichelli, G.; et al. New materials and diagnostic techniques to prevent and control calcarenite degradation. In Proceedings of the XII International Conference Diagnosis, Conservation and Enhancement of the Cultural Heritage, Naples, Italy, 9–10 December 2021.
15. Campanella, L.; Cardellicchio, F.; Dell'Aglio, E.; Reale, R.; Salvi, A.M. A green approach to clean iron stains from marble surfaces. *Herit. Sci.* **2022**, *10*, 14. [CrossRef]
16. Little, B.J.; Gerke, T.L.; Lee, J.S. Mini-review: The morphology, mineralogy and microbiology of accumulated iron corrosion products. *Biofouling* **2014**, *30*, 941–948. [CrossRef]
17. Nabika, H. Liesegang phenomena: Spontaneous pattern formation engineered by chemical reactions. *Curr. Phys. Chem.* **2015**, *5*, 5–20. [CrossRef]
18. Reale, R.; Campanella, L.; Sammartino, M.P.; Visco, G.; Bretti, G.; Ceseri, M.; Notarnicola, F. A mathematical, experimental study on iron rings formation in porous stones. *J. Cult. Herit.* **2019**, *38*, 158–166. [CrossRef]
19. Merino, E.; Banerjee, A. Terra Rossa Genesis, Implications for Karst, and Eolian Dust: A Geodynamic Thread. *J. Geol.* **2008**, *116*, 62–75. [CrossRef]
20. Loeppert, R.H.; Hossner, L.R. Reaction of Fe^{2+} and Fe^{3+} with Calcite. *Clays Clay Miner.* **1984**, *32*, 213–222. [CrossRef]
21. Magalhaes, M.S.; Brandao, P.R.G.; Tavares, R.P. Types of goethite from Quadrilátero Ferrífero's iron ores and their implications in the sintering process. *Miner. Process. Extr. Metall.* **2007**, *116*, 54–64. [CrossRef]
22. Castle, J.E.; Chapman-Kpodo, H.; Proctor, A.; Salvi, A.M. Curve-fitting in XPS using extrinsic and intrinsic background structure. *J. Electron Spectrosc. Relat. Phenom.* **2000**, *106*, 65–80. [CrossRef]
23. Castle, J.E.; Salvi, A.M. Chemical state information from the near-peak region of the X-ray photoelectron background. *J. Electron Spectrosc. Relat. Phenom.* **2001**, *114–116*, 1103–1113. [CrossRef]
24. Graat, P.C.J.; Somers, M.A.J. Simultaneous determination of composition and thickness of thin iron oxide films from XPS Fe2p spectra. *Appl. Surf. Sci.* **1996**, *100–101*, 36–40. [CrossRef]
25. Briggs, D.; Grant, J.T. *Surface Analysis by Auger and X-ray Photoelectron Spectroscopy*; IM Publications: Chichester, UK, 2003.
26. Briggs, D.; Seah, M.P. *Practical Surface Analysis*; John Wiley & Sons: Chichester, UK, 1990; ISBN 0-471-26279-X.
27. Yamashita, T.; Hayes, P. Analysis of Fe^{2+} and Fe^{3+} ions in oxide materials. *Appl. Surf. Sci.* **2008**, *254*, 2441–2449. [CrossRef]
28. NIST X-ray Photoelectron Spectroscopy Database 20, Version 4.1. Available online: https://srdata.nist.gov/xps/default.aspx (accessed on 30 September 2022).
29. Kloprogge, J.T.; Ponce, C.P.; Ortillo, D.O. X-ray Photoelectron Spectroscopy Study of some Organic and Inorganic Modified Clay Minerals. *Materials* **2021**, *14*, 7115. [CrossRef] [PubMed]
30. Li, Y.; Li, Y.; Li, H.; Fan, X.; Yan, H.; Cai, M.; Zhu, M. Insights into the tribological behavior of choline chloride—Urea and choline chloride—Thiourea deep eutectic solvents. *Friction* **2021**, *11*, 76–92. [CrossRef]
31. Ciotonea, C.; Averlant, R.; Rochard, G.; Mamede, A.S.; Giraudon, J.M.; Alamdari, H.; Lamonier, J.F.; Royer, S. A simple and green procedure to prepare efficient manganese oxide nanopowder for the low temperature removal of formaldehyde. *ChemCatChem* **2017**, *9*, 2366–2376. [CrossRef]
32. Wang, H.; Liang, Y.; Gong, M.; Li, Y.; Chang, W.; Mefford, T.; Zhou, J.; Wang, J.; Regier, T.; Wei, F.; et al. An ultrafast nickel–iron battery from strongly coupled inorganic nanoparticle/nanocarbon hybrid materials. *Nat. Commun.* **2012**, *3*, 917. [CrossRef]
33. Tian, Z.; Wang, C.; Yue, J.; Zhang, X.; Ma, L. Effect of a potassium promoter on the Fischer–Tropsch synthesis of light olefins over iron carbide catalysts encapsulated in graphene-like carbon. *Catal. Sci. Technol.* **2019**, *9*, 2728–2741. [CrossRef]
34. Wang, X.; Zhang, H.; Lin, H.; Gupta, S.; Wang, C.; Tao, Z.; Fu, H.; Wang, T.; Zheng, J.; Wu, G.; et al. Directly converting Fe-doped metal-organic frameworks into highly active and stable Fe-N-C catalysts for oxygen reduction in acid. *Nano Energy* **2016**, *25*, 110–119. [CrossRef]
35. Martín-García, L.; Bernal-Villamil, I.; Oujja, M.; Carrasco, E.; Gargallo-Caballero, R.; Castillejo, M.; Marco, J.F.; Gallego, S.; de la Figuera, J. Unconventional properties of nanometric FeO (111) films on Ru (0001): Stoichiometry and surface structure. *J. Mater. Chem. C* **2016**, *4*, 1850–1859. [CrossRef]
36. Reale, R. Alterazioni Cromatiche di Materiali Lapidei Carbonatici: Studio Delle Macchie Indotte Dalla Coesistenza Con Materiali Ferrosi e Metodi Innovativi per la Loro Rimozione. Ph.D. Thesis, Sapienza University, Rome, Italy, 2017.
37. Cornell, R.M.; Schwertmann, U. *The Iron Oxides: Structure, Properties, Reactions, Occurrences, and Uses*; Wiley-Vch: Weinheim, Germany, 2003; Volume 664.
38. Leidheiser, H., Jr.; Czakó-Nagy, I. A Mössbauer spectroscopic study of rust formed during simulated atmospheric corrosion. *Corros. Sci.* **1984**, *24*, 569–577. [CrossRef]
39. Joos, A.; Rumenapp, C.; Wagner, F.; Gleich, B. Characterisation of iron oxide nanoparticles by Mossbauer spectroscopy at ambient temperature? *J. Magn. Magn. Mater.* **2016**, *399*, 123–129. [CrossRef]
40. Beltran, M.; Playà, E.; Artigau, M.; Arroyo, P.; Guinea, A. Iron patinas on alabaster surfaces (Santa Maria de Poblet Monastery, Tarragona, NE Spain). *J. Cult. Herit.* **2016**, *18*, 370–374. [CrossRef]
41. Grassini, S.; Angelini, E.; Parvis, M.; Bouchar, M.; Dillmann, P.; Neff, D. An in situ corrosion study of Middle Ages wrought iron bar chains in the Amiens Cathedral. *Appl. Phys. A* **2013**, *113*, 971–979. [CrossRef]
42. Fassina, V. Environmental pollution in relation to stone decay. *Durab. Build. Mater.* **1988**, *5*, 317–358.

43. Gaylarde, C.; Little, B. Biodeterioration of stone and metal- Fundamental microbial cycling processes with spatial and temporal scale differences. *Sci. Total Environ.* **2022**, *823*, 153193. [CrossRef] [PubMed]
44. Lagarec, K.; Rancourt, D.G. *Recoil-Mössbauer Spectral Analysis Software for Windows*; University of Ottawa: Ottawa, ON, USA, 1998.

Disclaimer/Publisher's Note: The statements, opinions and data contained in all publications are solely those of the individual author(s) and contributor(s) and not of MDPI and/or the editor(s). MDPI and/or the editor(s) disclaim responsibility for any injury to people or property resulting from any ideas, methods, instructions or products referred to in the content.

Article

Iron Forms Fe(II) and Fe(III) Determination in Pre-Roman Iron Age Archaeological Pottery as a New Tool in Archaeometry

Lidia Kozak [1], Andrzej Michałowski [2,3], Jedrzej Proch [1,3], Michal Krueger [2,3], Octavian Munteanu [4] and Przemyslaw Niedzielski [1,3,*]

1. Department of Analytical Chemistry, Faculty of Chemistry, Adam Mickiewicz University in Poznań, 8 Uniwersytetu Poznanskiego Street, 61-614 Poznań, Poland; lkozak@amu.edu.pl (L.K.); jedrzej.proch@amu.edu.pl (J.P.)
2. Faculty of Archaeology, Adam Mickiewicz University in Poznań, 7 Uniwersytetu Poznanskiego Street, 61-614 Poznań, Poland; misiek@amu.edu.pl (A.M.); krueger@amu.edu.pl (M.K.)
3. Interdisciplinary Research Group Archaeometry, Faculty of Archaeology and Faculty of Chemistry, Adam Mickiewicz University in Poznań, 7–8 Uniwersytetu Poznanskiego Street, 61-614 Poznań, Poland
4. World History Department, State Pedagogical University, 1 Ion Creanga Street, MD-2069 Chisinau, Moldova; ocmunteanu@gmail.com
* Correspondence: pnied@amu.edu.pl

Abstract: This article presents studies on iron speciation in the pottery obtained from archaeological sites. The determination of iron forms Fe(II) and Fe(III) has been provided by a very simple test that is available for routine analysis involving the technique of molecular absorption spectrophotometry (UV–Vis) in the acid leachable fraction of pottery. The elemental composition of the acid leachable fraction has been determined by inductively coupled plasma optical emission spectrometry (ICP-OES). Additionally, the total concentration of the selected elements has been determined by X-ray fluorescence spectrometry with energy dispersion (EDXRF). The results of the iron forms' determinations in archaeological pottery samples have been applied in the archaeometric studies on the potential recognition of the pottery production technology, definitely going beyond the traditional analysis of the pottery colour.

Keywords: pottery; iron; speciation; archaeometry; spectrophotometry; spectrometry; pre-Roman Iron Age

Citation: Kozak, L.; Michałowski, A.; Proch, J.; Krueger, M.; Munteanu, O.; Niedzielski, P. Iron Forms Fe(II) and Fe(III) Determination in Pre-Roman Iron Age Archaeological Pottery as a New Tool in Archaeometry. *Molecules* 2021, 26, 5617. https://doi.org/10.3390/molecules26185617

Academic Editors: Maria Luisa Astolfi, Maria Pia Sammartino and Emanuele Dell'Aglio

Received: 25 July 2021
Accepted: 13 September 2021
Published: 16 September 2021

Publisher's Note: MDPI stays neutral with regard to jurisdictional claims in published maps and institutional affiliations.

Copyright: © 2021 by the authors. Licensee MDPI, Basel, Switzerland. This article is an open access article distributed under the terms and conditions of the Creative Commons Attribution (CC BY) license (https://creativecommons.org/licenses/by/4.0/).

1. Introduction

Clay is a natural rock material that has been used as a building material by people for over 10,000 years due to its availability and properties [1]. From a chemical perspective, it combines one or more silicate minerals, also known as phyllosilicates, as well as traces of metals, metal oxides, sand, and organic matter. Clay minerals are usually hydrated, and they exhibit plasticity due to their water content. After drying and/or firing, they lose their water content, and clay itself becomes hard and non-plastic [2]. Crafting vessels from clay was definitely a large step in the development of prehistoric communities. Ceramics and their properties, such as quality, element or water content, or even shapes and patterns, are very important indicators in archaeometric research that can lead us to various conclusions about the past [3].

Archaeometry is an important branch of archaeology that applies various scientific techniques in order to acquire more information about archaeological samples, such as pottery [4], metal artefacts [5] or building materials [6,7]. The archaeometric study of ceramics is especially important because the pottery was used daily by settled farming communities [8]. Applying scientific methods to ubiquitous ceramic samples found at various excavation sites helps us to find answers about the date and place of the pottery production, the likely technique and parameters used during firing, or the purpose of their creation [9–11]. Admixtures (and also clay firing parameters) can be markers providing information about connections between samples found in different places [12]. Redox and phase speciation (for solid samples) can be used to determine the parameters in which the studied object was formed [13]. Moreover, it can be used to understand the processes that changed the chemical composition of an object from the state when it was formed [14]. Ceramics as works of art are often painted on the surface. There are numerous examples of element speciation studies in ceramics in which the oxidation state of the elements was the subject of determination: iron and manganese in Sicilian "proto-majolica" pottery [15], iron in black glaze of Chinese pottery [16], cobalt in Chinese porcelains from the 16th to 17th century [17], and iron in ceramics from Brazil [18].

The pigments used in coatings or glazes are often metal oxides in different oxidation states, which differ in colour [13]. Moreover, different parameters during clay firing are responsible for elements' oxidation state distribution, which is visible as a change in the appearance of the ceramic form. Iron is an element that is the component of many silicate minerals, such as illite, chlorite, glauconite, or biotite [19]. The determination of iron speciation in pottery is especially important because various types of clay, just like different clay firing techniques and temperatures, have a major influence on the oxidation state and—furthermore—on the colour of ceramics [18]. Fe(III) is responsible for an orange-red colour, whereas Fe(II) imparts a dark grey colour. In practice, both forms can simultaneously exist in ceramics and their coatings, so the determination of the Fe(III)/Fe(II) ratio is of great importance (Orecchio, 2011). The simplest way to quantitatively determine Fe(II) and Fe(III), is to carry out a mineralisation of a powdered sample by hydrofluoric acid and to use UV–Vis spectrophotometric methods in order to determine Fe(II) and Fe(III). A similar method was proposed for studies of iron speciation in ancient pottery from Sicily [20]. The sample was mineralised, and then the method for determining Fe(II) by the reaction of Fe^{2+} with 1,10-phenanthroline was applied [21]. Moreover, the author determined the total iron using ICP-OES, and the results were in agreement with those of the phenanthroline method [20]. There are many other analytical methods for determining iron speciation, but they require more advanced equipment. One of them uses Mössbauer spectroscopy, which is based on the Mössbauer effect. It was applied in the recognition of the oxidation state of iron in Brazilian ceramic samples, along with Scanning Electron Microscope (SEM), X-ray fluorescence (XRF), and X-ray diffraction (XRD) [18].

A non-destructive X-ray absorption spectroscopy (XAS) is also used for archaeological artifacts analysis. Compared with XRD, it does not require ordered structures [22]. Therefore, it can be applied to amorphous samples with no restrictions on their type or size [23]. In addition, the Extended X-ray Absorption Fine Structure (EXAFS) range shows the local structure around the iron sites. The whole study demonstrates numerous possibilities for analysing samples by XAS [20].

In this article, the studies on iron speciation in archaeological pottery are described. The results of the determination of iron forms: Fe(II), Fe(III) and chemical composition (the occurrence of the selected elements, both total concentration and occurrence in the acid leachable fraction of pottery) are compared for different archaeological sites.

2. Experimental

2.1. Instrumentation

For determination of the total concentration of selected elements (As, Ba, Ca, Co, Cr, Cu, Fe, Mn, Mo, Nb, Ni, Pb, Rb, Sb, Sn, Sr, Th, Ti, U, Y, Zn, Zr) in pottery samples, the portable XRF spectrometer Tracer III Handheld XRF Bruker (Billerica, MA, USA) was used. The spectrometer has worked in quantitate mode (determination limits on the level of 1 mg kg^{-1}, uncertainty level below 15%) elaborated by the manufacturer for geochemical analysis with two built-in calibrations: Bruker Mudrock Major (Al, Ba, Ca, Fe, K, Mg, Mn, and Ti determination; instrumental parameters: 15 keV, 25, µA, vacuum < 17 Torr); and Bruker Mudrock Trace (As, Co, Cr, Cu, Mo, Nb, Ni, Pb, Rb, Sb, Sn, Sr, Th, U, Y, Zn, Zr determination; instrumental parameters: filter 0.3048 mm Al and 0.0254 mm Ti, 40 kV, 12 µA). The problem of accuracy in studies was discussed in the Supplementary Data.

ICP-OES spectrometer Agilent 5110 ICP-OES Agilent (Santa Clara, CA, USA) was used in selected elements (Al, As, B, Ba, Bi, Ca, Cd, Ce, Co, Cr, Cu, Dy, Er, Eu, Fe, Ga, Gd, Ge, Ho, In, K, La, Li, Lu, Mg, Mn, Mo, Na, Nd, Ni, Pb, Pd, Pr, Re, Rh, Sb, Sc, Se, Sm, Sr, Tl, Tm, Y, Yb, Zn) for determination in simultaneous mode. The simultaneous axial and radial view of plasma was allowed by the synchronous vertical dual view (SVDV). For multi-elemental determination, the common conditions were used: 3 replicates, measuring time 5 s, plasma gas flow 12.0 L min^{-1}, auxiliary gas flow 1.0 L min^{-1}, nebulizer gas flow 0.7 L min^{-1}, Radio Frequency (RF) power 1.2 kW. For determination of higher level of the selected elements, the alternative (less sensitive) wavelengths were used (indicated by bold in Table S1). Spectrometer build-in method of background correction (fitted) was used. The detection limits were calculated based on the standard deviation value of multiple (n = 10) calibration blank analysis: 3-sigma criteria. The detection limits for all determined elements were in the range of 0.01–0.09 mg kg^{-1} (Table S1 in Supplementary Data).

The uncertainty budget was estimated for the complete analytical procedure, including preparation of samples, instrument calibration, and determination of the content of elements. The propagated uncertainty (a coverage factor k = 2 for approximate 95% confidence) was at a level below 20%.

Due to the lack of access to standard reference materials (CRM) for the multi-elemental pottery analysis, the soil and sediments matrix CRM was used because of its geological similarity to the raw materials of ceramics (post-glacial and sedimentary materials). For the traceability studies, the following CRMs were selected: NIST 2709—soil; IAEA 405—estuarine sediments; CRM S-1—loess soil; BCR 667- estuarine sediments (Table S2 in Supplementary Data). Due to the fact that the information about elements' concentration of acid extractable fraction was available only for NIST 2709a, the analysis of CRMs were provided using two procedures: (i) with sample digestion using the mixture of concentrated HCl and HNO$_3$ (aqua regia (AR)); (ii) with sample extraction by HCl following the procedure described below. The first step allowed the calibration and interferences correction to be checked (using spectrometer build-in background correction method); the second step allowed the matrix-dependent interferences to be controlled. Additionally, the standard addition method was applied for HCl extracts. In all procedures, the acceptable recovery (in the range 80–120%) was found (Table S3 in Supplementary Data). In colorimetric analyses, the photometer Slandi LF300 Slandi (Michalowice, Poland) was used (measurements of absorbance at 470 and 520 nm). The detection limits for both iron forms were determined by dilution of calibration standards, and the levels of 10 mg kg^{-1} were determined for Fe(II) and Fe(III), respectively, with uncertainty below 15%. To check the accuracy, the standard addition method was applied with good recovery (in the range 80–120%, Table S4 in Supplementary Data). Additionally, the reference procedures were applied in accuracy studies (Table S5 and procedures description in Supplementary Data).

For homogenisation of the samples, the Pulverisette agate laboratory grinder Fritsch (Idar-Oberstein, Germany) was used.

2.2. Reagents

Only analytical purity reagents and deionised water from a Milli-Q device Millipore (Burlington, VT, USA) were used. Standard solutions (1.00 g L^{-1}) of iron forms: Fe(III) and Fe(II) were prepared from ferric ammonium sulphate dodecahydrate and ferrous ammonium sulphate hexahydrate Acros-Thermo Fisher Scientific (Geel, Belgium), respectively. The commercial standards (1.000 g L^{-1}) were used for ICP-OES analysis Romil (Cambridge, UK). The following reagents POCh (Gliwice, Poland) were used: 2.0 mol L^{-1} solution of hydrochloric acid (HCl), 0.5% (m/m) solution of 2,2'-Bipirydyl ($C_{10}H_8N_2$), acetate buffer (sodium acetate and acetic acid), 5% (m/m) solution of potassium thiocyanate (KSCN).

2.3. Samples

The 78 fragments of pottery chosen for chemical analysis came from three archaeological sites from western Poland. Sites were located in close proximity to each other: Borzejewo

(B) and Poznań-Nowe Miasto (P), located in central Wielkopolska, and Grabkowo (G) in the eastern part of Kujawy. The reference material for the mentioned sites was the archaeological site of Poieneşti-Lukaševka Culture from Orcheiul Vechi (M) from Moldova. The whole collection of analysed samples can be combined with a unified time horizon—associated with the early phases of the younger pre-Roman Iron Age and an approaching cultural component visible in the material, initially referred to as Jastorf Culture. This also applies to the areas of Moldova. The collection of this reference material was the result of an attempt to capture potentially similar cultural factors that could be manifested in radically different raw materials. Furthermore, 126 pottery fragments from Pławce (Pl) in western Poland were selected for comparative analysis. In the case of this latter site, the selection criteria were the area of origin and cultural phenomena readable in archaeological material analogous to other previously mentioned sites. An important factor was to implement an analysis of all samples taken from the archaeological site.

2.4. Methodology

The acid leaching procedure was described for geochemical studies [24]. The established analytical procedure of the speciation analysis [25,26] was optimised and applied for pottery samples. The analysis by XRF technique was described in previous work [27]. The sample preparation procedure is identical to the one applied for metals determination described in previous work [28]. The accuracy of the studies regarding all of the analytical procedures has been described in the Supplementary Data.

2.5. Total Iron and Selected Elements Determination

The XRF analysis was provided in laboratory using desktop spectrometer-stand. The pottery sample was placed on the spectrometer stand, oriented in correspondence with the original external surface of the ceramic vessel, and analysed. After analysis, the sample was rotated and the analysis was repeated. The acquisition time was 15 s. The mean value of concentration of selected elements (As, Ba, Ca, Co, Cr, Cu, Fe, Mn, Mo, Nb, Ni, Pb, Rb, Sb, Sn, Sr, Th, Ti, U, Y, Zn, Zr) and relative standard deviation were calculated from three repetitions ($n = 3$).

2.6. Acid Leachable Fraction Analysis

2.6.1. Sample Extraction by Hydrochloric Acid

The extraction by hydrochloric acid (methodology of acid leaching) was prepared following the previous studies [24]. The ceramic material was homogenised by grinding; the coarse material was removed using a plastic sieve (diameter of particle > 0.02 mm). Samples weighed to be 2.00 ± 0.01 g were put into a flask and 20 mL of hydrochloric acid solution (2 mol L^{-1}) was added. The flask (with a reflux condenser) was heated to approximately 80 °C for 30 min. After cooling, the solution was drained through a paper filter (rinsed previously using 200 mL of water) into a test tube; finally, water was added to a volume of 50.0 mL.

2.6.2. Elemental Analysis

Hydrochloric acid extracts of samples were analysed using ICP-OES technique. The selected elements were determined (indicated in the Supplementary Data).

2.6.3. Iron Chemical Forms Determination

The content of Fe(III) was determined using reaction of Fe(III) with thiocyanate in pH < 2.0 (in the hydrochloric acid environment). The intensity of light absorption (absorbance) by red complex was measured at wavelength 470 nm by UV–Vis spectrophotometer and compared with the calibration curve prepared using Fe(III) standard. The content of Fe(II) was determined in reaction of Fe(II) with 2,2'-bipirydyl (in the acetate buffer pH 4.5). The intensity of light absorption (absorbance) by red complex was measured at wavelength

520 nm by UV–Vis spectrophotometer and compared with the calibration curve prepared using Fe(II) standard.

2.7. Statistical Analysis

The analysis of the experimental data was performed using computer software Statistica 13.1 StatSoft—Dell (Round Rock, TX, USA). The multidimensional statistical analysis (principal components analysis PCA) was provided for the results of XRF and ICP-OES analysis to indicate the individual differences in the elemental composition of the pottery samples. For all statistical tests, the probability value $p = 0.05$ was applied [29,30].

3. Results and Discussion

3.1. Analysis of the Pottery Fragments from Four Sites

3.1.1. Chemical Composition of Pottery

The results (n = 4964 of single results for 78 fragments of pottery from sites B, G, P and M) of the total concentration of the elements (obtained in the XRF analysis of the raw material) and the concentration of elements in the acid leachable fraction (obtained in the ICP-OES analysis of the HCl extracts) were analysed using exploratory analysis (PCA). The 95.5% variability of the results was described by two components (Figure 1).

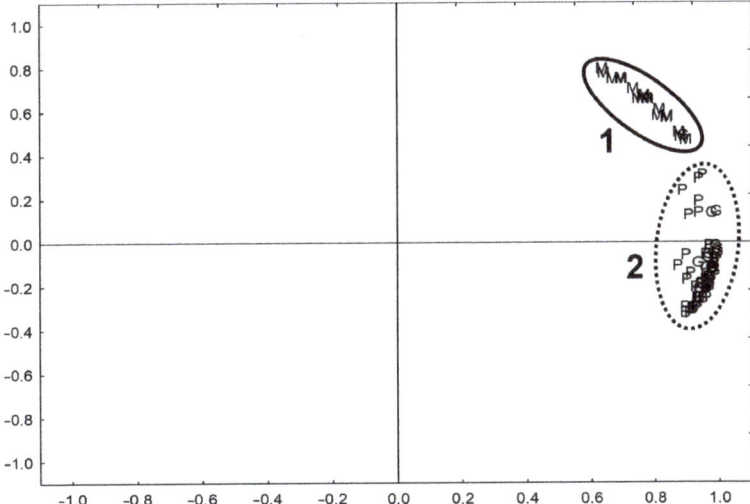

Figure 1. Results of Principal Components Analysis for elements concentration in pottery samples from four archaeological sites (Borzejewo (B), Poznań-Nowe Miasto (P), Grabkowo (G), Moldova(M)); 1—the group of samples from site M (Moldova), 2—the group of samples from the other sites.

It is clearly indicated that the elemental composition of the studied samples is different for the pottery from Moldova and Poland. The difference is based on the geology of the regions: the Moldovan clays were marine deposits, while, in Poland, the formation of the clays was connected with the glacial and post-glacial processes. Different raw materials (clay) were used for the production of the pottery, and they were indicated in the composition of the material.

3.1.2. Iron Speciation in Pottery

The results of the determination of iron forms (Fe(II) and Fe(III)) were put together in Figure 2. Two groups of pottery samples were formed: the first for pottery from archaeological sites in Bozejewo (B) and Grabkowo (G)—group 1 in Figure 2—and the second one for pottery from Poznan (P) and Moldavia (M)—group 2.

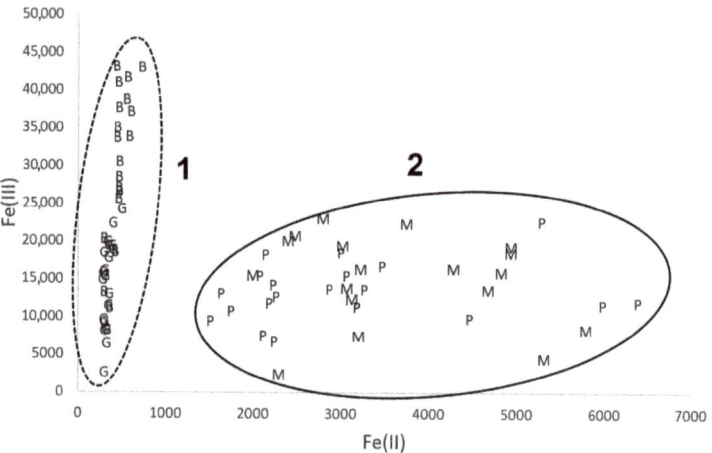

Figure 2. The iron speciation in pottery samples: 1—the group of samples from sites Borzejewo (B) and Grabkowo (G); 2—the group of samples from sites Poznań-Nowe Miasto (P) and Moldova(M).

The similarities of the iron speciation are different to the shapes of the chemical compositions of the samples. Due to the different origin of the raw material (marine clay versus glacial or post-glacial clay) used for pottery production, the pottery chemical composition of the samples from the Poznan archaeological site is different to the pottery chemical composition of the samples from Moldavia. The differences in the material from which the ceramics are made are not reflected in the iron speciation. According to the literature data, the presence of Fe (III) and Fe (II) forms and their Fe(III)/Fe(II) ratio reflect the technology of the ceramics production process [18], particularly the temperature and conditions of the ceramics firing process [31]. Thus, it can be concluded that, regardless of the origin of the ceramics tested, and, therefore, the material (clay) from which it was made, two groups of ceramic objects (Figure 3) stand out on the basis of iron speciation, probably as a result of similar technological processes.

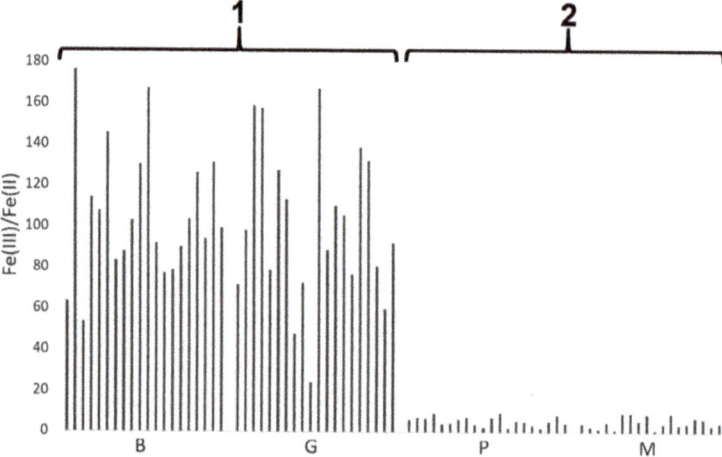

Figure 3. The Fe(III)/Fe(II) ratio for samples from sites Borzejewo (B), Poznań-Nowe Miasto (P), Grabkowo (G), Moldova(M); 1—the group of samples with Fe(III)/Fe(II) ratio greater than 20 (sites (Borzejewo (B) and Grabkowo (G))); 2—the group of samples with Fe(III)/Fe(II) ratio less than 20 (sites (Poznań-Nowe Miasto (P) and Moldova(M))).

The first group consists of ceramics from two archaeological sites (B and G), characterized by high values of the Fe(III)/Fe(II) ratio, exceeding the range from 24 to 176 (median values 101 and 95 for site B and site G, respectively). The predominant form of iron was that of Fe(III), which indicates both a higher ceramic firing temperature and the provision of an oxidizing environment during firing [20]. On the other hand, for the second group of samples (P and M), the value of the ratio Fe(III)/Fe(II) was definitively lower, ranging between 1 and 9 (median values 5 and 4 for site P and site M, respectively). This indicates a definitively higher concentration of Fe(II) than for the first group, which indicates technology using a lower ceramic firing temperature, with a definitively limited supply of oxidizing agent [32]. Importantly, the ceramic firing techniques, although characteristic of a given archaeological site, were unrelated to both the origin of the source material (clay) and the location of the site.

3.2. Analysis of Pottery Fragments from One Site

In the second step of the research, the mass pottery material obtained from one archaeological site (Plawce—Pl) was analysed. The 126 pottery samples represented all of the samples collected at the given site without any pre-selection or elimination. This creates the opportunity to compare fragments of ceramics made of a similar material but resulting from the actions of various pottery manufacturers.

3.2.1. Chemical Composition of Pottery

The results (n = 9198 of single results for 126 fragments of pottery from sites Pl) of the total concentration of the elements (obtained in the XRF analysis of the raw material) and the concentration of elements in the acid leachable fraction (obtained in the ICP-OES analysis of the HCl extracts) were analysed using exploratory analysis (PCA). The 96.0% variability of the results was described by two components (Figure 4).

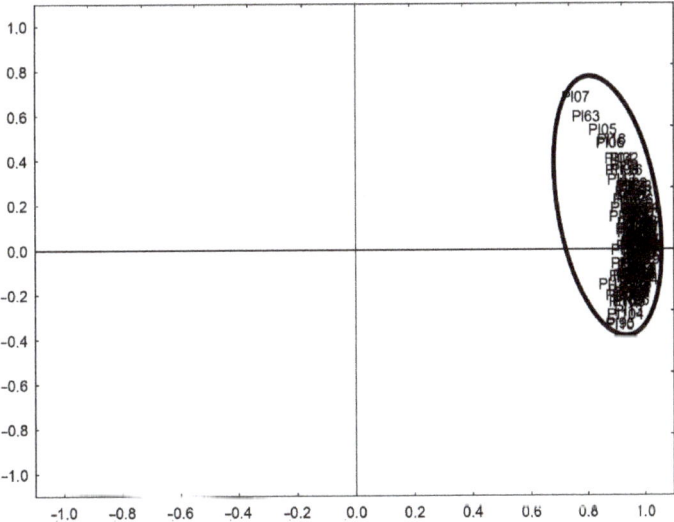

Figure 4. Results of Principal Components Analysis for elements concentration in pottery samples from Pl archaeological site.

The results of the exploratory analysis indicated that the actually tested ceramics were characterized by a very similar chemical composition, which allows the hypothesis of the local exploitation of one source of clay by ceramic manufacturers at the time to be considered. Thus, we were dealing with a homogeneous set of ceramic fragments in terms of chemical properties.

3.2.2. Iron Speciation in Pottery

The results of the determinations of iron forms (Fe(II) and Fe(III)) were presented together in Figure 5. Two groups of pottery samples were indicated based on Figure 2: the first for pottery similar to pottery from archaeological sites in Bozejewo (B) and Grabkowo (G)—group 1 in Figure 5, the second one for pottery "similar to" pottery from Poznan (P) and Moldavia (M)—group 2.

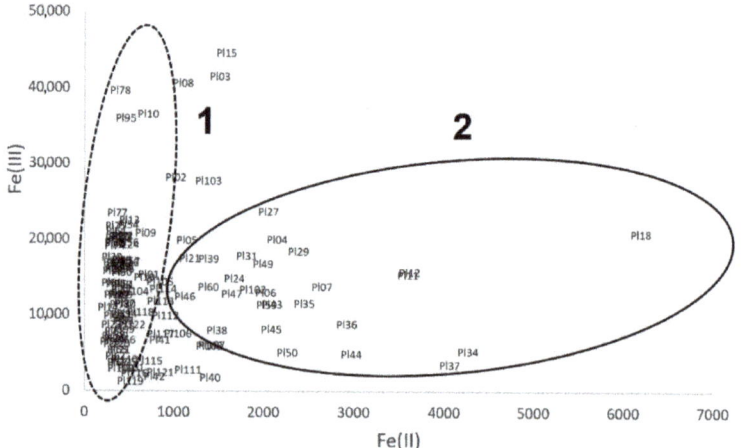

Figure 5. The iron speciation in pottery samples: 1—the group of samples "similar to" pottery from sites B and G; 2—the group of samples "similar to" pottery from sites M and P.

The factor Fe(III)/Fe(II) for samples from site Pl was shaped in a very broad range: from 1 to 485 (Figure 6).

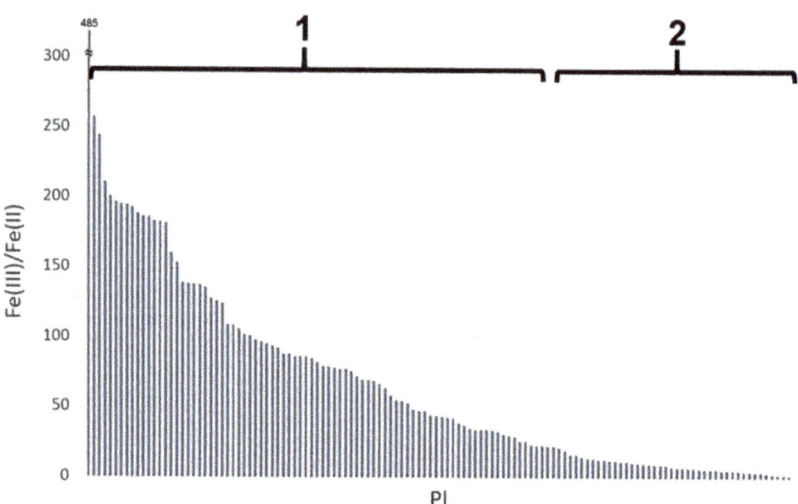

Figure 6. The Fe(III)/Fe(II) ratio for samples from site Pl: 1—the group of samples with Fe(III)/Fe(II) ratio greater than or equal to 20; 2—the group of samples with Fe(III)/Fe(II) ratio less than 20.

For most (85) samples, the value of the Fe(III)/Fe(II) ratio was greater than or equal to 20. As stated above [20], this indicated a higher ceramic firing temperature, as well as the presence of an oxidizing atmosphere during firing. For a smaller number of samples (41), the value of the Fe(III)/Fe(II) ratio was less than 20, indicating different firing conditions: a lower temperature and a reducing atmosphere [32].

Despite the similar chemical composition of the examined ceramic fragments, suggesting the origin of the raw materials from the same or a very geologically similar source, the iron speciation showed a wide variation within the considered sample collection. The reason for this could be the different technology used in different pottery workshops operating in the same place. In addition, a large variation in the value of the Fe(III)/Fe(II) ratio may indicate the variability of the firing conditions even within one workshop. The high variability of the results for the Fe(III)/Fe(II) coefficient may arise for two reasons: (i) the inhomogeneity of the distribution of iron forms in single vessels; (ii) the high variability of the firing conditions even within one pottery workshop. However, the inhomogeneity indicated in point (i) may also result from the instability of the firing process and the lack of its full repeatability. Although the group of analyzed pottery fragments was not large, the results allow for the formulation of a hypothesis about the large differentiation and randomness of the conditions of ceramics firing within one site.

4. Conclusions

The speciation analysis has been demonstrated to be a useful tool for archaeometrical studies that make it possible to obtain new information about the characteristics of pottery. However, the speciation studies require the destruction of a part of a sample (although this is not problematic for mass pottery materials). The very simple, well-known, and, for most researchers, readily available analytical procedures enable analysis of a great number of samples. A large dataset is the ideal condition to provide a statistical analysis that can be used to find the similarities and differences in the pottery as it pertains to the results of, e.g., pottery manufacturing. It was found that the values of the Fe(III)/Fe(II) coefficient do not depend on the origin of the original material from which the ceramics were made. On the one hand, similar values of the Fe(III)/Fe(II) coefficient were obtained for pottery from regions with different geology (clays from marine deposits versus clays whose formation is connected with the glacial and post-glacial processes). On the other hand, the coefficient values were different for pottery samples with a similar chemical composition coming from the same area. Based on the literature, the reason for the differentiation of the Fe(III)/Fe(II) coefficient can be indicated by the different technological processes used in the manufacture of ceramics.

The most important novelty of the work is the development of simple, routine analytical tools that enable the indication of the technology of producing ceramics. The possibility of applying the developed procedures in the analysis of a large number of samples allows the use of statistical analysis to interpret the results, which is not possible with the traditional classification of ceramics based on their colour.

Supplementary Materials: The following are available online. Description of XRF accuracy studies, ICP-OES analytical figures of merit: Table S1. Wavelengths used in analysis, detection limits (DL), calibration range (range) and precision for ICP-OES determination, ICP-OES accuracy studies: Table S2. Results of certified reference materials analysis, Table S3. Results of acid extractable fraction of certified reference material NIST 2709a analysis and spike recovery in standard addition method, Colorimetric analysis accuracy studies: Table S4. Spike recovery in colorimetric procedures, Table S5. Results of iron forms determination using different colorimetric procedures and hyphenated analytical systems References to previous works.

Author Contributions: Conceptualization, L.K. and P.N.; methodology, L.K., A.M. and P.N.; validation, J.P. and P.N.; investigation, J.P. and M.K.; resources, O.M.; data curation, J.P. and M.K.; writing—original draft preparation, L.K. and A.M.; writing—review and editing, P.N.; funding acquisition, A.M. All authors have read and agreed to the published version of the manuscript.

Funding: This research was funded by National Science Centre, Poland, under the program OPUS 8: UMO-2014/15/B/HS3/02279.

Institutional Review Board Statement: Not applicable.

Informed Consent Statement: Not applicable.

Data Availability Statement: All data presented in the article are available from the corresponding author.

Conflicts of Interest: The authors declare no conflict of interest.

References

1. Shelach, G. On the invention of pottery. *Science* **2012**, *336*, 1644–1645. [CrossRef] [PubMed]
2. Wagner, U.; Gebhard, R.; Grosse, G.; Hutzelmann, T.; Murad, E.; Riederer, J.; Shimada, I.; Wagner, F. Clay: An important raw material for prehistoric man. *Hyperfine Interact.* **1998**, *117*, 323–335. [CrossRef]
3. Foster, W. Chemistry and Grecian archaeology. *J. Chem. Educ.* **1933**, *10*, 270. [CrossRef]
4. Quarta, G.; Giorgia, A.; Ingravallo, E.; Tiberi, I.; Calcagnile, L. Radiocarbon dates and XRF analyses from two prehistoric contexts in the Badisco area (Otranto–Le). *Measurement* **2018**, *125*, 279–283. [CrossRef]
5. Gójska, A.; Miśta-Jakubowska, E.; Banaś, D.; Kubala-Kukuś, A.; Stabrawa, I.; Miśta, E. Archaeological applications of spectroscopic measurements. Compatibility of analytical methods in comparative measurements of historical Polish coins. *Measurement* **2018**, *135*, 869–874. [CrossRef]
6. Ergenç, D.; Fort, R. Multi-technical characterization of Roman mortars from Complutum, Spain. *Measurement* **2019**, *147*, 106876. [CrossRef]
7. Lezzerini, M.; Raneri, S.; Pagnotta, S.; Columbu, S.; Gallello, G. Archaeometric study of mortars from the Pisa's Cathedral Square (Italy). *Measurement* **2018**, *126*, 322–331. [CrossRef]
8. Rosado, L.; Van Pevenage, J.; Vandenabeele, P.; Candeias, A.; Lopes, M.D.C.; Tavares, D.; Alfenim, R.; Schiavon, N.; Mirão, J. Multi-analytical study of ceramic pigments application in the study of Iron Age decorated pottery from SW Iberia. *Measurement* **2018**, *118*, 262–274. [CrossRef]
9. Cano, N.F.; Ribeiro, R.B.; Munita, C.S.; Watanabe, S.; Neves, E.G.; Tamanaha, E.K. Dating and determination of firing temperature of ancient potteries from São Paulo II archaeological site, Brazil by TL and EPR techniques. *J. Cult. Herit.* **2015**, *16*, 361–364. [CrossRef]
10. Ricciardi, P.; Nodari, L.; Gualtieri, S.; De Simone, D.; Fabbri, B.; Russo, U. Firing techniques of black slipped pottery from Nepal (12th–3rd century B.C.): The role of Mossbauer spectroscopy. *J. Cult. Herit.* **2008**, *9*, 261–268. [CrossRef]
11. Venkatachalapathy, R.; Sridharan, T.; Dhanapandian, S.; Manoharan, C. Determination of firing temperature of ancient potteries by means of infrared and Mossbauer studies. *Spectrosc. Lett.* **2002**, *35*, 769–779. [CrossRef]
12. Meloni, S.; Oddone, M.; Genova, N.; Cairo, A. The production of ceramic materials in Roman Pavia: An archaeometric NAA investigation of clay sources and archaeological artifacts. *J. Radioanal. Nucl. Chem.* **2000**, *244*, 553–558. [CrossRef]
13. Bardelli, F.; Barone, G.; Crupi, V.; Longo, F.; Maisano, G.; Majolino, D.; Mazzoleni, P.; Venuti, V. Iron speciation in ancient Attic pottery pigments: A non-destructive SR-XAS investigation. *J. Synchrotron Radiat.* **2012**, *19*, 782–788. [CrossRef] [PubMed]
14. Doménech-Carbó, A.; Doménech Carbó, M.; Costa, V. *Electrochemical Methods in Archaeometry, Conservation and Restoration*; Springer: Berlin/Heidelberg, Germany, 2009.
15. Bardelli, F.; Barone, G.; Crupi, V.; Longo, F.; Majolino, D.; Mazzoleni, P.; Venuti, V. Combined non-destructive XRF and SR-XAS study of archaeological artefacts. *Anal. Bioanal. Chem.* **2011**, *399*, 3147–3153. [CrossRef] [PubMed]
16. Nakai, I.; Taguchi, I.; Yamasaki, K. Chemical speciation of archaeological objects by XRF/XANES analysis using synchrotron radiation. *Anal. Sci.* **1991**, *7*, 365–368. [CrossRef]
17. Figueiredo, M.; Silva, T.; Veiga, J. A XANES study of cobalt speciation state in blue-and-white glazes from 16th to 17th century Chinese porcelains. *J. Electron Spectrosc. Relat. Phenom.* **2012**, *185*, 97–102. [CrossRef]
18. Floresta, D.L.; Ardisson, J.D.; Fagundes, M.; Fabris, J.D.; Macedo, W.A.A. Oxidation states of iron as an indicator of the techniques used to burn clays and handcraft archaeological Tupiguarani ceramics by ancient human groups in Minas Gerais, Brazil. *Hyperfine Interact.* **2013**, *224*, 121–129. [CrossRef]
19. Stewart, J.D.; Adams, K.R. Evaluating visual criteria for identifying carbon- and iron-based pottery paints from the four corners region using SEM-EDS. *Am. Antiq.* **1999**, *64*, 675–696. [CrossRef]
20. Orecchio, S. Speciation studies of iron in ancient pots from Sicily (Italy). *Microchem. J.* **2011**, *99*, 132–137. [CrossRef]
21. Begheijn, L.T. Determination of iron(II) in rock, soil and clay. *Analyst* **1979**, *104*, 1055–1061. [CrossRef]
22. Frankel, D.; Webb, J.M. Pottery production and distribution in prehistoric Bronze Age Cyprus. An application of pXRF analysis. *J. Archaeol. Sci.* **2012**, *39*, 1380–1387. [CrossRef]
23. Forster, N.; Grave, P.; Vickery, N.; Kealhofer, L. Non-destructive analysis using PXRF: Methodology and application to archaeological ceramics. *X-Ray Spectrom.* **2011**, *40*, 389–398. [CrossRef]
24. Kozak, L.; Niedzielski, P. The evolution of December 2004 tsunami deposits: Temporal and spatial distribution of potentially toxic metalloids. *Chemosphere* **2013**, *93*, 1856–1865. [CrossRef] [PubMed]
25. Niedzielski, P.; Kozak, L. Iron's fingerprint of deposits-iron speciation as a geochemical marker. *Environ. Sci. Pollut. Res.* **2017**, *25*, 242–248. [CrossRef]
26. Niedzielski, P.; Zielińska-Dawidziak, M.; Kozak, L.; Kowalewski, P.; Szlachetka, B.; Zalicka, S.; Wachowiak, W. Determination of iron species in samples of iron-fortified food. *Food Anal. Methods* **2014**, *7*, 2023–2032. [CrossRef]

27. Niedzielski, P.; Kozak, L.; Wachelka, M.; Jakubowski, K.; Wybieralska, J. The microwave induced plasma with optical emission spectrometry (MIP–OES) in 23 elements determination in geological samples. *Talanta* **2015**, *132*, 591–599. [CrossRef]
28. Niedzielski, P.; Michałowski, A.; Teska, M.; Krzyżanowska, M.; Jakubowski, K.; Kozak, L.; Krueger, M.; Żółkiewski, M. The analysis of variability of chemical composition of ceramics for archaeometrical studies. In *Kulturkonzepte und Konzipierte Kulturen*; Michałowski, A., Schuster, J., Eds.; Dr. Rudolf Habelt GmbH: Bonn, Germany, 2018; pp. 199–211.
29. Electronic Statistics Textbook. StatSoft 2020. Available online: http://www.statsoft.com/textbook/ (accessed on 20 January 2020).
30. Moore, D.S. *The Basic Practice of Statistics*; Palgrave Macmillan: London, UK, 2010.
31. Doménech-Carbó, A.; Sánchez-Ramosa, S.; Doménech-Carbó, M.; Gimeno-Adelantado, J.V.; Bosch-Reig, F.; Yusá-Marco, D.J.; Saurí-Peris, M.C. Electrochemical determination of the Fe(III)/Fe(II) ratio in archaeological ceramic materials using carbon paste and composite electrodes. *Electroanalysis* **2002**, *14*, 685–696. [CrossRef]
32. Mangueira, G.; Toledo, R.; Teixeira, S.; Franco, R. Evaluation of archeothermometric methods in pottery using electron paramagnetic resonance spectra of iron. *Appl. Clay Sci.* **2013**, *86*, 70–75. [CrossRef]

Article

Visual and Physical Degradation of the Black and White Mosaic of a *Roman Domus* under Palazzo Valentini in Rome: A Preliminary Study

Claudia Colantonio [1], Paola Baldassarri [2], Pasquale Avino [3,*], Maria Luisa Astolfi [1,4] and Giovanni Visco [1]

1. Department of Chemistry, University of Rome "La Sapienza", p.le Aldo Moro 5, I-00185 Rome, Italy
2. Città Metropolitana di Roma Capitale, U.C. 2, Servizio 3, Palazzo Valentini, Via Quattro Novembre 119a, I-00187 Rome, Italy
3. Department of Agricultural, Environmental and Food Sciences (DiAAA), University of Molise, Via De Sanctis, I-86100 Campobasso, Italy
4. Research Center for Applied Sciences to the safeguard of Environment and Cultural Heritage (CIABC), University of Rome "La Sapienza", p.le Aldo Moro 5, I-00185 Rome, Italy
* Correspondence: avino@unimol.it; Tel.: +39-0874-404634

Abstract: Palazzo Valentini, the institutional head office of Città Metropolitana di Roma Capitale, stands in in a crucial position in the Roman archaeological and urban contexts, exactly between the Fora valley, Quirinal Hill slopes, and Campus Martius. It stands on a second-century A.D. complex to which belong, between other archeological remains, two richly decorated aristocratic domus. One of these buildings, the *domus A*, presents an outward porticoed room with a fourth-century AD central *impluvium* (open air part of the atrium designed to carry away rainwater) with a black/white tiled mosaic pavement, the preservation status of which is compromised by an incoherent degradation product that has caused gradual detachment of the mosaic tiles. To identify the product and determine the causes of degradation, samples of the product were taken and subjected to SEM-EDS, XRF, NMR, FT-IR and GC-MS analyses. The findings reported in this study can help restorers, archaeologists and conservation scientists in order to improve knowledge about the Roman mosaic, its construction phases, conservation problems and proper solutions.

Keywords: Roman domus; mosaic restoration; multi-method diagnostic; NMR; FT-IR; GC-MS

Citation: Colantonio, C.; Baldassarri, P.; Avino, P.; Astolfi, M.L.; Visco, G. Visual and Physical Degradation of the Black and White Mosaic of a *Roman Domus* under Palazzo Valentini in Rome: A Preliminary Study. *Molecules* **2022**, *27*, 7765. https://doi.org/10.3390/molecules27227765

Academic Editor: Antonella Curulli

Received: 29 September 2022
Accepted: 7 November 2022
Published: 11 November 2022

Publisher's Note: MDPI stays neutral with regard to jurisdictional claims in published maps and institutional affiliations.

Copyright: © 2022 by the authors. Licensee MDPI, Basel, Switzerland. This article is an open access article distributed under the terms and conditions of the Creative Commons Attribution (CC BY) license (https://creativecommons.org/licenses/by/4.0/).

1. Introduction

Roman mosaic is one of the most beautiful expressions of ancient art that has survived to this day. The etymology of the word "mosaic" comes from "Muse", which was related to the parietal decorations of the caves consecrated to the Muses that were built in Roman gardens. The art of mosaic [1], as a decorative technique, found its first expression in the Aegean culture in productions of natural colored pebble pavements, initially grounded without stone processing and later carved to realize more refined geometries.

The retaining structure (*substructio*) for a mosaic pavement installation was made up of three layers: a primer coat called *statuminatio*, consisting of a conglomerate of big pieces of rocks with a fist-like size; a second layer, almost 25-cm deep, called *rudus*, made up of three parts of gravel and one part of lime; a *nucleus*, almost half the depth of the previous layer, composed of three parts of hydraulic lime mixed to crushed pottery and one part of lime. This use of crushed pottery together with lime was defined *opus signinum* by Romans, so-called from the town of Signia, renowned for its earthenware ("terracotta") shingles, and represented an innovation, compared with Greek mosaic pavements cemented with lime, thanks to which mosaic basements gained major resistance and stability.

Tiles were inserted into a superficial plaster layer and their surface was made perfectly smooth using appropriate tools to level them off. The last step involved applying a mixture of marble powder, sand and lime to produce a compact and tough surface.

Some beautiful examples of late Roman mosaic art, belonging to the floors of two residential rooms, were found in 2005 in the underground excavations of Palazzo Valentini in Rome [2–4], together with a private thermal complex, both pertaining to the fourth-century A.D. phase, and most significantly offering two rich aristocratic domus.

Palazzo Valentini (Figure 1a), rising at a crucial point regarding Roman archaeological and urban contexts, was found exactly between the Fora valley, Quirinal Hill slopes and Campus Martius, and was built by Cardinal Michele Bonelli, starting from 1585, over a preexisting building, Palazzo Zambeccari, from the middle of the sixteenthth century, that he bought from Giacomo Boncompagni [5–7]. The great enlargement and renovation work commissioned by Cardinal Bonelli were within the perspective of the supremacy of a nerve center in the town [8], as confirmed by the Roman cartography of the time, i.e., Antonio Tempesta's *"Veduta di Roma"*, 1593 (Figure S1 of the Supplementary Material). In 1827, the palace was bought by a banker, Vincenzo Valentini, to be his house, fostering the conclusion of works towards the Fora.

Figure 1. In the red circle is evidenced Palazzo Valentini (source: Google Earth [9]) (**a**). The paviment mosaic and the immediately facing wall (**b**).

The most significant and fascinating findings were the rooms pertaining to at least two buildings of the middle and late Imperial Age, which were surely configurated as *domus A* and *domus B* in the fourth century A.D.: indeed, their construction must be dated between the reigns of Hadrian and Septimius Severus (117–211 A.D.), but their story continues until the end of the fifth century A.D. [10].

The prestige of these two residences in the first half of the fourth century A.D. is immediately confirmed by the richness of decorative setup, consisting of great mosaic pavements in the *peristilium* (Figure S2a of the Supplementary Material) and the *triclinium* (Figure S2b of the Supplementary Material) of *domus A*, and in the marble covering of a staircase and an apsidal Aula in the *domus B*. The walls and pavements were covered by refined opus sectile with different polychrome marbles, which provided the room with the aspect of a boardroom.

The mosaic room is located at a depth of about 5.5 m under the modern street level and is enclosed on three sides by two walls dating to about 1865, and one perimetral wall of the palace, which has a basement window at its upper end, opening onto the road. The state of decay is clearly visible in the photographs and is not present in any other area of the excavation (Figure 2).

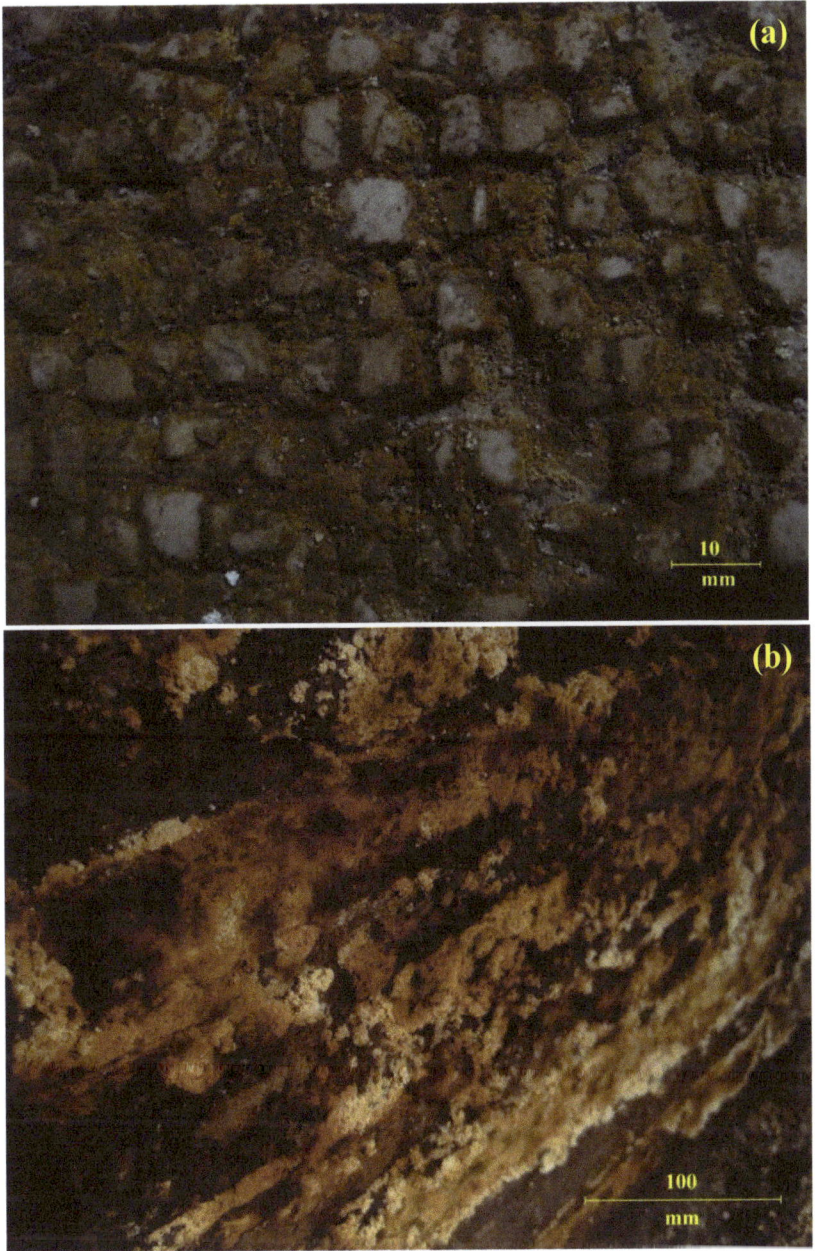

Figure 2. *Cont.*

Figure 2. Photos of the degradation (**a**) of the floor and (**b**) of the wall and optical microscope image (**c**) on the surface of the black and white mosaic pavement (10× and 40× magnification, respectively; width focusing 25 mm, 2 mm increments; 6 V/20 W halogen light). The sample of the degradation product of the black and white pavement mosaic of *domus B* was analyzed under an optical microscope in transmittance and reflectance).

Although the excavation areas investigated up to this day are still limited, compared with the total extension of the area of interest, the quality and quantity of archaeological remains allow us to assert with confidence that they belonged to an important residential quarter of senators or dignitaries of the imperial court, which occupied the most part of the NW and NE areas of the Trajan's Forum.

The paper focuses on the state of chemical degradation of the mosaic's right part and the part immediately facing the wall (Figure 1b–d). The need to identify the causes of chemical degradation is important for the historical reconstruction of the use of this environment over time and for the correct restoration of the wall and, essentially, the mosaic itself. This paper shows the preliminary results in the possible identification of the compounds responsible for such degradation.

Further, it would be advantageous that all the operations were coordinated among restorers, archaeologists, and conservation scientists in order to improve our knowledge about the Roman mosaic, its construction phases, conservation problems, and proper solutions to the problems.

Finally, it should be underlined that the archaeological excavation campaign started eight years ago and has assumed unexpected dimensions and the perfect state of conservation of the structures has made it possible to undertake work to enhance the site and open it to the public with the offer of a sophisticated multimedia system as virtual support to the museum itinerary.

2. Results

During a study led by one of the authors to contrast micro-biological and micro-flora growth, typical of hypogeum settings, by different enlightenment systems (the authors apologize but the procedure and data cannot be displayed because they are under international patent process/review, although some information can be found at ref. [11]), the widespread presence of brown spots was observed and ascribed to some biodegradation products but later assessed differently. As a result, the geometric black and white mosaic of *domus A* was completely restored in 2007 (though without replenishing the detached tiles, still present on the surface) to allow a clear reading for visitors.

By accurate examination, it was possible to observe that the mosaic tiles were intact, and the degradation product was lying over the surface of the tile and in the interstices between one tile and another; this suggested that the degradation mainly concerned the mosaic mortar layer, essentially made up of calcium carbonate ($CaCO_3$).

Degradation also extended to the room walls (Figure 2a,b) and seemed similar to that affecting the pavement. Over the centuries (especially after the sixth century), the functional use of these underground settings, located at between 3.5 and 5.5 m of depth in respect to the current ground level, is unclear because of fragmented or missing documentation.

By OM, it has been possible to highlight a substantial homogeneity of the degradation product, consisting of small grain aggregate with porous surface (Figure 3). Nevertheless, the mosaic tiles are entirely covered by a yellowish-brown degradation product of incoherent substance (Figure 2a,c), more damp and compact at some pavement points. In addition to the visual and aesthetic damage, there is the implication of a loss of mechanical–structural properties and, therefore, of binding power.

The samples taken from the pavimental mosaic and the wall were observed by SEM at magnifications from 50× up to over 1000× to highlight the crystalline structure. The SEM analysis confirmed the porous structure observed by means of optical microscope. EDS microprobe analyses were then used to identify the elemental composition of the samples. Figure 4a,b show the SEM images of the sample from the pavimental mosaic, whereas Figure 4c,d show the SEM images of the sample from the wall. Table 1 reports the EDS results obtained for both the floor and wall samples.

The difference between gypsum morphology and $CaSO_4$ and $CaCO_3$ morphologies are important issues important at this point, requiring some consideration. Basically, it should be remembered that the relationship between the thermodynamic concentrations of Ca for pure $CaCO_3$ and $CaSO_4$ solutions is dependent on the pH of the $CaCO_3$ solution. $CaSO_4$ precipitated in the form of gypsum has a needle shaped structure, while $CaCO_3$ has a spiral growth and precipitates in the form of calcite. The precipitate structure is affected by the co-existence of salts. Their co-precipitation results in $CaCO_3$ crystals interwoven with $CaSO_4$ crystals. This tends to result in a co-precipitate that is stronger than pure $CaSO_4$ and weaker than pure $CaCO_3$ precipitate [12]. EDS spectra showed that the sample from the pavement was almost entirely made up of gypsum (calcium sulphate, $CaSO_4$) with well grown and homogeneous crystals of small dimensions. A small quantity of calcium carbonate ($CaCO_3$) was also revealed, the occurrence of which can be ascribed to the mosaic mortar, since the tile surfaces and bodies look intact, although moved along from their inclusion points in the mosaic pavement. On the basis of such observation, it can be supposed that the tiles have slipped from their location because of pressure exerted by the degradation product (of a chemical or biological nature), which caused a physical degradation with structural failure of the mortar [13,14].

The sample from the wall (Figure 4c,d) looks similar to the pavement one but contains some fragments of a more compact aspect (zone A in Figure 4d). EDS spectra (Figure 5a,b, Table 1) revealed a lower gypsum content and a higher one of calcium carbonate. There is also a component, albeit minimal, of Mg (0.3%), Al (0.3%), and Si (0.8%), feldspars probably imputable to grains of earth.

Figure 3. (a) Photo of the samples of the degradation product of both the floor and the wall (glass Petri plate 80 mm); (b) photo of the degradation product of the pavimental mosaic taken by the optical microscope (1000× magnification; width focusing 25 mm, 2 mm increments; 6 V/20 W halogen light).

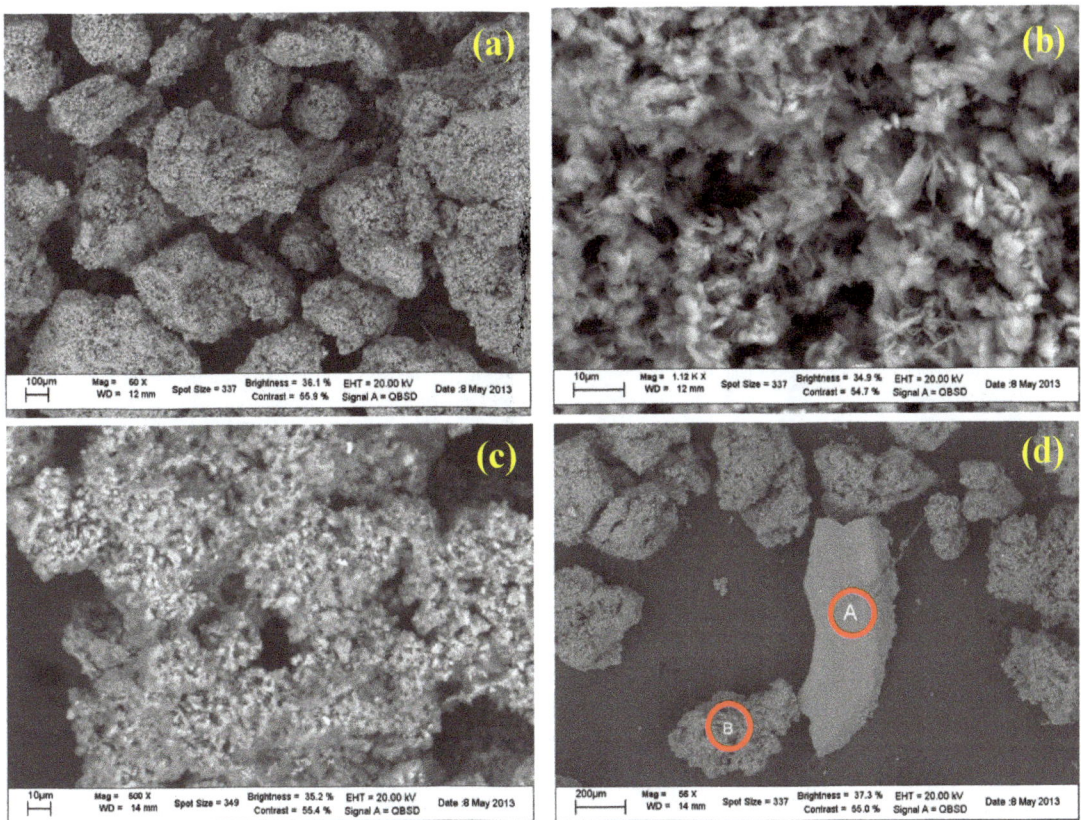

Figure 4. SEM micrographs of the samples collected from pavimental mosaic (**a**,**b**) and from the wall (**c**,**d**).

Table 1. EDS analysis (% w/w) of the wall and floor samples.

Spectrum	C	O	S	Ca	Mg	Al	Si	Total
Floor	16.5	47.9	14.6	21.0				100.0
Wall	25.9	51.1	4.9	16.7	0.3	0.3	0.8	100.0
A [1]	24.7	47.6	11.1	16.6				100.0
B [1]	25.9	56.1	7.8	9.8	0.4			100.0

[1] Portions (particles "A" and "B") of the wall sample as shown in Figure 4d.

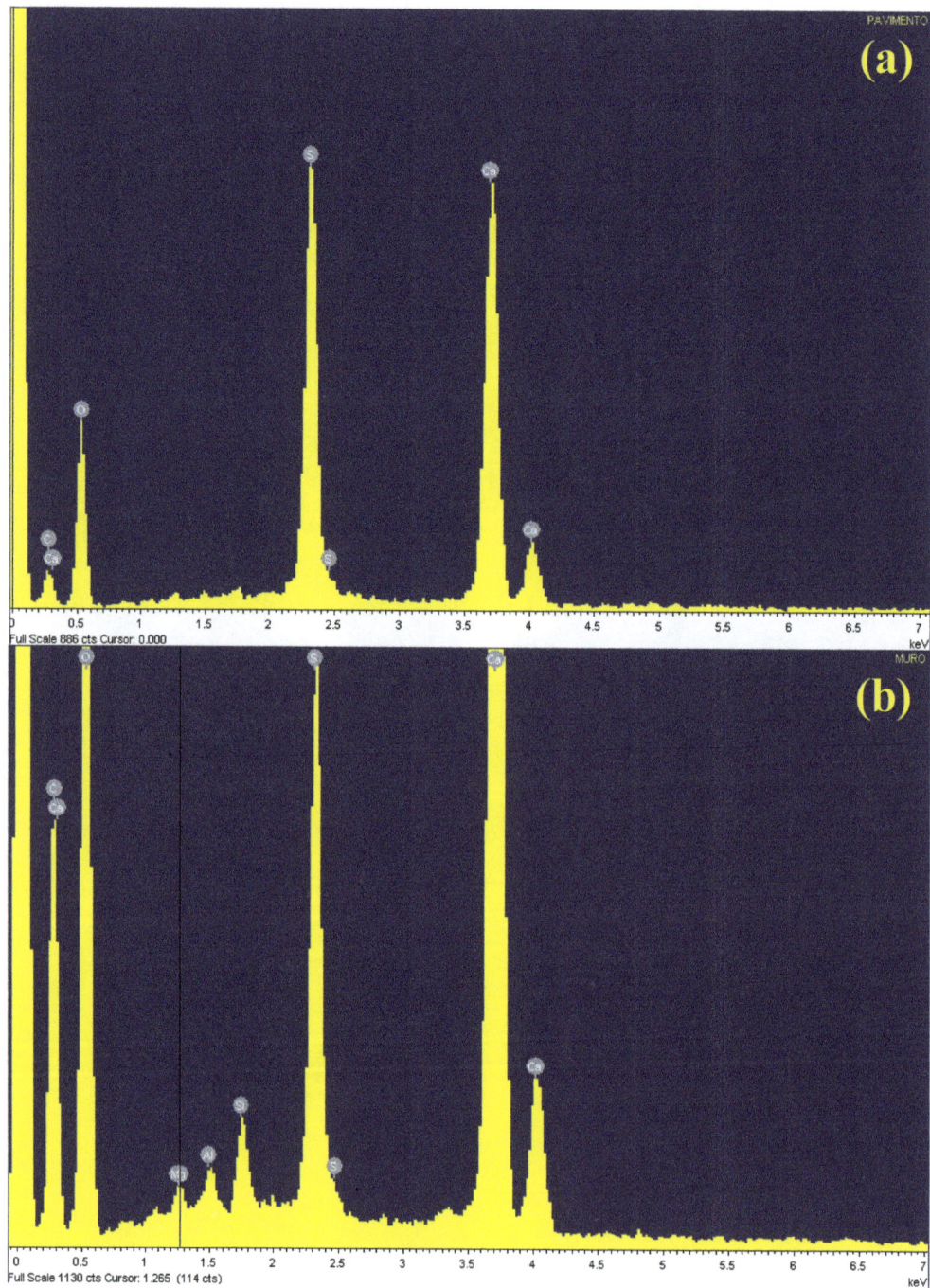

Figure 5. EDS spectra of (**a**) floor sample and (**b**) wall sample. For experimental condition: see Section 4 (Materials and Methods).

The more compact portion (particle "A") differs from the others (particle "B"), with a higher gypsum content and higher calcium carbonate (Figure 6b and Table 1). At higher magnification (1000×) (Figure 6a), as expected very small grains became visible. Another significant difference between the pavement and wall samples was the higher content of Carbon, which cannot be fully imputed to the higher carbonate content because of the lower Calcium (which, as for the pavement sample also cannot satisfy both gypsum and carbonate stoichiometry), so revealing a higher content of organic compounds (Table 1).

Indeed, the yellowish-brown color of the degradation product, and the possible use of the room as a deposit for materials, whose composition or consequences of their use or abandonment were not clear, strengthens the presence of hydrocarbon compounds in the samples collected from the pavimental mosaic and the wall. Following this hypothesis, NMR analyses of pavement and wall samples were performed to identify contingent organic compounds (Figure 7a,b) [15].

Analysis of the NMR spectra showed the presence of aliphatic hydrocarbons in the samples (shift at 7.3 ppm is deuterated chloroform used as solvent). In particular, Figure 7b clearly shows the presence of aliphatic hydrocarbons (shifts between 0.8 and 1.7 ppm). On the other hand, Figure 7a shows a broad signal. The presence of some -OH group caused a shift broadening. Basically, protons on carbon adjacent to the alcohol oxygen showed up in the region of 3.4–4.5 ppm. The electronegativity of the alcohol oxygen de-shielded these protons causing them to appear downfield, when compared to alkane protons. Further, protons directly attached to the alcohol oxygen often appeared in the region of 2.0 to 2.5 ppm. These peaks tended to appear as short, broad singlets [16,17]. Actually, there was also another reasonable interpretation of these broad signals in the NMR spectrum:. They may also be the product of a larger molecular size of the compound or the presence of some paramagnetic impurities in the sample. In any case, the identification of the substance was carried out through a comparison with the online database of ^{1}H-NMR spectra of organic substances in $CDCl_3$ provided by the National Institute of Advanced Industrial Science and Technology (AIST). From the analysis of the ^{1}H-NMR spectra, the presence of aliphatic hydrocarbons seemed to emerge, some of which also contained amino and sulfur groups, such as octane, 2-methylhexane, 2-mthylnonane, decylamine, 1-methyleptylamine, methylpentylsulfide, sodium 1-octanesulfonate (Figure S3 of the Supplementary Material), and, mainly, a more complex composition in the sample coming from the wall (Figure 7a) than that coming from the pavement (Figure 7b).

The FT-IR analyses were conducted by putting the samples from the wall (Figure 8a) and floor (Figure 8b) on potassium bromide (KBr) tablets. The spectra (recorded in the range 4000–400 cm^{-1}) confirmed the presence of calcium sulfate (by diagnostic peaks at 607, 1117, 1620 cm^{-1}) in the product lying on the mosaic pavement (Figure 8b). The calcium carbonate peak (at 875 and 1384 cm^{-1}) was attributed to mosaic mortar, while the double peak at 3400–3500 cm^{-1} was due to a N–H simple bond in the compound.

The FT-IR analysis of the samples coming from the wall (Figure 8a) and floor (Figure 8b) showed that their hydrocarbon compositions were essentially similar. The possible reason for this was common contamination. In addition, the study of the FT-IR spectra also indicated that the sample coming from the floor contained only gypsum, whereas that coming from the wall also contained calcium carbonate, which was associated with the binder of the mortar that constituted it. Further, the FT-IR spectra also demonstrated the presence of possible compounds containing N–H: these compounds could be the same as those identified by NMR analysis and reported in the Supplementary Material, namely 1-methylheptylamine and decylamine.

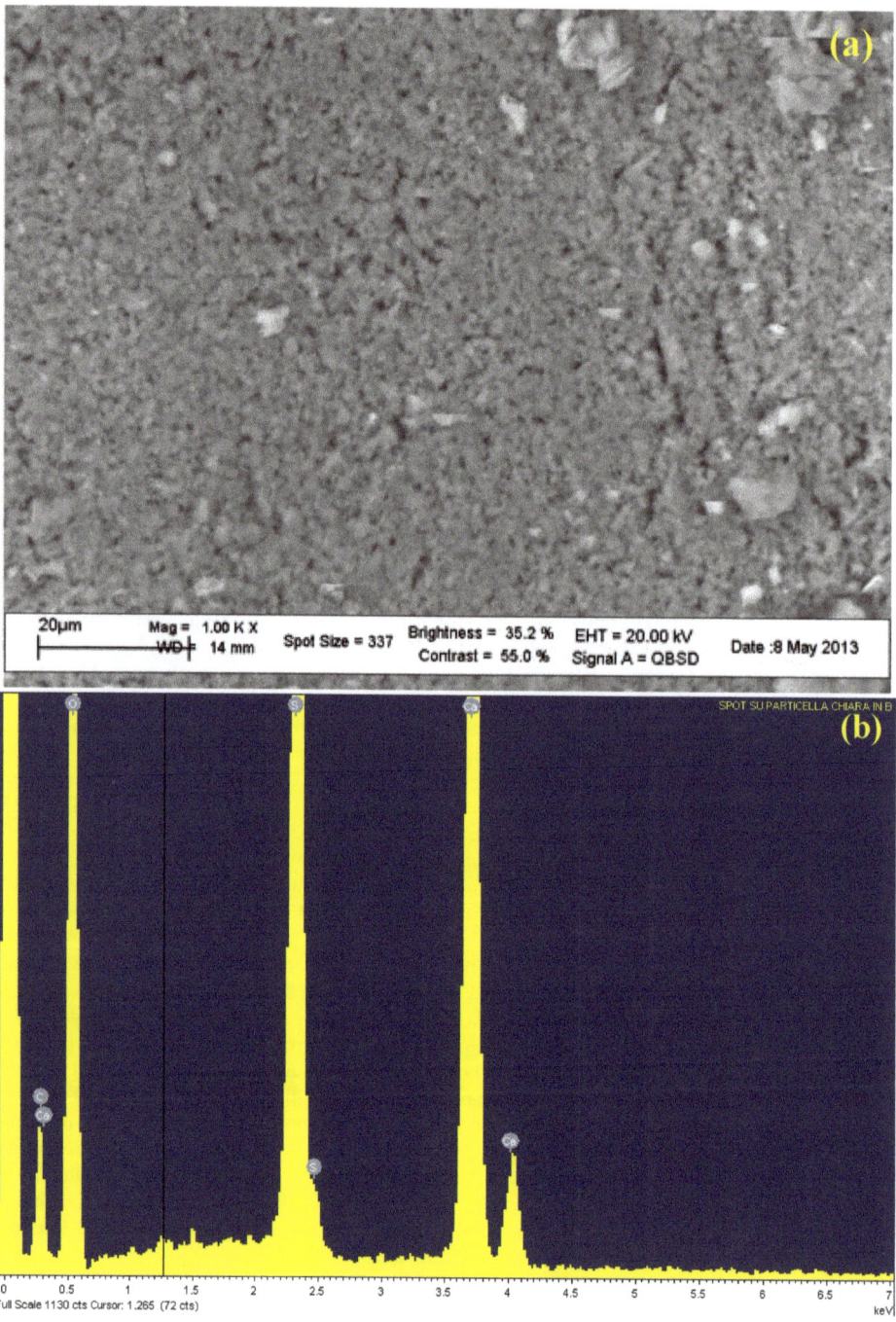

Figure 6. SEM photo (**a**) and EDS spectrum (**b**) of the spot of the particle "A" of the wall sample. For experimental condition: see Section 4 (Materials and Methods).

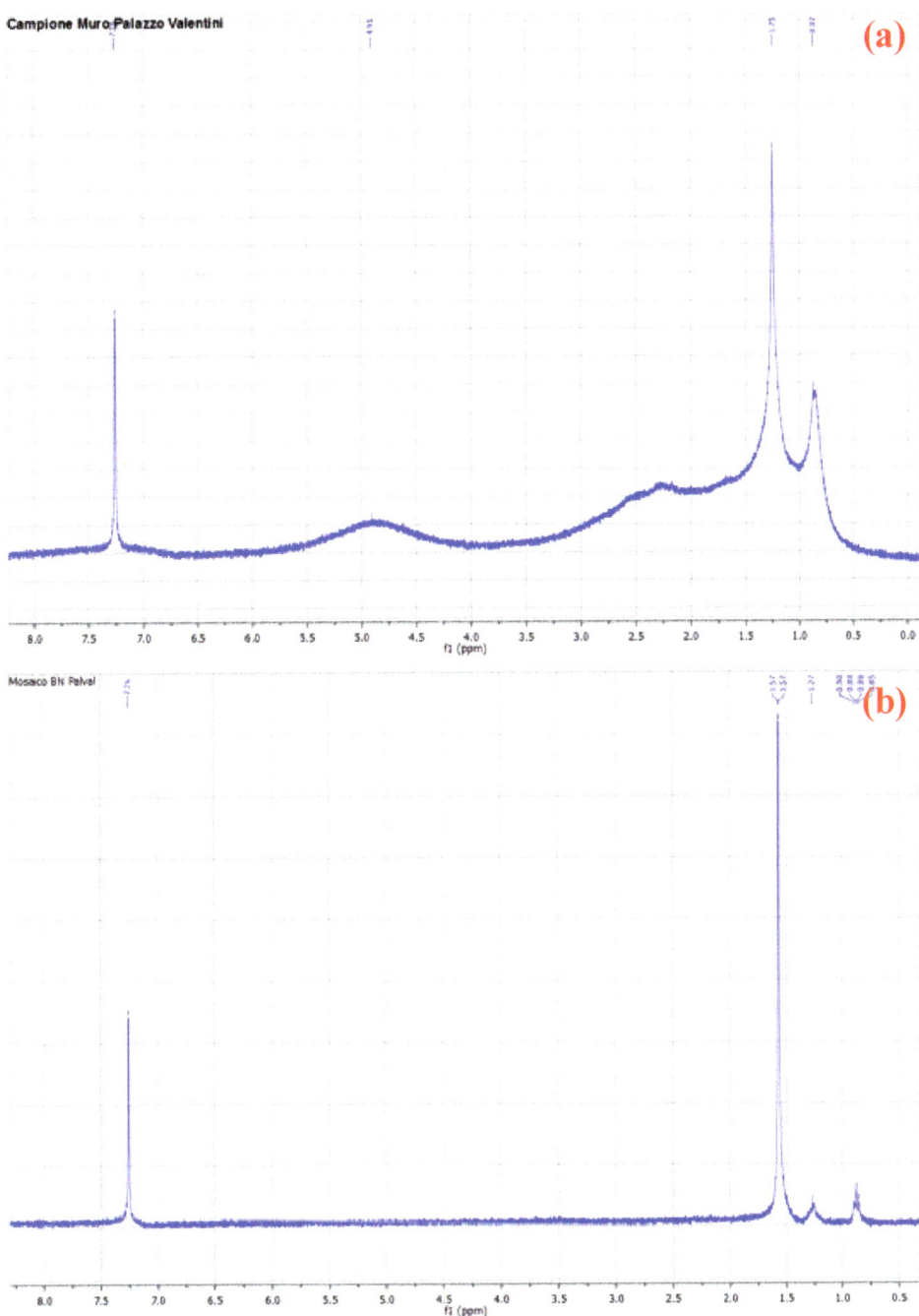

Figure 7. ^{1}H-NMR spectra (300 MHz, CDCl$_3$) of (**a**) wall and (**b**) pavement samples.

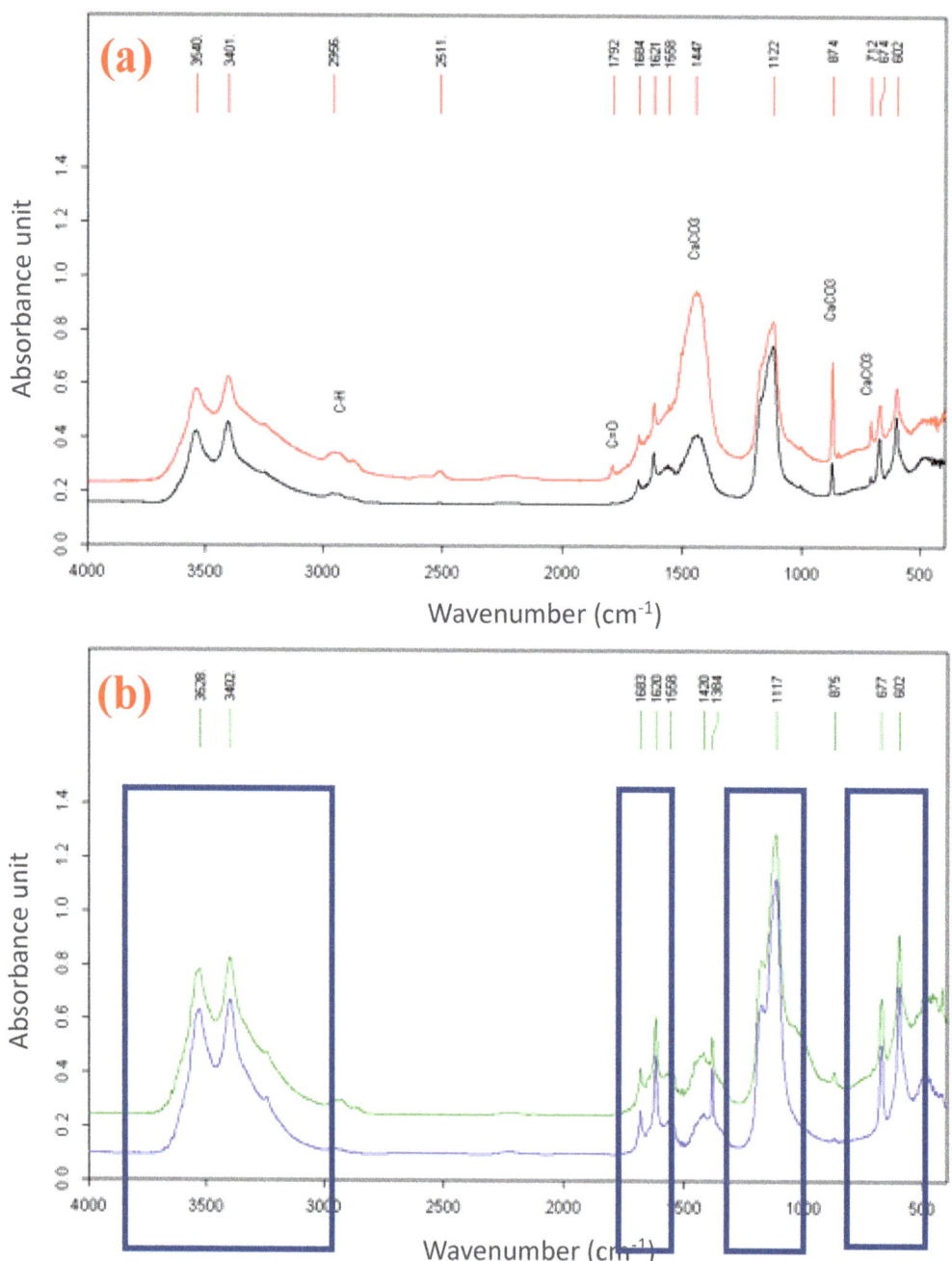

Figure 8. FT-IR spectra (4000–400 cm^{-1}; Globar source and deuterated triglycine sulfate, DTGS, detector; 200 scans; resolution 2 cm^{-1}) of samples taken from (**a**) wall and (**b**) floor (CaCO$_3$ 875 cm^{-1}; CaSO$_4$ 607 cm^{-1}). The red/black and green/blue spectra in figure (**a**) and figure (**b**), respectively, are related to two different samples from wall (**a**) and floor (**b**). The blue boxes show the signals due to calcium sulfate.

Finally, the GC-MS analysis on a floor sample subjected to chemical degradation made it possible to identify the organic composition responsible for the yellow/brown color with greater accuracy. Figure 9 shows the Total Ion Current (TIC) chromatogram, whereas Figure S4 of the Supplementary Material shows the mass spectra of the identified compounds. The MS spectra confirmed the NMR results, namely the presence of low molecular weight hydrocarbons as octadecane and hexadecane, and the presence of sulfur as sulfurous acid, and the identification of complex compounds, such as ethyl 5-chloro-2-nitrobenzoato and tributyl-chloro-stannane.

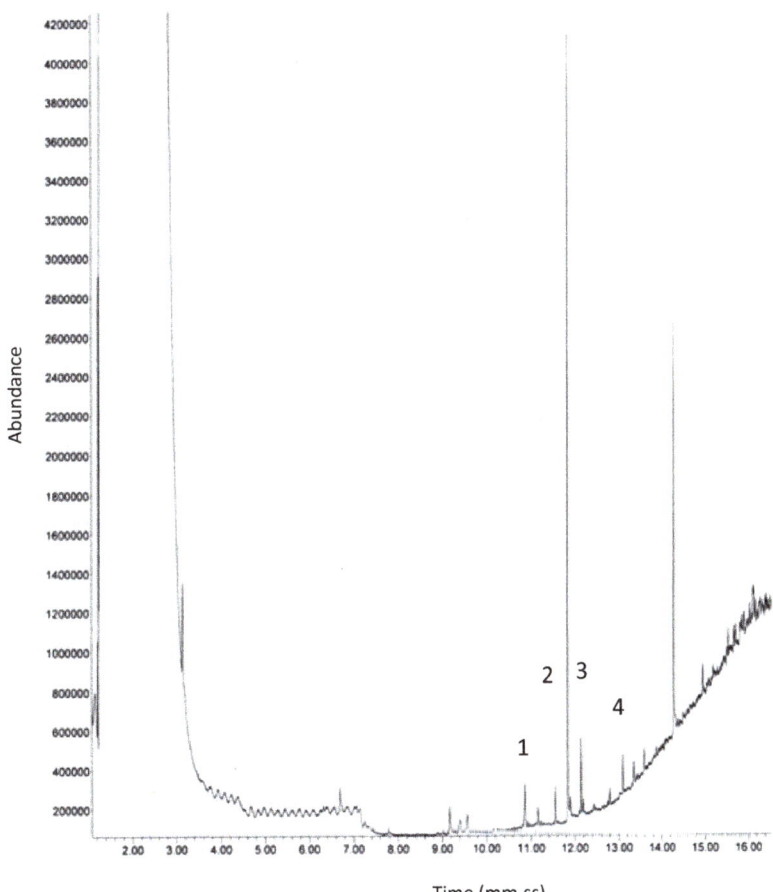

Figure 9. GC-MS chromatogram (fused-silica capillary column, HP-5MS; stationary phase SE54; 30 m × 0.25 mm I.D. and 0.25 µm d_f; T_{inj} 300 °C; splitless mode 30 s; T transfer line 200 °C; program: T_0 120 °C, isothermal for 1 min, 30 °C min^{-1} up to 200 °C, 5 °C min^{-1} up 230 °C, then 30 °C min^{-1} up to T final 290 °C; carrier gas He; m/z from 50 to 700 at 70 eV.) of a pavement sample. Peaks: 1 hexadecane, 2 octadecane, 3 nonyl 2-propyl ester sulfurous acid, 4 tributylchloro-stannane. Mass spectra are shown in Figure S4 of the Supplementary Material.

3. Discussion

Diagnostic analyses demonstrated that the black and white mosaic of the domus A was affected by a diffuse degradation of chemical origin, with the formation of gypsum crystals, due to the presence of sulfur in the topsoil beneath the mosaic bedding layer [18]. An aesthetic degradation was also evident because of organic compounds with 10–20 carbon atoms that could be traces of some unrefined combustive oil used in the past, but not yet present in the bedding layer, like the remains of up-to-date chemical pollutants used in pharmaceutical and agricultural industries [19].

These modern pollutants, as well as the sulfur traces, could have arisen from the aquifer beneath the mosaic room, permeating the overlying topsoil and consequently filtering through the mortar in the form of salts, or could be present in the soil bordering the mosaic room wall, which is still to be investigated. Another possible source of sulfur in the urban context could be sulfur oxide air pollution, in particular, due to wood heating. Still this system has by now almost completely disappeared in big cities such as Rome. Considering the wide micro-flora colonization of Palazzo Valentini underground settings, another hypothesis for the presence of sulfur could be special anoxygenic photosynthetic bacteria, such as Thiobacillus sp., able to oxidize sulfur (from sulfuric acid or other sulfur compounds produced during mineralization processes) to sulphates using the energy produced to fix organic carbon (CO_2) in the photosynthetic process. Anaerobic sulfur bacteria can also partially oxidize sulfuric acid to elemental sulfur, which tends to form sediment in soils and takes part in long cyclic processes [20].

Based on the results of this study, three different hypotheses can be drawn on the reason for the deterioration. The first hypothesis is based on documents of the archaeological excavation where it is reported that, in the seventeenth century, some underground rooms of Palazzo Valentini, but at a higher level than the room of the *domus B* with the floor mosaic, were used as a printing house. Therefore, it can be hypothesized that the traces of hydrocarbons are attributable to printing oil percolating over time. A second hypothesis, on the other hand, is based on the presumed presence, reported in the chronicles of the seventeenth century, of a boiler in the room of the *domus B* and, therefore, on the fuel used for the same. The critical point of this hypothesis is that, although we have news of this boiler, it has never been found, much less traces of its possible presence. Finally, a third hypothesis on the origin of the degradation, perhaps more concrete and real, is based on more recent historical facts. Documentary research made during the excavation campaigns reports the use of these underground rooms as fuel oil storage for a boiler installed in proximity to the mosaic room. In addition, it is important to report that these subterranean settings were changed into a bunker during the Second World War, still existing as part of the exhibition itinerary, which extended to under *Palazzo delle Assicurazioni Generali di Venezia* and whose frequentation probably implied the use of some combustive substance for lighting and heating. This last hypothesis was confirmed in the finding, during the excavation, of metallic remains (lids, etc.) of drums of traction/heating fuel.

A possible definitive solution could be the isolation of mosaic pavement (or at a still unknown depth) from the underlying soil soaked with the organic compound. This kind of expensive intervention should involve the surviving mosaic's detachment, the excavation of all back-up layers and an additional one to gain a tank far larger than the mosaic surface extension for the new waterproofing layer coating, without binder and of suitable thickness. The purpose of these steps would be reconstructing every bedding layer discovered during the detachment operation and re-installing the pre-existing mosaic, eventually replenishing it with tiles discovered in the new excavations (a similar approach has been chosen for the Giotto *frescoes* in the Upper Basilica of Assisi).

Finally, it would be really important to use compatible materials, as similar as possible to the original ones, such as ten-year aged hydrated lime, pozzolana, white marble powder, washed river sand and the like, recovering old recipes mentioned by Romans or during the Renaissance.

4. Materials and Methods

The sampling was performed at a few different points, in very small amounts, of course. After this, the samples were collected together and homogenized in a mortar and two representative samples were analyzed.

The experimental approach involved the use of multiple analytical methods for better identification of the compounds responsible for the degradation, namely Optical Microscope (OM), Scanning Electron Microscope with Energy Dispersion Spectroscopy (SEM-EDS), X-ray fluorescence (XRF), Nuclear Magnetic Resonance (NMR), Gas Chromatography coupled with Mass Spectroscopy (GC-MS) and Fourier Transform Infrared Spectroscopy (FT-IR). The authors opted for an analytical study to identify the substances involved and their common role in the degradation action. For this purpose, samples of the degradation product were taken from both the wall and the floor. Two samples of the degradation product were taken with tweezers and spatula, one from the floor and one from the wall (Figure 3a). The samples were observed under the Motic BA200 binocular optical microscope (Figure 3b) and in reflection mode with the Motic MLC-150C optical fiber. The instrument features were: 4×, 10×, 40×, and 100× quadruple nosepiece; focusing with a width of 25 mm, in 2 mm increments; 6 V/20 W halogen lighting system with light intensity control.

Samples were observed with SEM coupled with EDS to better investigate the crystal structure. The use of EDS made it possible to get a non-destructive qualitative and semi-quantitative sample analysis, identifying the chemical elements in the investigated spots unequivocally [21]. SEM analysis, over information on topography and crystal structure of the samples collected, also gave a grey-scale map of the surface, revealing zones with different medium atomic density (where the heaviest elements, which in our spectra was calcium, appeared clearer and brighter than those of lighter ones). As these were non-metallic samples, they needed to be coated with a conductive material during SEM sample preparation to make them compatible with SEM: a thin layer of gold was used by means of a sputter coater. Microscope analyses were performed by an SEM instrument (mod. Leica LEO 440 S, Wetzlar, Germany) equipped with an energy dispersive spectrometer for X-ray EDS (mod. INCA Energy 400, Oxford Instruments, Abingdon-on-Thames, UK). SEM allowed high magnification (beyond 100,000×) digital images of different kinds of samples with an up to 5-nm resolution while EDS allowed identification of the elemental composition, both of the entire sample and on more or less restricted areas. Our main experimental conditions were 10^{-6} hPa vacuum and 20 keV accelerating voltage. Diffractometry instead revealed the presence of crystalline compounds. This was carried out using a Bruker AXS D8 Focus automatic diffractometer for powders, operating in a reflection in Bragg Brentano θ/2θ geometry with an exposure of 12 h.

In addition, NMR analysis was performed over the degradation product samples from both pavement and wall of the mosaic room to analyze a possible organic component [22]. For NMR measurements, the samples taken from the pavimental mosaic and the wall were dissolved in deuterated chloroform ($CDCl_3$), and after being centrifuged and filtered, they were subjected to ^{1}H-NMR analysis. NMR analysis for the degradation product of the pavimental mosaic [23] was carried out using the mod. 300 MHz Varian Mercury (Varian, Palo Alto, CA, USA) operating at 300 MHz (superconducting magnet at 7.05 T). Chemical shifts in ^{1}H-NMR spectra were reported in parts per million (ppm) on the δ scale using the solvent signal as an internal standard.

The Fourier transform infrared (FT-IR) spectroscopy was used to identify the organic compound structure by using characteristic vibrations of molecules producing spectra with characteristic bands in the infrared region (from 400 cm^{-1} to 4000 cm^{-1}), where single bands can be attributed to vibrations of specific chemical groups. FT-IR spectra were carried out using an Alpha FT-IR instrument (Bruker Optics, Ettingen, Germany) equipped with a Globar source (i.e., a microscope) and a deuterated triglycine sulfate (DTGS) detector [24,25]. Each sample was analyzed, collecting 200 scans or more at a resolution of 2 cm^{-1} in the spectroscopic range 4000–400 cm^{-1}. The spectra collected showed excellent reproducibility [26].

Finally, a floor sample was analyzed by GC-MS, an analytic method by which complex mixtures of chemical compounds may be separated, and the compounds were identified and quantified. Each sample was dissolved in acetone and the analysis was performed by means of a HP-5890 Series II (HP, Rome, Italy) gas chromatograph coupled to a HP-5972 mass selective detector (HP). Chromatographic separations were achieved on a fused-silica capillary column (HP-5MS), stationary phase SE54 (5% phenyl-95% methylpolysiloxane), 30 m × 0.25 mm I.D. and 0.25 μm d_f [27–29]. The chromatographic conditions were as follows: injector temperature 300 °C; injections made in splitless mode (30 s delay before opening the splitter); transfer line temperature 200 °C; initial oven temperature 120 °C, isothermal for 1 min; 30 °C min^{-1} up to 200 °C; 5 °C min^{-1} up 230 °C; then 30 °C min^{-1} up to the final temperature of 290 °C. The carrier gas was helium, constant inlet gas pressure of 5 psi. The mass spectrometer was scanned from m/z 50 to 700 at 70 eV.

5. Conclusions

This paper is a preliminary study on a black and white Roman mosaic. In particular, the authors wished to suggest a methodology for studying the state of degradation of an archaeological find. Such degradation, due to the presence of hydrocarbons, which to date cannot be dated with certainty, risk irreversibly damaging the mosaic, as well as the rooms and the walls in such *domus*. Different analytical methodologies were used to understand the organic composition involved in this contamination and the composition of the wall and pavement. The contamination, as well as the sulfur traces, could have arisen from the aquifer beneath the mosaic room, permeating the overlying topsoil and, consequently, filtering through the mortar in the form of salts, or could be present in the soil bordering the mosaic room wall, still not investigated. Another possible source of sulfur in the urban context could be sulfur oxide air pollution, in particular due to wood heating, but this system has by now almost completely disappeared in big cities like Rome. The authors believe that all these findings are fundamental for preliminary studies but they could also be very important for developing protocols of restoration and conservation. It should also be considered that during the writing of this paper, a cleaning intervention was made on the mosaic surface, but more invasive solutions should be applied in the event of a new occurrence of the above-mentioned degradation problems. A possibly definitive solution could be the isolation of mosaic pavement, or a still unknown depth, from the underlying soil soaked with the organic compound to ascertain it. This kind of expensive intervention should involve the surviving mosaic's detachment, the excavation of all back-up layers and of an additional one to gain a tank far larger than the mosaic surface extension for a new waterproofing layer coating, without binder and of suitable thickness, with the purpose of then reconstructing every bedding layer discovered during the detachment operation and re-installing the pre-existing mosaic, and eventually replenishing it with tiles discovered in the new excavations (a similar approach has been chosen for the Giotto frescoes in the Upper Basilica of Assisi).

Supplementary Materials: The following supporting information can be downloaded at: https://www.mdpi.com/article/10.3390/molecules27227765/s1, Figure S1: "Veduta di Roma" (1593) by Antonio Tempesta: in the red box Palazzo Valentini is highlighted; Figure S2: Palazzo Valentini: (a) *domus A*: rests of the *peristilium* with black and white mosaic pavement and bases for columns or pilasters; (b) *domus A*: *triclinium* with colored geometric mosaic pavement; Figure S3. NMR spectra of the aliphatic hydrocarbons identified in the investigated samples; Figure S4. Selected Ion Monitoring (SIM) spectra of the compounds (octadecane, hexadecane, nonyl 2-propyl ester sulfurous acid, tributylchloro-stannane) investigated in the floor sample.

Author Contributions: Conceptualization, P.B., P.A. and G.V.; methodology, P.A.; validation, P.A. and G.V.; investigation, C.C. and G.V.; resources, P.A. and G.V.; data curation, P.A.; writing—original draft preparation, C.C. and P.A..; writing—review and editing, P.B., P.A., M.L.A. and G.V.; visualization, M.L.A.; supervision, P.A. and G.V. All authors have read and agreed to the published version of the manuscript.

Funding: This research received no external funding.

Institutional Review Board Statement: Not applicable.

Informed Consent Statement: Not applicable.

Data Availability Statement: Not applicable.

Acknowledgments: The authors thank Maria Pia Sammartino for the scientific support of this study. The authors would like to thank the Spectral Database for Organic Compounds (SDBS) for the spectra.

Conflicts of Interest: The authors declare no conflict of interest.

Sample Availability: Samples are available from the authors.

References

1. Maltese, C. *Le Tecniche Artistiche*, 7th ed.; Ugo Mursia Editore: Milan, Italy, 1991; pp. 361–366, ISBN 978-8842508823.
2. Baldassarri, P. Indagini archeologiche a Palazzo Valentini. La campagna 2005–2007. In *Palazzo Valentini. L'area tra Antichità ed età Moderna: Scoperte Archeologiche e Progetti di Valorizzazione*; Del Signore, R., Ed.; Ediart: Rome, Italy, 2008; pp. 29–80, ISBN 978-88-95759-02-9.
3. Quattrocchi, M. I mosaici della domus A. In *Palazzo Valentini. L'area tra Antichità ed età Moderna: Scoperte Archeologiche e Progetti di Valorizzazione*; Del Signore, R., Ed.; Ediart: Rome, Italy, 2008; pp. 81–93, ISBN 978-88-95759-02-9.
4. Baldassarri, P. Archaeological excavations at Palazzo Valentini: A residential area in the shade of the Trajan's Forum. In Proceedings of the 11th International Colloquium on Ancient Mosaics, Bursa, Turkey, 16–20 October 2009; Şahin, M., Ed.; Zero Books: Istanbul, Turkey, 2011; pp. 43–67, ISBN 978-605-5607-81-4.
5. Acconci, A.; Baldassarri, P.; Nuzzo, M.; Tommasi, F.M. Profilo storico, artistico e archeologico di Palazzo Valentini. In *'La Provincia Capitale'. Storia di una Istituzione e dei suoi Presidenti*; Del Signore, R., Ed.; Provincia di Roma: Rome, Italy, 2005; pp. 153–198.
6. Cicconi, M. La 'Fabbrica' di Palazzo Bonelli-Valentini, residenza cardinalizia del Cinquecento. Il punto di partenza. In *Palazzo Valentini. L'area tra Antichità ed età Moderna: Scoperte Archeologiche e Progetti di Valorizzazione*; Del Signore, R., Ed.; Ediart: Rome, Italy, 2008; pp. 1–27, ISBN 978-88-95759-02-9.
7. Cola, M.C. *Palazzo Valentini in Roma: La committenza Zambeccari, Boncompagni, Bonelli tra Cinquecento e Settecento*; Gangemi: Rome, Italy, 2012; 270p, ISBN 9788849219357.
8. Passigli, S. Urbanizzazione e topografia di Roma nell'area dei Fori Imperiali tra XIV e XVI secolo. *Mélanges L'école Française Rome* **1989**, *101*, 273–325. [CrossRef]
9. Google Earth. Available online: https://earth.google.com/web/@41.896258,12.48398616,35.16175745a,910.69330694d,35y,0h,0t,0r (accessed on 29 November 2002).
10. Napoli, L.; Baldassarri, P. Palazzo Valentini: Archaeological discoveries and redevelopment projects. *Front. Archit. Res.* **2015**, *4*, 91–99. [CrossRef]
11. Sammartino, M.P.; Visco, G. Led Lighting Installation and Related Process for Inhibiting the Development of Photosynthetic Biodeteriogen Organisms in Hypogeal Environments. Italian Patent RM2014R000823, 15 May 2014.
12. Sudmalis, M.; Sheikholeslami, R. Precipitation and co-precipitation of $CaCO_3$ and $CaSO_4$. *Can. J. Chem. Eng.* **2000**, *78*, 21–31. [CrossRef]
13. Faulstich, F.R.L.; Schnellrath, J.; De Oliveira, L.F.; Scholz, R. Rockbridgeite inclusion in rock crystal from Galileia region, Minas Gerais, Brazil. *Eur. J. Mineral.* **2013**, *25*, 817–823. [CrossRef]
14. Sudoł, E.; Małek, M.; Jackowski, M.; Czarnecki, M.; Strąk, C. What makes a floor slippery? A brief experimental study of ceramic tiles slip resistance depending on their properties and surface conditions. *Materials* **2021**, *14*, 7064. [CrossRef] [PubMed]
15. Mason, J. Nitrogen nuclear magnetic resonance spectroscopy in inorganic, organometallic, and bioinorganic chemistry. *Chem. Rev.* **1981**, *81*, 205–227. [CrossRef]
16. Fulmer, G.R.; Miller, A.J.M.; Sherden, N.H.; Gottlieb, H.E.; Nudelman, A.; Stoltz, B.M.; Bercaw, J.E.; Goldberg, K.I. NMR chemical shifts of trace impurities: Common laboratory solvents, organics, and gases in deuterated solvents relevant to the organometallic chemist. *Organometallics* **2010**, *29*, 2176–2179. [CrossRef]
17. Babij, N.R.; McCusker, E.O.; Whiteker, G.T.; Canturk, B.; Choy, N.; Creemer, L.C.; De Amicis, C.V.; Hewlett, N.M.; Johnson, P.L.; Knobelsdorf, J.A.; et al. NMR chemical shifts of trace impurities: Industrially preferred solvents used in process and green chemistry. *Org. Process Res. Dev.* **2016**, *20*, 661–667. [CrossRef]
18. Van Driessche, A.E.S.; Stawski, T.; Kellermeier, M. Calcium sulfate precipitation pathways in natural and engineering environments. *Chem. Geol.* **2019**, *530*, 119274. [CrossRef]
19. Stepanova, A.Y.; Gladkov, E.A.; Osipova, E.S.; Gladkova, O.V.; Tereshonok, D.V. Bioremediation of Soil from Petroleum Contamination. *Processes* **2022**, *10*, 1224. [CrossRef]
20. Waluś, K.L.; Warguła, Ł.; Wieczorek, B.; Krawiec, P. Slip risk analysis on the surface of floors in public utility buildings. *J. Build. Eng.* **2022**, *54*, 104643. [CrossRef]
21. Nasrazadani, S.; Hassani, S. Modern analytical techniques in failure analysis of aerospace, chemical, and oil and gas industries. In *Handbook of Materials Failure Analysis with Case Studies from the Oil and Gas Industry*; Makhlouf, A.S.H., Aliofkhazraei, M., Eds.; Butterworth-Heinemann: Oxford, UK, 2016; pp. 39–54.
22. Bifulco, G.; Dambruoso, P.; Gomez-Paloma, L.; Riccio, R. Determination of relative configuration in organic compounds by NMR spectroscopy and computational methods. *Chem. Rev.* **2007**, *107*, 3744–3779. [CrossRef] [PubMed]

23. Lazzari, M.; Reggio, D. What fate for plastics in artworks? An overview of their identification and degradative behavior. *Polymers* **2021**, *13*, 883. [CrossRef] [PubMed]
24. Tarquini, G.; Nunziante-Cesaro, S.; Campanella, L. Identification of oil residues in Roman amphorae (Monte Testaccio, Rome): A comparative FTIR spectroscopic study of archeological and artificially aged samples. *Talanta* **2014**, *118*, 195–200. [CrossRef] [PubMed]
25. Nunziante-Cesaro, S.; Lemorini, C. The function of prehistoric lithic tools: A combined study of use-wear analysis and FTIR microspectroscopy. *Spectrochim. Acta A* **2012**, *86*, 299–304. [CrossRef] [PubMed]
26. Nucara, A.; Nunziante-Cesaro, S.; Venditti, F.; Lemorini, C. A multivariate analysis for enhancing the interpretation of infrared spectra of plant residues on lithic artefacts. *J. Archaeol. Sci. Rep.* **2020**, *33*, 102526. [CrossRef]
27. Russo, M.V.; Avino, P.; Cinelli, G.; Notardonato, I. Sampling of organophosphorus pesticides at trace levels in the atmosphere using XAD-2 adsorbent and analysis by gas chromatography coupled with nitrogen-phosphorus and ion-trap mass spectrometry detectors. *Anal. Bioanal. Chem.* **2012**, *404*, 1517–1527. [CrossRef]
28. Russo, M.V.; Notardonato, I.; Avino, P.; Cinelli, G. Fast determination of phthalate ester residues in soft drinks and light alcoholic beverages by ultrasound/vortex assisted dispersive liquid-liquid microextraction followed by gas chromatography-ion trap mass spectrometry. *RSC Adv.* **2014**, *4*, 59655–59663. [CrossRef]
29. Avino, P.; Notardonato, I.; Perugini, L.; Russo, M.V. New protocol based on high-volume sampling followed by DLLME-GC-IT/MS for determining PAHs at ultra-trace levels in surface water samples. *Microchem. J.* **2017**, *133*, 251–257. [CrossRef]

Article

The Use of UV-Visible Diffuse Reflectance Spectrophotometry for a Fast, Preliminary Authentication of Gemstones

Maurizio Aceto [1,*], Elisa Calà [1], Federica Gulino [1], Francesca Gullo [1], Maria Labate [2], Angelo Agostino [2] and Marcello Picollo [3]

1 Dipartimento per lo Sviluppo Sostenibile e la Transizione Ecologica (DiSSTE), Università degli Studi del Piemonte Orientale, Piazza Sant'Eusebio, 5-13100 Vercelli, Italy; elisa.cala@uniupo.it (E.C.); federica.gulino@uniupo.it (F.G.); francesca.gullo@uniupo.it (F.G.)
2 Dipartimento di Chimica, Università degli Studi di Torino, Via P. Giuria, 7-10125 Torino, Italy; maria.labate@unito.it (M.L.); angelo.agostino@unito.it (A.A.)
3 Istituto di Fisica Applicata "Nello Carrara" del Consiglio Nazionale delle Ricerche (IFAC-CNR), Via Madonna del Piano, 10-50019 Sesto Fiorentino, Italy; m.picollo@ifac.cnr.it
* Correspondence: maurizio.aceto@uniupo.it

Abstract: The identification of gemstones is an important topic in the field of cultural heritage, given their enormous value. Particularly, the most important precious stones, namely diamond, emerald, ruby and sapphire, are frequently subjected to counterfeit by substitution with objects of lesser value with similar appearance, colour or shape. While a gemmologist is able to recognise a counterfeit in most instances, more generally, it is not easy to do this without resorting to instrumental methods. In this work, the use of UV-visible diffuse reflectance spectrophotometry with optic fibres (FORS) is proposed as a fast and easy method for the preliminary identification of gemstones, alternative to the classical methods used by gemmologists or to Raman spectroscopy, which is by far the instrumental method with the best diagnostic potential, but cannot be used in situations of problematic geometric hindrance. The possibilities and the limitations given by the FORS technique are critically discussed together with the spectral features of the most important gemstones. Finally, the application of chemometric pattern recognition methods is described for the treatment of large sets of spectral data deriving from gemstones identification.

Keywords: FORS; reflectance; non-invasive; gemstones; colour

1. Introduction

Due to their enormous value, the identification of gemstones is an important topic in the field of cultural heritage. Counterfeit is rather common, and particularly, the most precious stones, namely diamond, emerald, ruby and sapphire, are frequently substituted with gemstones of lesser value which have the same appearance, colour or shape. While a skilful gemmologist with expertise may be able to differentiate between authentic and fake gemstones, it can be challenging without resorting to instrumental methods.

The classical methods used by gemmologists are based on the measurement of the refraction index, of the birefringence or double refraction and of specific gravity [1,2]. Apart from these methods, instrumental techniques are presently available which allow even non-experts to correctly identify gemstones. Raman spectroscopy is one of the techniques with the most accurate diagnostic potential due to the fact that it can provide a fingerprint of nearly every known gemstone [3,4]; moreover, the process is completely non-invasive and non-destructive. When a portable Raman system is available, this analysis can be performed in situ, i.e., inside museums where precious artworks are kept, without the need to remove the objects from their original location in order to analyse them in laboratories [5–8]. Fourier Transform-Infrared (FT-IR) spectroscopy can be used as well, exploiting the various configurations available, i.e., absorbance, reflectance, Attenuated Total Reflection (ATR), etc. [9].

Despite the reliability of Raman spectroscopy in the identification of gemstones, in certain situations, it cannot be used because of geometric hindrances (e.g., an artwork located inside a cabinet). The same holds when a jewellery artwork is to be analysed using classical methods that need gemstones to be studied detached from the jewel frame.

Another technique commonly used in gemmological laboratories is UV-Visible-NIR absorption spectrophotometry. It is well known that most gemstones owe their colour to the presence of small amounts of transition metal ions occurring as impurities inside their structure [10,11]: such gemstones are called *allochromatic*, in contrast to *idiochromatic* gemstones in which the chromophore is a main chemical constituent (an example is turquoise -$CuAl_6(PO_4)_4(OH)_8 \cdot 4H_2O$-in which the chromophore is Cu^{2+}). Particularly relevant is the incomplete set of 3d electrons of transition metal ions. Since most of the d–d transitions occur in the visible region, UV-Visible-NIR absorption spectrophotometry is suitable for the identification of coloured gemstones. Other known phenomena causing the appearance of colour in gemstones, such as charge-transfer and colour centres, generate spectral features [12–14] that can be appreciated as well. However, this technique has two main drawbacks in the analysis of gemstones: (1) when used in absorbance mode, it functions on transparent gemstones only; (2) it can be used only on gemstones analysed in a laboratory. The first drawback can be addressed by means of an integration sphere, a sampling geometry that enables the collecting of reflectance spectra even from opaque or translucid gemstones, thus allowing to obtain *apparent absorbance* spectra; the second drawback cannot be addressed.

One additional drawback of UV-Visible-NIR absorption spectrophotometry is the fact that many gemstones are *pleochroic*, that is, they show two or three different colours when viewed from different angles or irradiated with different lights. This means that although such gemstones can be identified with Raman spectroscopy because the vibrational behaviour does not change even when the angle is changed, the absorption response—and therefore the possibility to reliably identify these gemstone—will vary because of both the different chromophore system present and the angle of collection of the response itself.

Finally, it is well known that artificial treatments, such as heating or irradiation, can cause colour changes due to induction (or improvement) of the charge transfer mechanism or creation of colour centres. In such cases, again, the absorption spectrum is changed while the vibrational spectrum is not.

In such cases, a good alternative could be the use of a preliminary technique such as UV-Visible diffuse reflectance spectrophotometry with optic fibres (FORS). The FORS technique, due to the use of a small probe, can be employed in situ nearly anywhere without steric constraints; moreover, the technique works on transparent, translucent and opaque objects. A patented method has been recently issued by Takahashi and Perera [15]. When examining pleochroic gemstones, it is relatively easy to change the angle of measurement in order to verify the different responses. The diagnostic issues generated by the artificial treatments of gemstones are obviously like those encountered in the absorption mode. Ultimately, the FORS technique can be an advantageous alternative to classical methods used in gemmology, in particular in cases where jewellery artworks cannot be moved from their natural locations, such as museums.

In this work, the possibilities and the limitations given by the FORS technique in the identification of gemstones are critically discussed.

2. Results

The spectral response yielded by the FORS analysis is mainly related to the chromophore system of the gemstone. In most cases, this involves the presence of one or more metal ions in specific oxidation states. The spectrum can therefore provide information useful for (a) the identification of the gemstone and (b) its geographical or geological provenance that can be related to the presence of specific elements.

The FORS technique is particularly useful in cases where the analysis is carried out on opaque material. In fact, these are the cases in which a larger amount of light is diffused by the sample because of scattering and therefore can reach the detection system. Most of the gemstones, however, are transparent or translucent; hence, the amount of reflected radiation is low and poor spectra must be expected. Nevertheless, the spectral features necessary for the identification (reflectance minima/apparent absorbance maxima) can usually be detected with instruments of good sensitivity.

In FORS analysis of gemstones, the influence of ambient light (LED, neon lights, direct sunlight) on the spectral response must be taken into account. To exploit the advantages of this technique, the measurements are usually carried out in open systems, that is, presenting the probe directly to the gemstone without covers, contrarily to the measurements carried out inside spectrophotometers such as in transmission mode or with an integration sphere. Ambient light can generate undesired spectral artefacts, which are sometimes easy to recognise because they occur as sharp bands. A proper way to avoid this drawback is to cover the tip of the probe with a small cylindrical sheath cut into a slope (Figure 1) to exclude external sources of light.

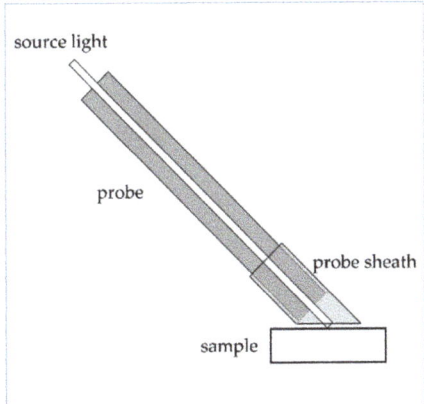

Figure 1. FORS probe with a small cylindrical sheath to exclude external source of light.

In the following paragraphs, the possibility of identifying the most important gemstones with FORS will be evaluated. The discussion is arranged according to the colour of gemstones. Spectra are shown mostly in Log(1/R) coordinates, i.e., in *apparent absorbance*, in order to better appreciate the absorption features, except for cases in which luminescence bands must be highlighted (e.g., sapphire, ruby), and so, the corresponding spectra are shown in the usual reflectance coordinates. In the figures where multiple spectra are presented, spectra are offset for clarity.

The identification of all the gemstones analysed has been previously confirmed by means of Raman spectroscopy.

2.1. Blue Gemstones

The blue gemstone par excellence is the *sapphire*, a variety of corundum—Al_2O_3. The mechanism of the blue colour generation in sapphires has long been debated [16]. Bristow et al. [17] provided spectroscopic evidence that the mechanism responsible is an intervalence charge transfer (IVCT) between Ti^{4+} and Fe^{2+}; Palanza et al. [18,19], besides the IVCT mechanism, cited overlapping crystal field transitions of Cr^{2+}, Cr^{3+}, Ti^{3+}, V^{2+} and V^{3+} ions. In Figure 2, the FORS spectrum of a blue sapphire is shown: it is characterised by main absorption bands at ca. 390, 456 and 706 nm due to Fe^{3+} and a band at 570 nm due to Fe^{2+}-Ti^{4+} IVCT.

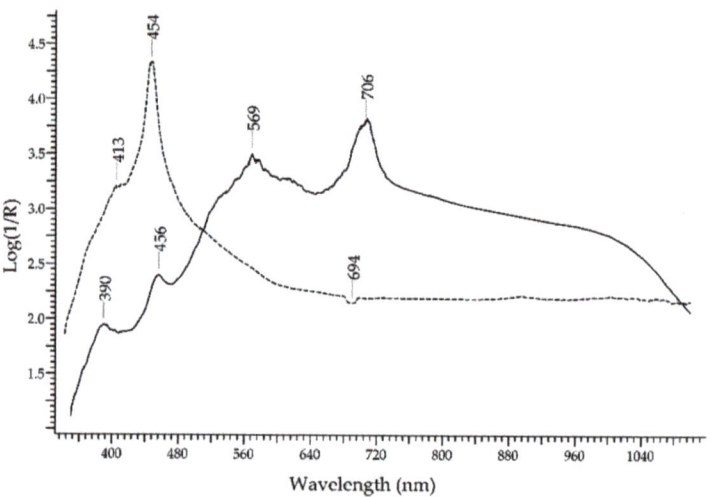

Figure 2. FORS spectrum in reflectance coordinates of a blue sapphire from Cambodia (solid line) and a yellow synthetic sapphire (dashed line).

Sapphires, which usually are blue, can also be yellow–orange due to the presence of Fe^{3+}. The corresponding spectrum shows the absorption band at ca. 450 nm more prominent and a shoulder at 413 nm (see Figure 2). Sapphires may also appear green when both yellow and blue chromophores are present.

Additional spectral features in sapphires can include two luminescence bands at 693 and 694 nm, due to Cr^{3+} ions [20,21], that appear as a single sharp negative band in the FORS spectrum (see Figure 1).

Recently [22,23], it has been demonstrated that UV-visible spectrophotometry, besides other techniques [24], can differentiate treated and non-treated corundum, which is an important issue in the gemstones market, based on the presence of a strong, wide absorption band at ca. 555 nm due to the formation of the blue colour $[FeTi]^{6+}$ complex.

The presence of Fe^{2+} and Fe^{3+} ions, instead, causes the typical blue–green colour of *aquamarine*, which is due to a variety of beryl—$Be_3Al_2Si_6O_{18}$. The spectrum (Figure 3) shows two main absorption bands at 425 nm, due to Fe^{3+} in octahedral sites, and at ca. 820 nm, due to Fe^{2+}-Fe^{3+} intervalence charge transfer [25].

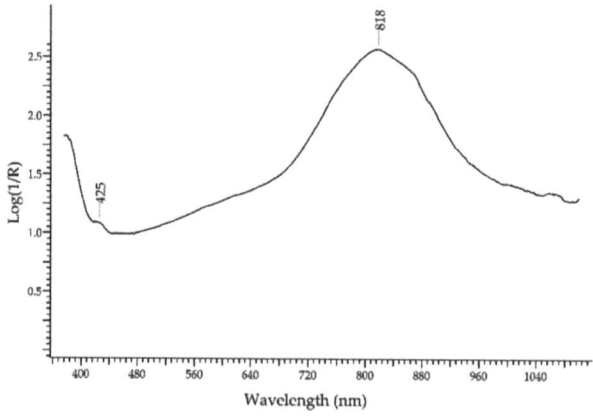

Figure 3. FORS spectrum in Log(1/R) coordinates of aquamarine.

Another precious blue stone, known by mankind since at least 5 millennia, is the *lapis lazuli*. The typical blue colour is due to the lazurite phase—$Na_6Ca_2(Al_6Si_6O_{24})(SO_4,S,S_2,S_3,Cl,OH)_2$—while other accessory phases (e.g., diopside, calcite, pyrite) are present but do not contribute significantly to the absorption features. The FORS spectrum (Figure 4) is dominated by a main band at 600 nm and a second band at ca. 400 nm, which are both due to the intervalence charge transfer mechanism of absorption between HS_3^-, S_2^- and S_3^- radicals entrapped in the lazurite cage. The second band has been considered as distinctive for samples of Afghan origin [26].

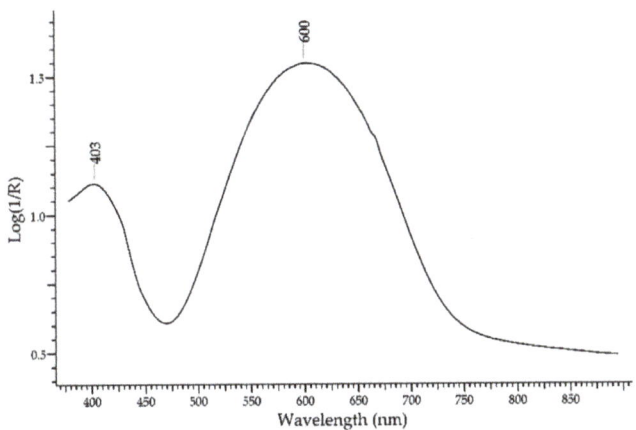

Figure 4. FORS spectrum in Log(1/R) coordinates of lapis lazuli.

The *turquoise* is considered a step below the most precious gemstones. It is a hydrated Cu and Al phosphate with formula $CuAl_6(PO_4)_4(OH)_8 \cdot 4H_2O$. The colour is mainly due to Cu2+, with contribution by substituted elements such as Fe^{2+}, Fe^{3+} and Zn^{2+}. The FORS spectrum of turquoise (Figure 5) shows a broad band at ca. 680 nm due to Cu^{2+} and a sharp peak at ca. 420 nm due to Fe^{3+}, accounting for a greener hue [27]. The presence of broad absorption bands at ca. 620 and 680 nm usually indicates dyed turquoises [28].

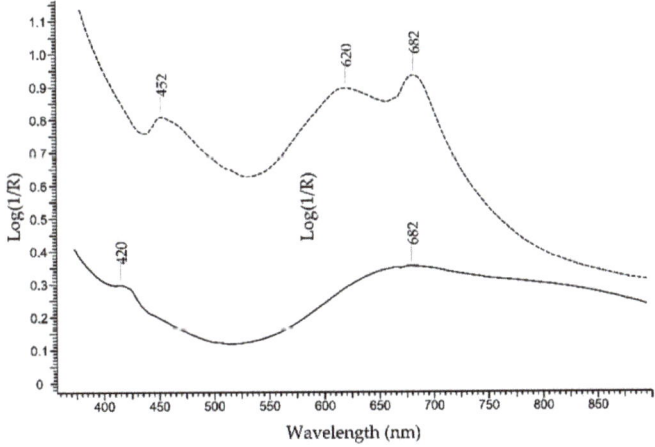

Figure 5. FORS spectra in Log(1/R) coordinates of a natural turquoise (solid line) and a dyed turquoise (dashed line).

2.2. Green Gemstones

The most important green gemstone is the *emerald*, which is the most precious variety of beryl—$Be_3Al_2(Si_6O_{18})$. The UV-visible absorption spectrum of emerald obtained by FORS is well known and characterised by two main broad bands occurring at 440 and 616 nm, due to Cr^{3+} and V^{3+} ions, and two sharp bands occurring around 682 nm, due to Cr^{3+} ions [29]. Spectral features due to Fe^{2+} (a broad band at 843 nm) and Fe^{3+} ions can be present as well; the presence or absence of these features can provide information on the geographic provenance of emeralds. In Figure 6, the spectrum of an emerald is shown in comparison with the spectra of an emerald-like green glass coloured with Cu^{2+} and of an emerald-like green glass coloured with Cr^{3+} and Cu^{2+} (Kremer Pigmente 39132, Colored Glass, Emerald Green, transparent): the possibility of distinguishing emerald from the two glasses according to the absorption features is apparent, even in the case of glass containing Cr^{3+}, which is the same ion that generates the colour of emerald but inside a ligand field of different strength.

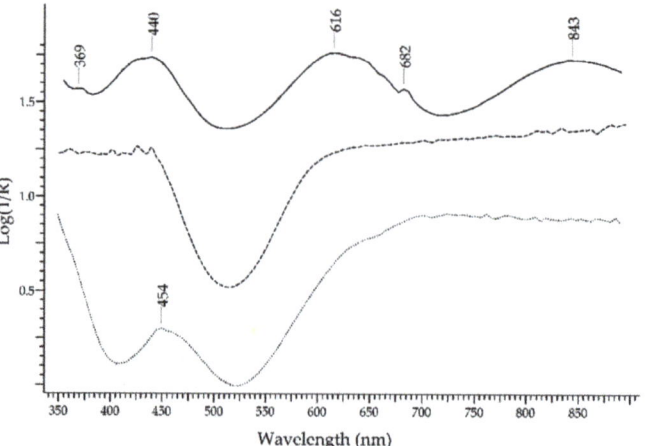

Figure 6. FORS spectra in Log(1/R) coordinates of an emerald (solid line), a Cu^{2+} green glass (dashed line) and a Cu^{2+}/Cr^{3+} green glass (dotted line).

Another gemstone containing beryllium is the *alexandrite*, which is a variety of chrysoberyl—$BeAl_2O_4$. It is a rare pleochroic gemstone that appears green in daylight but turns red in incandescent light [30]. The colour is due to impurities of the Cr^{3+} ion substituting Al^{3+} in the structure, which generate a main band at ca. 580 nm and a shoulder at ca. 680 nm; another band at 446 and minor bands between 380 and 400 nm are due to Fe^{3+}. Figure 7 shows a comparison of the spectra of alexandrite, a green heliodor (another gemstone of the beryl family) and of a yellow chrysoberyl (see below).

The *chrome-chalcedony* or *mtorolite* is a rare variety of chalcedony. Its aspect is green due to Cr^{3+} impurities, and its reflectance spectrum (Figure 8) is dominated by a single band at ca. 610 nm. It was commonly employed in glyptic art of the Roman age [31], possibly as a substitute of emerald, and in medieval precious bindings such as the *Pace di Chiavenna* [32] and the binding of the C Codex of Vercelli [33].

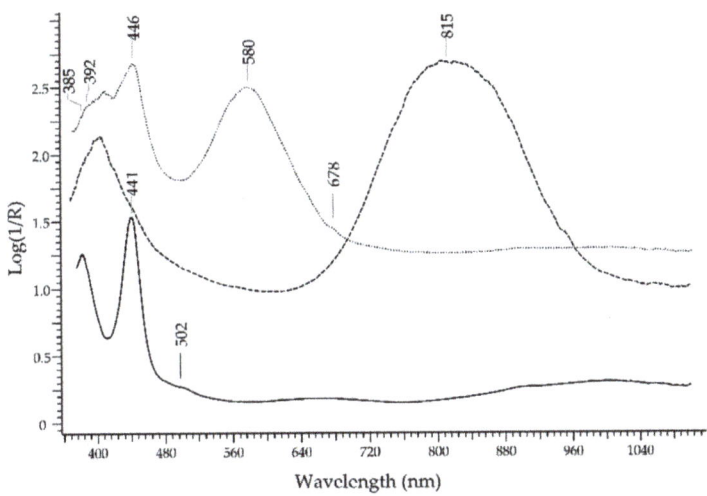

Figure 7. FORS spectrum in Log(1/R) coordinates of alexandrite (dotted line), a green heliodor (dashed line) and a yellow chrysoberyl (solid line).

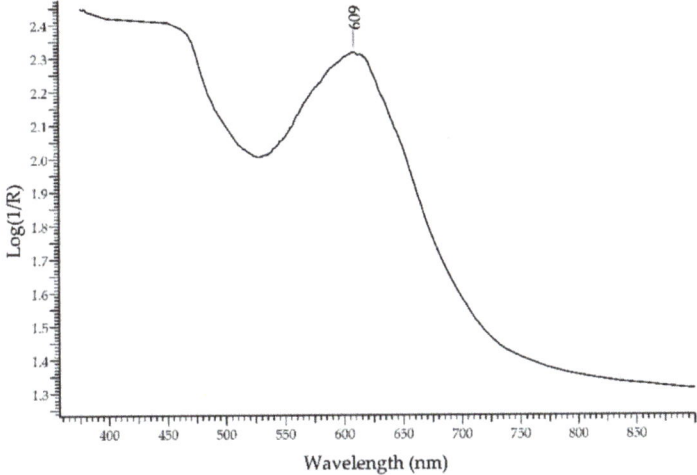

Figure 8. FORS spectrum in Log(1/R) coordinates of chrome-chalcedony.

A very common gemstone is the *malachite*. Its absorption spectrum is dominated by a peak at ca. 800 nm due to Cu^{2+} [34]; because this gemstone is opaque, it can be distinguished from a transparent green glass but not from an opaque green glass coloured with Cu^{2+}, which will show similar absorption features. *Chrysocolla* is coloured by Cu^{2+} too, but the absorption maximum occurs at ca. 690 nm [35].

The *peridot* is a green-to-yellow variety of the mineral olivine, with formula $Mg_2SiO_4 \cdot Fe_2SiO_4$. (Figure 9). The colour is due to Fe^{2+} ion [36] (weak bands between ca. 450 and 490 nm); other features at 513 and 653 nm can possibly be due to another chromophore such as Cr^{3+}.

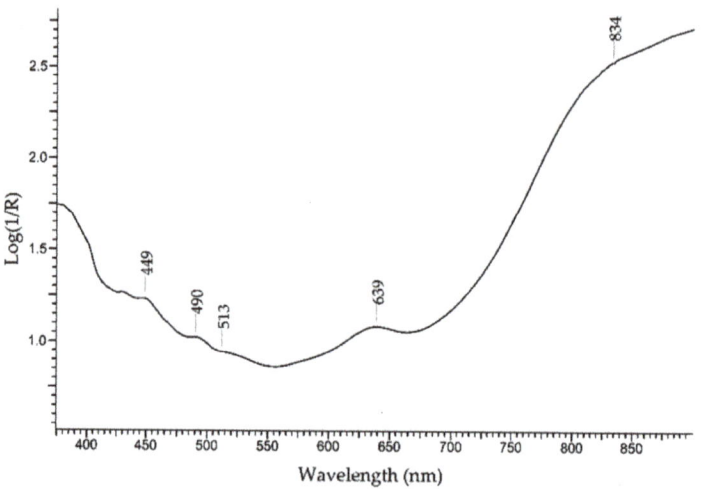

Figure 9. FORS spectrum in Log(1/R) coordinates of peridot.

2.3. Pink Gemstones

The *rhodochrosite*, a Mn(II) carbonate with formula $MnCO_3$, is one of the pink gemstones. The colour is due to the three bands of Mn^{2+} occurring at 407, 445 and 547 nm (Figure 1). Another Mn-containing pink gemstone is the *rhodonite*, a silicate with formula $MnSiO_3$. The spectral features are very similar to those of rhodochrosite, with Mn^{2+} bands occurring at 409, 456 and 549 nm (Figure 10). Therefore, FORS systems with the 350–1100 nm spectral range will generally be unable to distinguish between rhodochrosite and rhodonite, while systems with an extended range up to 2500 nm will reveal spectral features of the anions (carbonate and silicate in this case), thus enabling a more accurate identification.

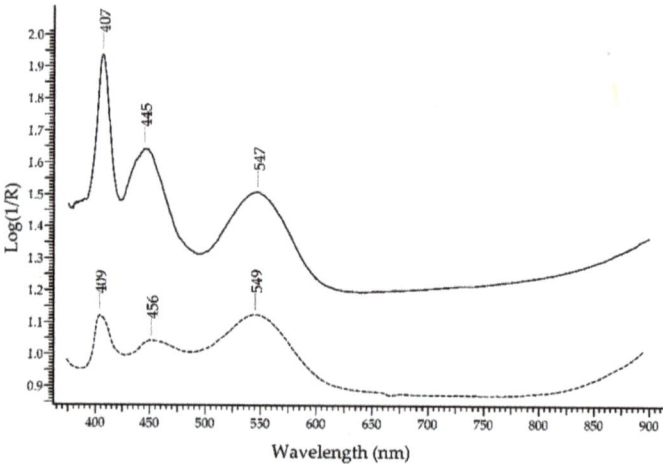

Figure 10. FORS spectra in Log(1/R) coordinates of rhodochrosite (solid line) and rhodonite (dashed line).

2.4. Red and Purple Gemstones

Ruby, the most precious variety of corundum—Al_2O_3—owes its colour to Cr^{3+} [16]. It is relatively easy to identify and discriminate from its substitutes, such as ruby-like glasses, thanks to two luminescence bands occurring at 693 and 694 nm, due to Cr^{3+}; these bands can be clearly seen upon reflectance measurements, despite not being true reflectance bands.

In Figure 11 (top), the FORS spectra of ruby and a ruby-like glass coloured with selenium (Kremer Pigmente 39224, Colored glass, Gold ruby extra, transparent) are shown in reflectance coordinates: the spectrum of ruby (solid line) is clearly dominated by the two luminescence bands. In Log(1/R) coordinates (Figure 11, bottom), the spectrum of ruby shows two absorption bands at ca. 413 and 550 nm, which were again due to Cr^{3+}.

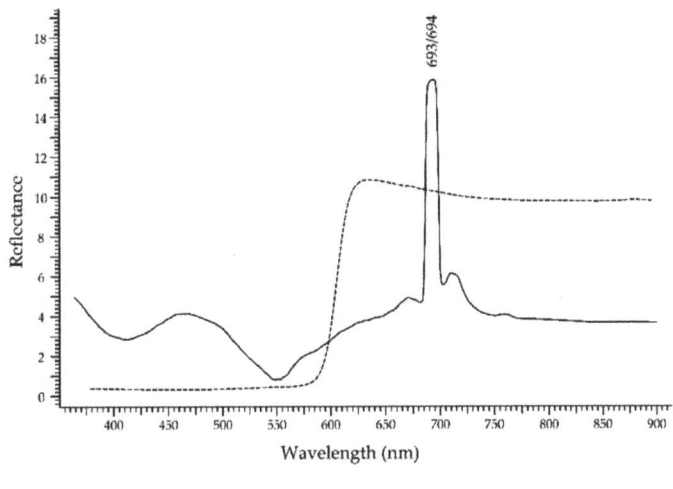

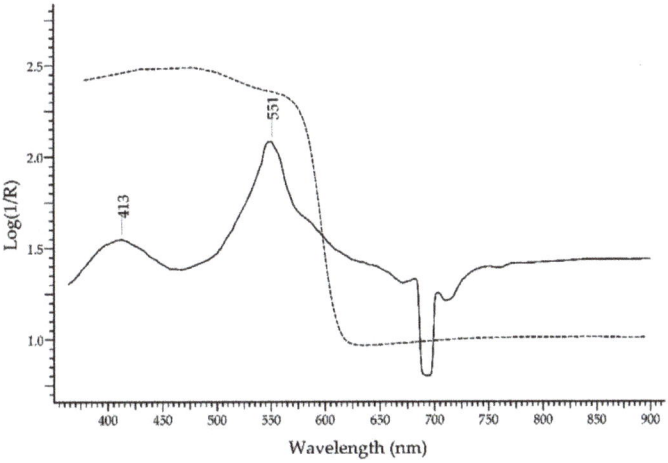

Figure 11. Top: FORS spectra in reflectance coordinates of ruby (solid line) and a ruby-coloured glass (dashed line). **Bottom**: FORS spectra in Log(1/R) coordinates (same legend).

The difference between the spectral features of ruby and a ruby glass are evident. In the former, the chromophore system is the Cr^{3+} ion in the network of corundum (Al_2O_3) which causes the crystal-field splitting of the energy levels of Cr^{3+} [37]. In the latter, the ruby colour can be obtained by adding Se, metallic Cu or metallic Au, but whatever the chromophore is, the spectral features are widely different from those of Cr^{3+} ion in corundum.

The garnet group of minerals has a broad range of chemical composition and colours from the most common red to orange, purple, brown, up to colourless [38]. The spectrum of garnet is depending on the transition metal ions present in the structure, mostly Cr^{3+}, Fe^{2+}, Fe^{3+}, Mn^{2+} and V^{3+} [39,40] and on Fe^{2+}-Ti^{4+} intervalence charge transfer, and it is usually rich in features. In the spectrum in Figure 12, taken from a purple pyrope-almandine garnet (dashed line), absorption bands at 397, 464, 505, 521, 618 and 697 nm can be attributed to Fe^{2+}, with the last band attributable also to Fe^{3+}-Fe^{2+} IVCT; bands at 426 and 464 nm can be attributed to Mn^{2+}. The band at 573 nm may be attributed to Fe^{3+}, Cr^{3+} and/or V^{3+}. The spectrum of an orange spessartite (solid line) is dominated by the features of Mn^{2+}. at 410, 432 and 480 nm.

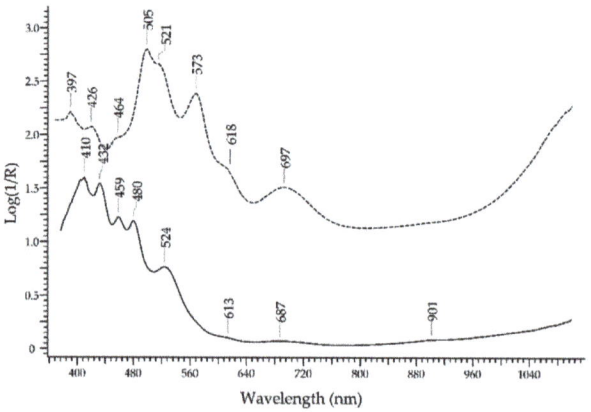

Figure 12. FORS spectrum in Log(1/R) coordinates of a spessartite garnet (solid line) and a purple pyrope–almandine garnet (dashed line).

A particular case of red gemstone is the *coral*, one among the very few of organic nature. The red-to-pink colour of corals is not due to inorganic chromophores but to carotenoids; therefore, it is produced by electronic transitions among delocalised molecular orbitals [41]. Natural corals can be counterfeit by bleaching and dyeing the surface in order to obtain a more homogeneous and rich coloration. In natural corals, the absorption spectrum is dominated by a main band structured in three sub-bands at ca. 465, 498, and 525 nm, with minor spectral features in the UV region. Dyed coral samples do not show these features. In Figure 13, the FORS spectrum of a natural coral is shown.

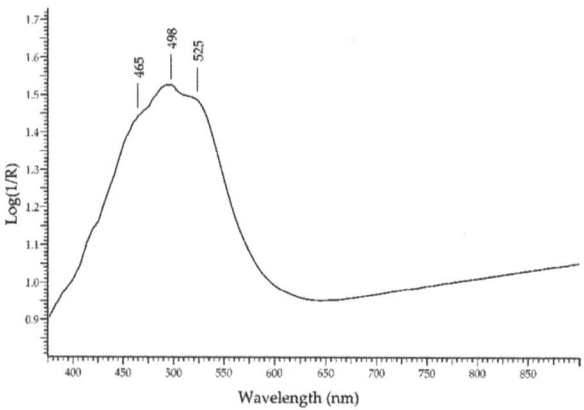

Figure 13. FORS spectrum in Log(1/R) coordinates of a natural coral.

2.5. Violet Gemstones

A violet–blue gemstone is the *tanzanite*, which is a pleochroic variety of zoisite—$(Ca_2Al_3[Si_2O_7][SiO_4]O(OH))$. The colour is mainly due to the V^{3+} ion [42]. Due to the rarity of natural high-quality gemstones, lower-quality products are generated by means of heat treatment. The absorption spectrum is dominated by the features of V^{3+} with two main bands occurring at 600 and 750 nm and a shoulder at 540 nm (Figure 14).

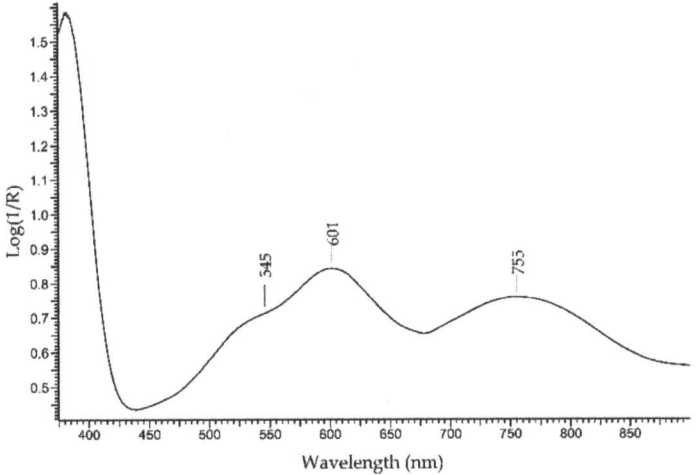

Figure 14. FORS spectrum in Log(1/R) coordinates of tanzanite.

2.6. Yellow Gemstones

The *chrysoberyl* is a gemstone containing beryllium as the beryl family, of various colours, although yellow and yellow–green are considered the most valuable. The FORS spectrum is dominated by a sharp band at ca. 440 nm due to Fe^{3+}, with a minor band at 502 nm (Figure 7).

Heliodor is another variety of beryl with golden–green to yellow–green hue. Its colour is due to a mechanism of charge transfer between Fe^{3+} ions and the surrounding oxygen ions [43], generating an absorption band at ca. 815 nm (Figure 7).

2.7. Multicoloured Gemstones

These gemstones represent a challenge for FORS analysis. In such gemstones, the presence of metal ions impurities or other mechanisms of colour generation can vary extensively, rendering the absorption response highly variable (although not the vibrational behaviour), and it is difficult to identify specific spectral features. In some cases, the same gemstone may include areas with different colours [44].

A very common group of multicoloured gemstones is that of *quartz*—crystalline SiO_2. It includes macro-, micro- and crypto-crystalline varieties, with a wide range of colours arising from colour centres, from optica effects and from inclusions. [45]. The set of varieties has recently been reviewed by Jovanovski et al. [46]. As to the most valuable quartz gemstones, that is the macro-crystalline varieties, the main mechanism generating colour is that of colour centres associated with ions external to the structure of quartz, mostly Fe^{3+} and Al^{3+} [47]. *Amethyst*, the most precious variety of quartz, owes its violet colour to Fe^{3+} impurities exposed to ionising radiation, arising from the natural decay of ^{40}K nuclides or from artificial irradiation; irradiation causes the oxidation of substitutional Fe^{3+} to Fe^{4+} and reduction in interstitial Fe^{3+} to Fe^{2+}. [48]. Its FORS spectrum (Figure 1) is characterised by a broad absorption band at ca. 540 nm; further bands are present at ca. 350 and 950 nm. It is not possible to distinguish between naturally or artificially irradiated amethysts by means of the FORS response only.

Another variety is *citrine quartz*, with a yellow to brown colour. While natural citrines are rare, most of them are obtained by heat treatment of amethysts between 350 and 450 °C: this will increase the number of substitutional and interstitial sites filled with Fe^{3+}. The FORS spectrum (Figure 15) has no specific features, showing only a generic decrease towards NIR.

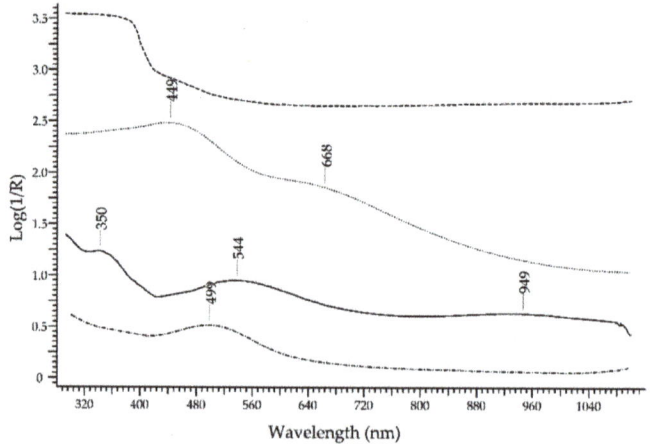

Figure 15. FORS spectrum in Log(1/R) coordinates of amethyst (solid line), citrine quartz (dashed line), smoky quartz (dotted line) and rose quartz (dashed–dotted line).

Smoky quartz is a variety in which the colour centres, created by natural or artificial irradiation, are associated with impurities of Al^{3+}. The FORS spectrum (Figure 15) will present a main band at ca. 450 nm with a shoulder at ca. 670 nm. Additional colours can be generated by heating, either natural or artificial.

Finally, *rose quartz*, the well-known variety of pink colour, owes its chromatic features to fibrous inclusions of dumortierite and in particular to IVCT between Fe^{2+} and Ti^{4+} that occur as impurities inside this aluminoborosilicate mineral [49]. The FORS spectrum (Figure 15) contains a main band centred at ca. 500 nm and assigned to the above-mentioned IVCT.

The *opal* is an unusual gemstone. It is a hydrous silica material—$SiO_2 \cdot nH_2O$—with different degrees of crystallinity and crystal structure. It can be colourless, white, yellow, orange, or red, besides other minor colour varieties. The particular aspect of precious opals, also called *noble opals*, is due to the diffraction of light by the regular stacking of the silica microspheres forming the body of the gemstone. The colour can be caused by specific mineral phases; red–orange hues are usually associated with iron oxides. Hydrophane gemstones from Ethiopia, which are very porous and can thus easily absorb water and be subjected to dyeing or impregnation processes, can result in opals with an aspect similar to fire opals from Mexico [50]. Wu et al. [51] have recently studied the possibility to distinguish natural fire opals from dyed opals. Figure 16 shows a natural fire opal from Mexico (solid line), with two inflection points at 462 and 563 nm that can be assigned to hydrated iron oxides rather than to hematite [34], and an Ethiopian dyed opal (dashed line) with features due to the impregnating solution.

The family of *spinel* comprises different members that can be defined as multiple oxides with a highly variable composition [52]. The simplest composition is AB_2O_4 with A representing a divalent ion and B representing a trivalent ion. This structure can accommodate different transition element ions that act as chromophores, such as Co^{2+}, Cr^{3+}, Cu^{2+}, Fe^{2+}, Fe^{3+}, Mn^{2+}, Mn^{3+}, and V^{3+}. Consequently, there is a wide range of displayed colours due to diverse absorption features determined by the presence of different transition elements. The most common spinels are red, coloured by Cr^{3+}; blue, coloured by Fe^{2+} or Co^{2+};

and pink, coloured by both Cr^{3+} and Fe^{2+}. Figure 17 shows two examples of spectra. In the spectrum of a red spinel, the band at 390 and the shoulder at 560 nm are attributed to V^{3+}, while the bands at ca. 410 and 535 nm are attributed to Cr^{3+}. In the spectrum of a blue spinel, the typical spectral signature of Co^{2+} in the tetrahedral site can be detected.

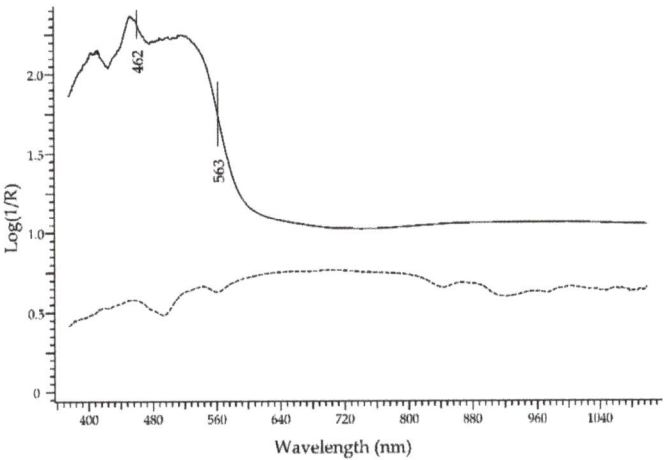

Figure 16. FORS spectrum in Log(1/R) coordinates of a Mexican fire opal (solid line) and an Ethiopian dyed opal (dashed line).

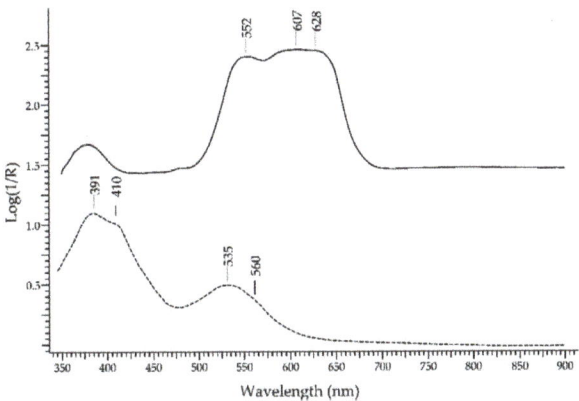

Figure 17. FORS spectrum in Log(1/R) coordinates of a red spinel (solid line) and a blue spinel (dashed line).

The *topaz* is a particularly challenging case for the FORS technique. Its formula is $Al_2SiO_4(F,OH)_2$, but its colour can vary from blue to green, yellow, pink, brown, and it can even be colourless. In addition, it is a *pleochroic* gemstone. Blue is the most common colour of topaz on the market, but natural topazes are rarely blue; indeed, the hue is obtained artificially by means of heat treatment and irradiation. The spectrum shows a main band at 620–650 nm depending on the measurement angle (Figure 18), which is possibly due to irradiation-induced defects [53] or to Cr^{3+}, Fe^{2+} and Mn^{2+} ions [54]. A pink colour characterises the so-called *imperial topaz* but is instead caused by Cr^{3+} ion according to the bands at 395, 418, 536 and 687 nm. Green-irradiated topaz shows features at 618 and 658 nm.

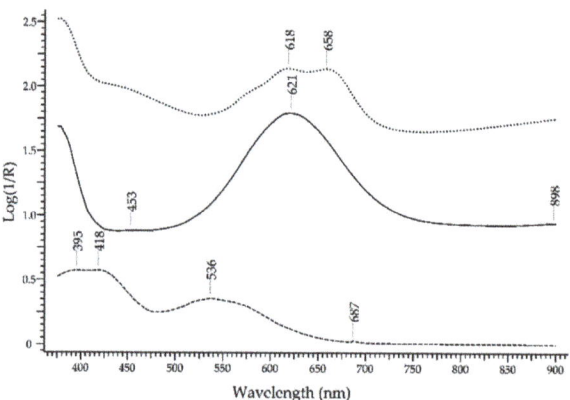

Figure 18. FORS spectrum in Log(1/R) coordinates of a blue topaz (solid line), a pink natural topaz (dashed line) and a green irradiated topaz (dotted line).

Tourmaline is the name of a large group of gemstones that share a common crystal structure (hexagonal) but have different compositions. The basic formula is $XY_3Z_6(T_6O_{18})\cdot(BO_3)_3 V_3W$, with X, Y, Z, T, V, and W representing different elements and, hence, different chromophore systems are possible. The main ions generating colour are Cr^{3+}, Cu^{2+}, Fe^{2+}, Fe^{3+}, Mn^{2+}, Mn^{3+}, Ti^{4+} and V^{3+}. The transition mechanisms can be due to the ligand field and/or to IVCT. Therefore, as in the case of topaz, the tourmaline group is highly challenging for the FORS technique, since it contains members of nearly all colours. Figure 19 shows a very limited example of the many varieties: a red *rubellite*, with its colour ascending from the absorption band at ca. 530 nm due to Mn^{3+} [55], and a green tourmaline with an intense absorption band at ca. 710 nm due to Fe^{2+} ion and Fe^{2+}-Ti^{4+} IVCT.

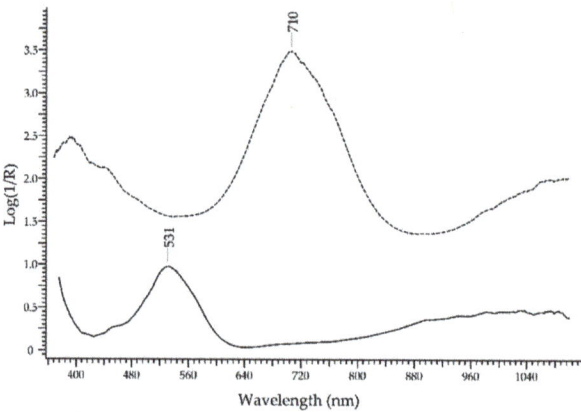

Figure 19. FORS spectrum in Log(1/R) coordinates of a rubellite variety (solid line) and a green tourmaline (dashed line).

The *zircon* family comprises gemstones with the same formula—$ZrSiO_4$—showing different colours according to the chromophores. The origin of the colour is not entirely clear [56]: pure zircon is colourless (it can a substitute of diamond), but more frequently, the natural content of U^{4+} or Th^{4+} ions substituting Zr^{4+} in the structure generates blue gemstones. The radioactive decay of these ions causes radiation damages that in turn generate colour centres and the increase in red–brown and amber colours. The absorption features of zircons can be highly variable; as an example, in Figure 20, the spectra of a green, a pink and a yellow zircon are shown. The features are mainly due to U^{4+} ion and to colour centres.

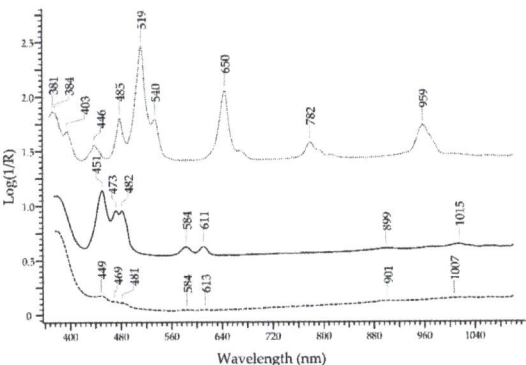

Figure 20. FORS spectrum in Log(1/R) coordinates of a green (solid line), a pink (dotted line) and a yellow (dashed line) zircon.

2.8. Uncoloured Gemstones

This group of gemstones constitutes a clear limit to the possibilities of FORS in their identification. Uncoloured transparent gemstones, such as *diamond* or *rock crystal* (uncoloured quartz), yield very poor—if any—spectral responses in the analysis in reflectance mode. Whether cut gemstones or rough stones, the exciting light enters the material, undergoes several refractions inside it and does not exit or exits very faintly; the result is a nearly flat line at 0% reflectance. The well-known N3 centre of diamonds, a lattice defect constituted by 3 nitrogen atoms bonded to a vacancy, causes an absorption band at 415.2 nm, but this cannot be seen by FORS. As a comparison, Drift-FT-IR spectroscopy can differentiate diamond from cubic zirconia and synthetic moissanite which resemble it [57]. Lipatov et al. [58] claimed that optical absorption spectroscopy combined with cathodoluminescence spectroscopy can be used for identifying natural and synthetic diamond, but they exploited spectra obtained in transmittance mode, not in reflectance.

2.9. Glassy Materials

Glasses and vitreous pastes were commonly used in medieval jewellery artworks, possibly as substitutes of authentic gemstones [33]. FORS analysis cannot highlight the glassy nature of a gemstone, of course, but it can provide indirect identification by yielding information on the chromophore system. This is particularly true as far as vitreous pastes are concerned, being them opaque materials. The main metal ions that impart colour to glass, i.e., Co^{2+}, Cu^{2+}, Fe^{2+} and Fe^{3+}, Mn^{3+} and Mn^{4+}, etc. can be identified in the FORS spectrum according to their typical absorption bands [59], therefore suggesting the presence of glassy gemstones.

2.10. Comparison of FORS with other Techniques

The diagnostic potential of FORS in the correct identification of gemstones has been tested by comparison of the FORS responses with those obtained with Raman spectroscopy and with refractometry in the analysis of three precious medieval bindings: the *Pace di Ariberto* o *Evangeliario di Ariberto*, held in the Museo del Tesoro del Duomo at Milan (Italy), the *Pace di Chiavenna*, held in the Museo del Tesoro di San Lorenzo at Chiavenna (Lumbardy, Italy) and the *Legatura di Vercelli*, held in the Museo del Tesoro del Duomo at Vercelli (Piedmont, Italy). These notable jewellery artworks are datable to the 11th century and are decorated with rich and various gemstone goods. In particular, the gemstones on the *Pace di Ariberto* [60] and the *Legatura di Vercelli* [33] have been previously analysed with

Raman spectroscopy, while the gemstones on the *Pace di Chiavenna* have been analysed with refractometry by an expert gemmologist [32]. Table 1 shows the results of the comparison.

Table 1. Comparison of the results obtained with FORS, Raman spectroscopy and refractometry (Ref) in the analysis of three precious bindings.

Gemstones	Pace di Ariberto		Pace di Chiavenna		Legatura di Vercelli	
	FORS	Raman	FORS	Ref	FORS	Raman
agate	-	1				
amethyst	24	24	6	6	5	5
carnelian	-	1				
chalcedony	-	2				
doublet	-	4				
emerald	16	16	7	7	3	3
garnet	11	11	55	56	14	14
glass/vitreous paste	-	23	-	2	-	31
mtorolite			-	1	-	1
pearl	-	21	-	93	-	23
rock crystal	-	18			-	2
sapphire	10	10	19	19	5	5
turquoise	2	2				
other stones		1		6		
total identified	63	134	87	190	27	84
unidentified by FORS	71		103		57	
total excluding pearls, rock crystals and glassy materials	63	72	87	95	27	28
unidentified by FORS excluding pearls, rock crystals and glassy materials	9		8		1	

It is apparent that the diagnostic performances of FORS are satisfying in all three cases: if pearls, rock crystals and glassy materials are not considered, between 88 and 96% of the identification of the gemstones is correct.

2.11. Chemometric Treatment of Data

Among the greatest advantages of the FORS technique is the speed of analysis: spectra can be collected in as low as 1 s, so that several spectra can be acquired in a short time. This justifies the fact that FORS can be proposed as a survey technique in the identification of gemstones on a complex jewellery artwork. After collecting several spectra, it can be useful to treat them with multivariate analysis in order to identify groups of gemstones with similar features. Using a well-known chemometric pattern recognition method, Hierarchical Cluster Analysis (HCA), it is possible to discriminate gemstones, colour by colour, according to their composition—or better to their chromophore system—which is reflected inside the FORS spectrum as minima or luminescence peaks. As an example, this approach is shown in the discrimination of the green gemstones contained in the three above cited medieval bindings. The gemstones were the following:

26 emeralds (em);
10 emerald-like glasses coloured with Cu^{2+} (gg em);
6 green glasses coloured with Ni^{2+} (gg);
2 chrome-chalcedony gemstones (cc).

A total of 44 green gemstones have been included in the analysis. The FORS spectra have been pre-treated by selecting the range 250–900 nm with a 1 nm path; this yielded 650 variables. Then, range scaling has been applied along the spectrum. After HCA, the dendrogram shown in Figure 21 was obtained. The result highlights the differences among the three main types of gemstones, arising from the spectral features of their FORS responses. The two chrome-chalcedony gemstones were classified among the group of green glasses.

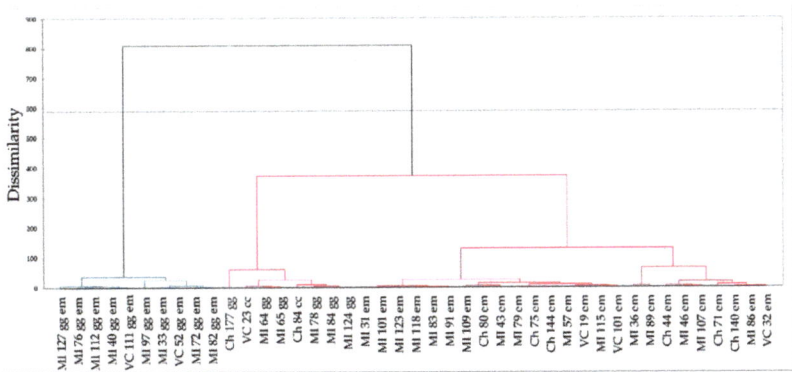

Figure 21. Dendrogram obtained by means of HCA on the FORS spectra of green gemstones.

3. Discussion

As it has been described in the previous paragraphs, one clear disadvantage of the FORS technique in the identification of gemstones is its strict dependence from the chromophore system and not from the structure of the gemstone. Nevertheless, apart from diamond, the most important gemstones, i.e., ruby, sapphire and emerald, can be easily discriminated from other gemstones with similar hues. This topic will be discussed in the next paragraphs.

3.1. Ruby vs. Red Gemstones

The possibility of discriminating ruby from red–purple gemstones (Figure 22) relies mostly on the typical luminescence bands of ruby, due to Cr^{3+}, that do not even occur in glasses containing Cr^{3+}. The two absorption bands occurring at 413 and 550 nm are not selective enough to allow a reliable identification. The red spinel, in fact, has a similar chromophore, i.e., Cr^{3+}, which generates absorption bands at 410 and 535 nm in addition to a band at 390 nm due to V^{3+}; however, the spectrum of red spinel generally lacks the strong luminescence bands at 693/694 nm. Purple garnets, though showing a somewhat similar hue, have a totally different spectral fingerprint with several bands (397, 426, 464, 505, 521, 573, 618 and 697 nm) due to Cr^{3+}, Fe^{2+}, Fe^{3+}, Mn^{2+} and V^{3+}. The red tourmaline variety called rubellite shows a single absorption band at 530 nm due to Mn^{3+}. Finally, ruby glasses, regardless of the chromophore, have generally sigmoid-like spectra.

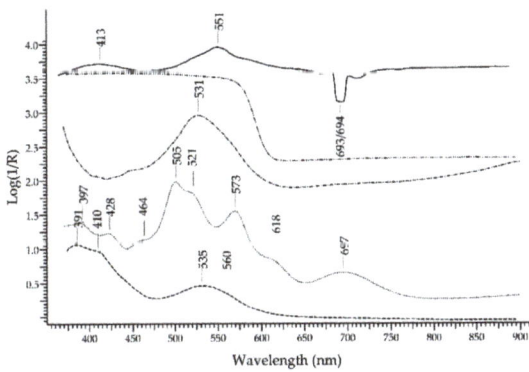

Figure 22. FORS spectrum in Log(1/R) coordinates of ruby (solid line), red spinel (dashed line), ruby garnet (dotted line), red tourmaline (dashed–dotted line) and a ruby-like glass (dashed–dotted–dotted line).

3.2. Sapphire vs. Blue Gemstones

The blue sapphire can be easily discriminated from blue aquamarine, blue spinel, blue topaz and blue zircon (lapis lazuli and turquoise are of course not considered, being opaque gemstones). The spectral features of sapphire, i.e., the bands at 390, 456 and 706 nm due to Fe^{3+} and the band at 570 nm due to Fe^{2+}-Ti^{4+} IVCT, are selective enough to allow a reliable identification (Figure 23). Aquamarine, despite having Fe^{2+} and Fe^{3+} as the ions generating colour, shows a main band at ca. 820 nm. A blue spinel coloured by Co^{2+} will show the typical signature of the ion with three sub-bands between 550 and 650 nm, while a blue spinel coloured by Fe^{2+} will show bands at 459, 655 and 902 nm. The blue topaz shows a main large band at ca. 620 nm, due to irradiation-induced defects or to Cr^{3+}, Fe^{2+} and Mn^{2+} ions. The blue zircon has a complex spectrum with sharp bands due to U^{4+}, so it can be easily recognised. Of course, blue glasses coloured with Co^{2+}, Cu^{2+} or Fe^{2+} have spectral features quite different from those of sapphire.

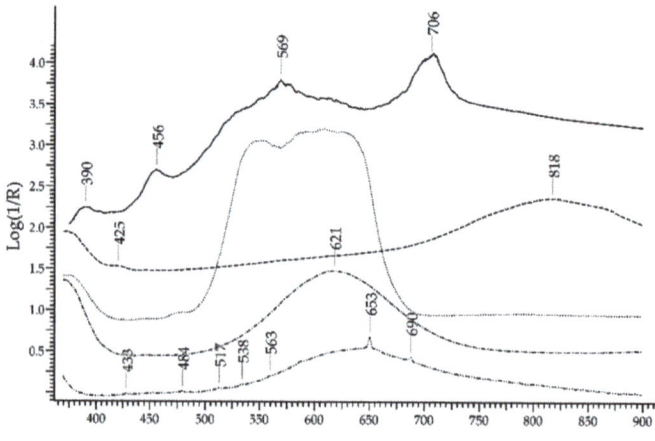

Figure 23. FORS spectrum in Log(1/R) coordinates of sapphire (solid line), a blue aquamarine (dashed line), a blue spinel (dotted line), a blue topaz (dashed–dotted line) and a blue zircon (dashed–dotted–dotted line).

3.3. Emerald vs. Green Gemstones

The discrimination between emerald and other green gemstones can be easily obtained despite the fact that some of the potential substitutes have a similar chromophore, i.e., the Cr^{3+} ion (Figure 24, top). The characteristic spectral features of emerald are two main bands occurring at 440 and 616 nm and two sharp bands occurring at 682 nm, which are all due to Cr^{3+} ion. Alexandrite, which has the same chromophore, shows a main band at ca. 580 nm and sharp bands at 682 nm, plus another band at 446 due to Fe^{3+}. Chrome-chalcedony has a main band at 610 nm due to Cr^{3+} ion. Among the potential substitutes with Fe^{2+}/Fe^{3+} chromophores (Figure 24, bottom), heliodor has a single band at 815 nm; chrysoberyl has a sharp band at ca. 440 nm, due to Fe^{3+}, and a minor band at 502 nm; peridot shows only weak bands between 450 and 490 nm, due to Fe^{2+}, and between 513 and 653 nm, due to Cr^{3+}; green tourmaline has a main intense band at ca. 710 nm, due to Fe^{2+} ion and Fe^{2+}-Ti^{4+} IVCT. Green topaz, a version obtained by irradiation, shows two bands at 618 and 658 nm. Finally, green glasses can be obtained by adding Cr^{3+}, Cu^{2+}, Ni^{2+} or V^{3+}/V^{5+} [59], but in no case, even in that of a Cr^{3+}-containing glass, is the resulting spectrum comparable to the one of emerald.

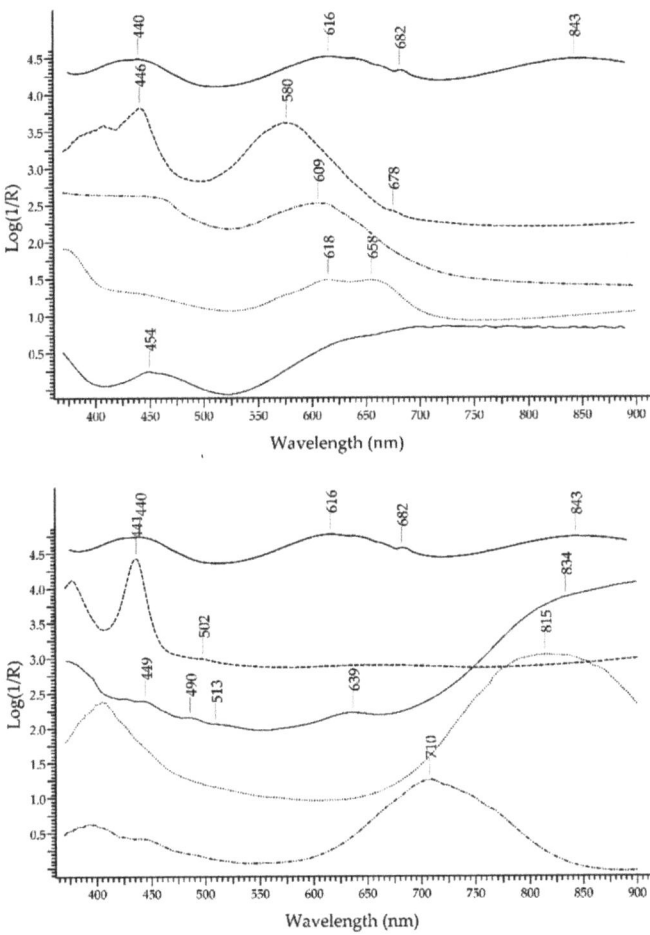

Figure 24. Top: FORS spectrum in Log(1/R) coordinates of emerald (solid line), alexandrite (dashed line), a green topaz (dotted line), a Cr^{3+} containing green glass (dashed–dotted line) and a chrome–chalcedony (dashed–dotted–dotted line). Bottom: FORS spectrum in Log(1/R) coordinates of emerald (solid line), chrysoberyl (dashed line), heliodor (dotted line), peridot (dashed–dotted line) and a green tourmaline (dashed–dotted–dotted line).

3.4. Final Considerations

The results described above show that the FORS technique have clear limits in the identification of gemstones but also clear advantages. A large number of gemstones can be identified; in cases where a gemstone has different varieties (i.e., topaz, tourmaline, quartz, etc.), the availability of a proper database can counteract the relative low diagnostic power of the technique.

The main advantage of the technique lies in its ease and speediness of use, that allows analysing jewellery artworks in a very safe way, without need of moving them outside their natural locations. The building of complete databases, that include the largest number of varieties, is a prerequisite for a proper use of this technique. In this view, apart from accessing databases available in the literature and on the web [61], the best choice is to build your own spectral database that is fully compatible with your own instrumental setup.

4. Materials and Methods

4.1. Samples of Gemstones

The gemstones analysed in this work, listed in Table 2, were provided by Effeffe Preziosi di Gilberto Faccaro & C. at Valenza (city, Italy). Their identification was confirmed by means of Raman spectroscopy.

Table 2. List of gemstones analysed in this work.

Gemstone	Provenance	Colour	Notes
alexandrite	Brazil	green	
amethyst	Brazil	violet	
aquamarine	Brazil	blue	
chrome-chalcedony	unknown	green	[1]
chrysoberyl	Brazil	yellow	
citrine quartz	Brazil	yellow	
coral	Italy	red	
emerald	Colombia	green	
garnet	India	purple	Pyrope–almandine
garnet	Kenya	orange	spessartite
glass with Cr^{3+}		green	Artificial [2]
glass with Se		red	Artificial [2]
heliodor	Brazil	green-yellow	
lapis lazuli	Afghanistan	blue	
opal	Mexico	various	
opal	Ethiopia	various	
peridot	Sri Lanka	green	
rhodochrosite	Romania	pink	
rhodonite	Tanzania	pink	
rose quartz	Brazil	pink	
ruby	Myanmar	red	
sapphire	Cambodia	blue	
sapphire		yellow	artificial
smoky quartz	Brazil	grey	
spinel	Russia	blue	
spinel	Myanmar	red	
tanzanite	Tanzania	violet	
topaz	Brazil	blue	
topaz		green	artificial
topaz	Brazil	pink	
tourmaline	Brazil	green	
tourmaline	Brazil	red	rubellite
turquoise	China	turquoise	
turquoise	China	turquoise	dyed
zircon	Myanmar	blue	
zircon		green	artificial
zircon		pink	artificial
zircon		yellow	artificial

[1] From the binding of the *C Codex* of Vercelli [33]. [2] Kremer Pigmente.

4.2. UV-Visible Diffuse Reflectance Spectrophotometry with Optic Fibres (FORS)

FORS analysis was carried out with two instruments. Most of the measurements were performed with an Avantes (Apeldoorn, The Netherlands) AvaSpec-ULS2048XL-USB2 model spectrophotometer and an AvaLight-HAL-S-IND tungsten halogen light source; the detector and light source are connected with fibre optic cables to a 1.5 mm diameter FCR-7UV200-2-1,5 × 100 probe, which contains cables for both illumination and detection; therefore, incident and detecting angles were respectively 45° and −45° from the surface normal in order to exclude specular reflectance. The spectral range of the detector was 200–1160 nm; considering the range of emission of the light source, the optimal range of acquisition of spectra was 350–1100 nm. The best spectral resolution of the system, calculated as FWHM, was 2.4 nm. Diffuse reflectance spectra of the samples were referenced against the WS-2 reference tile, guaranteed to be reflective at 98% or more in the spectral range investigated. The investigated area on the sample was 1 mm diameter. In all measurements, the distance between probe and sample was 2 mm. The instrumental parameters were as follows: 10 ms integration time, 100 scans for a total acquisition time of 1 s for each spectrum. The reproducibility of the system, as far as the conditions (distance between probe and sample, angle of the probe) are kept constant, is better than 5% in terms of band position and height. The system was managed by means of AvaSoft 8 software running under Windows 10™.

4.3. Raman Spectroscopy

In order to confirm the identification of the gemstones analysed in this work, all of them were previously subjected to Raman analysis. For this task, a high-resolution dispersive Horiba (Villeneuve d'Ascq, France) LabRAM HR Evolution model spectrometer coupled with a confocal microscope was used. The instrument was equipped with 532, 633 and 785 nm excitation lasers, an 1800 lines/mm dispersive grating, an 800 mm focal length achromatic flat field monochromator and a multichannel air-cooled CCD detector. The spectral resolution was 2 cm^{-1}. Spectra were taken with long working distance 50x and 80x objectives. All spectra were recorder at full laser power. Exposure time was 1–10 s according to needs (3 accumulations). The system was managed with LabSpec 6 software running under Windows 10™.

Author Contributions: Conceptualization, M.A. and M.P.; methodology, M.A. and M.P.; software, E.C.; validation, A.A. and M.L.; formal analysis, F.G. (Federica Gulino) and F.G. (Francesca Gullo); investigation, F.G. (Federica Gulino), F.G. (Francesca Gullo) and M.L.; resources, M.A., M.P. and A.A.; data curation, E.C.; writing—original draft preparation, M.A. and M.P.; writing—review and editing, A.A.; visualization, E.C.; supervision, M.A.; project administration, M.P.; funding acquisition, A.A. All authors have read and agreed to the published version of the manuscript.

Funding: This research received no external funding.

Institutional Review Board Statement: Not applicable.

Informed Consent Statement: Not applicable.

Data Availability Statement: The spectral database generated in this study will be soon rendered available on a dedicated web site.

Acknowledgments: The authors would like to thank Effeffe Preziosi di Gilberto Faccaro & C. at Valenza (Italy) for allowing access to the gemstones.

Conflicts of Interest: The authors declare no conflict of interest.

Sample Availability: Samples of the compounds are not available from the authors.

References

1. O'Donoghue, M.; Joyner, L. *Identification of Gemstones*; Butterworth-Heinemann: Oxford, UK, 2003; ISBN 9780750655125.
2. Riccardi, M.P.; Prosperi, L.; Tarantino, S.C.; Zema, M. Gemmology in the Service of Archaeometry. In *EMU Notes in Mineralogy*; Artioli, G., Oberti, R., Eds.; Mineralogical Society of Great Britain & Ireland: Twickenham, UK, 2019; Volume 20, pp. 345–366. ISBN 978-0903056-61-8.
3. Bersani, D.; Lottici, P.P. Applications of Raman Spectroscopy to Gemology. *Anal. Bioanal. Chem.* **2010**, *397*, 2631–2646. [CrossRef] [PubMed]
4. Kiefert, L.; Karampelas, S. Use of the Raman Spectrometer in Gemmological Laboratories: Review. *Spectrochim. Acta Part A Mol. Biomol. Spectrosc.* **2011**, *80*, 119–124. [CrossRef]
5. Jehlička, J.; Culka, A.; Baštová, M.; Bašta, P.; Kuntoš, J. The Ring Monstrance from the Loreto Treasury in Prague: Handheld Raman Spectrometer for Identification of Gemstones. *Philos. Trans. R. Soc. A Math. Phys. Eng. Sci.* **2016**, *374*, 20160042. [CrossRef]
6. Osterrothová, K.; Minaříková, L.; Culka, A.; Kuntoš, J.; Jehlička, J. In Situ Study of Stones Adorning a Silver Torah Shield Using Portable Raman Spectrometers. *J. Raman Spectrosc.* **2014**, *45*, 830–837. [CrossRef]
7. Barone, G.; Bersani, D.; Crupi, V.; Longo, F.; Longobardo, U.; Lottici, P.P.; Aliatis, I.; Majolino, D.; Mazzoleni, P.; Raneri, S.; et al. A Portable versus Micro-Raman Equipment Comparison for Gemmological Purposes: The Case of Sapphires and Their Imitations. *J. Raman Spectrosc.* **2014**, *45*, 1309–1317. [CrossRef]
8. Barone, G.; Bersani, D.; Jehlička, J.; Lottici, P.P.; Mazzoleni, P.; Raneri, S.; Vandenabeele, P.; Di Giacomo, C.; Larinà, G. Nondestructive Investigation on the 17–18th Centuries Sicilian Jewelry Collection at the Messina Regional Museum Using Mobile Raman Equipment. *J. Raman Spectrosc.* **2015**, *46*, 989–995. [CrossRef]
9. Hainschwang, T. Gemstone Analysis by Spectroscopy. In *Encyclopedia of Spectroscopy and Spectrometry*; Elsevier: Amsterdam, The Netherlands, 2017; pp. 18–24.
10. Fritsch, E.; Rossman, G.R. An Update on Color in Gems. Part 1: Introduction and Colors Caused by Dispersed Metal Ions. *Gems Gemol.* **1987**, *23*, 126–139. [CrossRef]
11. Nassau, K. The Origin of Color in Minerals. *Am. Mineral.* **1978**, *63*, 219–229.
12. Nassau, K. *The Physics and Chemistry of Color: The Fifteen Causes of Color*; John Wiley & Sons Ltd.: New York City, NY, USA, 1983.
13. Fritsch, E.; Rossman, G.R. An Update on Color in Gems. Part 2: Colors Involving Multiple Atoms and Color Centers. *Gems Gemol.* **1988**, *24*, 3–15. [CrossRef]
14. Fritsch, E.; Rossman, G.R. An Update on Color in Gems. Part 3: Colors Caused by Band Gaps and Physical Phenomena. *Gems Gemol.* **1988**, *24*, 81–102. [CrossRef]
15. Takahashi, H.; Perera, P.N. Ultraviolet-Visible Absorption Spectroscopy for Gemstone Identification. U.S. Patent 17/382,317, 3 February 2022.
16. Dubinsky, E.V.; Stone-Sundberg, J.; Emmett, J.L. A Quantitative Description of the Causes of Color in Corundum. *Gems Gemol.* **2020**, *56*, 2–28. [CrossRef]
17. Bristow, J.K.; Parker, S.C.; Catlow, C.R.A.; Woodley, S.M.; Walsh, A. Microscopic Origin of the Optical Processes in Blue Sapphire. *Chem. Commun.* **2013**, *49*, 5259–5261. [CrossRef] [PubMed]
18. Fontana, I.; Le Donne, A.; Palanza, V.; Binetti, S.; Spinolo, G. Optical Spectroscopy Study of Type 1 Natural and Synthetic Sapphires. *J. Phys. Condens. Matter* **2008**, *20*, 125228. [CrossRef]
19. Palanza, V.; Chiodini, N.; Galli, A.; Lorenzi, R.; Moretti, F.; Paleari, A.; Spinolo, G. Updating of the Interpretation of the Optical Absorption and Emission of Verneuil Synthetic and Natural Metamorphic Blue Sapphire: The Role of V^{2+}, V^{3+} and Cr^{2+}. *IOP Conf. Ser. Mater. Sci. Eng.* **2010**, *15*, 012087. [CrossRef]
20. Marfunin, A.S. Luminescence. In *Spectroscopy, Luminescence and Radiation Centers in Minerals*; Springer: Berlin/Heidelberg, Germany, 1979; pp. 141–222.
21. Gaft, M.; Reisfeld, R.; Panczer, G. Interpretation of Luminescence Centers. In *Modern Luminescence Spectroscopy of Minerals and Materials*; Springer: Berlin/Heidelberg, Germany, 2005; pp. 119–251.
22. Jaliya, R.G.C.; Dharmaratne, P.G.R.; Wijesekara, K.B. Characterization of Heat Treated Geuda Gemstones for Different Furnace Conditions Using FTIR, XRD and UV–Visible Spectroscopy Methods. *Solid Earth Sci.* **2020**, *5*, 282–289. [CrossRef]
23. Huang, R.R.; Yin, Z.W. Spectroscopy Identification of Untreated and Heated Corundum. *Spectrosc. Spectr. Anal.* **2017**, *37*, 80–84.
24. Calvo Del Castillo, H.; Deprez, N.; Dupuis, T.; Mathis, F.; Deneckere, A.; Vandenabeele, P.; Calderón, T.; Strivay, D. Towards the Differentiation of Non-Treated and Treated Corundum Minerals by Ion-Beam-Induced Luminescence and Other Complementary Techniques. *Anal. Bioanal. Chem.* **2009**, *394*, 1043–1058. [CrossRef]
25. Bunnag, N.; Kasri, B.; Setwong, W.; Sirisurawong, E.; Chotsawat, M.; Chirawatkul, P.; Saiyasombat, C. Study of Fe Ions in Aquamarine and the Effect of Dichroism as Seen Using UV–Vis, NIR and X-ray. *Radiat. Phys. Chem.* **2020**, *177*, 109107. [CrossRef]
26. Bacci, M.; Cucci, C.; Del Federico, E.; Ienco, A.; Jerschow, A.; Newman, J.M.; Picollo, M. An Integrated Spectroscopic Approach for the Identification of What Distinguishes Afghan Lapis Lazuli from Others. *Vib. Spectrosc.* **2009**, *49*, 80–83. [CrossRef]
27. Qiu, J.-T.; Qi, H.; Duan, J.-L. Reflectance Spectroscopy Characteristics of Turquoise. *Minerals* **2016**, *7*, 3. [CrossRef]
28. Han, W.; Lu, T.; Dai, H.; Su, J.; Dai, H. Impregnated and Dyed Turquoise. *Gems Gemol.* **2015**, *51*, 3.
29. Wood, D.L.; Nassau, K. The Characterization of Beryl and Emerald by Visible and Infrared Absorption Spectroscopy. *Am. Mineral.* **1968**, *53*, 777–800.

30. Schmetzer, K.; Hyršl, J.; Bernhardt, H.J.; Hainschwang, T. Purple to Reddish Purple Chrysoberyl from Brazil. *J. Gemmol.* **2014**, *34*, 32–40. [CrossRef]
31. Butini, E.; Aliprandi, R. Le Gemme Di Oplontis: Aspetto Gemmologico. In *Gli ori di Oplontis Gioielli Romani dal Suburbio Pompeiano*; D'Ambrosio, A., Ed.; Bibliopolis—Edizioni di Filosofia e Scienze, Soprintendenza Archeologica di Pompei: Napoli, Italy, 1987.
32. De Michele, V.; Aceto, M.; Agostino, A.; Fenoglio, G. La Gemmatura Nella Pace Di Chiavenna. Considerazioni Gemmologiche. Indagini Archeometriche Non Invasive. *Arte Lomb.* **2019**, *185*, 72–80. [CrossRef]
33. Agostino, A.; Aceto, M.; Fenoglio, G.; Operti, L. Caratterizzazione Chimica Della Coperta Del Codice, C. In *Tabula Ornata Lapidibus Diversorum Colorum. la Legatura Preziosa del Codice C nel Museo del Tesoro del Duomo di Vercelli*; Lomartire, S., Ed.; Viella: Roma, Italy, 2015; pp. 125–162.
34. Aceto, M.; Agostino, A.; Fenoglio, G.; Idone, A.; Gulmini, M.; Picollo, M.; Ricciardi, P.; Delaney, J.K. Characterisation of Colourants on Illuminated Manuscripts by Portable Fibre Optic UV-Visible-NIR Reflectance Spectrophotometry. *Anal. Methods* **2014**, *6*, 1488. [CrossRef]
35. Ravikumar, R.V.S.S.N.; Madhu, N.; Chandrasekhar, A.V.; Reddy, B.J.; Reddy, Y.P.; Rao, P.S. Cu(II), Mn(II) in Tetragonal Site in Chrysocolla. *Radiat. Eff. Defects Solids* **1998**, *143*, 263–272. [CrossRef]
36. Adamo, I.; Bocchio, R.; Pavese, A.; Prosperi, L. Characterization of Peridot from Sardinia, Italy. *Gems Gemol.* **2009**, *45*, 130–133. [CrossRef]
37. Tilley, R.J.D. *Colour and the Optical Properties of Materials: An Exploration of the Relationship Between Light, the Optical Properties of Materials and Colour*; John Wiley & Sons: Hoboken, NJ, USA, 2011; ISBN 9780470746967.
38. Deer, W.A.; Howie, R.A.; Zussman, J. *An Introduction to the Rock-Forming Minerals*, 2nd ed.; Longman: London, UK, 1992.
39. Manning, P.G. The Optical Absorption Spectra of the Garnets; Almandine-Pyrope, Pyrope and Spessartine and Some Structural Interpretations of Mineralogical Significance. *Can. Mineral.* **1967**, *9*, 237–251.
40. Izawa, M.R.M.; Cloutis, E.A.; Rhind, T.; Mertzman, S.A.; Poitras, J.; Applin, D.M.; Mann, P. Spectral Reflectance (0.35–2.5 μm) Properties of Garnets: Implications for Remote Sensing Detection and Characterization. *Icarus* **2018**, *300*, 392–410. [CrossRef]
41. Smith, C.P.; McClure, S.F.; Eaton-Magaña, S.; Kondo, D.M. Pink-to-Red Coral: A Guide to Determining Origin of Color. *Gems Gemol.* **2007**, *43*, 4–15. [CrossRef]
42. Koziarska, B.; Godlewski, M.; Suchocki, A.; Czaja, M.; Mazurak, Z. Optical Properties of Zoisite. *Phys. Rev. B* **1994**, *50*, 12297–12300. [CrossRef] [PubMed]
43. Andersson, L.O. The Yellow Color Center and Trapped Electrons in Beryl. *Can. Mineral.* **2013**, *51*, 15–25. [CrossRef]
44. Wang, G.-Y.; Yu, X.-Y.; Liu, F. Genesis of Color Zonation and Chemical Composition of Penglai Sapphire in Hainan Province, China. *Minerals* **2022**, *12*, 832. [CrossRef]
45. Rossman, G.R. Colored Varieties of the Silica Minerals. *Rev. Mineral. Geochem.* **1994**, *29*, 433–463.
46. Jovanovski, G.; Šijakova-Ivanova, T.; Boev, I.; Boev, B.; Makreski, P. Intriguing Minerals: Quartz and Its Polymorphic Modifications. *ChemTexts* **2022**, *8*, 14. [CrossRef]
47. Henn, U.; Schultz-Güttler, R. Review of Some Current Coloured Quartz Varieties. *J. Gemmol.* **2012**, *33*, 29–43. [CrossRef]
48. Lehmann, G. On the Color Centers of Iron in Amethyst and Synthetic Quartz: A Discussion. *J. Earth Planet. Mater.* **1975**, *60*, 335–337.
49. Kibar, R.; Garcia-Guinea, J.; Çetin, A.; Selvi, S.; Karal, T.; Can, N. Luminescent, Optical and Color Properties of Natural Rose Quartz. *Radiat. Meas.* **2007**, *42*, 1610–1617. [CrossRef]
50. Renfro, N.; McClure, S.F. Dyed Purple Hydrophane Opal. *Gems Gemol.* **2011**, *47*, 260–270. [CrossRef]
51. Wu, J.; Ma, H.; Ma, Y.; Ning, P.; Tang, N.; Li, H. Comparison of Natural and Dyed Fire Opal. *Crystals* **2022**, *12*, 322. [CrossRef]
52. Andreozzi, G.B.; D'Ippolito, V.; Skogby, H.; Hålenius, U.; Bosi, F. Color Mechanisms in Spinel: A Multi-Analytical Investigation of Natural Crystals with a Wide Range of Coloration. *Phys. Chem. Miner.* **2019**, *46*, 343–360. [CrossRef]
53. Krambrock, K.; Ribeiro, L.G.M.; Pinheiro, M.V.B.; Leal, A.S.; Menezes, M.D.B.; Spaeth, J.M. Color Centers in Topaz: Comparison between Neutron and Gamma Irradiation. *Phys. Chem. Miner.* **2007**, *34*, 437–444. [CrossRef]
54. Skvortsova, V.; Mironova-Ulmane, N.; Trinkler, L.; Chikvaidze, G. Optical Properties of Natural Topaz. *IOP Conf. Ser. Mater. Sci. Eng.* **2013**, *49*, 012051. [CrossRef]
55. Phichaikamjornwut, B.; Pongkrapan, S.; Intarasiri, S.; Bootkul, D. Conclusive Comparison of Gamma Irradiation and Heat Treatment for Color Enhancement of Rubellite from Mozambique. *Vib. Spectrosc.* **2019**, *103*, 102926. [CrossRef]
56. Kempe, U.; Trinkler, M.; Pöppl, A.; Himcinschi, C. Coloration of Natural Zircon. *Can. Mineral.* **2016**, *54*, 635–660. [CrossRef]
57. *Diamonds—Characterized by FT-IR Spectroscopy*; Application Note no. AN#81; Bruker: Ettlingen, Germany, 2010; pp. 1–3. Available online: https://www.optikinstruments.hr/runtime/cache/an-m81-diamonds-en-daa1b24eb7b29aa45a60e6d8355a40d/.pdf (accessed on 21 June 2022).
58. Lipatov, E.I.; Burachenko, A.G.; Avdeev, S.M.; Tarasenko, V.F.; Bublik, M.A. Identification of Natural and Synthetic Diamonds from Their Optical Absorption and Cathodoluminescence Spectra. *Russ. Phys. J.* **2018**, *61*, 469–483. [CrossRef]
59. Aceto, M.; Fenoglio, G.; Labate, M.; Picollo, M.; Bacci, M.; Agostino, A. A Fast Non-Invasive Method for Preliminary Authentication of Mediaeval Glass Enamels Using UV–Visible–NIR Diffuse Reflectance Spectrophotometry. *J. Cult. Herit.* **2020**, *45*, 33–40. [CrossRef]

60. Superchi, M. Le Gemme Dell'Evangeliario Di Ariberto. In *Evangeliario di Ariberto*; Tomei, A., Ed.; Silvana Editore: Milano, Italy, 1999; pp. 149–157.
61. California Institute of Technology Mineral Spectroscopy Server. Available online: http://minerals.gps.caltech.edu/FILES/Visible/Index.html (accessed on 21 June 2022).

Article

Inside the History of Italian Coloring Industries: An Investigation of ACNA Dyes through a Novel Analytical Protocol for Synthetic Dye Extraction and Characterization

Ilaria Serafini [1,*], Kathryn Raeburn McClure [1], Alessandro Ciccola [1], Flaminia Vincenti [1], Adele Bosi [1,2], Greta Peruzzi [1], Camilla Montesano [1], Manuel Sergi [1], Gabriele Favero [3] and Roberta Curini [1]

1. Department of Chemistry, Sapienza University of Rome, P. le Aldo Moro 5, 00185 Rome, Italy; k.mcclure.1@research.gla.ac.uk (K.R.M.); alessandro.ciccola@uniroma1.it (A.C.); flaminia.vincenti@uniroma1.it (F.V.); adele.bosi@uniroma1.it (A.B.); peruzzi.1957090@studenti.uniroma1.it (G.P.); camilla.montesano@uniroma1.it (C.M.); manuel.sergi@uniroma1.it (M.S.); roberta.curini@uniroma1.it (R.C.)
2. Department of Earth Sciences, Sapienza University of Rome, P. le Aldo Moro 5, 00185 Rome, Italy
3. Department of Environmental Biology, Sapienza University of Rome, P. le Aldo Moro 5, 00185 Rome, Italy; gabriele.favero@uniroma1.it
* Correspondence: ilaria.serafini@uniroma1.it

Abstract: The introduction of synthetic dyes completely changed the industrial production and use of colorants for art materials. From the synthesis of the first synthetic dye, mauveine, in 1856 until today, artists have enjoyed a wider range of colors and selection of chemical properties than was ever available before. However, the introduction of synthetic dyes introduced a wider variety and increased the complexity of the chemical structures of marketed dyes. This work looks towards the analysis of synthetically dyed objects in heritage collections, applying an extraction protocol based on the use of ammonia, which is considered favorable for natural anthraquinone dyes but has never before been applied to acid synthetic dyes. This work also presents an innovative cleanup step based on the use of an ion pair dispersive liquid–liquid microextraction for the purification and preconcentration of historical synthetic dyes before analysis. This approach was adapted from food science analysis and is applied to synthetic dyes in heritage science for the first time in this paper. The results showed adequate recovery of analytes and allowed for the ammonia-based extraction method to be applied successfully to 15 samples of suspected azo dyes from the Azienda Coloranti Nazionali e Affini (ACNA) synthetic dye collection, identified through untargeted HPLC-HRMS analyses.

Keywords: synthetic dyes; ACNA; ammonia extraction protocol; dLLME; HPLC-HRMS; cultural heritage

1. Introduction

Historians and conservators consider 1856 a key year in the history of industry. The work of the young British chemist William Henry Perkin introduced mauveine (the first synthetic dye) [1], and this discovery started a new era in fabric dyeing. The synthesis of dyes in laboratory environments opened up the possibility of exploring new methodologies to obtain new hues and shades, which no longer relied upon the intrinsic variability of natural dye resources [2]. For this reason, the price of production was reduced and the dependency of the textile industry on natural dyes declined [3–5]. This synthesis completely changed the face of the industry and allowed for rapid scaling up of textile dye production and an increase in market access to textile dyes, leading to the launch of new large chemical plants and companies across the world [6,7].

Feverish experimentation, encouraged by the new potential in terms of possible substances, colors and synthesis routes was the result of this new phase, which invested not only the world of industry but also the world of art materials. Colored objects originating from this period of rapid expansion offer important information that can shape our understanding of the histories of industries and art that shaped the development, use and production of these materials.

This importance explains why synthetic dye studies, which represent a relatively new field in cultural heritage analyses due to their modernity, are rapidly increasing in importance, taking into account that the artistic productions of the 19th–20th century now require conservation interventions and raise curatorial questions [2,8].

While scientific analysis of synthetically dyed heritage objects is carried out in a similar way to the analysis of natural dyes, chemical variabilities between the two groups introduce different complexities. In general, synthetic dyes possess greater molecular uniformity than natural dyes, which are often a mixture of different chromophore compounds (i.e., we can count 68 anthraquinone dye molecules for madder roots). This is because they are produced under controlled conditions in laboratory environments, whilst natural dyes are influenced by a significant number of "uncontrolled" natural variables [4,5]. In contrast, however, a vast range of molecular classes exist for synthetic textile dyes compared to natural dyes. This reflects the extensive variability of the dye molecules able to be synthesized in the laboratory compared to those that are derived from the natural world [6]. Due to this incredibly high diversity in chemical structure, substituents, etc., it is difficult to develop analytical protocols that are well suited to the identification of all of the different chemical classes and typologies of synthetic dye; for example, charged dyes (e.g., reactive dyes) may require different extraction methods than uncharged dyes.

This variability means that different processes may be required for the extraction of different types of dyes from fibers. Furthermore, the variability introduces additional challenges for the heritage scientist in identifying the chemical structures of unknown dyes, as reference spectra do not exist for the thousands of potential commercial dyes sold, and minor variations (e.g., substituent position) make even mass spectrometry data, which do not require reference data, difficult to interpret. Reference spectra databases currently only contain a small proportion of the dyes made and sold on the market, and the vast range of available dyes make it difficult to develop, navigate and update databases—making it challenging to use techniques that rely upon comparison with known compounds. Moreover, inconsistencies in nomenclature—where different manufacturers refer to the same dye molecules by their own brand names, or use similar names for chemically different dyes—make it challenging to identify dyes even when their commercial names are listed [2,9,10]. Together, these factors make the identification of these types of dyes from historical and artistic matrices a highly complex matter; for this reason, it is desirable to develop ad hoc methodologies and (re-)organize bibliographical sources.

However, the improved molecular uniformity within a single sample of synthetic dyes makes some aspects of analysis simpler than for natural dyes, which are generally made up of several low-concentration chromophores. For example, in spectroscopic analysis such as Raman spectroscopy, synthetic dyes often produce more intense spectral peaks. This can allow for spectra to be obtained without enhancements such as surface-enhanced Raman spectroscopy, which is required for natural dyes [2,9].

The development of new protocols for synthetic dye analysis represents a relatively new research area that has the potential to lead to significant improvements in how we research and understand modern heritage objects. Toward this goal, some interesting work has been published in recent years. For example, the excellent potential of nondestructive or minimally invasive analyses using Raman and SERS for providing new information on synthetically dyed heritage materials has been researched [9,11–14].

High-pressure liquid chromatography, coupled with mass spectrometry analyses (HPLC-MS)—generally considered the "gold-standard for dye analysis"—require extraction of the dye from a sample of the object. The extraction methods used for synthetic dyes are generally adapted from natural dye studies, such as oxalic acid or organic solvent, such as pyridine, at high temperatures [3,15], and applied directly to synthetically dyed textiles [12]. However, if the chemistries of the synthetic dyes under analysis are considered, more effective extraction protocols can be developed for specific objects. Furthermore, unlike other fields that employ synthetic dye analysis (e.g., food science) current methods in heritage science do not usually consider the use of cleanup steps, which are used to purify the sample before analysis. This lack of purification introduces impurities to the MS spectra that can at times produce signals that are higher than dyes or overlap their signals. These issues are even more pronounced in high-resolution mass spectrometry, which is a technique the field is increasingly moving toward. Furthermore, as synthetic dyes come from multistep synthesis and, therefore, may contain various side products, performing any separation or purification on the final compounds may not be commercially feasible for industries. Therefore, commercial synthetic dyes commonly contain a medium-high purity, and the remainder comprises side products [16]. For these reasons, the application of well-suited extraction methods (decided upon by the historical context and literary information available for the object under analysis), the use of an effective cleanup step (which isolates and purifies the molecules of interest before analysis), and analysis with high-sensitivity techniques such as UHPLC-MS is a beneficial approach that avoids sample losses, decreases spectral interference, and maximizes signal intensity [17].

Furthermore, historical textiles have occasionally been subjected to more than one dyeing process, sometimes both natural and synthetic, as demonstrated previously [18]. In these cases, two separate samples must be taken from a textile artefact that is likely to contain both natural and synthetic dyes. These factors work in opposition to the cultural heritage goal of achieving minimal destructiveness and maximum information. Recent papers are therefore starting to present methodologies that look at both components [12,19], but these methods still employ acid conditions or organic solvents at high temperatures, which have been demonstrated to be less effective in preserving the molecular pattern of natural dyes (e.g., madder, cochineal dyes) compared to ammonia methodologies [20,21].

For this reason, in this paper, the authors propose a new extraction protocol and novel clean up strategy for the recovery of synthetic dyes, starting from the innovative application of the ammonia–EDTA extraction methodology to such compounds. In particular, the method is focused on the development of a cleanup protocol to be used in the extraction of acid dyes, starting from the azo class specifically. This class of dyes, characterized by a N=N bond, was one of the earliest developed, with the first dye, Bismarck brown, commercially synthesized after 1861, shortly after mauveine synthesis [2,22]. Different colors can be obtained by modifications to the chemical structure, but the azo class is most typically associated with red, yellow and orange hues [22]. Since their first appearance on the market, azo dyes have been widely used in historical objects, representing one of the largest classes of synthetic dyes; it is therefore common to find them in early synthetic dye collections [22]. For this reason, the authors considered them a relevant class with which to start research with this specific focus.

Several previous analytical methodologies for azo dyes have been investigated in the field of food science, as azo dyes were largely employed, such as in edible products, until the emergence of scientific evidence related to the carcinogenic effects of some compounds in this class [23]. In response, several governments banned their use and a rapid development in analytical methods for the detection and identification of azo dyes in foods became necessary for enforcement of these regulations [24–26]. Despite the extensive study of these dyes in food science, heritage science has not yet utilized this significant body of work in improving its methods for the identification of azo dyes.

This work therefore uses the methods developed for food science as a springboard to propose a new methodology suited to the analysis of synthetic dyes in the heritage field. The method was developed using three specific azo dyes, broadly representative of the azoic acid dye class. The standards chosen were Acid Yellow 25 (CI: 18835), Congo Red (CI: 22120) and Red 2G (CI: 18050). The choice of these three dyes lies not only in their chemical properties (good water solubility and possessing negative charges), which allows for the use of the ammonia extraction, but also because they are potential reference standards for the historical samples chosen as a case study to evaluate the effectiveness of the methodology in the art–historical context. These historical samples were taken from the Azienda Coloranti Nazionali e Affini (ACNA) collection at the Museum of Chemistry, Sapienza University, and the links to these compounds were predicted through literature research into the commercial names listed beside the samples selected for analysis.

The Museum of Chemistry, located within Sapienza University of Rome's Department of Chemistry, holds an extensive collection of early synthetic dyes from several different dye companies and a significant group is represented by the dyes from the ACNA—an Italian chemical company active from 1882 until 1999 [7]. The collection of dyes analyzed contains dye powders in glass jars and card-backed sheets holding samples of dyed wool fibers. These samples came directly from the ACNA laboratories, and this collection is likely to date from the 1930s.

Researching this type of collection represents a precious opportunity to deepen historical knowledge about the production of dyes in a precise historical period, which is still understudied. The data produced also contributes to the published reference spectra available for other researchers studying unknown dyes. Moreover, samples coming from the ACNA laboratories can provide useful information to reconstruct the synthetic processes followed and history of the industrial process. The nature of these samples also means that many of these dyes may have been synthesized but then discarded due to their performance on textiles not considered suitable for large-scale production or due to a lack of chemical–physical properties.

2. Results and Discussion

The ammonia–EDTA protocol represents the first method of dye extraction in a basic environment at room temperature. It was initially developed for natural anthraquinone dyes due to the sensitivity of many dye components to the acid environment. The method showed that, even in comparison with organic solvent methods, it could better preserve highly sensitive glycosylated moieties [20,21]. Never before applied to synthetic dyes, the present work arises from a desire to evaluate the performance of this methodology for a different dye group. One motivation for the development of a method applicable to both dyes is that it is not unusual to find textile artifacts that contain both natural and synthetic dyes, especially in the years immediately following the synthesis of mauveine [18]. In cases such as these, if this protocol was found to be applicable to synthetic dyes, it would be possible to minimize the quantity of materials and maximize the information obtained from a single extraction.

Recently, the ammonia protocol has been successfully applied for the microgel extraction of natural anthraquinone dyes [27]. This research found that the application of a cleanup protocol strongly improves the quality of the spectra obtained, as mentioned in the introduction. Traditional liquid–liquid extraction (LLE) approaches cannot be used for the cleanup of azoic acid dyes, which are charged species stored as powders with cationic counter-ions before application to textiles to which they bind directly through their ionic group [28]. When extracted from the textiles, they revert to their anionic form and hence have a high affinity for water. This charged characteristic means they have an extremely low affinity for less polar solvents; so, it is unlikely for them to be recovered from an organic extracting solvent during liquid–liquid extraction. This is a significant gap in the literature that this paper addresses through the application of dLLME with the addition of an ion pair reagent (IP-dLLME).

2.1. Development of IP-dLLME Protocol

For the development of a cleanup protocol for synthetic dyes, solvent ratios were decided by referring to a study on dLLME extractions of azo dyes from ice cream samples presented by Faraji et al. [24]. dLLME has recently been applied for the first time to heritage dye analysis, and this represents one of the first well-suited cleanup methods applied in this field [29]. The method was first developed in 2006 for environmental science [30], and nowadays is widely used in analytical chemistry fields including forensics, food science and environmental biology [30,31]. Based on a three-solvent system, dLLME involves the rapid injection of an organic extracting solvent and a disperser solvent into an aqueous solution. This forms a cloudy solution that maximizes the contact between phases, increasing the opportunities for analytes of interest to move into the extracting phase. This promotion of extraction means that smaller quantities of extraction solvent can be used with improved recoveries. This has the benefit of combining purification and extraction into a single step [29–33]. For synthetic azo dyes, the aim of using an ion pair reagent is to overcome their very high affinity for water and allow for transition to a less polar extracting phase during liquid–liquid extractions. The ion pair reagent tetra-n-butylammonium bromide (TBAB) was used. The quaternary character of TBAB means that it has significant steric hindrance and through this can form an "ion associate" with anionic molecules—such as azoic acid dyes [34]. These ion associate pairs are bound by the steric effects of the TBAB, and act in a similar way to nonpolar molecules. This apolar behavior allows the ion associate pair to transition into the organic extraction solvent. The use of this ion pair reagent enables the application of the dLLME cleanup protocol to synthetic dyes, and whilst it can also be used with traditional LLE, the use of dLLME enhances extraction recovery, efficiency and precision. TBAB was specifically chosen, as it is reported to be significantly more efficient than the chloride and iodide countered quaternary ammonium salts [35]. It was also successfully applied to azo dyes from food samples by Faraji et al. [24] in the protocol used as the basis for this research.

The results of the disperser tests for synthetic dyes are displayed in Figure 1.

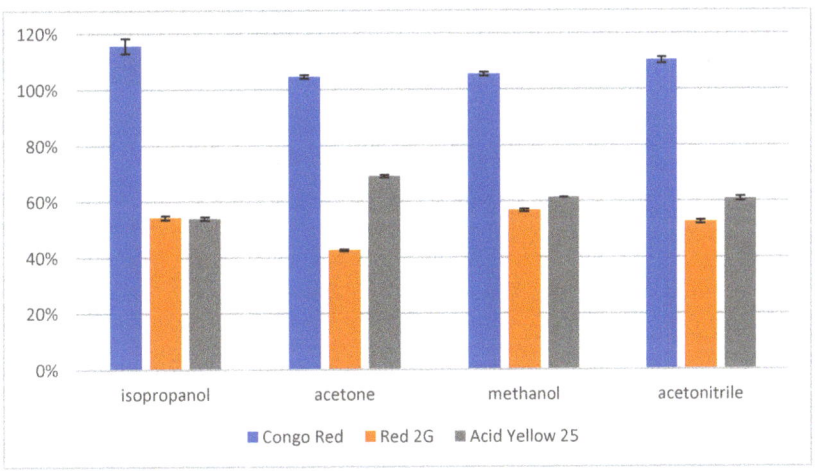

Figure 1. Comparison of the recoveries of three azo dye standards after the performance of dLLME with chloroform as the extracting solvent trialing four dispersers: isopropanol, acetone, methanol and acetonitrile.

Under the IP-dLLME protocol, all dispersers display good recoveries for all analytes. The recovery of Congo Red—over 100% for all dispersers—is explained by a combination of the error margins, and the matrix effect of the TBAB ion pair reagent. Despite the errors in recovery attributed to matrix effects, the disperser trials were sufficient to evaluate that the best recovery was achieved with methanol as the disperser. Methanol was therefore used as the disperser for further tests.

To counter the matrix effects, further analyses used a spiked sample as a reference—which takes into account the matrix. The spiked sample was prepared identically to the experimental conditions described but using 100 µL methanol instead of 500 ppb reference mix. After drying, the residue was reconstituted with the 500 ppb reference mix before analysis.

After deciding the best disperser conditions for the extraction of the synthetic azo dyes, the state-of-the-art ammonia–EDTA protocol was applied alongside dLLME to assess its effects on the recovery of the samples. The results of this analysis are presented in Figure 2.

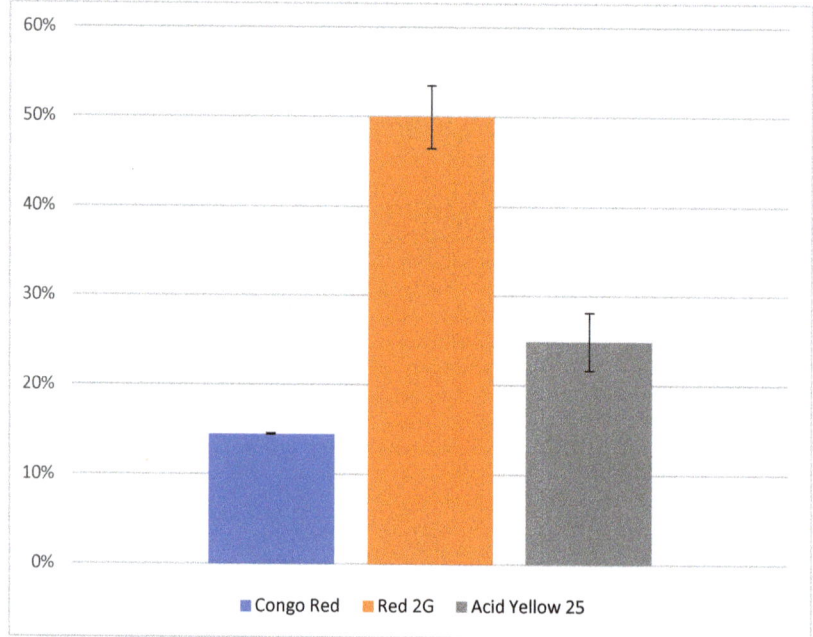

Figure 2. Recoveries of the three trialed synthetic dye analytes with the complete analytical protocol combining the ammonia and Na2EDTA extraction method with the novel dLLME cleanup protocol developed in this research.

These results display a more significant reduction in recovery than was observed when dLLME was performed alone—particularly with reference to Congo Red. This reduction is likely to be attributed to the reference in this trial accounting for the matrix effect, but also could be related to the application of the extraction protocol. However, despite this reduction, recoveries are sufficient to verify the coherence and effectiveness of the complete analytical protocol. This means that the ammonia–EDTA extraction method can now be applied to synthetic dyes as well as natural dyes and brings together a coherent extraction methodology that requires only one sample to be taken from artefacts suspected to contain both natural and synthetic dyes. While the methodologies for cleanup for natural and synthetic dyes are divergent, the improved recoveries of the ammonia–EDTA extraction method are sufficient to allow for the extracted sample to be divided into two when both natural and synthetic dyes are extracted.

2.2. Application to the Case Study: ACNA Industries Samples

This study marks a first step toward understanding the full collection of synthetic dyes held by the Museum of Chemistry and provides insight into the naming conventions of the ACNA, an understudied dye manufacturer. Information on ACNA dye naming conventions provides a useful understanding that could aid further studies related to their dyestuffs, both within the Museum of Chemistry and elsewhere. The study of these dyes is an excellent opportunity to understand the behaviors and synthetic procedures of a company that was active throughout the 20th century. Studying this through the lens of a collection obtained directly from the manufacturer, such as the collection in Sapienza University's Museum of Chemistry, is uniquely useful for two reasons: It offers the chance to obtain information about dyes that were available on the market—which have the potential to offer a synthetic dye database that can be referred to in the analysis of unknown objects. It also provides the opportunity to study dyes that may never have been placed on the market—which could provide insight into the internal testing methods and motivations of the ACNA. In particular, results from dyes such as Rosso Amidonaftolo 2G and Giallo Luce Solido 2G for which both powders and dyed fibers were analyzed provides interesting information regarding the consistency of naming.

The samples were first studied using nondestructive Raman spectroscopy before proceeding to extraction [8] and untargeted HPLC-HRMS analysis, presented in this paper. These results were combined to obtain information regarding the identities of the dyes studied.

Fifteen samples were taken from the collection—11 fiber samples from one card-backed sheet (Figure 3a), and 4 powder samples from glass jars (Figure 3b: photo of a part of the collection, where the jars are visible too). The samples were chosen after an initial visit, during which names were recorded to allow for literary research into their commercial names. This particular group of dyes was then chosen due to research indicating a high likelihood that a majority of the dyes in the group are of the azo class.

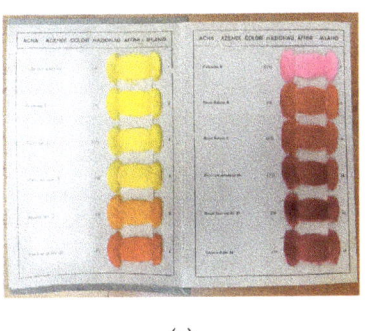
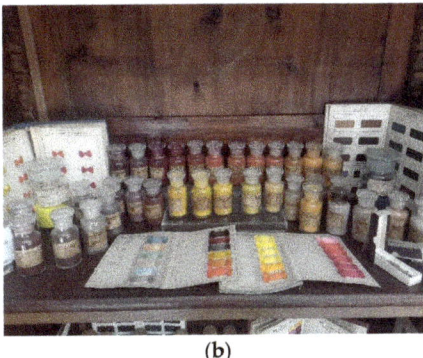

(a) (b)

Figure 3. (a) Photograph of the card-backed sheet containing the sampled dyes. Note: all dyes were sampled except Rodamina B, which was excluded due to predictions from the name that it was likely to be a Rhodamine basic dye and hence would not be suitable for the new methodology; (b) photograph of part of the collection of the dyes on display from the ACNA synthetic dye collection based at Sapienza University of Rome.

2.3. HRMS Analyses—Analytical Challenges

To pursue the scope of identification of the dye compounds presents on the museum samples, untargeted mass analysis was performed using an Orbitrap high-resolution mass spectrometer. The mass spectrometry results contained several peaks that were present in all analytes and hence likely to be related to the matrix and these were therefore discounted when analyzing the data.

Where assignments agreed with the predictions formulated from the Raman spectroscopy interpretations [9], it was concluded that the molecules were highly likely to correspond to the projected molecular structures. In cases where Raman spectroscopy was not sufficient to obtain formal predictions about the molecular structures (particularly when there were no corresponding spectra available in the literature) but the mass analyses were able to provide possible predictions, Raman spectroscopy was utilized as further confirmation of possible assignments. In some cases, characterization of the specific species was not obtained and will require additional research.

The reasons for the difficulties associated with this characterization are the lack of databases available for the identification of synthetic dye molecules, and sometimes the dyes come from one specific company and therefore may not have been widely used or indeed ever been commercially available. Further challenges with the mass spectrometry of azoic acid dyes most likely to be present in this collection are the fact that these species contain varying degrees of charged character, meaning that m/z ratios may refer to several different molecular masses.

Alongside this, where databases exist, they generally include the counter-ion mass when reporting the overall molecular mass, so possible variations in the cationic species must be considered. A summary table (Table 1) of evidence found and/or hypothesized is presented and a case-by-case discussion, also in connection with the Raman data discussed in [9], immediately follows.

Table 1. A summary table of evidences found and/or hypothesized for each sample.

Sample	Possible Match or Possible Chemical Features	Chemical Structure
Arancio Luce G	Acid Orange 31	
Giallo Eliaminia RL	Further studies ongoing	
Giallo Italana 2G	Structural similarities with Acid Orange 31	
Giallo Luce Solido 2G—powder	Acid Yellow 11	
Giallo Luce Solido 2G—fiber	Further studies ongoing	

Table 1. Cont.

Sample	Possible Match or Possible Chemical Features	Chemical Structure
Giallo Novamina 2G	Acid Yellow 25	
Rosso Amidonaftolo 2G	Red 2G	
Tartrazine J	Further studies ongoing	
Rosso Italana B	Further studies ongoing	
Rosso Luce Solido BL; Rosso Italana R	Further studies ongoing	
Rosso Naftolo SJ	A naphthalene group (as suggested by the name) as well as at least one sulfonate group and one azo group	
Rosso Novamina 2G	Acid Orange 19	

Table 1. *Cont.*

Sample	Possible Match or Possible Chemical Features	Chemical Structure
SEII Azoico Acido Pag	Acid Orange 7	(structure of Acid Orange 7 with HO-naphthalene, N=N azo linkage, phenyl-SO₃⁻ Na⁺)

- Assignments of "Arancio Luce G"

Preliminary research for "Arancio Luce G" suggested a possible association of this name to Acid Orange 10 and Food Orange 4, which have the same chemical structure (C.I. 16230) [8,36]. Mass spectral data of the compound were fairly weak, with three unique m/z peaks observed, all of which had relatively low intensity chromatographic peaks. The peaks observed were for m/z 360.3131, 361.2608 and 526.0877. The highest intensity of these peaks was observed at m/z 361.2608 and this species was therefore used to make some possible projections (Figure 4). Projections were made considering that "Arancio Luce G" is likely to be an azoic acid dye, and that the counter-ion is likely to be Na+ (as is usually the case for this class). Regarding the species, which seems to be a singly charged azoic dye based on the isotopic pattern observed, the m/z is likely to represent [M-Na]⁻, where M would represent the molecular mass and be equal to 384.2506 u. When this information was searched on chemical databases, a tentative possible assignment to Acid Orange 31 was made [37].

- Assignment of "Giallo Eliaminia RL"

Preliminary investigations into the identification of "Giallo Eliamina RL" were strongly based on the nomenclature. Several chemical databases listed Yellow Eliamina as a synonym for a variety of dyes: Direct Yellow 29, Direct Yellow 44, Direct Yellow 49, and Direct Yellow 50. Except for Direct Yellow 29, the other molecules share some features. Specifically, they are diazo structures with the presence of a central carbamide group.

Mass spectral data of the "Giallo Eliamina RL" powder revealed several species present in the sample; however, a very intense peak corresponding to m/z 388.7797 at a retention time of 2.25 min was by far the most prominent. The species seems to be a singly charged azoic dye; the m/z is likely to represent [M-Na]⁻, where M would represent the molecular mass and be equal to 411.7695u. Molecular weight searches into the projected mass were performed but no yellow dyes were found to correspond.

- Assignment of "Giallo Italana 2G"

In the case of Giallo Italana 2G, preliminary studies were not indicative for identification.

The mass spectral data for "Giallo Italana 2G" presented three m/z species, and by far the most intense was m/z 236.9883, which had a retention time of 3.35 min (Figure S1 in Supplementary Materials). The species, based on isotopic pattern observed, is a singly charged azoic dye; thus, the m/z is likely to represent [M-Na]⁻, where M would represent the molecular mass and be equal to 259.9781u. Molecular weight searches into the projected mass were performed but no yellow dyes were found to correspond. However, it is possible to infer from the Raman spectrum [9] that the spectral template for "Giallo Italana 2G" is likely to share structural details with Acid Orange 31.

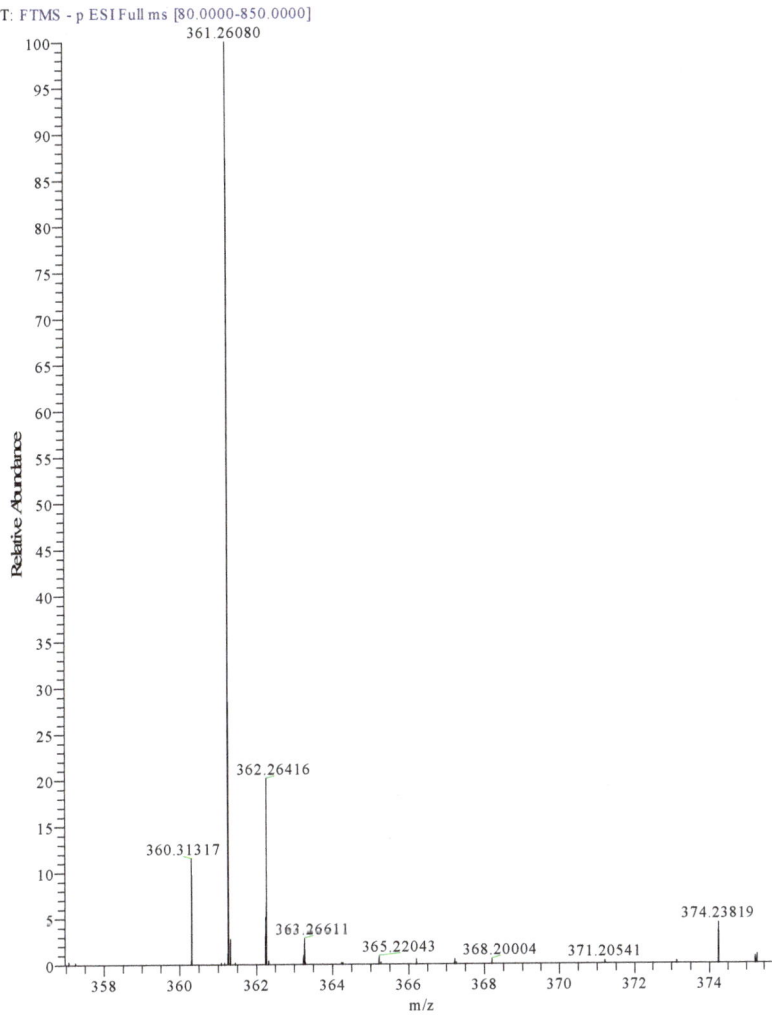

Figure 4. Diagnostic peak at 361.2608 from Arancio Luce G.

- Assignment of "Giallo Luce Solido 2G"

Preliminary research on both the nomenclature and Raman spectra agreed that it was highly likely that that "Giallo Luce Solido 2G" powder was likely to be the dye Acid Yellow 11. From the interpretation of the Raman spectrum, whilst there were some spectral similarities between the fiber sample and the powder sample, several of the peaks did not correspond strongly [9]. Preliminary analyses considered that this may be due to the dyeing process, but also introduced the possibility that the fiber and powder may have different molecular structures despite sharing the same name.

Powder: The mass spectrometry analysis of the powder sample corroborated the prediction that the dye was Acid Yellow 11. The mass spectrometry results indicated the presence of a very intense peak corresponding to m/z 357.0572 between retention times 2.59–2.99 min (Figure S2). The m/z corresponds to the following species: [M-Na]-, where M is the mass of Acid Yellow 11 and is equal to 380.0555u. The spectrum of the "Giallo Luce Solido 2G" powder was therefore considered highly likely to be Acid Yellow 11 [38].

Fiber: For the fiber species, there is no significant peak present corresponding to the m/z 357.0572. This agrees with the preliminary Raman data in which the spectra did not appear to completely match with the powder spectrum despite some similarities [9]. This indicates that the powder and fiber samples named "Giallo Luce Solido 2G" are different molecular species despite having the same commercial names.

The fiber sample did however present a very intense chromatographic peak corresponding to an m/z 417.3234 at retention time 3.88 min. From this m/z, it is possible to draw several interpretations, but if it is considered likely that the fiber is an azoic acid (which the similarities in the Raman spectrum to the powdered sample would point towards) and that the counter-ion is likely to be Na+, some projections for possible masses can be put forward: the molecule, if singly charged, is characterized by a m/z ratio likely representative of [M-Na]-, and the molecular mass would likely be to be 440.3132 u. Molecular weight searches into the projected mass were performed but no yellow dyes were found to correspond.

- Assignment of "Giallo Novamina 2G"

Three molecules were proposed as possible identifications of the dye "Giallo Novamina 2G" from preliminary research on nomenclature: Acid Yellow 61, Acid Yellow 39 and Acid Yellow 25. Whilst no reference Raman spectra were available for Acid Yellow 61 and 39 for comparison, the Raman analysis on an analytical standard of Acid Yellow 25 performed in the laboratory was found to correspond strongly to the peaks of the "Giallo Novamina 2G" Raman spectrum.

Upon mass spectrometric analysis, an intense chromatographic peak was observed corresponding to m/z 526.0876 at retention time 2.86 min, which is exactly as observed for the Acid Yellow 25 analytical standard. The m/z corresponds to the following species: [M-Na]-, where M is the mass of Acid Yellow 25 with a sodium counter-ion and is equal to 549.0774u.

The chemical structure of "Giallo Novamina 2G" is therefore understood as highly likely to be that of Acid Yellow 25 [39].

- Assignment of "Rosso Amidonaftolo 2G"

The predictions proposed by the preliminary research on both the nomenclature of the dye "Rosso Amidonaftolo 2G" and the comparison of the Raman spectra obtained meant that Red 2G was predicted as a likely candidate for the identification of the molecular structure of both the powder and fiber samples.

This identification was corroborated by the presence of an intense chromatographic peak corresponding to the m/z 464.0233 at retention times 2.17 min for the powder (Figure S3) and 2.24 min for the fiber, which are close to those observed for the Red 2G standard. Furthermore, another the peak at m/z 358.9780 was observed, as reported in the literature [40].

Both the powder and fiber samples of Rosso Amidonaftolo 2G are hence identified as highly likely to be Red 2G [41].

- Assignment of "Tartrazina J"

The preliminary predictions for "Tartrazina J" were tartrazine (based on nomenclature; however, significant spectral differences were observed between the Raman spectra for these compounds) and Acid Yellow 17 (based on a very strongly similar Raman spectrum) [9,42].

For Acid Yellow 17, the expected m/z peak is likely to exist at m/z 251.6546, corresponding to the species [M-2Na]$^{2-}$, where M is the mass of Acid Yellow 17 and is equal to

551.2888u. This peak was not present in the spectrum of Tartrazina J, and it was therefore inferred that the dye molecule is unlikely to correspond to Acid Yellow 17.

For tartrazine, there could be significant problems in detection with mass analysis owing to a triple charge on a very small molecule and the existence of the molecule in several states. In trials undertaken as part of this work, even an analytical standard of tartrazine was could not be detected in targeted analysis, so it is unlikely that it would be possible to detect the species in untargeted analysis—which is less sensitive for specific compounds. Alongside this, the Raman spectrum of "Tartrazina J" showed some significant spectral peaks that did not correspond to the peaks observed in the analytical standard of tartrazine.

However, one of the recorded m/z peaks in the literature for tartrazine is m/z 233.1, owing to the following species: $[M-3Na+H]^+$ [43], and this peak was present in the mass spectrum of "Tartrazina J", alongside another equally intense peak at m/z 228.9509 (Figure S4). Due to problems with analyzing Tartrazine, it was not possible to conclude whether tartrazine may have been present in the sample, but if it is present, it is possible that the dyed fiber may contain a mixture of dyes, which would account for the extra peaks on the Raman spectrum [9].

- Assignment of "Rosso Italana B"

Preliminary research on the nomenclature of the "Italana" dyes yielded no results. Raman spectral comparisons with databases indicated possible correlations with structures similar to Acid Red 26 [9]. Upon corroboration with the mass spectral data, however, this possible attribution was found to be unlikely, as the following predictions were made based on the only diagnostic peak that appeared on the spectra, which had an m/z of 236.9884 and a retention time of 3.36 min. If the species is a singly charged azoic dye, the m/z is likely to represent $[M-Na]^-$, where M would represent the molecular mass and be equal to 259.9782 u. If the species is a doubly charged azoic dye, the m/z is likely to represent $[M-2Na]^{2-}$, where M would represent the molecular mass and be equal to 519.9564u.

Molecular weight searches into the two projected masses were performed but no red dyes were found to correspond.

- Assignment of "Rosso Luce Solido BL" and "Rosso Italana R"

For both "Rosso Luce Solido BL" and "Rosso Italana R", m/z peaks corresponding to the major peaks in the chromatogram were all present in a wide range of the spectra acquired from the whole set of the museum's dyes and were hence not considered to be indicative of the dye compounds present in the samples. It is possible that, upon further analysis of the chromatograms in the laboratory, other chromatographic peaks may be identified, with more diagnostic m/z values. For "Rosso Italana R" in particular, a very broad peak was acquired from 1.45–2.41 min, which unfortunately includes the retention time of an extremely intense peak at m/z 360.3130, which appears in every spectrum. It is therefore likely that if the diagnostic peak elutes within this range that its signal may be overwhelmed by the intensity of the m/z 360.3130 peak in the mass spectrum and therefore not be visible. For both "Rosso Luce Solido BL" and "Rosso Italana R", the tentative data obtained from preliminary predictions were not adequate for making any informed decision about the identification of dyes without a defined mass peak.

- Assignment of "Rosso Naftolo SJ"

Naphthol reds are a very common and significantly varied range of azoic acid dyes, so the name "Rosso Naftolo SJ" is not particularly indicative for obtaining preliminary predictions from. Unfortunately, no preliminary predictions were possible based on Raman comparisons [9].

This meant that the interpretation of the mass spectral data was approached only with the understanding that the dye was likely to be a species containing a naphthalene group. For interpretation of the chromatogram, the only intense peak considered likely to be indicative of the molecular structure of the compound was the peak recorded at a m/z

of 200.9741 and retention time of 3.08 min. This was therefore used to make a tentative projection for the possible mass of the molecule as follows: if the species is a singly charged azoic dye, the m/z is likely to represent [M-Na]$^-$, where M would represent the molecular mass and be equal to 223.9308u; if the species is a doubly charged azoic dye, the m/z is likely to represent [M-2Na]$^{2-}$, where M would represent the molecular mass and be equal to 447.8616u.

Molecular weight searches into the two projected masses were performed but no red dyes were found to correspond. The authors contend that due to the likelihood that the dye molecule contains a naphthalene group (as suggested by the name) as well as at least one sulfonate group and one azo group (as is the case for the other dyes identified from the sample set), it is suspected that a molecule of mass 223.9308u is unlikely to correspond as these components have a cumulative mass > 223.9308u.

- Assignment of "Rosso Novamina 2G"

Preliminary research on the nomenclature of "Rosso Novamina 2G" found two related azoic acid dye species—however, a historical document [44] strongly indicated that "Rosso Novamina 2G" is likely to be Acid Orange 19 [45]. No Raman spectra were available in the literature for the standard for spectral comparisons to be performed, and the Raman spectrum of "Rosso Novamina 2G" was also very strongly affected by the signals of the wool compared to the other red dyes, meaning that only very few peaks were visible.

As such, the prediction was made solely on the nomenclature for this sample and it was predicted that if the sample corresponded to Acid Orange 19, which has a mass of 519.0535u, then the major m/z peak in the mass spectral data should correspond to the following species: [M-Na]$^-$ = 496.0655. Indeed, an intense peak corresponding to this m/z was observed at a retention time of 2.88 min (Figure S5).

It is hence proposed that the identification of the dye "Rosso Novamina 2G" is likely to be Acid Orange 19.

- Assignment of "SEII Azoico Acido Pag"

For "SEII Azoico Acido Pag", unlike the other samples, the label on the glass jar was simply handwritten. Due to this (alongside the fact that the naming format was very different to the others in the set), it was deemed likely that the label may have simply corresponded to an internal sample management system and not the commercial dye nomenclature. Research on the name found no bibliographic references, and a Raman spectrum of this sample was not obtained due to the spectrum being overwhelmed by fluorescence.

This meant that the mass spectrometry data were observed without any prior knowledge or corroborative data about the possible molecular structure and hence were difficult to interpret. There were several very high-intensity m/z peaks observed in the data. A fairly intense chromatographic peak was observed corresponding to the m/z 231.5072 associated with Red 2G at a retention time of 2.14 (as observed with the standard), as well as intense peaks at m/z 217.01257, 253.1350, 327.0452, 355.0767 and 577.4851. The peak at m/z 327.0452—recorded at a retention time of 2.68 min—was by far the highest peak observed; however, all peaks were of a significant magnitude, indicating that it is possible that the sample is a mixture. This peak showed two diagnostic fragments at m/z 170.9987 and 155.98758 (Figure 5); for this reason, taking into account the literature data [46], it was identified as Acid Orange 7 [47].

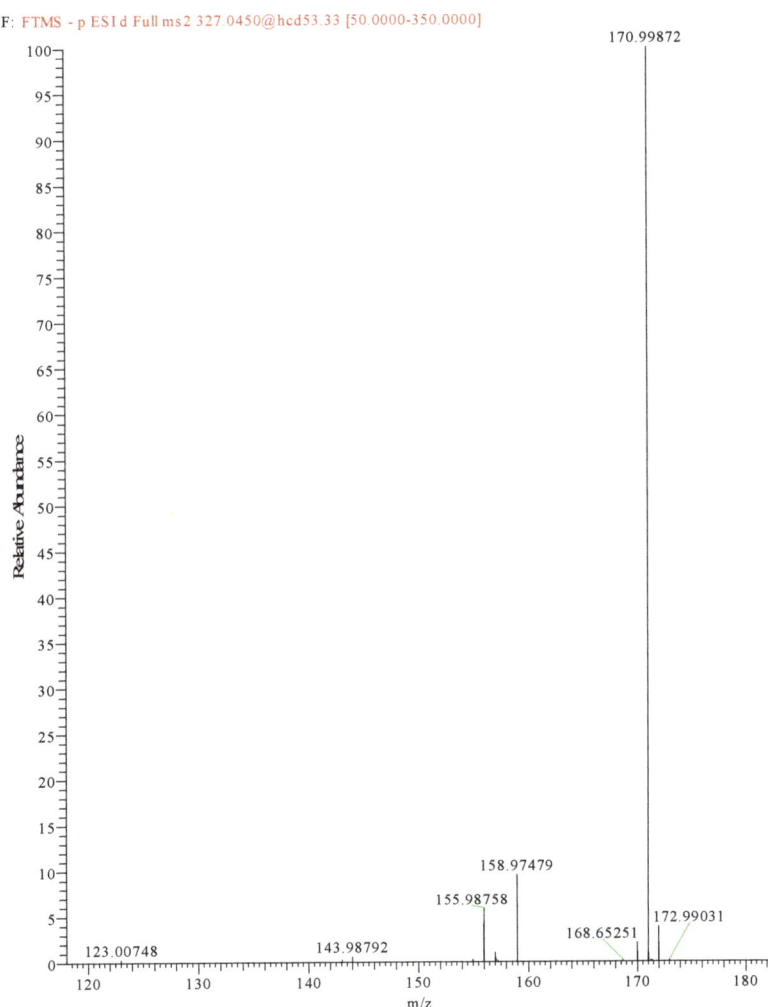

Figure 5. Fragmentation of peak at m/z 327.0450, 2.73 min. The two diagnostic fragments at m/z 170.9987 and 155.9875 are visible.

3. Materials and Methods

3.1. Solvents and Reagents

High-purity analytical standards of Congo Red and Red 2G were purchased from Sigma Aldrich. A $\geq 40\%$ purity standard of Acid Yellow 25 was also purchased from Sigma Aldrich. Solvents, acids and bases were purchased from Sigma Aldrich and used without further purification. $Na_2EDTA \cdot 2H_2O$ was purchased from Carlo Erba while TBAB and other salts were purchased from Sigma Aldrich.

3.2. Development of Clean Up for Synthetic Dyes: dLLME

Development of the cleanup protocol was performed on a mixed reference sample containing 500 ppb of Acid Yellow 25, Congo Red, and Red 2G. The analytes were dissolved to 500 ppm and diluted to 100 ppm in Millipore water and the final dilution to 500 ppb was then performed in methanol. All experiments were repeated three times and each replicate was analyzed twice using mass spectrometry to obtain an average.

3.2.1. Evaluation of Disperser for Synthetic Dyes

The dLLME method was adapted from a protocol outlined by Faraji et al. [24] for the determination of azo dyes from ice cream samples. Chloroform was used as the extracting solvent for the tests and methanol, isopropanol, acetonitrile and acetone were all trialed as dispersers.

The disperser trials were carried out as follows: 500 µL of 100 ppb mixed standard was placed together in a tube. The sample was then made up to 5 mL with Millipore water, and 150 µL previously made up 2M tetra-n-butylammonium bromide (TBAB) in water was added as the ion pair reagent. The tube was lightly swirled to ensure homogeneity. Quantities of 750 µL of the desired disperser and 100 µL of chloroform were then drawn up into a syringe and injected rapidly into the aqueous phase, forming a cloudy solution. The mixture was then vortexed for 10 s and sonicated for 10 min before undergoing 5 min of centrifugation at 4200 rpm. The bottom organic layer was removed with a syringe and placed in a vial where it was dried under N_2 flow. The extract was reconstituted with 100 µL methanol for analysis using HPLC-MS.

3.2.2. Trial of the Complete Analytical Protocol for Synthetic Dyes

After the dLLME procedure was set up, the chosen method was paired with the initial extraction protocol to ensure functionality of the whole protocol when joined together. The initial extraction method was exactly as described in the literature for natural dyes in Serafini et al. 2017 [21]. After that, the dLLME protocol was applied, employing methanol as the disperser, as described above.

3.2.3. Analysis of Historical Textile and Powder Samples

The fiber samples were extracted using the method described above. Specifically, the fiber sample was placed in a vial containing 4.4 mg NaCl, 0.8 mL 30% NH_3 and 0.8 mL 1 mM Na_2EDTA. The samples were left in the extraction mixture and covered in aluminum foil for 2 days and left to extract at room temperature. The solution was then pipetted out. The sample was then placed under N_2 flow to facilitate the evaporation of the ammonia, and this was performed until a neutral solution was obtained.

The neutral solution was then placed in a tube and made up to 5 mL with Millipore water to which 150 µL 2M TBAB was added. The tube was lightly swirled to ensure homogeneity. Quantities of 750 µL methanol and 100 µL chloroform were then drawn up into a syringe and injected rapidly into the aqueous phase, forming a cloudy solution. The mixture was then vortexed and sonicated for 10 min before undergoing 5 min of centrifugation at 4200 rpm. The bottom organic layer was removed and placed in vial, where it was dried under N_2 flow.

For the powder samples, it was not necessary to perform the initial extraction step and therefore the samples were dissolved in 5 ml water in a tube and the dLLME protocol was then performed directly on the sample.

HPLC-HRMS was then carried out using untargeted analysis by following the instrumental setup reported below.

3.3. HPLC-MS Analyses

3.3.1. Targeted HPLC-MS Analyses

Recovery analysis during method development was carried out using the results from targeted HPLC-MS. For the chromatographic analysis, a Series 200 Perkin Elmer micro-LC system equipped with an autosampler was used. The system was coupled to a PE-Sciex API 2000 triple quadrupole mass spectrometer equipped with a TurboIon-Spray ionization source, operating in negative ionization mode. The columns tested were a Kinetex XB-C18 2.6 µm core–shell particle column and a Luna-C18 5 µm column, and the Luna-C18 was chosen. The system was used in SIM. To individuate the best chromatographic conditions for synthetic dyes, three trials were performed on the mobile phases:

i. Phase A: 0.1% formic acid in acetonitrile; Phase B: 0.1% formic acid in water;

ii. Phase A: 5 mM ammonium acetate in acetonitrile; Phase B: 5 mM ammonium acetate in water;
iii. Phase A: methanol; Phase B: 5 mM ammonium acetate in water.

The mobile phases chosen were methanol and 5 mM ammonium acetate in Millipore water in a gradient elution, as shown in Table 2.

Table 2. Gradient elution for HPLC-MS analyses.

Time (min)	Phase A (MeOH)	Phase B (5 mM Ammonium Acetate in H_2O)
0.0	0%	100%
3.0	100%	0%
4.5	100%	0%
5.0	0%	100%
7.0	0%	100%

The SIM targets are listed below, in Table 3:

Table 3. Synthetic dye standard mass spectrometry targets.

Dye	Exact Mass	Parent Ion (m/z)	DP (V)	FP (V)	EP (V)	CEP (V)
Acid Yellow 25	549.5537	526.2	−101	−140	−9	−29.66
Congo Red	696.6622	325.3	−38	−365	−9	−24.79
Red 2G	509.4200	231.5	−15	−326	−7	−22.52

Recoveries were calculated by performing peak area integrations with Sciex Analyst software, and these areas were compared to results from the original standards in Microsoft Excel.

3.3.2. Untargeted HPLC-MS Analyses

Untargeted mass spectrometric data of the unknown museum samples were acquired using a Thermo Fisher Scientific DionexTM UltiMateTM 3000 (RSLC) UHPLC system equipped with an RS autosampler and coupled with a high-resolution Q-Exactive Orbitrap mass spectrometer equipped with a heated electrospray ionization source (H-ESI).

The H-ESI source operated in negative ionization mode with tuning parameters set at sheath gas flow rate (nitrogen) = 45 units, auxiliary gas flow rate (nitrogen) = 20 units, spray voltage = −3.00 kV, capillary temperature = 350 °C, source temperature = 350 °C. MS experiments were carried out in full scan–data dependent acquisition mode (Full-dds). A full scan was conducted with a scan range between 100 and 800 m/z with a resolution of 70,000 FWHM; automatic gain control (AGC) was 1×10^6, maximum injection time was 100 ms. For MS/MS experiments, resolution was 17,500 FWHM, AGC was 5×10^6, maximum injection time was 80 ms, loop count and TopN were 5, isolation window was set to 2.0 m/z, the fixed first mass was 50 m/z. Minimum AGC target was 8×10^3 and the intensity threshold was 1×10^6. The dissociation of molecular ions was induced in a high-energy collision cell (HCD) by means of nitrogen; simultaneous experiments were conducted at three different normalized collision energies: 10, 30 and 50%.

The Luna-C18 column used for the setting of the methods and the mobile phase conditions individuated for the method development were used for analysis on historical samples.

3.4. Sampling from ACNA Collection

As mentioned in the introduction, the collection partially studied in this work comes from the Museum of Chemistry, located within Sapienza University of Rome's Department of Chemistry, which holds an extensive collection of early synthetic dyes from several different dye companies. ACNA industry, an acronym for Azienda Coloranti Nazionali e Affini (ACNA), was an Italian chemical company active from 1882 until 1999 [6]. The company was extremely controversial throughout the entirety of its history. It was first founded as an explosives factory that operated under the name Dinamitificio Barbieri and subsequently the Italian Society of Explosive Products. The company later moved away from the production of explosives and retooled as a colorant manufacturer after being acquired by Italgas in 1925. The ACNA received significant investment from the fascist regime in the years immediately following 1925, to promote Italian manufacturing industries, and its name was once again changed, this time to the Associated National Chemistry Companies—at which point it obtained the acronym ACNA. It was then acquired by the larger companies IG Farben and Montecatini, who gave it its final name: Azienda Coloranti Nazionali e Affini [48]. Under these companies, the ACNA manufactured colorants, working until 1999, during which time their production caused extensive pollution of the surrounding areas, inflicting severe damage to both the environment and the health of nearby residents [48].

Sampling of the ACNA Museum's dye collection was carried out in July 2020 after a preliminary viewing of the collection. The preliminary visit was used to take photographs and note commercial names of dyes in the collection, which were then researched in the subsequent months to identify possible assignments for the dyes based on the nomenclature. Fibers (less than 1 mg) and powdered samples (about 1 mg) were then selected as the case study for this research.

4. Conclusions

The primary aim of this work was the development and application of a novel extraction strategy, based on the ammonia–EDTA protocol, adding a novel cleanup protocol for the purification and enrichment of the analytes for the analysis of synthetic dyes from historical and artistic matrices. As a secondary aim, the protocol developed for synthetic dyes was applied to a real case study of historical synthetic dyes from the Museum of Chemistry, Sapienza University to verify the effectiveness of this protocol in historical sample analyses. This research adapted a methodology from the analytical science of food and successfully developed a protocol well-suited to cultural heritage. For the dLLME protocol, the best results were obtained with the use of methanol as the disperser solvent with chloroform as the extracting solvent.

One of the main benefits of this application is that it allows for synthetic dyes to be extracted alongside natural dyes in a single extraction step, with the ammonia extraction protocol, and the best-suited dLLME protocol can be employed to purify the analytes. In this way, both types of dye now only require the acquisition of a single sample instead of two, minimizing the destructiveness of analysis.

For application to synthetic dye samples from the ACNA industrial collection, the application of the novel analytical protocol for the extraction and preconcentration of synthetic dyes was effective for this research, and it can therefore be considered a significant contribution to the study of synthetically dyed artefacts. Furthermore, the results obtained confirmed the importance of scientific study in improving understanding of naming conventions in the context of the synthetic dye collections. In ACNA dyes, particularly interesting was the case of "Giallo Italana 2G", where it appears that, despite being labelled with the same name and being from the same company, the dyed fiber and powdered sample are likely to have different chemical structures. In contrast, in "Rosso Amidonaftolo 2G", the chemical structure was consistent between the powder sample and the dyed wool. This is a strong exemplification of the complexity of the study of synthetic dye collections—where, even within a single company, naming conventions can vary widely.

The case study also perfectly illustrates the benefits and necessity of a multi-analytical approach to the analysis of unknown artefacts. In some cases, Raman spectroscopy was highly indicative of the likely assignment of the structure, but when compared with the mass spectral results, the preliminary assignment was shown to be incorrect. This was illustrated with the "Tartrazina J" sample, which appeared strongly similar to the spectrum for Acid Yellow 17, but the expected m/z was not observed in the mass spectrum. Contrastingly, for some samples, the indicative nature of the Raman meant that mass spectra could be rapidly interpreted to corroborate the data, for example, with "Rosso Amidonaftolo 2G". In other cases, such as for "Arancio Luce G", the preliminary data were not indicative of a particular compound but could be used to corroborate the data obtained from the mass spectrum.

Moreover, the preliminary investigations using Raman analyses once again demonstrated the importance of a multi-analytical approach, which meant that several samples from the collection could be analyzed with only a very small textile sample.

It is also hoped that further studies will be conducted in the future, and hopefully provide more indicative data for the unidentified samples, thus implementing the database of such dyes.

Supplementary Materials: The following supporting information can be downloaded at: https://www.mdpi.com/article/10.3390/molecules28145331/s1, Figure S1: Diagnostic peak at 236.9883 from Giallo Italana 2G; Figure S2: Chromatograms of Giallo luce solido 2G, powder, with the diagnostic peak at 357.0572 at 2.59–2.99 min; Figure S3: Diagnostic peak at 464.0233 from Rosso Amidonaftolo-powder. Figure S4. Chromatograms of Tartrazine J, fiber; Figure S5: Chromatograms of Rosso Novamina 2G, fiber, with the diagnostic peak at m/z 496.0655, 2.88 min.

Author Contributions: Conceptualization, I.S., K.R.M., F.V. and C.M.; methodology, I.S., F.V. and A.C.; validation, K.R.M., A.B. and G.P.; formal analysis, F.V., I.S., K.R.M., A.B., A.C. and G.P.; investigation, K.R.M., A.C., F.V. and I.S.; resources, A.C., G.F. and I.S.; data curation, I.S., F.V. and K.R.M.; writing—original draft preparation, K.R.M. and I.S.; writing—review and editing, F.V. and I.S.; visualization, K.R.M.; supervision, C.M., M.S., G.F. and R.C.; project administration, G.F. and R.C. All authors have read and agreed to the published version of the manuscript.

Funding: This research received no external funding.

Institutional Review Board Statement: Not applicable.

Informed Consent Statement: Not applicable.

Data Availability Statement: Not applicable.

Acknowledgments: The authors would like to thank Vittoria Primiceri and Andrea Zanotti from the ancient RICO-ACNA-Research Center of Coulorants, and Alessandro Bacaloni and Donato Monti, past and present directors, respectively, of the Museum of Chemistry, located within Sapienza University of Rome's Department of Chemistry, for providing access to the collection, and their continuous collaboration to the research.

Conflicts of Interest: The authors declare no conflict of interest.

References

1. Holme, I. Sir William Henry Perkin: A review of his life, work and legacy. *Color. Technol.* **2006**, *122*, 235–251. [CrossRef]
2. Mcclure, K.R. Development of New Extraction Methods for Analysis of Natural and Synthetic Organic Colourants from Historical and Artistic Matrices. Master's Thesis, Sapienza University of Rome, Rome, Italy, 2020.
3. Liu, J.; Zhou, Y.; Zhao, F.; Peng, Z.; Wang, S. Identification of early synthetic dyes in historical Chinese textiles of the late nineteenth century by high-performance liquid chromatography coupled with diode array detection and mass spectrometry. *Color. Technol.* **2016**, *132*, 177–185. [CrossRef]
4. Chequer, F.D.; de Oliveira, G.A.; Ferraz, E.A.; Cardoso, J.C.; Zanoni, M.B.; de Oliveira, D.P. Textile dyes: Dyeing process and environmental impact. *Eco-Friendly Text. Dye. Finish.* **2013**, *6*, 151–176. [CrossRef]
5. Shahid, M.; Wertz, J.; Degano, I.; Aceto, M.; Khan, M.I.; Quye, A. Analytical methods for determination of anthraquinone dyes in historical textiles: A review. *Anal. Chim. Acta* **2019**, *1083*, 58–87. [CrossRef] [PubMed]
6. Khatri, A.; Peerzada, M.H.; Mohsin, M.; White, M. A review on developments in dyeing cotton fabrics with reactive dyes for reducing effluent pollution. *J. Clean. Prod.* **2015**, *87*, 50–57. [CrossRef]
7. Colombo, S. American Institute of Chemical Engineers. Global Outlook: Innovation and Italy's Chemicals Industries. 2014. Available online: https://www.aiche.org/sites/default/files/cep/20140454_2.pdf (accessed on 1 November 2020).

8. Souto, C.S. Analysis of Early Synthetic Dyes with HPLC-DAD-MS: An Important Database for Analysis of Colorants Used in Cultural Heritage. Ph.D. Thesis, Faculdade de Ciências e Tecnologia Department of Conservation and Restoration, Universidade Nova de Lisboa, Lisbon, Portugal, 2010.
9. Ciccola, A.; McClure, K.R.; Serafini, I.; Vincenti, F.; Montesano, C.; Gentili, A.; Curini, R.; Favero, G.; Postorino, P. The XXth Century and its new colours: Investigating the molecular structures of historical synthetic dyes using Raman spectroscopy, submitted to Journal of Raman Spectroscopy, Rome, Italy, 2023. Available online: https://www.mdpi.com/journal/molecules/instructions#references (accessed on 1 November 2020).
10. Cooksey, C.J. Quirks of dye nomenclature. 2. Congo red. *Biotech. Histochem.* **2014**, *89*, 384–387. [CrossRef]
11. Fremout, W.; Saverwyns, S. Identification of synthetic organic pigments: The role of a comprehensive digital Raman spectral library. *J. Raman Spectrosc.* **2012**, *43*, 1536–1544. [CrossRef]
12. Smith, G.D.; Esson, J.M.; Chen, V.J.; Hanson, R.M. Forensic dye analysis in cultural heritage: Unraveling the authenticity of the earliest Persian knotted-pile silk carpet. *Forensic. Sci. Int. Synergy* **2021**, *3*, 100130. [CrossRef]
13. Leona, M.; Decuzzi, P.; Kubic, T.A.; Gates, G.; Lombardi, J.R. Nondestructive identification of natural and synthetic organic colorants in works of art by surface enhanced Raman scattering. *Anal. Chem.* **2011**, *83*, 3990–3993. [CrossRef] [PubMed]
14. Sessa, C.; Weiss, R.; Niessner, R.; Ivleva, N.P.; Stege, H. Towards a Surface Enhanced Raman Scattering (SERS) spectra database for synthetic organic colourants in cultural heritage. The effect of using different metal substrates on the spectra. *Microchem. J.* **2018**, *138*, 209–225. [CrossRef]
15. Pirok, B.W.; Moro, G.; Meekel, N.; Berbers, S.V.; Schoenmakers, P.J.; Van Bommel, M.R. Mapping degradation pathways of natural and synthetic dyes with LC-MS: Influence of solvent on degradation mechanisms. *J. Cult. Herit.* **2019**, *38*, 29–36. [CrossRef]
16. Weisz, A.; Lazo-Portugal, R.; Perez-Gonzalez, M.; Ridge, C.D. Identification, synthesis, and quantification of a novel carbazole derivative as an impurity in the anthraquinone dye Ext. D&C Violet No. 2 (Acid Violet 43). *Dyes Pigment.* **2022**, *200*, 110161.
17. Sandstrom, E.; Wyld, H.; Mackay, C.L.; Troalen, L.G. An optimised small-scale sample preparation workflow for historical dye analysis using UHPLC-PDA applied to Scottish and English Renaissance embroidery. *Anal. Methods* **2021**, *13*, 4220–4227. [CrossRef]
18. Serafini, I.; Lombardi, L.; Fasolato, C.; Sergi, M.; Di Ottavio, F.; Sciubba, F.; Montesano, C.; Guiso, M.; Costanza, R.; Nucci, L.; et al. A new multi analytical approach for the identification of synthetic and natural dyes mixtures. The case of orcein-mauveine mixture in a historical dress of a Sicilian noblewoman of nineteenth century. *Nat. Prod. Res.* **2019**, *33*, 1040–1051. [CrossRef]
19. Chen, V.J.; Smith, G.D.; Holden, A.; Paydar, N.; Kiefer, K. Chemical analysis of dyes on an Uzbek ceremonial coat: Objective evidence for artifact dating and the chemistry of early synthetic dyes. *Dyes Pigment.* **2016**, *131*, 320–332. [CrossRef]
20. Lombardi, L.; Serafini, I.; Guiso, M.; Sciubba, F.; Bianco, A. A new approach to the mild extraction of madder dyes from lake and textile. *Microchem. J.* **2016**, *126*, 373–380. [CrossRef]
21. Serafini, I.; Lombardi, L.; Vannutelli, G.; Montesano, C.; Sciubba, F.; Guiso, M.; Curini, R.; Bianco, A. How the extraction method could be crucial in the characterization of natural dyes from dyed yarns and lake pigments: The case of American and Armenian cochineal dyes, extracted through the new ammonia-EDTA method. *Microchem. J.* **2017**, *134*, 237–245. [CrossRef]
22. Christie, R. *Colour Chemistry*, 2nd ed.; Royal Society of Chemistry: London, UK, 2014.
23. Piccinini, P.; Senaldi, C.; Buriova, E. *European Survey on the Presence of Banned Azodyes in Textiles*; JRC Scientific and Technical Reports; JRC European Commission: Brussels, Belgium, 2008. Available online: https://publications.jrc.ec.europa.eu/repository/handle/JRC44198 (accessed on 1 July 2023).
24. Faraji, M.; Nasiri Sahneh, B.; Javanshir, R. An ion-pair dispersive liquid-liquid microextraction for simultaneous determination of synthetic dyes in ice cream samples by HPLC. *Anal. Bioanal. Chem. Res.* **2017**, *4*, 213–225.
25. Unsal, Y.E.; Tuzen, M.; Soylak, M. Spectrophotometric determination of Sudan Blue II in environmental samples after dispersive liquid-liquid microextraction. *Quim. Nova* **2014**, *37*, 1128–1131. [CrossRef]
26. Long, C.; Mai, Z.; Yang, X.; Zhu, B.; Xu, X.; Huang, X.; Zou, X. A new liquid-liquid extraction method for determination of 6 azo-dyes in chilli products by high-performance liquid chromatography. *Food Chem.* **2011**, *126*, 1324–1329. [CrossRef]
27. Germinario, G.; Ciccola, A.; Serafini, I.; Ruggiero, L.; Sbroscia, M.; Vincenti, F.; Fasolato, C.; Curini, R.; Ioele, M.; Postorino, P.; et al. Gel substrates and ammonia-EDTA extraction solution: A new non-destructive combined approach for the identification of anthraquinone dyes from wool textiles. *Microchem. J.* **2020**, *155*, 104780. [CrossRef]
28. Hunger, K.; Mischke, P.; Rieper, W.; Raue, R.; Kunde, K.; Engel, A. *Ullmann's Encyclopedia of Industrial Chemistry*; Wiley-VCH: Weinheim, Germany, 2000; Volume 4.
29. Serafini, I.; Bosi, A.; Vincenti, F.; Peruzzi, P.; McClure, K.R.; Ciccola, A.; Hamza, N.M.; Moricca, C.; Sadori, L.; Montesano, C.; et al. Overcoming the limit of in situ gel supported liquid microextraction: Development of the new InGeL-LC-MS analyses, a smart methodology for the identification of natural dyes from Tutankhamun tomb relics. *Anal. Chim. Acta* **2023**, to be resubmitted.
30. Rezaee, M.; Assadi, Y.; Hosseini, M.R.M.; Aghaee, E.; Ahmadi, F.; Berijani, S. Determination of organic compounds in water using dispersive liquid–liquid microextraction. *J. Chromatog A* **2006**, *1116*, 1–9. [CrossRef]
31. Yan, H.; Wang, H. Recent development and applications of dispersive liquid–liquid microextraction. *J. Chromatogr. A* **2013**, *1295*, 1–15. [CrossRef] [PubMed]
32. Vincenti, F.; Montesano, C.; Cellucci, L.; Gregori, A.; Fanti, F.; Compagnone, D.; Curini, R.; Sergi, M. Combination of pressurized liquid extraction with dispersive liquid liquid micro extraction for the determination of sixty drugs of abuse in hair. *J. Chromatogr. A* **2019**, *1605*, 360348. [CrossRef]

33. Quigley, A.; Cummins, W.; Connolly, D. Dispersive liquid-liquid microextraction in the analysis of milk and dairy products: A review. *J. Chem-NY* **2016**, *2016*, 4040165. [CrossRef]
34. El-Sheikh, A.H.; Al-Degs, Y.S. Spectrophotometric determination of food dyes in soft drinks by second order multivariate calibration of the absorbance spectra-pH data matrices. *Dyes Pigments* **2013**, *97*, 330–339. [CrossRef]
35. Brändström, A. Principles of phase-transfer catalysis by quaternary ammonium salts. In *Advances in Physical Organic Chemistry*; Academic Press: New York, NY, USA, 1977; Volume 15, pp. 267–330.
36. Colour Index, 2nd ed., The Society of Dyers and Colorists. Available online: https://www.worldcat.org/title/colour-index/oclc/1171036 (accessed on 2 November 2022).
37. World Dye Variety. Acid Orange 31. Available online: http://www.worlddyevariety.com/acid-dyes/acid-orange-31.html (accessed on 2 November 2020).
38. World Dye Variety. Acid Yellow 11. Available online: http://www.worlddyevariety.com/acid-dyes/acid-yellow-11.html (accessed on 2 November 2020).
39. PubChem. Compound Summary Acid Yellow 25. Available online: https://pubchem.ncbi.nlm.nih.gov/compound/Acid-Yellow-25 (accessed on 2 November 2020).
40. Martin, F.; Oberson, J.-M.; Meschiari, M.; Munari, C. Determination of 18 water-soluble artificial dyes by LC–MS in selected matrices. *Food Chem.* **2016**, *197*, 1249–1255. [CrossRef] [PubMed]
41. PubChem. Compound Summary Red 2G. Available online: https://pubchem.ncbi.nlm.nih.gov/compound/Acid-red-1 (accessed on 2 November 2020).
42. PubChem. Compound Summary Acid Yellow 17. Available online: https://pubchem.ncbi.nlm.nih.gov/compound/22842 (accessed on 2 November 2020).
43. Gao, H.-G.; Gong, W.-J.; Zhao, Y.-G. Rapid Method for quantification of seven Synthetic Pigments in Colored Chinese Steamed Buns using UFLC-MS/MS without SPE. *Anal. Sci.* **2015**, *31*, 205–210. [CrossRef]
44. Giustiniani, P.; Natta, G.; Mazzanti, G.; Crespi, G. Composizioni Poliofiniche Adatte Alla Produzione di Fibre Tessili Crystalline Aventi una Migliorata Affinita per Coloranti. 1959. Available online: http://www.giulionatta.it/pdf/brevetti/img157.pdf (accessed on 2 November 2020).
45. World Dye Variety. Acid Orange 19. Available online: http://www.worlddyevariety.com/acid-dyes/acid-orange-19.html (accessed on 2 November 2020).
46. Liu, X.; Yang, J.L.; Li, J.H.; Li, X.L.; Li, J.; Lu, X.Y.; Shen, J.Z.; Wang, Y.W.; Zhang, Z.H. Analysis of water-soluble azo dyes in soft drinks by high resolution UPLC–MS. *Food Addit. Contam—Chem. Anal. Control Expo. Risk Assess.* **2011**, *28*, 1315–1323. [CrossRef]
47. PubChem. Compound Summary Acid Orange 7. Available online: https://pubchem.ncbi.nlm.nih.gov/compound/Acid-orange-7 (accessed on 2 November 2020).
48. Scripo Museum. Aziende Chimiche Nazionali Associate. Available online: https://scripomuseum.com/aziende-chimiche-nazionali-associate/ (accessed on 2 November 2020).

Disclaimer/Publisher's Note: The statements, opinions and data contained in all publications are solely those of the individual author(s) and contributor(s) and not of MDPI and/or the editor(s). MDPI and/or the editor(s) disclaim responsibility for any injury to people or property resulting from any ideas, methods, instructions or products referred to in the content.

Article

New Advances in Dye Analyses: In Situ Gel-Supported Liquid Extraction from Paint Layers and Textiles for SERS and HPLC-MS/MS Identification

Adele Bosi [1,2], Greta Peruzzi [3], Alessandro Ciccola [1,*], Ilaria Serafini [1,*], Flaminia Vincenti [1], Camilla Montesano [1], Paolo Postorino [4], Manuel Sergi [1], Gabriele Favero [5] and Roberta Curini [1]

1 Department of Chemistry, Sapienza University of Rome, P. le Aldo Moro 5, 00185 Rome, Italy; adele.bosi@uniroma1.it (A.B.)
2 Department of Earth Sciences, Sapienza University of Rome, P. le Aldo Moro 5, 00185 Rome, Italy
3 Institute for Complex System, National Research Council, Sapienza University, Piazzale Aldo Moro 5, 00185 Rome, Italy; peruzzi.1957090@studenti.uniroma1.it
4 Department of Physics, Sapienza University of Rome, P. le Aldo Moro 5, 00185 Rome, Italy
5 Department of Environmental Biology, Sapienza University of Rome, P. le Aldo Moro 5, 00185 Rome, Italy
* Correspondence: alessandro.ciccola@uniroma1.it (A.C.); ilaria.serafini@uniroma1.it (I.S.)

Abstract: To date, it is still not possible to obtain exhaustive information about organic materials in cultural heritage without sampling. Nonetheless, when studying unique objects with invaluable artistic or historical significance, preserving their integrity is a priority. In particular, organic dye identification is of significant interest for history and conservation research, but it is still hindered by analytes' low concentration and poor fastness. In this work, a minimally invasive approach for dye identification is presented. The procedure is designed to accompany noninvasive analyses of inorganic substances for comprehensive studies of complex cultural heritage matrices, in compliance with their soundness. Liquid extraction of madder, turmeric, and indigo dyes was performed directly from paint layers and textiles. The extraction was supported by hydrogels, which themselves can undergo multitechnique analyses in the place of samples. After extraction, Ag colloid pastes were applied on the gels for SERS analyses, allowing for the identification of the three dyes. For the HPLC-MS/MS analyses, re-extraction of the dyes was followed by a clean-up step that was successfully applied on madder and turmeric. The colour change perceptivity after extraction was measured with colorimetry. The results showed ΔE values mostly below the upper limit of rigorous colour change, confirming the gentleness of the procedure.

Keywords: dyes; in situ extraction; hydrogels; SERS; dLLME; HPLC-MS/MS; colorimetry

1. Introduction

While sample reduction and procedure miniaturisation are generally desirable in analytical chemistry, the principle of minimal invasiveness is imperative when analysing cultural heritage and, where possible, completely noninvasive analyses are preferred.

Nonetheless, when sampling is forbidden the study of organic components is disadvantaged. Dyes, in particular, represent one of the most complex challenges because of their low concentration and tendency to fade. When they are in complex matrices, care must be taken not to completely lose information about their presence. Noninvasive techniques, like fibre optic reflectance spectroscopy (FORS) or fluorescence spectroscopy, were proved to be effective, in some instances, for dye identification in rather simple matrices [1–5], but they generally lack specificity. Because of their strong fluorescence, dyes are not suited for Raman spectroscopy identification. However, fluorescence can be quenched when signals are enhanced by means of metal nanoparticles (NPs), and very good dye spectra can be obtained by means of surface enhanced Raman spectroscopy (SERS) [6–10]. Nonetheless, the phenomenon occurs when the analyte interacts with metal NPs, which cannot be applied directly on the artifact and, thus, micro sampling is required anyways [6,11]. To overcome this problem, different kinds of solid substrates have been proposed over the last decade. For instance, cellulose films incorporated with NPs can be applied directly on the artifact's surface and, following SERS measurements, can be removed, therefore limiting the residues left behind [12].

Gels, on the other hand, can be loaded with extraction solvents and used to perform extractions directly from the artifacts. Gels can be prepared with NPs inside or covered with NPs after removal from the artifact, just before undergoing SERS analysis in the place of samples [13–18]. Hydrogels, in particular, because of their water retention and release properties, are appreciated for in situ extractions by means of aqueous solutions. Agar gel prepared with Ag NPs has repeatedly been used to obtain a Ag-agar gel support for SERS analyses [14–18]. A Ag-agar gel loaded with an extraction solution was applied on an artifact surface for dye extraction and then left to dry for SERS analyses. While drying, an agar gel shrinks consistently, promoting the close interaction of Ag colloids with the dyes, thus enhancing the SERS signals [19].

However, in spite of SERS efficacy, when it comes to dye analyses, it must be said that high-performance liquid chromatography (HPLC) coupled with diode array detector (DAD) and mass spectrometry (MS) specificity is unrivalled. Hence, when possible, the common procedure involves removing a paint sample larger than 200 µm, or 1–2 mm of thread [6], for the dye extraction from the matrix. But complications with dyes are yet to end; extraction is a very critical and debated step, and useful information can be lost even with a consistent sample in hand. This is because traditional extraction methods imply the use of HCl and high temperatures. These harsh conditions are needed to break the chemical linkage between dyes and their metal substrate. Nonetheless, this way the linkage between glycosylated dyes and their sugar moiety becomes broken, too. Even if dye identification is possible based exclusively on aglycones, glycosylated molecules can be extremely precious to obtain specific information [20]. Taking this into account, several alternatives to HCl have been proposed [21–26], from mild acids to organic solvents, mostly supported by high temperatures. In 2015, a completely new basic approach, based on ammonia, was proposed for anthraquinone dyes [27]. The extraction solution, composed of NH_3, Na_2EDTA, and NaCl, is able to break the linkage between dyes and their metal substrates, preserving the entire glycosylated molecules at ambient temperature. The resultant analytical composition of ammonia extracts matches the source composition reported in the literature [27]. In 2020, the ammonia–EDTA extraction method was combined with hydrogel-supported microextraction in an attempt to coalesce SERS with the HPLC-MS/MS identification of dyes for the first time [19]. This procedure was designed to obtain as much information as possible on the analytes while preserving their molecular pattern together with the artifact's soundness. The ammonia solution, being aqueous, is very suited to hydrogel extraction, and its ability to work at ambient temperature makes direct application on the object possible. After extraction from wool, anthraquinone dyes were re-extracted from the gel support, and an additional liquid–liquid extraction was performed to purify the analytes and eliminate Na_2EDTA and NaCl, both of which are not compatible with the HPLC-MS/MS system. To test the procedure, agar gel and Nanorestore Gel® HWR (high water retention), designed and produced by the Italian Center for Colloids and Surface Science (CSGI), were compared. The outcomes revealed that while SERS analyses provided good results, traditional liquid–liquid extraction was not able to recover the small amount of analyte extracted, and a dispersive liquid–liquid microextraction (dLLME) clean-up procedure was thus developed and validated for this purpose. The development and validation study, which comprehends the application of this methodology to archaeological textiles, is under submission to another journal by the same authors. During the presented work, the whole procedure was also adapted to paint layers and extended to additional dye classes. Three textile mock-ups were prepared dying wool using madder, indigo, and a direct dye: turmeric. Furthermore, madder lake pigment and indigo in powder were mixed with egg yolk to obtain tempera paint mock-ups. While the ammonia–EDTA extraction and dLLME clean-up procedure was used on madder and turmeric, indigo gels were imbibed into an aqueous solution of NaOH:Na$_2$S$_2$O$_4$ 1:2, able to reduce indigotin into its water-soluble form, leuco indigo [16,28] (Figure 1). Once in the hydrogel pores, leuco indigo re-oxides into indigotin when in contact with the atmosphere. A dLLME clean-up procedure was then tested to re-extract indigoids from the gels for the HPLC-MS/MS analyses. All details are listed in Section 3 (Materials and Methods).

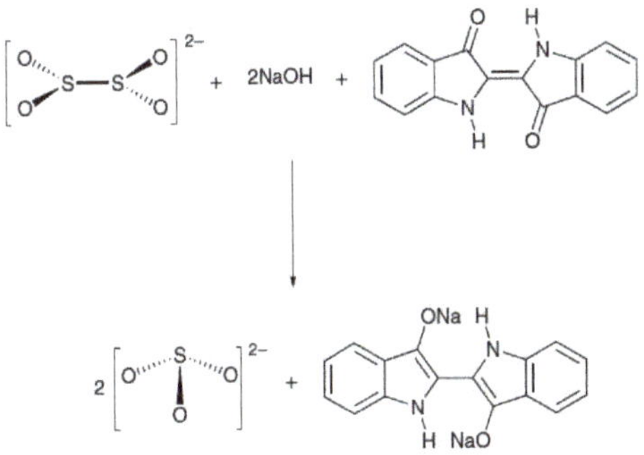

Figure 1. Indigo reduction reaction into leucoindigo using sodium hydrosulphite in alkaline solution: sodium hydrosulphite is oxidised into sodium sulphite (S^{III} → S^{IV}), releasing two electrons [28].

2. Results and Discussion

2.1. Extraction

2.1.1. Madder and Turmeric

The ammonia–EDTA extraction solution was visibly able to extract madder and turmeric dyes both from wool and paint layers. As already observed by Germinario et al. [19], agar gel was homogeneously coloured after extraction (Figure 2a), while for the Nanorestore Gel® HWR, the dyes appeared more concentrated on the contact surface (Figure 2b), in accordance with its high retention power.

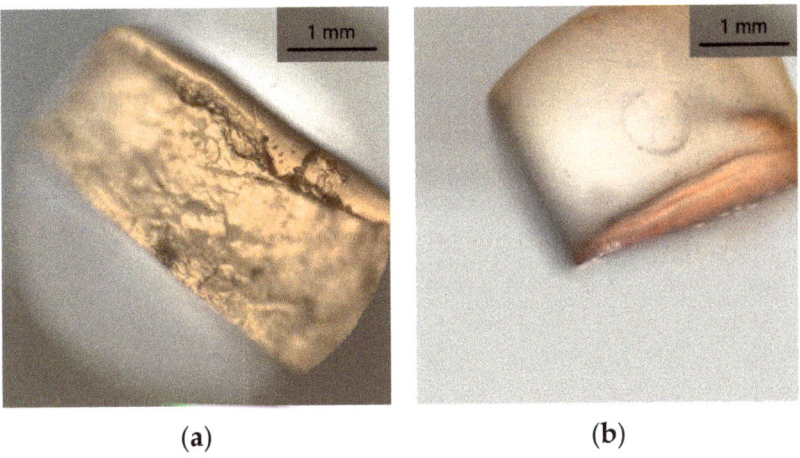

(a)   (b)

Figure 2. (a) Agar gel (3% w/v) after 3 h extraction from wool dyed with madder; (b) Nanorestore Gel® HWR after 3 h extraction from wool dyed with madder.

2.1.2. Indigo

Agar gel was extremely effective for indigoid dye extraction from both wool and paint mock-ups. The gel appeared greenish coloured right after extraction; however, after contact with the atmosphere, the oxidation of leucoindigo back to indigotin slowly turned the colour back to blue. Conversely, the Nanorestore Gel® HWR showed not to be compatible with the reducing solution. After soaking, the gel appeared altered by look and by touch. A substance absorbed from the solution or degradation product was observable right at the centre of the gel cylinder (Figure 3), and the Nanorestore Gel® HWR consistency was hardened.

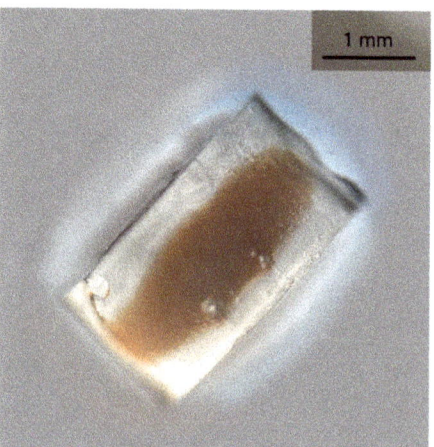

Figure 3. Nanorestore Gel® HWR after 90 min soaking in aqueous solution containing NaOH:Na$_2$S$_2$O$_4$ 1:2.

2.2. SERS

2.2.1. Madder and Turmeric

Spectra were acquired directly on the gels after the addition of Ag colloids. The colloids were made in pastes to prevent excessive absorption into the gel cylinders (see Section 3 (Materials and Methods)). The spectra recorded show SERS scattering peaks attributable to madder and turmeric dyes (Table 1). Peaks at approximately 1270–1280 and 1320 cm^{-1}, observed in the spectra of the agar gel after madder extraction from textile and paint mock-ups (Figure 4a), are diagnostic for the presence of alizarin and are attributed to the C-C stretching, H-C-C, and C-C-C bending modes of the anthraquinone ring [15–19,29–34].

Table 1. Main SERS scattering peaks observed in the spectra of agar gel (3% w/v) and Nanorestore Gel® HWR after extraction of madder and turmeric from wool and tempera mock-ups. The vibrational modes are reported as ν (stretching) and δ (bending).

Peaks from Madder				Attribution	Peaks from Turmeric		Attribution	Peaks from the Gel Matrix	
Agar Gel (3% w/v)		Nanorestore Gel® HWR			Agar Gel (3% w/v)	Nanorestore Gel® HWR		Agar Gel (3% w/v)	Nanorestore Gel® HWR
Tempera	Textile	Tempera	Textile		Textile	Textile			
1144 cm^{-1}	1155 cm^{-1}	1156 cm^{-1}	1160 cm^{-1}	ν C-C, δ C-H	1139 cm^{-1}		skeleton vibr.	838 cm^{-1} 1000 cm^{-1} 1034 cm^{-1}	839 cm^{-1} 999 cm^{-1} 1249 cm^{-1}
1270 cm^{-1}	1283 cm^{-1}	1290 cm^{-1}	1292 cm^{-1}	ν C-C, δ H-C-C	1166 cm^{-1}	-	skeleton vibr.	1268 cm^{-1} 1332 cm^{-1}	1321 cm^{-1} 1355 cm^{-1}
1320 cm^{-1}	1321 cm^{-1}	1324 cm^{-1}	1315 cm^{-1}	ν C-C, δ H-C-C	1290 cm^{-1}		δ C-C-C, C-C-H and C=CH	1383 cm^{-1} 1461 cm^{-1}	1422 cm^{-1} 1448 cm^{-1} 1760 cm^{-1}
1615 cm^{-1}	1614 cm^{-1}	1622 cm^{-1}	1611 cm^{-1}	ν C=O	1600 cm^{-1}		δ C-OH		

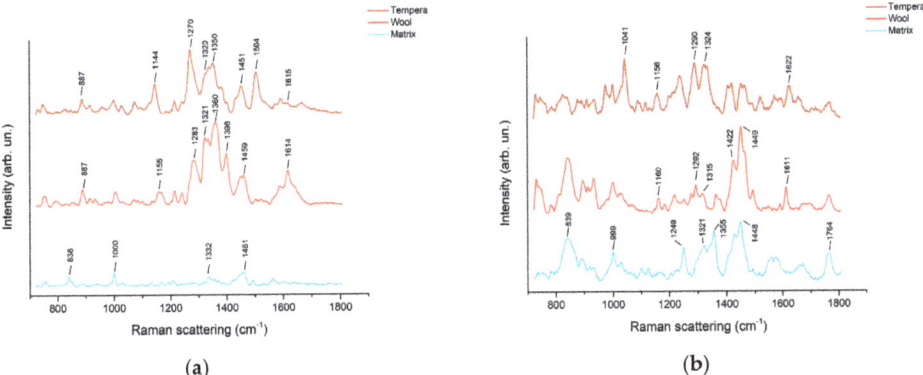

Figure 4. (a) SERS spectra of agar gel (3% w/v) after 15 min extraction of madder from tempera paint and 3 h extraction of madder from wool; (b) SERS spectra of Nanorestore Gel® HWR cylinders after 15 min extraction of madder from tempera paint and 3 h extraction of madder from wool.

The peak at approximately 1615 cm^{-1}, mainly visible in the spectrum from wool extraction, is attributed to the C=O stretching of the anthraquinone ring [14,15,29,30]. The peak at approximately 1150 cm^{-1} (1144 for tempera and 1155 cm^{-1} for wool) can be ascribed to the C-C stretching and C-H bending modes of the dye molecules [29]. Additional medium and weak intensity peaks between 800 and 1000 cm^{-1}, such as the one at 887 cm^{-1}, are attributed mainly to anthraquinone skeletal vibration [14,15,29]. The peaks at approximately 1450 cm^{-1} are generally attributed to alizarin C-O stretching and C-O-H and C-H bending [14,15,30]. The scattering band at 1398 cm^{-1}, on the other hand, is diagnostic for the presence of purpurin [30,34,35]. This peak is reproducibly more pronounced after extraction from wool than from tempera and could be due to the different preparation recipes. Additional peaks, such as the ones at approximately 1360 and 1505 cm^{-1}, could not be assigned to specific modes of anthraquinone dyes and can, hence, be related to the complex molecular pattern extractable from madder roots. Scattering peaks at approximately 1330–1350 cm^{-1}, however, were already observed in the agar gel [19,35], and its attribution, hence, remains uncertain.

Madder spectra acquired on Nanorestore Gel® HWR (Figure 4b) similarly exhibit peaks at approximately 1155, 1290, 1320, and 1620 cm^{-1}. The peaks were less intense in comparison to the ones from the gel matrix, especially after extraction from wool. In this latter case, peaks at approximately 1422 and 1449 cm^{-1}, with a shoulder at 1465 cm^{-1}, were preferentially enhanced. These peaks are generally attributed to alizarin C-O stretching and C-O-H and C-H bending [14,15,29]. Nonetheless, peaks in the same positions were also observed in the gel matrix. In addition, the Nanorestore Gel® HWR matrix presents some peaks at approximately 1320–1350 cm^{-1} that overlap with the ones typical of alizarin. A peak at 1041 cm^{-1}, visible after extraction from tempera, is attributed in the literature to the alizarin ring C-C-C in-plane bending [29]. The intensity of this specific peak in this spectrum could be due to the geometry of the interaction between the analytes and the AgNPs.

In general, on the paint layers as on the textile, the typical anthraquinone dye SERS scattering peaks were more recognizable after agar-gel-supported extraction. This is in accordance with already published results [19]. This can be related to the high water retention of Nanorestore Gel® HWR, which could, in some instances, hinder the extraction solution release and, hence, the analytes collection.

In the case of turmeric, too, typical SERS scattering peaks could be observed on the agar gel after extraction from wool (Figure 5). Signals at 1139 and 1166 cm^{-1} can be assigned to the molecular skeleton vibration, while the signal at 1290 cm^{-1} is due to the phenolic ring C-C-C, C-C-H, and C=CH bending [36]. Lastly, the band at 1600 cm^{-1} can be assigned to the C-OH bending and its shoulder at 1630 cm^{-1} to the C=C and C=O stretching [36,37]. The peak at 999 cm^{-1} is likely due to the gel matrix. On the contrary, no peaks related to curcumin could be observed on Nanorestore Gel® HWR after the dye extraction. As mentioned for madder, the difference in the extraction capabilities can be explained by taking into account the difference in water retention.

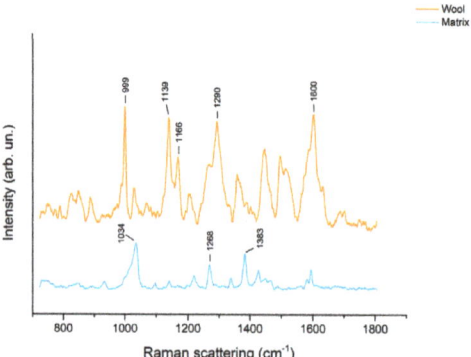

Figure 5. SERS spectra of agar gel (3% w/v) after 3 h extraction of turmeric from wool.

In general, the gel matrices exhibited rather reproducible signals. While the agar gel peak positions are reproducible, anyways, their intensities can change consistently. This fact is a consequence of the agar gel macromolecular structure, which can have variable interaction with AgNPs.

The signals observed on the Nanorestore Gel® HWR blank were also quite reproducible during the experiments that were carried out, and the main scattering peaks were at 839 cm^{-1}, 999 cm^{-1}, 1355 cm^{-1}, 1422, and 1448 cm^{-1}.

To avoid any attribution uncertainty, however, especially in the region around 1320 and 1450 cm^{-1}, SERS analyses of dyes re-extracted from the gels will be attempted in the future.

2.2.2. Indigo

Because of the degradation that occurred in the reducing solution, the Nanorestore Gel® HWR was not able to extract indigo. Spectra were, thus, acquired only on the agar gel after indigo extraction and Ag colloid pastes addition.

In the spectra recorded, intense and sharp peaks related to indigoids were visible (Figure 6 and Table 2).

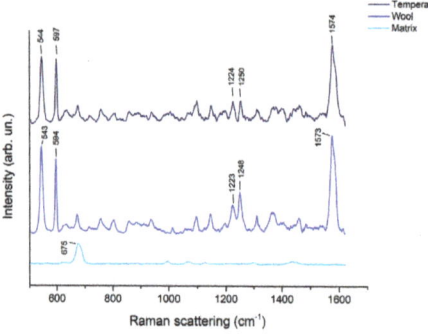

Figure 6. Spectra of agar gel (3% w/v) cylinders after 5 min extraction of indigo from tempera and from wool.

Table 2. Main scattering peaks observed in the spectra of agar gel 3% (w/v) after extraction of indigo from wool and tempera mock-ups. The vibrational modes are reported as ν (stretching) and δ (bending).

Peaks from Indigo		Attribution	Peaks from the Gel Matrix
Tempera	Textile		
544 cm^{-1}	543 cm^{-1}	δ C=C-CO-C and C-N	675 cm^{-1}
597 cm^{-1}	594 cm^{-1}	δ C-C and C-N	1000 cm^{-1}
1224 cm^{-1}	1223 cm^{-1}	δ N-H and C-H	1440 cm^{-1}
1250 cm^{-1}	1248 cm^{-1}	δ C-H, C=C, and N-H	
1574 cm^{-1}	1573 cm^{-1}	ν C=C and C=O	

The intense peaks at 544 and 597 cm^{-1} are related to the indigotin C=C-CO-C and C-N bending modes and to the C-C and C-N bending modes, respectively. The peak at 1224 cm^{-1} is due to the molecule N-H and C-H in-plane bending, while the one at 1250 cm^{-1} is due to the C-H, C=C, and N-H in-plane bending [16]. The strong signal at 1574 cm^{-1} can be attributed to the molecule's C=C and C=O stretching modes [16,38].

The SERS peaks obtained are consistent with the ones published by Platania et al. [16], who extracted indigo from cotton using agar gel loaded with the same reducing solution (NaOH:Na$_2$S$_2$O$_4$ 1:2). Nonetheless, in comparison to the spectrum obtained by Platania et al., the spectra reported here exhibit a pronounced enhancement of the peaks at 545, 597, and 1573 cm^{-1}. Interestingly, the spectra obtained after extraction from wool and after extraction from tempera are very similar in spite of the different preparation recipes and are highly reproducible.

2.3. HPLC-MS/MS

2.3.1. Madder and Turmeric

The chromatograms obtained show the presence of the target analytes selected. The chromatographic peaks related to alizarin fragmentation (Rt = 4.17 min) and purpurin fragmentation (Rt = 4.52) could be observed after re-extraction from agar gel applied on wool (Figures 7 and 8).

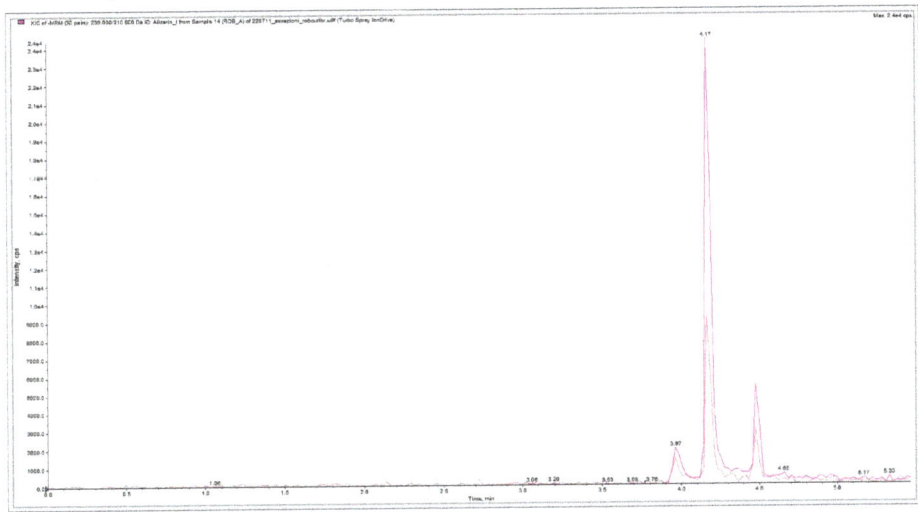

Figure 7. Alizarin chromatogram relative to agar gel (3% w/v) madder extraction from wool. The chromatogram in darker pink refers to the transition from 239 to 210 m/z, while the chromatogram in lighter pink refers to the transition from 239 to 167 m/z.

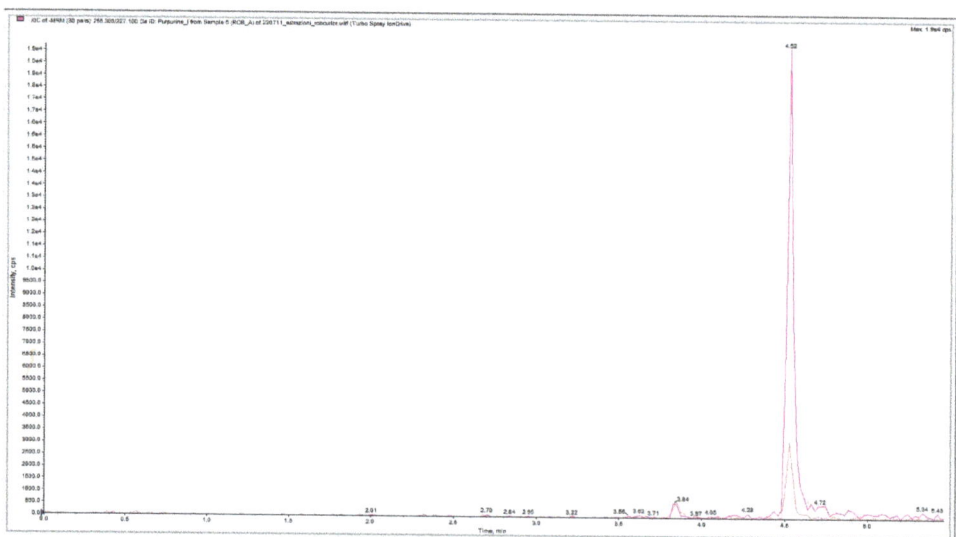

Figure 8. Purpurin chromatogram relative to agar gel (3% w/v) madder extraction from wool. The darker chromatogram refers to the transition from 255 to 227 m/z, while the lighter chromatogram refers to the transition from 255 to 171 m/z.

On the contrary, re-extraction from Nanorestore Gel® did not produce satisfying results.

The chromatographic peak areas of the analytes are greater for agar gel (3% w/v) extraction than for Nanorestore Gel® HWR extraction. Again, this is in accordance with their different water retention.

Regarding the extraction from tempera paint layers, no peaks related to alizarin and purpurin were observed. Nonetheless, considering the SERS results obtained, the extraction times from the paint mock-ups could be adjusted to further increase the amount of analyte collected.

The dLLME clean-up procedure, successfully tested and validated for aglycones and glycosylated dyes, demonstrated efficacy also with turmeric. The chromatographic peaks related to curcumin transitions (Rt = 4.68) could be observed after re-extraction from agar gel and Nanorestore Gel® HWR. No peaks related to the analyte were observed in the reference blanks.

2.3.2. Indigo

Because of the extraction solution complexity, the clean-up procedure developed for indigoids (see Section 3 (Materials and Methods)) could not be perfected. The formation of a sulphur salt was observed both in the aqueous phase and organic phase after dLLME. For this reason, HPLC-MS/MS analyses of indigoid dyes with this procedure requires further studies, and the protocol is still under refinement. Considering the good results obtained by means of SERS, however, the reducing solution could be further diluted to hinder the salt precipitation.

2.4. Colorimetry

2.4.1. Madder and Turmeric

The colour variations induced using gel-supported liquid extraction are mainly around the colour tolerance applied in the industrial field ($\Delta E = 3$ CIELAB units) considered the upper limit of rigorous colour tolerance (Table 3) [39].

Table 3. Colour variation (ΔE_{00}) measured for every mock-up before and after dye extraction.

Sample	Name	Description	Gel	Extraction Time	ΔE_{00}
	R1R	Tempera paint, madder lake pigment	Agar gel (3% w/v)	15 min	2.97
			Nanorestore Gel® HWR	15 min	4.96
	ROB	Wool dyed with madder	Agar gel (3% w/v)	3 h	2.09
			Nanorestore Gel® HWR	3 h	2.76
	CUR	Wool dyed with turmeric	Agar gel (3% w/v)	3 h	3.87
			Nanorestore Gel® HWR	3 h	5.24
	I1AR	Tempera paint, indigo over azurite	Agar gel (3% w/v)	5 min	4.14
	IND	Wool dyed with indigo	Agar gel (3% w/v)	5 min	1.52

On madder, only Nanorestore Gel® HWR extraction from tempera paint induced a colour variation higher than three. However, no mark was visible on the paint to the authors' eyes. Agar gel, on the other hand, induced a colour variation of 2.97 CIELAB units, and showed better results when analysed by means of SERS. On wool, both Nanorestore Gel® HWR and agar gel extraction was invisible, leaving a large margin for further developments.

Turmeric colour measurements showed that the dye is sensitive to the extraction solution. After 3 h, a light shift to a reddish colour was effectively visible under careful observation. Nevertheless, the presence of turmeric could be detected using both SERS and HPLC-MS/MS, which means there could be a margin for a reduction in the extraction time.

2.4.2. Indigo

The reducing solution proved to be harsher than the ammonia–EDTA solution. The extraction time was reduced from 3 h to 5 min on wool, providing very good SERS results and a colour variation of only 1.52 CIELAB units (Table 3). Conversely, the extraction on tempera induced a colour variation of 4.14 CIELAB after 5 min. Considering the spectra obtained, the reducing solution could be further diluted and the extraction times adjusted.

3. Materials and Methods

3.1. Materials

Madder roots (*Rubia tinctorum* L.) and alum ($KAl(SO_4)_2$) were purchased from "Chroma" srl. Genuine indigo in pieces (Indigofera Tinctoria), rabbit skin glue, and $CaSO_4$ were purchased from Kremer Pigmente (Germany). Bricks were purchased from "Mattone Romano" srl (Italy). $KC_4H_5O_6$ was purchased from a grocery shop. Turmeric was purchased from Chroma. Agar in powder (ash 2.0–2.4%), solvents, and salts, such as ammonia (30–33%), $NaCO_3$ ($\geq$99.0%), $Na_2S_2O_4$ ($\geq$82.5%), K_2CO_3 (with impurities $\leq$55.0 ppm), NaOH ($\geq$95%), hydroxylamine hydrochloride (99.9%), NaCl (with impurities $\leq$0.005% as insoluble matter), Na_2EDTA (with impurities $\leq$0.005% as insoluble matter), silver nitrate ($\geq$99.0%), HCl (37%), HCOOH ($\geq$95%), 2-propanol, and 1-pentanol, were purchased from Sigma-Aldrich. Nanorestore Gel® High Water Retention was purchased from CSGI (Center for Colloid and Surface Science).

3.2. Lake Pigments Preparation

Red madder lake pigment was prepared following a recipe from Daniels et al. [40]. Briefly, 5 g of madder roots were soaked in 150 mL distilled water and left overnight. The roots in water were heated up until 70 °C for 30 min. After filtration, 2.5 g potassium alum was added to the solution, and the temperature was brought to 80 °C. Meanwhile, 0.94 g K_2CO_3 was dissolved in 25 mL water and gently poured into the dye bath under continuous stirring. The lake pigment formed was left to precipitate overnight. Once precipitated, the lake pigment was filtered and finely ground.

3.3. Paint Mock-Ups Preparation

Paint mock-ups were made on bricks prepared with eight layers of $CaSO_4$ and rabbit skin glue and applied with a brush [41]. The organic pigments were mixed with a solution of 2 mL water and an egg yolk. The pigments were mixed with the binder in a rough proportion 2:1 (w/v) using a spatula and then applied by means of a brush, and then left to age naturally for six months.

3.4. Wool Dyeing

Textile mock-ups were prepared from wool dyed using madder, indigo, and turmeric. First, 2 g wool yarn was mordanted as follows: 620 mg alum and 120 mg potassium bitartrate ($KC_4H_5O_6$) were dissolved in 250 mL of distilled water. The solution was warmed to 40 °C and kept for ten minutes. Wool was soaked in the solution once cooled down. The temperature was increased again to 80 °C for 40 min and kept for 1 h under gentle magnetic stirring [27]. Once at ambient T, the wool was removed, squeezed, and left to dry.

To dye wool using madder, 1 g madder roots was soaked in 400 mL distilled water and warmed to 35 °C. One gram of mordanted wool was soaked, the temperature was brought to 80 °C over 40 min, and then kept for 1 h under gentle magnetic stirring [27].

For direct wool dyeing using turmeric, 0.5 g turmeric in powder was soaked in 400 mL distilled water, brought to 90 °C, and then kept for 15 min under magnetic stirring. Successively, 1 g unmordanted wool was soaked and left for 30 min at the same temperature.

To dye wool with indigo, 0.6 g ground indigo was added to 10 mL distilled water at 45 °C. Successively, a solution of 0.6 g $NaCO_3$ in 6 mL distilled water and a solution of 1.5 g $Na_2S_2O_4$ in 50 mL distilled water at 45 °C were added one after the other. The mixture obtained was brought to 55 °C and kept for 20 min. Afterwards, 3 g unmordanted wool was soaked in the bath and left for 10 min. The wool was then removed and left in contact with the atmosphere to allow the dye to re-oxidise.

After the dyeing was performed, distilled water was always used to wash the wool until the rinse was transparent; then, it was left to dry and age naturally for six months. The textile mock-ups were made by winding dyed wool yarns around microscope glass slides.

3.5. Agar Gel Preparation

For the agar gel preparation, 0.24 g agar in powder was added to 8 mL distilled water in a 100 mL beaker. The beaker was then gently shaken in a boiling water bath for 10 min so that the polymer reached its melting temperature. The solution was cooled for 30 min at ambient temperature and stored in a refrigerator overnight [19,35].

3.6. Ag Colloids Pastes Preparation

The Ag colloids were prepared following the protocol developed by Leopold and Lendl [19,42]. Specifically, 20 mg NaOH was added to 5 mL MilliQ water, and 21 mg $NH_2OH \cdot HCl$ was added to another 5 mL MilliQ water. The two solutions were mixed together and poured into a solution of 17 mg $AgNO_3$ in 90 mL MilliQ water under constant magnetic stirring. The obtained Ag colloid solution was stored in the fridge. To obtain colloidal pastes, 10 mL colloids were centrifuged at 4500 RPM for 20 min, and the supernatant was discarded [43,44].

3.7. Gel-Supported Liquid Extraction

Agar gel 3% (*w/v*) and Nanorestore Gel® HWR, as a commercial product, were cut into cylinders of approximately 4 mm in diameter. The cylinders were soaked for 90 min in a solution of NH_3 (30–33%) and Na_2EDTA 1 mM (1:1). NaCl was added until a final concentration of 4.7 mM [19,20,27]. After 90 min, the gel cylinders were removed using tweezers and left to lose 5% of their weight. In the case of indigo extraction, after soaking in the reducing solution, the gel cylinders were quickly dried on absorbent paper and applied directly on the mock-ups to avoid prolongated contact with O_2 in the atmosphere. The gel cylinders were applied on paint and textiles with glass slides on the top to prevent solution's evaporation. The extraction time varied, from 5 min on the paint layers to 3 h on the textiles. The whole procedure workflow is summarised in Figure 9 as a schematic diagram.

Figure 9. Schematic diagram of the whole gel-supported liquid extraction workflow for SERS and HPLC-MS/MS identification.

3.8. SERS Analysis

For the SERS analyses, 20 µL of colloidal pastes were poured on the gel face, which was in contact with the mock-ups. The gels were then left to dry for 12 h at ambient temperature. The SERS analyses were carried out directly on the dry gel using a Horiba Jobin-Yvon HR Evolution micro-Raman spectrometer equipped with a He-Ne laser (λ = 633 nm) coupled with a microscope with a set of interchangeable objectives. The spectra were collected using a 100× magnification objective, and the laser intensity was varied from 0.15 to 0.75 mW. The acquisition time was varied from 5 to 10 s and the scan number from 30 to 60, depending on the sample, in order to obtain the best signal-to-noise ratio. A minimum of three spectra were collected for every sample, both on the reference gel soaked into the extraction solution (blank gel matrix) and the gel after the dyes' extraction. All spectra were processed using OriginPro 9 software (©OriginLab): the background was subtracted fitting a polynomial baseline to the power of five, the spectra were normalised, and "adjacent averaging" smoothing was applied to reduce noise.

3.9. dLLME

The dLLME clean-up procedure was developed and validated to enhance dyes recovery. The development and validation study is under submission to another journal by the same authors. After the gel-supported liquid extraction from the mock-ups, the dyes were re-extracted from the gels for the HPLC-MS/MS analyses using the same aqueous solution. The gels loaded with madder, and the turmeric dyes were soaked in 0.8 mL NH_3 (30–33%), 0.8 mL Na_2EDTA (1 mM), and 4.4 mg NaCl. After 24 h, 495.6 mg NaCl was added together with 1 mL HCl 6 M and 0.8 mL HCOOH ($\geq$95%) to bring the solution pH to 3. The dyes were then extracted from the aqueous phase into the organic solvent:

250 µL 2-propanol was added to every sample and, subsequently, together in the same syringe, 200 µL 1-pentanol and 100 µL 2-propanol were vigorously injected to obtain a highly dispersed thee-phasic system, known as a cloudy solution [28,45,46].

For indigo dyes, the gels were re-extracted in 1.5 mL distilled water containing NaOH:Na$_2$S$_2$O$_4$ 1:2 (w/w). After 24 h, 530 µL HCl 6 M was added, and the solution volume was brought to 5 mL with distilled water. For the extraction into organic solvent, 750 µL 2-propanol and 100 µL chloroform were vigorously injected together to obtain the cloudy solution [46]. All samples were successively sonicated for 10 min and centrifuged for 10 min at 10,000 RPM and 5 °C. The aqueous solution was discarded. For samples containing anthraquinone dyes, 1-pentanol was washed using 3.1 mL of a solution of 2.8 M NaCl. Lastly, the organic solvent was evaporated under N$_2$ flow. All samples were reconstituted with 100 µL MeOH:H$_2$O 1:1 for the HPLC-MS/MS analyses.

3.10. HPLC-MS/MS Analyses

For the chromatographic analysis, a SCIEX Exion LC AD System was used. The system was coupled to a Sciex QTRAP 6500 mass spectrometer system with electrospray ionisation (ESI) and multicomponent IonDrive Technology.

The column chosen was a reversed phase BEH C18 (2.1 mm × 50 mm) with 1.7 µm silica particles. The injected volume was 3 µL. The mobile phases chosen were 0.1% formic acid in Millipore water (phase A) and 0.1% formic acid in acetonitrile (phase B). The gradient programme is shown in Table 4. Table 5 reports the MRM transitions used for the main target analytes' identification (in negative) based on optimisation using certified standards. Identifications were made relying on retention times and two fragmentation transitions.

Table 4. Chromatographic gradient used during for HPLC-MS/MS analyses.

Time (min)	Phase A: 0.1% HCOOH in H$_2$O	Phase B: 0.1% HCOOH in ACN
0.00	95%	5%
1.00	95%	5%
5.00	35%	65%
5.20	0%	100%
6.50	0%	100%
7.00	95%	5%
8.50	95%	5%

Table 5. Instrumental parameters optimised to detect every analyte basing on MRM transitions (negative mode). The parameters were optimised relying on certified standards analysis and literature reports [47–49].

Analyte	Parent Ion (m/z)	Fragments (m/z)	DP (V)	EP (V)	CE (V)	CXP (V)
Alizarin	239	210 167	−120	−8.1	−49 −40	−15 −10
Purpurin	255	227 171	−109	−8.3	−38 −42	−21 −13
Curcumin	367	134 158	−40	−2.5	−48 −44	−13 −14

3.11. Colorimetric Measurements

To assess the perceptivity of the procedure on paint layers and textiles, colorimetric measurements were performed before and after gel-supported extraction, and the ΔE values were calculated. Colorimetric coordinates were acquired using fibre optic reflectance spectroscopy (FORS) in the visible range. The spectrophotometer used was a BWTEK Exemplar LS (B&W Tek, Plainsboro, NJ, USA) with a tungsten lamp BWTEK (series BPS101, 5 W, emission spectrum from 350 to 2600 nm and 2800 K colour temperature). The acquisition range was from 180 to 1100 nm, with a resolution from 0.6 nm to 6.0 nm. The fibre optic, a THORLABS RP22, was used with a head to obtain a 45° inclination of the probe. The measurements were performed in the dark, before and after gel extraction, on the same point. Every measurement was repeated three times and mediated. The CIE L*a*b* parameters were extracted from the spectra using the software BWSpec. The colour variation ΔE00 was calculated using the formula reported in the CIEDE2000 guidelines [39,50].

4. Conclusions

In conclusion, gel-supported in situ extraction of madder, turmeric, and indigo was applied for the first time on wool and tempera paints for SERS and HPLC-MS/MS dyes identification.

The ammonia–EDTA solution proved to be able to extract turmeric, a direct dye, from textiles. Furthermore, the solution, already tested on water soluble paint layers [36], allowed anthraquinone dye extraction also from tempera paints that were naturally aged for six months.

The dyes extracted using the ammonia–EDTA solution were identified directly on the gels after Ag colloid pastes addition by means of SERS. While madder-related peaks were clearly distinguishable on both agar gel and Nanorestore Gel® HWR, no turmeric-related peaks could be detected on Nanorestore Gel® HWR.

A reducing solution, already tested by Platania et al. for gel-supported extraction from cotton [16], was selected for indigoids. Nanorestore Gel® HWR demonstrated to not be compatible with the reducing solution, while agar gel extraction was successful on wool and tempera paints. The SERS analysis on agar gel produced very good results, and indigoids-related peaks were clearly recognizable and highly reproducible.

A dLLME clean-up procedure was applied for the HPLC-MS/MS analyses. The procedure was necessary for the dye re-extraction from the gels and purification before injection in the instrument.

The dLLME clean-up workflow was already applied by Serafini et al. for the HPLC-MS/MS identification of anthraquinone dyes extracted from ancient textiles using the ammonia–EDTA solution. The results are under submission to another journal. The procedure demonstrated, for the first time, to be effective also for the HPLC-MS/MS identification of turmeric.

Moreover, a clean-up protocol for HPLC-MS/MS analysis of indigoids is still under development. Further dilution of the reducing solution could be effective in perfecting the dLLME procedure tested, which could not be applied because of a salt precipitation.

For both spectroscopic and chromatographic analyses, dye-related signals observed after Nanorestore Gel® HWR extraction were less intense when compared to the ones observed after agar gel (3% w/v) extraction.

This is in accordance with the Nanorestore Gel® HWR retention power. Hence, unless dealing with very water sensitive materials [36], agar gel use is preferable.

The colour change results on the mock-ups before and after gel-supported liquid extraction were very promising for all dyes tested. Considering the SERS and HPLC-MS/MS results, there is still a margin for improvement. The gel cylinder dimension could be consistently reduced and extraction times extended to make the procedure completely imperceptible.

Future research will be focused on artificially aged mock-ups to assess the effect of degradation phenomena on invasiveness and extraction efficiency.

Ultimately, the presented gel-supported in situ extraction is suited for multianalytical dye identification. Techniques traditionally used to study dyes, both invasively and noninvasively, can be combined without interferences from the matrix. During HPLC-MS/MS analyses, the addition of further transitions to the instrumental parameters related to the analytes of interest can allow for a more accurate portrayal of the dyes' molecular pattern, including glycosylated moieties [28]. Thus, the approach constitutes a valid alternative to accompany noninvasive analyses of inorganic substances for comprehensive studies of cultural heritage without posing a threat to their integrity.

Author Contributions: Conceptualisation, A.C. and I.S.; methodology, A.C., I.S., F.V., C.M. and A.B.; investigation, A.B., G.P. and A.C. resources, P.P., G.F. and R.C.; data curation, A.B., G.P. and I.S.; writing—original draft preparation, A.B.; writing—review and editing, A.B., G.P. and I.S.; supervision, P.P., M.S., G.F. and R.C.; project administration, R.C. All authors have read and agreed to the published version of the manuscript.

Funding: This research received no external funding.

Informed Consent Statement: Not applicable.

Data Availability Statement: Data will be made available on request.

Conflicts of Interest: The authors declare no conflict of interest.

Sample Availability: Not applicable.

References

1. Gueli, A.M.; Gallo, S.; Pasquale, S. Optical and colorimetric characterization on binary mixtures prepared with coloured and white historical pigments. *Dye. Pigment.* **2018**, *157*, 342–350. [CrossRef]
2. Peruzzi, G.; Cucci, C.; Picollo, M.; Quercioli, F.; Stefani, L. Non-invasive identification of dyed textiles by using VIS-NIR FORS and hyperspectral imaging techniques. *Cult. E Sci. Del Color. Color Cult. Sci.* **2021**, *13*, 61–69.
3. Degano, I.; Ribechini, E.; Modugno, F.; Colombini, M.P. Analytical Methods for the Characterization of Organic Dyes in Artworks and in Historical Textiles. *Appl. Spectrosc. Rev.* **2009**, *44*, 363–410. [CrossRef]
4. Melo, M.J.; Otero, V.; Vitorino, T.; Araújo, R.; Muralha, S.; Lemos, A.; Picollo, M. A Spectroscopic Study of Brazilwood Paints in Medieval Books of Hours. *Appl. Spectrosc.* **2014**, *68*, 434–443. [CrossRef] [PubMed]
5. Clementi, C.; Basconi, G.; Pellegrino, R.; Romani, A. *Carthamus tinctorius* L.: A Photophysical Study of the Main Coloured Species for Artwork Diagnostic Purposes. *Dye. Pigment.* **2014**, *103*, 127–137. [CrossRef]
6. Leona, M. Microanalysis of Organic Pigments and Glazes in Polychrome Works of Art by Surface Enhanced Resonance Raman Scattering. *Proc. Natl. Acad. Sci. USA* **2009**, *106*, 14757–14762. [CrossRef]
7. Melo, M.J.; Nabais, P.; Guimarães, M.; Araújo, R.; Castro, R.; Conceição Oliveira, M.; Whitworth, I. Organic Dyes in Illuminated Manuscripts: A Unique Cultural and Historic Record. *Philos. Trans. R. Soc. A Math. Phys. Eng. Sci.* **2016**, *374*, 20160050. [CrossRef]
8. Tamburini, D.; Dyer, J.; Davit, P.; Aceto, M.; Turina, V.; Borla, M.; Vandenbeusch, M.; Gulmini, M. Compositional and Micro-Morphological Characterisation of Red Colourants in Archaeological Textiles from Pharaonic Egypt. *Molecules* **2019**, *24*, 3761. [CrossRef]
9. Ciccola, A.; Serafini, I.; D'Agostino, G.; Giambra, B.; Bosi, A.; Ripanti, F.; Nucara, A.; Postorino, P.; Curini, R.; Bruno, M. Dyes of a Shadow Theatre: Investigating Tholu Bommalu Indian Puppets through a Highly Sensitive Multi-Spectroscopic Approach. *Heritage* **2021**, *4*, 1807–1820. [CrossRef]
10. Colantonio, C.; Lanteri, L.; Ciccola, A.; Serafini, I.; Postorino, P.; Censori, E.; Rotari, D.; Pelosi, C. Imaging Diagnostics Coupled with Non-Invasive and Micro-Invasive Analyses for the Restoration of Ethno-graphic Artifacts from French Polynesia. *Heritage* **2022**, *5*, 215–232. [CrossRef]
11. Pilot, R.; Signorini, R.; Durante, C.; Orian, L.; Bhamidipati, M.; Fabris, L. A Review on Surface-Enhanced Raman Scattering. *Biosensors* **2019**, *9*, 57. [CrossRef]
12. Doherty, B.; Brunetti, B.G.; Sgamellotti, A.; Miliani, C. A detachable SERS active cellulose film: A minimally invasive approach to the study of painting lakes. *J. Raman Spectrosc.* **2011**, *42*, 1932–1938. [CrossRef]
13. Leona, M.; Decuzzi, P.; Kubic, T.A.; Gates, G.; Lombardi, J.R. Nondestructive identification of natural and synthetic organic colorants in works of art by surface enhanced Raman scattering. *Anal. Chem.* **2011**, *83*, 3990–3993. [CrossRef] [PubMed]
14. Lofrumento, C.; Ricci, M.; Platania, E.; Becucci, M.; Castellucci, E. SERS detection of red organic dyes in Ag-agar gel. *J. Raman Spectrosc.* **2013**, *44*, 47–54. [CrossRef]
15. Platania, E.; Lombardi, J.R.; Leona, M.; Shibayama, N.; Lofrumento, C.; Ricci, M.; Becucci, M.; Castellucci, E. Suitability of Ag-agar gel for the micro-extraction of organic dyes on different substrates: The case study of wool, silk, printed cotton and a panel painting mock-up. *J. Raman Spectrosc.* **2014**, *45*, 1133–1139. [CrossRef]
16. Platania, E.; Lofrumento, C.; Lottini, E.; Azzaro, E.; Ricci, M.; Becucci, M. Tailored micro-extraction method for Raman/SERS detection of indigoids in ancient textiles. *Anal. Bioanal. Chem.* **2015**, *407*, 6505–6514. [CrossRef]
17. Becucci, M.; Ricci, M.; Lofrumento, C.; Castellucci, E. Identification of organic dyes by surface-enhanced Raman scattering in nano-composite agar-gel matrices: Evaluation of the enhancement factor. *Opt. Quantum Electron.* **2016**, *48*, 449. [CrossRef]
18. Ricci, M.; Lofrumento, C.; Castellucci, E.; Becucci, M. Microanalysis of Organic Pigments in Ancient Textiles by Surface-Enhanced Raman Scattering on Agar Gel Matrices. *J. Spectrosc.* **2016**, *2016*, 1380105. [CrossRef]
19. Germinario, G.; Ciccola, A.; Serafini, I.; Ruggiero, L.; Sbroscia, M.; Vincenti, F.; Fasolato, C.; Curini, R.; Ioele, M.; Postorino, P.; et al. Gel substrates and ammonia-EDTA extraction solution: A new nondestructive combined approach for the identification of anthraquinone dyes from wool textiles. *Microchem. J.* **2020**, *155*, 104780. [CrossRef]

20. Serafini, I.; Lombardi, L.; Vannutelli, L.; Montesano, C.; Sciubba, F.; Guiso, M.; Curini, R.; Bianco, A. How the extraction method could be crucial in the characterization of natural dyes from dyed yarns and lake pigments: The case of American and Armenian cochineal dyes, extracted through the new ammonia-EDTA method. *Microchem. J.* **2017**, *134*, 237–245. [CrossRef]
21. Zhang, X.; Laursen, R.A. Development of Mild Extraction Methods for the Analysis of Natural Dyes in Textiles of Historical Interest Using LC-Diode Array Detector-MS. *Anal. Chem.* **2005**, *77*, 2022–2025. [CrossRef] [PubMed]
22. Surowiec, I.; Quye, A.; Trojanowicz, M. Liquid chromatography determination of natural dyes in extracts from historical Scottish textiles excavated from peat bogs. *J. Chromatogr. A* **2006**, *1112*, 209–217. [CrossRef]
23. Sanyova, J. Mild extraction of dyes by hydrofluoric acid in routine analysis of historical paint micro-samples. *Microchim. Acta* **2008**, *162*, 361–370. [CrossRef]
24. Valianou, L.; Karapanagiotis, I.; Chryssoulakis, Y. Comparison of extraction methods for the analysis of natural dyes in historical textiles by high-performance liquid chromatography. *Anal. Bioanal. Chem.* **2009**, *395*, 2175–2189. [CrossRef]
25. Manhita, A.; Ferreira, T.; Candeias, A.; Barrocas Dias, C. Extracting natural dyes from wool-an evaluation of extraction methods. *Anal. Bioanal. Chem.* **2011**, *400*, 1501–1514. [CrossRef]
26. Lech, K.; Jarozs, M. Novel methodology for the extraction and identification of natural dyestuffs in historical textiles by HPLC–UV–Vis–ESI MS. Case study: Chasubles from the Wawel Cathedral collection. *Anal. Bioanal. Chem.* **2011**, *399*, 3241–3251. [CrossRef]
27. Lombardi, L.; Serafini, I.; Guiso, M.; Sciubba, F.; Bianco, A. A New Approach to the Mild Extraction of Madder Dyes from Lake and Textile. *Microchem. J.* **2015**, *126*, 373–380. [CrossRef]
28. Blackburn, R.S.; Bechtold, T.; John, P. The development of indigo reduction methods and pre-reduced indigo products. *Color. Technol.* **2009**, *125*, 193–207. [CrossRef]
29. Whitney, A.V.; Van Duyne, R.P.; Casadio, F. An innovative surface-enhanced Raman spectroscopy (SERS) method for the identification of six historical red lakes and dyestuffs. *J. Raman Spectrosc.* **2006**, *37*, 993–1002. [CrossRef]
30. Amato, F.; Micciché, F.; Cannas, M.; Gelardi, F.; Pignataro, B.; Li Vigni, M.; Agnello, S. Ag nanoparticles agargel nanocomposites for SERS detection of cultural heritage interest pigments. *Eur. Phys. J. Plus* **2018**, *133*, 74. [CrossRef]
31. Lofrumento, C.; Platania, E.; Ricci, M.; Mulana, F.; Becucci, M.; Castellucci, E. The SERS spectra of alizarin and its ionized species: The contribution of the molecular resonance to the spectral enhancement. *J. Mol. Struct.* **2015**, *1090*, 98–106. [CrossRef]
32. Retko, K.; Ropreta, P.; Cerc Korošec, R. Surface-enhanced Raman spectroscopy (SERS) analysis of organic colourants utilising a new UV-photoreduced substrate. *J. Raman Spectrosc.* **2014**, *45*, 1140–1146. [CrossRef]
33. Rambaldi, D.C.; Pozzi, F.; Shibayama, N.; Leona, M.; Preusser, F.D. Surface-enhanced Raman spectroscopy of various madder species on wool fibers: The role of pseudopurpurin in the interpretation of the spectra. *J. Raman Spectrosc.* **2015**, *46*, 1073–1081. [CrossRef]
34. Pozzi, F.; Zaleski, S.; Casadio, F.; Van Duyne, R.P. SERS Discrimination of Closely Related Molecules: A Systematic Study of Natural Red Dyes in Binary Mixtures. *J. Phys. Chem. C* **2016**, *120*, 21017–21026. [CrossRef]
35. Bosi, A.; Ciccola, A.; Serafini, I.; Peruzzi, G.; Nigro, V.; Postorino, P.; Curini, R.; Favero, G. Gel microextration from hydrophilic paint layers: A comparison between Agar-gel and Nanorestore Gel® HWR for spectroscopic identification of madder. *Microchem. J.* **2023**, *187*, 108447. [CrossRef]
36. Bruni, S.; Guglielmi, V.; Pozzi, F. Historical organic dyes: A surface-enhanced Raman scattering (SERS) spectral database on Ag Lee-Meisel colloids aggregated by NaClO$_4$: SERS database on historical organic dyes. *J. Raman Spectrosc.* **2011**, *42*, 1267–1281. [CrossRef]
37. Mollica Nardo, V.; Aliotta, F.; Mastelloni, M.A.; Ponterio, R.C.; Saija, F.; Trusso, S.; Vasi, C.S. A spectroscopic approach to the study of organic pigments in the field of cultural heritage. *Atti Della Accad. Peloritana Pericolanti* **2017**, *95*, A5.
38. Coralez, G.; Celis, F.; Gómez-Jeria, J.S.; Campos, M.; Cárcamo-Vega, J.J. Raman of indigo on a silver surface. Raman and theoretical characterization of indigo deposited on silicon dioxide-coated and uncoated silver nanoparticles. *Spectrosc. Lett.* **2017**, *50*, 316–321. [CrossRef]
39. Prieto, B.; Sanmartın, B.; Silva, B.; Martınez-Verdu, F. Measuring the Color of Granite Rocks: A Proposed Procedure. *Color Res. Appl.* **2010**, *35*, 368–375. [CrossRef]
40. Daniels, V.; Deviese, T.; Hacke, M.; Higgitt, C. Technological Insights into Madder Pigment Production in Antiquity. *Br. Mus. Tech. Res. Bull.* **2014**, *8*, 13–28.
41. Marconi, S. *Preparazione e finitura delle opere pittoriche. Materiale e metodi, preparazioni e imprimiture, leganti, vernici, cornice*; Mursia: Milan, Italy, 1993.
42. Leopold, N.; Lendl, B. A New Method for Fast Preparation of Highly Surface-Enhanced Raman Scattering (SERS) Active Siver Colloids at Room Temperature by Reduction of Silver Nitrate with Hydroxylamine Hydrochloride. *J. Phys. Chem. B* **2003**, *107*, 5723–5727. [CrossRef]
43. Idone, A.; Gulmini, M.; Henry, A.-I.; Casadio, F.; Chang, L.; Appolonia, L.; Van Duyned, R.P.; Shah, N.C. Silver colloidal pastes for dye analysis of reference and historical textile fibers using direct, extractionless, non-hydrolysis surface-enhanced Raman spectroscopy. *Analyst* **2013**, *138*, 5895. [CrossRef] [PubMed]
44. Saviello, D.; Alyami, A.; Trabace, M.; Giorgi, R.; Baglioni, P.; Mirabile, A.; Iacopino, D. Plasmonic colloidal pastes for surface-enhanced Raman spectroscopy (SERS) of historical felt-tip pens. *RSC Adv.* **2018**, *8*, 8365–8371. [CrossRef]

45. Vincenti, F.; Montesano, C.; Cellucci, L.; Gregori, A.; Fanti, F.; Compagnone, D.; Curini, R.; Sergi, M. Combination of pressurized liquid extraction with dispersive liquid liquid microextraction for the determination of sixty drugs of abuse in hair. *J. Chromatogr. A.* **2019**, *1605*, 360348. [CrossRef] [PubMed]
46. Rezaee, M.; Yamini, Y.; Faraji, M. Evolution of dispersive liquid–liquid microextraction method. *J. Chromatogr. A* **2010**, *1217*, 2342–2357. [CrossRef]
47. Lech, K.; Fornal, E. A Mass Spectrometry-Based Approach for Characterization of Red, Blue, and Purple Natural Dyes. *Molecules* **2020**, *25*, 3223. [CrossRef] [PubMed]
48. Rafaelly, L.; Heron, S.; Nowik, W.; Tchapla, A. Optimisation of ESI-MS detection for the HPLC of anthraquinone dyes. *Dye. Pigment.* **2008**, *77*, 191–203. [CrossRef]
49. Jiang, H.; Timmermann, N.B.; Gang, D.R. Use of liquid chromatography–electrospray ionization tandem mass spectrometry to identify diarylheptanoids in turmeric (*Curcuma longa* L.) rhizome. *J. Chromatogr. A* **2006**, *1111*, 21–31. [CrossRef] [PubMed]
50. Oleari, C. *Misurare il colore: Spettrofotometria, fotometria e colorimetria: Fisiologia e percezione*; Casa Editrice Libraria Ulrico Hoepli: Milan, Italy, 1998.

Disclaimer/Publisher's Note: The statements, opinions and data contained in all publications are solely those of the individual author(s) and contributor(s) and not of MDPI and/or the editor(s). MDPI and/or the editor(s) disclaim responsibility for any injury to people or property resulting from any ideas, methods, instructions or products referred to in the content.

Article

Reflectance Spectroscopy as a Novel Tool for Thickness Measurements of Paint Layers

Alice Dal Fovo [1,*], Marina Martínez-Weinbaum [2], Mohamed Oujja [2], Marta Castillejo [2] and Raffaella Fontana [1]

[1] Consiglio Nazionale delle Ricerche-Istituto Nazionale di Ottica (CNR-INO), Largo E. Fermi 6, 50125 Florence, Italy; raffaella.fontana@ino.cnr.it

[2] Instituto de Química Física Rocasolano, Spanish National Research Council (CSIC), C/Serrano 119, 28006 Madrid, Spain; mgmartinez@iqfr.csic.es (M.M.-W.); m.oujja@csic.es (M.O.); marta.castillejo@iqfr.csic.es (M.C.)

* Correspondence: alice.dalfovo@ino.cnr.it

Abstract: A major challenge in heritage science is the non-invasive cross-sectional analysis of paintings. When low-energy probes are used, the presence of opaque media can significantly hinder the penetration of incident radiation, as well as the collection of the backscattered signal. Currently, no technique is capable of uniquely and noninvasively measuring the micrometric thickness of heterogeneous materials, such as pictorial layers, for any painting material. The aim of this work was to explore the possibility of extracting stratigraphic information from reflectance spectra obtained by diffuse reflectance spectroscopy (DRS). We tested the proposed approach on single layers of ten pure acrylic paints. The chemical composition of each paint was first characterised by micro-Raman and laser-induced breakdown spectroscopies. The spectral behaviour was analysed by both Fibre Optics Reflectance Spectroscopy (FORS) and Vis-NIR multispectral reflectance imaging. We showed that there is a clear correlation between the spectral response of acrylic paint layers and their micrometric thickness, which was previously measured by Optical Coherence Tomography (OCT). Based on significant spectral features, exponential functions of reflectance vs. thickness were obtained for each paint, which can be used as calibration curves for thickness measurements. To the best of our knowledge, similar approaches for cross-sectional measurements of paint layers have never been tested.

Keywords: paintings; reflectance spectroscopy; OCT; LIBS; Raman spectroscopy; thickness measurements

1. Introduction

In recent decades, a wide variety of scientific techniques have been tested and optimized for the study of cultural heritage (CH) objects. The need to preserve the material integrity of works of art has directed the research toward defining non-invasive analytical approaches based on the combined application of methodologies that do not involve sampling or risk of damage to the object. A major challenge in heritage science is the non-invasive cross-sectional analysis of the pictorial stratigraphy in paintings. Thickness measurement of the micrometric layers is essential, for instance, to monitor the removal of surface materials during the cleaning operation [1] or to assess the compactness and adhesion between layers. In a non-invasive approach, when low-energy probes are used, the presence of opaque media can significantly hinder the penetration of the incident radiation, as well as the signal collection from within the examined materials.

One of the most widely used techniques for non-invasive stratigraphic measurements on paintings is Optical Coherence Tomography (OCT) [2,3]. Primarily applied in the field of ophthalmology, OCT is an interferometric method based on a Michelson interferometer, yielding 2- or 3-dimensional tomographic imaging that allows the visualization of the internal structure of pictorial layers with an axial resolution ranging from 1 to 10 µm (in air). The incident radiation is backscattered by the material and the optical interference is observed whenever the signal superposes with the reference beam, within the coherence length of the light source. The measurement is based on the detection of signals generated at the interfaces between different media—i.e., when the incident radiation experiences a refractive index (n) mismatch. OCT has proven particularly effective in probing materials that are semi-transparent in the near-infrared (NIR) spectral range. This includes most varnishes applied by artists on the painting surface with a protective and/or aesthetic function [4]. By combining an OCT setup with confocal microscope optics, which enables the beam focussing inside the material rather than on the outer surface, even highly reflecting coatings can be measured [5]. Pictorial layers, however, are often composed of pigments with dispersion and/or absorption properties that do not allow their thickness to be assessed by OCT. Moreover, the n-mismatch causes a delay in the optical path of the reference beam and, therefore, the optically measured distances must be corrected to geometrical distances by dividing them by the refractive index of the material. In the NIR, the refractive index of semi-transparent materials used in paintings is conventionally given as 1.5. However, pictorial layers (pigment dispersed in the binder) are often highly heterogeneous and exhibit variable optical properties that result in different n values. Therefore, if n is not known, the correct thickness of the painting layers is not achievable.

In recent decades, alternative methods to OCT have been proposed for the non-invasive in-depth analysis of paintings. Among others, Terahertz imaging [6,7] has proven effective in yielding 3D data sets of interfaces and projections of paintings in the presence of absorbing species. However, the low axial resolution achievable with THz radiation makes this method unfit for micrometric measurements of pictorial layers.

The use of Nuclear Magnetic Resonance (NMR) to obtain stratigraphic information on easel and wall paintings is also well-documented in the literature [8,9]. NMR profiling was tested to investigate both signal intensity and transverse relaxation time distribution, showing that the dependence of signal intensity on relaxation times makes the interpretation of the stratigraphic information difficult [10]. The NMR-sensitive volume averages the effect of irregularities in the layers and the signal from adjacent layers. In addition, the application of this method is hampered by the lack of application-specific operating software, while the low mass sensitivity resulting from low NMR frequencies results in long measurement times.

More recently, Nonlinear Optical (NLO) techniques [11] have been successfully used for cross-sectional analysis of a wide variety of artistic materials, including paint and varnish samples. The combined application of different NLO modalities allows for the acquisition of compositional and structural information based on the detection of fluorophores (by Multi-Photon Excitation Fluorescence, MPEF), crystalline or highly organized structures without inversion symmetry (by Second Harmonic Generation, SHG), or local differences in refractive index, i.e., interfaces (by Third Harmonic Generation, THG). While for MPEF and SHG, the main limitation in the stratigraphic analysis of paintings is the presence of highly diffusing and/or absorbing media (pigments) [12], the applicability of THG is confined to layers of transparent material, forward detection being the only possible configuration [13].

A cutting-edge methodology recently proposed for cross-sectional analysis in paintings is photoacoustics, which in a sense, can be considered complementary to OCT, as it takes advantage of the presence of non-transparent materials [14,15]. Acoustic waves are generated by the absorption of the radiation emitted by an intensity-modulated pulsed laser. The exponential attenuation of acoustic waves in the frequency domain, which depends on the absorption coefficient of the medium and the propagation path, can be exploited to measure the thickness of the examined material. Although early applications reveal the technique's potential [16], to date, its use is limited to specific cases only.

Given the above, it can be stated that, at present, there is no technique that can uniquely and non-invasively measure the thickness of heterogeneous and optically opaque materials such as pictorial layers.

In this work, we explored the feasibility of achieving stratigraphic information of painting layers from their reflectance spectra measured by Diffuse Reflectance Spectroscopy (DRS) [17]. In heritage science, DRS is typically applied for the analysis of paintings in multi- and hyper-spectral imaging modes or using fibre optics for point-wise measurements. The main objective is typically the identification and mapping of pigments and binders based on the absorption properties of electronic and vibrational transitions of molecules [18]. In the imaging mode, the use of the NIR spectral range allows for the visualization of hidden details underneath the painted surface related to the artistic working process, such as underdrawings and underpaintings. Diffuse reflectance is defined as the ratio of the irradiance of light reflected back to the detector to the irradiance on the surface of the object, as a function of wavelength. The measured light backscattered from the object includes contributions from the air/surface interface and varies with illumination and collection geometry.

The spectral reflectance behaviour of pigments and pictorial layers has been extensively studied [19–21]. It has been shown that the reflectance signal measured from pigment mixtures in paintings is the result of the nonlinear combination of the reflectance of the individual pigments. To cope with the complexity of spectral data interpretation, as well as to reduce the high dimensionality of DRS imaging datasets, new approaches based on artificial intelligence (AI), e.g., deep neural networks (DNN), have been recently explored [22].

In this preliminary study, we tested the proposed approach by examining single layers of pure paint, thus avoiding the use of optical models to predict the reflectance of pigments in mixtures [23] and not taking into account the influence of the surface roughness on the spectra [24]. A mock-up was created for this specific purpose: ten acrylic paints were laid with increasing thicknesses, ranging from 50 to 350 µm, on both white and black backgrounds. The chemical composition of each paint was first characterised by micro-Raman [25] and laser-induced breakdown spectroscopy (LIBS) [26]. The thickness of each layer was then measured by OCT, taking as a reference the portion of the visible substrate at the edge of the paint layer. The spectral behaviour was analysed by both Fibre Optics Reflectance Spectroscopy (FORS) and Vis-NIR multi-spectral reflectance imaging. Finally, for each acrylic paint, meaningful spectral features were identified to assess the dependence of the reflectance on the layer thickness, thus obtaining non-linear fitting curves.

To the best of our knowledge, no similar approaches have been explored before for cross-sectional measurements of paint layers.

2. Results

2.1. Chemical Characterization of the Acrylic Paints with LIBS and Micro-Raman Spectroscopy

The chemical composition of the ten acrylic paints was assessed by LIBS and micro-Raman spectroscopies (Table 1), using the information reported in the literature [27–32] and NIST [33] and IRUG [34] databases. Further information on the chemical composition of the phthalocyanine paints (PBC, PBL and PGL) can be found in our previous work [35].

Figure 1 shows the LIBS (top) and micro-Raman (bottom) spectra of cobalt blue (CB), cadmium red (CR), cadmium yellow (CY) and primary blue cyan (PBC) acrylic paints. The spectra measured on the other analysed paints are displayed in Figure S1 in the Supplementary Material. For all paints, the chemical composition (Table 1) agrees with what was declared by the manufacturer (Table 2), except for the absence of titanium dioxide in permanent green light (PGL) and the presence of additional components in most of the analysed paints, ascribed to the binder and fillers. Specifically, atomic emissions of Mg, Si, Ca, Al, Sr, Ba and Na detected by LIBS are ascribed to fillers such as kaolin ($Al_2Si_2O_5(OH)_4$), gypsum ($CaSO_4 \cdot 1/2H_2O$), carbonates ($CaCO_3$, $MgCO_3$), glass powder and barite ($BaSO_4$) [31]. The molecular bands of CN (Violet band), CH and C_2 (Swan bands) are due to the acrylic binder, as well as to the organic pigments. The LIBS results obtained in these findings agree with the ones by micro-Raman spectroscopy. Bands from calcium sulfate (1007 cm^{-1}), calcium carbonate (1085 cm^{-1}) and barium sulfate (985 cm^{-1}) are observed in each acrylic paint and attributed to the fillers. Additional bands at 482, 600, 620, 837, 841, 1106, 1150–1200, 1240, 1305, 1449, 1452, 1728, 2411 and 2800–3100 cm^{-1} are attributed to the constituents of the polymeric binder [27–32].

Table 1. Summary of elemental and molecular composition of the acrylic paints as found by LIBS and micro-Raman spectroscopies. The main elemental components and the Raman characteristic bands are indicated in bold.

Paint	Identified Elemental Components by LIBS	Identified Raman Bands [cm^{-1}] and Relative Intensities *. In Brackets the Excitation Wavelength
CB	Mg, Si, **Co**, **Al**, **CN**, Ca, Sr, **C$_2$**, Na	198 m, 408 w, **512 m**, **609 w**, **750 w**, 1007 m, 1150–1200 w, **2411 s**, **2800–3100 s** (λ_{exc} = 532 nm)
CR	Mg, Si, **Cd**, Al, **CN**, Ca, Sr, CH, **C$_2$**, **Ba**, Na	**136 s**, **200 s**, **269 s**, 488 w, **587 s**, 841 w, 985 w, 1007 w, 1150–1200 w, 1305 m, 1452 m (λ_{exc} = 632 nm)
CY	Mg, **Si**, **Cd**, Al, **CN**, Ca, Sr, CH, **C$_2$**, **Ba**, Na	**212 s**, **309 s**, **353 w**, 600 s, 841 w, 985 w, 1007 s, 1150–1200 w, 1305 m, 1449 s, 1728 w (λ_{exc} = 632 nm)
PBC	Mg, Si, Al, **Cu**, **CN**, Ca, **Ti**, CH, **C$_2$**, Na	231 w, 255 w, 482 w, 590 m, 680 m, 747 w, 837 w, 841 w, **951 w**, 1007 w, **1037 w**, 1106 w, **1143 w**, 1150–1200 w, 1305 m, **1341 w**, 1451 m, **1527 s**, **1595 w**, 2672 w, 2870 w, 2976 w, 3056 w (λ_{exc} = 532 nm)
PBL	Mg, Si, Al, **Cu**, CN, Ca, **Ti**, Sr, C$_2$, Na	142 w, 231 m, 255 m, **433 s**, 482 w, **590 s**, **609 s**, **680 s**, 747 w, 831 w, 841, **951 w**, 1007 w, **1037 w**, **1143 m**, 1150–1200, **1200 w**, **1341 s**, 1451 s, **1527 s**, 1595 w, 2870 w, 3056 w (λ_{exc} = 532 nm)
PGL	Mg, Si, **Cd**, Al, **Cu**, CN, Ca, Sr, CH, C$_2$, **Ba**, Na	162 w, 505 w, **620 m**, **685 s**, 818 m, **978 w**, 985 w, 1007 w, **1080 m**, 1150–1200 m, **1200 m**, 1284 s, 1340 m, 1388 s, **1503 s**, **1536 s** (λ_{exc} = 532 nm)
PRM	Mg, Si, Al, **CN**, Ca, CH, **C$_2$**, **Na**	841 m, 1007 w, 1150–1200, 1240 m, **1316 s**, **1570 s**, **1592 s**, **1645 s** (λ_{exc} = 632 nm)
PI	Mg, Si, Al, **Ti**, **CN**, **Ca**, **C$_2$**, Na	186 m, 223 w, 261 w, 318 w, 360 w, 401 w, 525 w, 600 w, 623 w, 646 w, 802 m, 922 w, 1066 w, 1090 m, 1162 m, 1150–1200 m, **1171 m**, **1266 s**, **1326 s**, **1351 s**, **1402 s**, **1489 s**, **1513 s**, 1500 m, 1593 s, **1667 w** (λ_{exc} = 532 nm)
TW	Mg, Si, **Ti**, CN, Ca, C$_2$, Na	138 m, 230 m, **445 s**, **609 s**, 841 w, 1007 w, 1452 w, 2900–3100 s (λ_{exc} = 532 nm)
ZW	Mg, **Zn**, **CN**, Ca, CH, **C$_2$**, **Na**	330 w, 381 w, **435 s**, 620 m, 841 m, 1007 w, **1075 w**, **1150 m**, 1150–1200 m, 1449 m, 1452 s, 1728 m, 2800–3100 s (λ_{exc} = 532 nm)

* s: strong; m: medium; w: weak.

LIBS results reported in Table 1 show the presence of the main paintings markers such as Co for cobalt blue (CB), Cd for cadmium red (CR) and cadmium yellow (CY), Cu and Ti for permanent blue light (PBL), Cu for permanent green light (PBC and PGL), CN and C$_2$ for primary red magenta (PRM), Ti, CN and C$_2$ for primary yellow (PI), Ti for titanium white (TW), and Zn for zinc white (ZW).

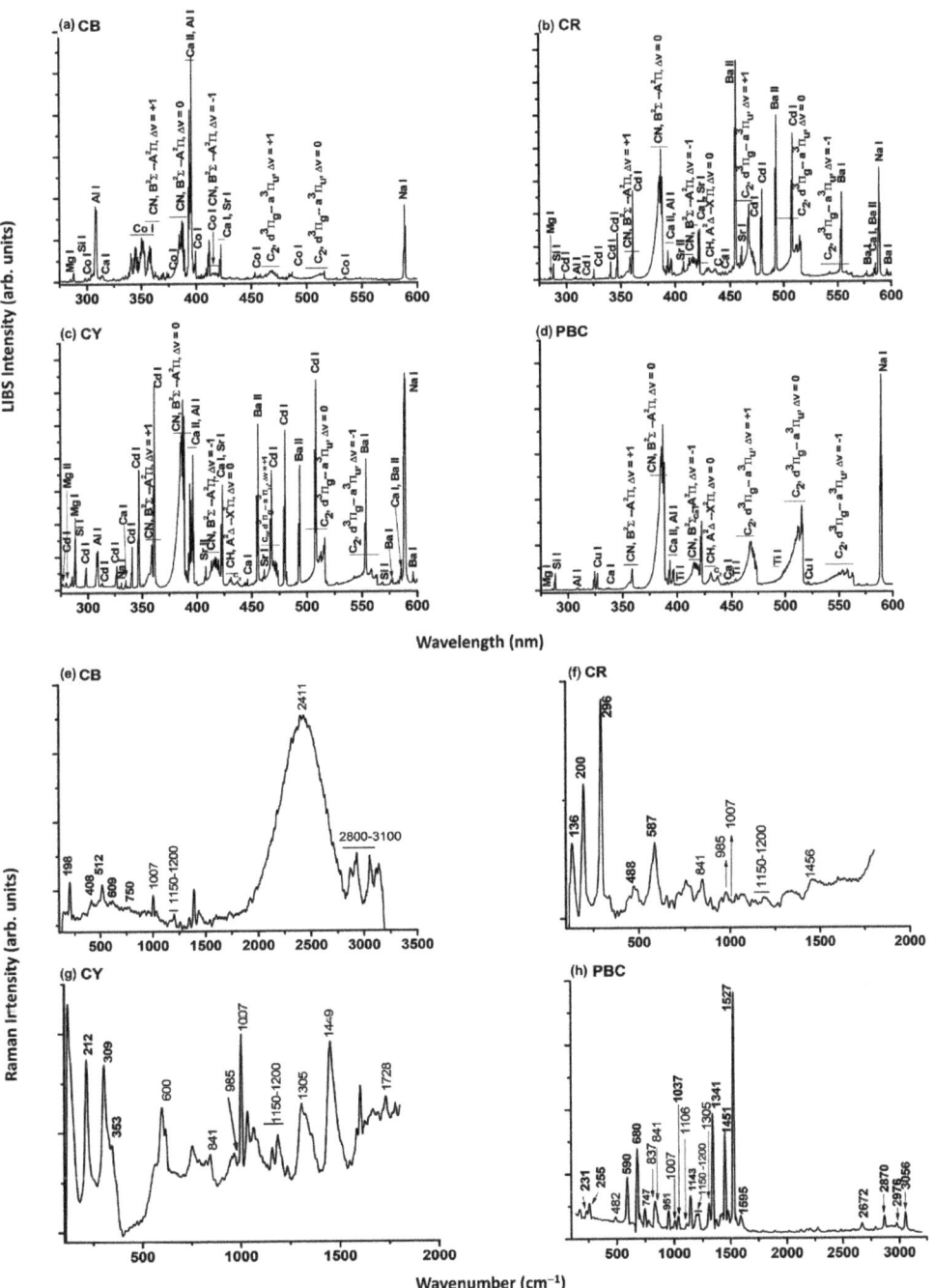

Figure 1. LIBS (top) and micro-Raman (bottom) spectra of cobalt blue (**a,e**), cadmium red (**b,f**), cadmium yellow (**c,g**) and primary blue cyan (**d,h**) acrylic paintings, respectively. The micro-Raman spectra are baseline subtracted.

Table 2. List of the ten analysed acrylic paints with their chemical composition and commercial code.

Paint with Acronym	Chemical Composition and Commercial Code (Maimeri Brera™)
Cobalt Blue (CB)	Cobalt(II) Aluminate [$CoAl_2O_4$], PB28—77346
Cadmium Red Medium (CR)	Cadmium Selenide Sulphide [Cd_2SSe], PR108—77202
Cadmium Yellow Medium (CY)	Cadmium Sulphide [CdS], PY35—77205
Primary Blue Cyan (PBC)	Copper Phthalocyanine β [$C_{32}H_{16}CuN_8$], PB15:3—74160
Permanent Blue Light (PBL)	Titanium Dioxide [TiO_2] PW6—77891, Chlorinated Phthalocyanine [$C_{32}HCl_{15}CuN$], PG7—74260, Copper Phthalocyanine β [$C_{32}H_{16}CuN_8$], PB15:3—74160
Permanent Green Light (PGL)	Arylide yellow, PY97—11767, Titanium Dioxide [TiO_2], PW6—77891, Chlorinated Phthalocyanine [$C_{32}HCl_{15}CuN$], PG7—74260
Primary Red Magenta (PRM)	Quinacridone [$C_{20}H_{12}N_2O_2$], PV19—73900
Primary Yellow (PI)	Arylide Yellow, PY97—11767
Titanium White (TW)	Titanium Dioxide [TiO_2], PW6—77891
Zinc White (ZW)	Zinc Oxide [ZnO], PW4—77947

2.2. Thickness Measurements with OCT

Four xz tomograms (8 × 0.6 mm^2, pixel size 3.5 μm^2) were acquired in each painted area. Given the low transparency of most of the analysed paints, the layer thickness was measured by taking the signal generated at the interface air-background (visible at the edges of each painted area) as a reference, as shown in Figure 2. The thickness of each pictorial layer was calculated as the average over 20 values, resulting in a range between a minimum of 45 μm (area 1) to a maximum of 350 μm (area 5) for all acrylics. For each thickness, the error, i.e., the standard deviation, resulted below 6 μm for all areas, demonstrating the micrometric homogeneity of the paint layers. Only three acrylics, namely PBC (Figure 3), PRM, and PY, showed sufficient transparency to enable the evaluation of their refractive index, which was calculated by dividing the thickness measured with OCT by the real one. The resulting n values at 1300 nm, i.e., at the OCT radiation wavelength, are in the range of 1.35–1.40 for both PBC and PRM, and 1.45–1.50 for PY.

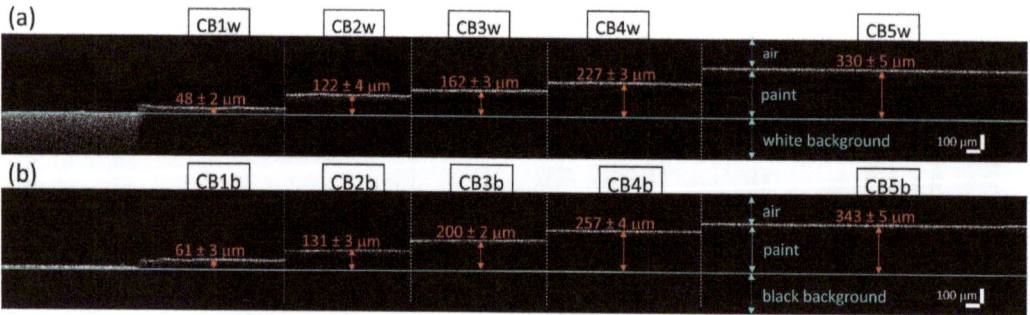

Figure 2. Assemblies of OCT tomograms acquired in Cobalt Blue (CB) paint laid on white (**a**) and black (**b**) backgrounds. For each area (1–5), thickness values are reported in red and calculated as the geometrical distance between the air–paint and the paint–background interfaces, with the latter highlighted by the light-blue line.

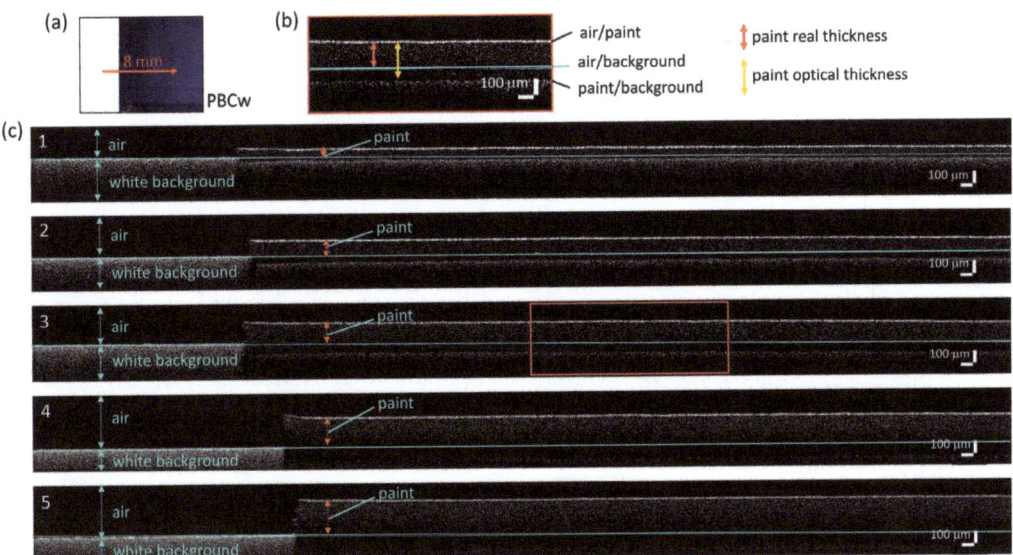

Figure 3. OCT results on Primary Blue Cyan paint laid on white background (PBCw); (**a**) microscope image of one of the five paint surfaces, with the red arrow indicating the location and length of the acquired section; (**b**) zoom-in of layer 3 delimited by the red rectangle in respective tomogram, enabling the assessment of the optical and real thicknesses used for calculating the refractive index of the acrylic paint; (**c**) OCT tomograms acquired on the five areas with increasing thickness (1–5), showing the transparency of the paint layer to the radiation probe.

2.3. Thickness Measurements with Reflectance Spectroscopy

FORS and multi-spectral data were compared for each paint, as shown in Figure 3a–d. In the graphs, the average spectra of each painted area (1→5) are plotted together with the spectrum of the underlying substrate (white or black background). First and second derivatives were computed for all spectra to facilitate the identification of the spectral feature best representing the reflectance dependence on the material thickness. For all paints, the trend of the reflectance as a function of the thickness is expressed by an exponential function, following the equation:

$$y = y_0 + Ae^{R_0 x} \qquad (1)$$

where y = R%, y_0 = offset, A = initial value, R_0 = growth constant, and x = layer thickness.

In the case of CB paint, the maximum reflectance R% values in the 808–811 and 710–760 nm ranges were selected for the white and the black background series, respectively, and plotted as a function of the thickness (Figure 4e,f). We noticed that the presence of the black background affects the position of the point of maximum reflectance, causing a blue shift as the thickness of the paint layer decreases and becomes gradually more transparent. In the presence of the white background, however, the point of maximum R remains around 810 nm regardless of the thickness of the paint layer. The resulting exponential fit curves (coefficient of determination $R^2 > 0.98$) show a good match between FORS and reflectance scanning results.

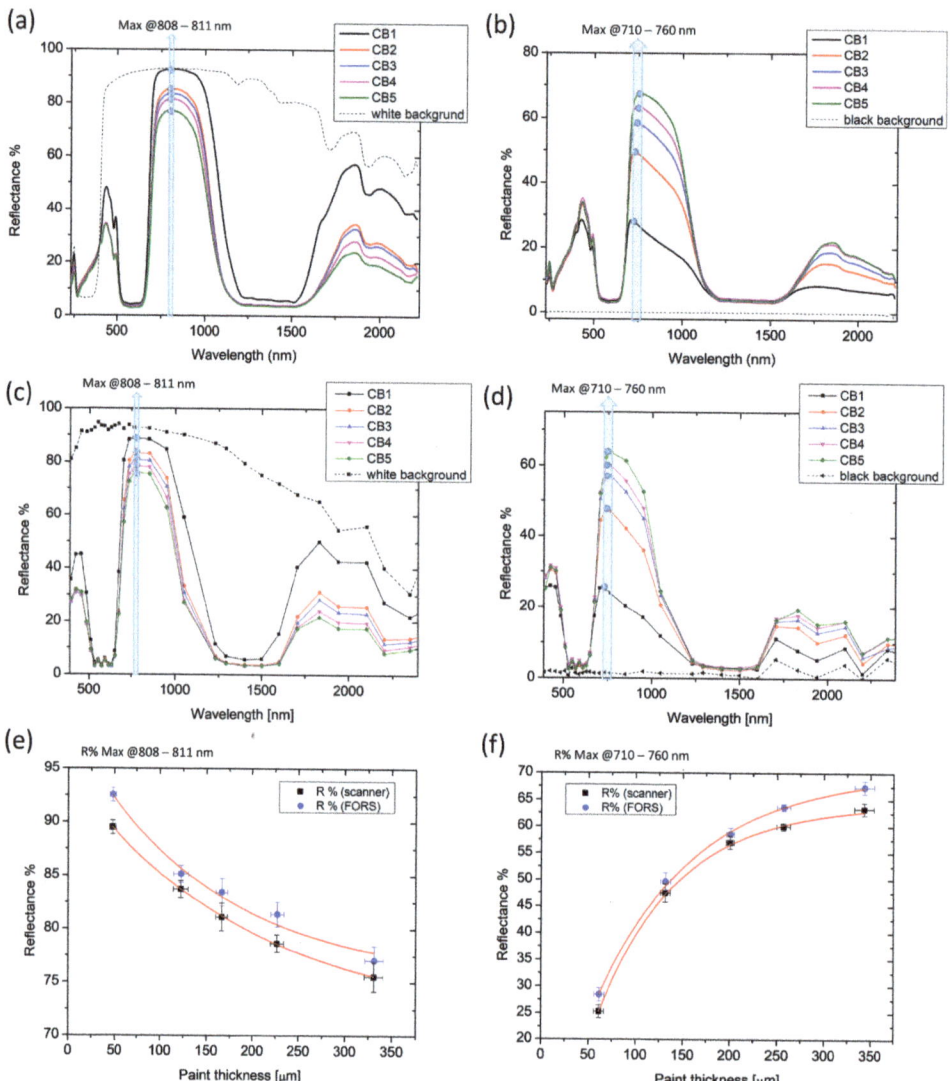

Figure 4. Results of diffuse reflectance spectroscopy on Cobalt Blue (CB) paint. Spectra acquired on each thickness layer (CB1-5) with FORS (**a**,**b**) and with the multi-spectral scanner (**c**,**d**) are reported with the spectra of the background (white or black). Maximum reflectance R% values in the 808–811 and 710–760 nm ranges are plotted as a function of the five OCT thicknesses (**e**,**f**). The length of the error bars is the standard deviation of each dataset. Red lines represent the fitting exponential functions.

The spectral feature selected for the analysis in the FORS spectra was not always identifiable in the spectra from the spectral cube due to the significantly lower spectral resolution of the multi-spectral scanner. Therefore, in order to evaluate the applicability of the proposed method in multi-spectral imaging mode, matching key points were found in the two datasets. With this aim, the results of PBC laid on the white background are shown in Figure 5 as an example. The spectral region between 600 and 1200 nm was chosen as significant for our computation: the multi-peak FORS spectra were fitted with a

5th-degree polynomial (Figure 5a) to reconstruct the shape of the multi-spectral spectrum. Maximum reflectance at 950 nm was then considered for both datasets. The resulting exponential fitting functions of the two DRS data show good accordance (Figure 5e). The same maximum was considered for the paint laid on the black background (Figure 5b,d,f) without fitting the FORS spectra to retrieve the same spectral feature. In this case, the reflectance measured on the thickest layer (PBC5 = 284 ± 10 μm) was excluded from the exponential fitting calculation (Figure 5f), since it clearly deviated from the increasing trend, being lower than that of PBC4. This measurable thickness threshold has also been found in other pigments for thicknesses exceeding 270 microns. Remarkably, this limit of detectability was exclusively found in the FORS spectra. This is possibly due to the different measurement configurations between the two DRS modalities, which results in a greater homogeneity of illumination and, therefore, depth of detection achievable with the multi-spectral scanner than with fibre optics.

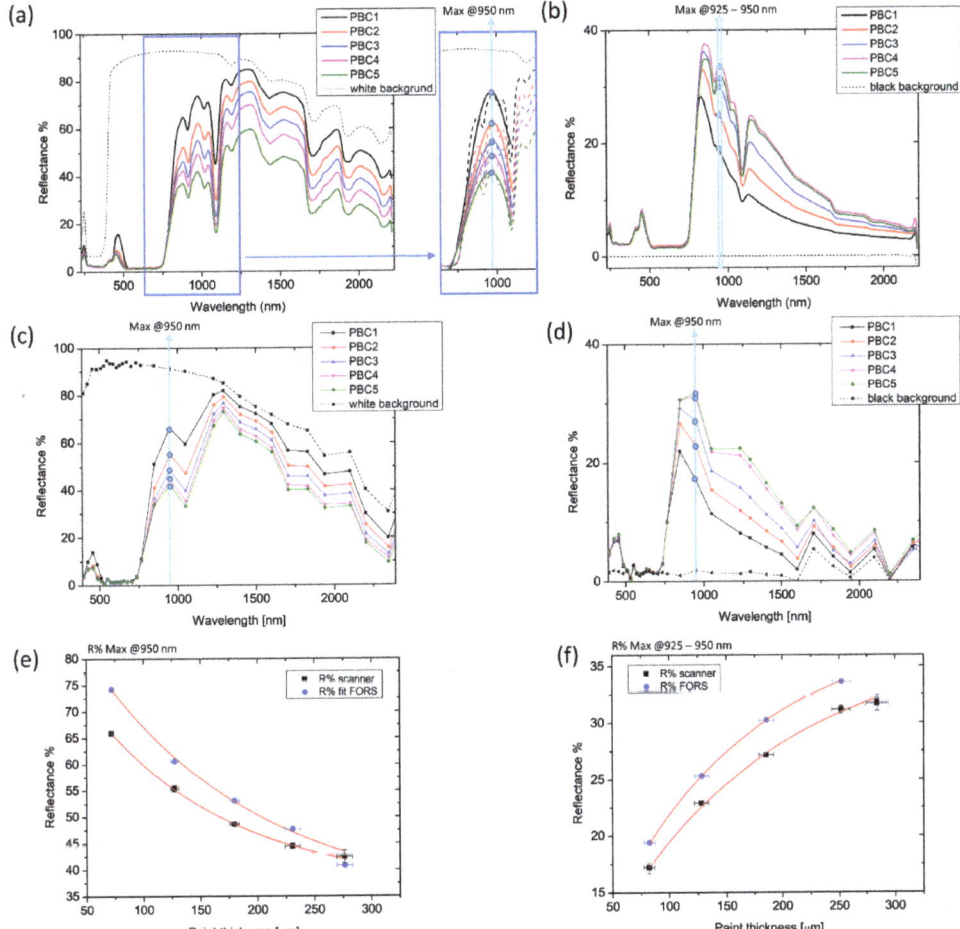

Figure 5. FORS (**a**,**b**) and multi-spectral scanner (**c**,**d**) results on Primary Blue Cyan (PBC) paint. The R% values at 950 nm are plotted with the five OCT thicknesses (**e**,**f**). The length of the error bars is the standard deviation of the dataset. Red lines represent the exponential fitting of the experimental points.

Optimal agreement between FORS and multispectral data was found in acrylics showing high transparency in the NIR range. As an example, results on PRM on a black background are shown in Figure 6a–c. As for highly scattering pigments, such as ZW shown in Figure 6d–f, an exponential fit curve could be derived only in the presence of the black background. Additionally, in this case, the detection limit found with FORS is 250 microns for both PRM and ZW, corresponding to the thickness of area 5.

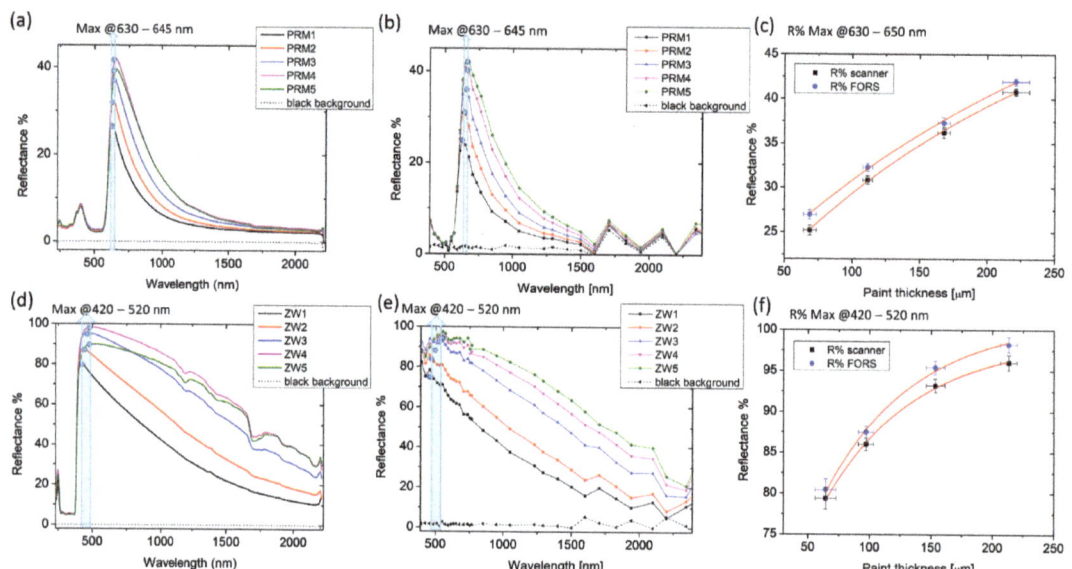

Figure 6. FORS (**a,d**) and multi-spectral scanner (**b,e**) results on Primary Red Magenta (PRM) and Zinc White (ZW) paints. R% values are plotted with the five OCT thicknesses (**c,f**). The length of the error bars is the standard deviation of the dataset. Red lines represent the exponential fitting of the experimental points.

As expected, in the presence of the white background, the reflectance of the identified spectral feature was found to decrease exponentially with increasing paint thickness for all acrylics. In contrast, in the presence of the black background, the reflectance was found to increase exponentially. The fitting parameters of function (1) for the ten acrylic paints, together with the considered spectral feature, are summarized in Table 3. In few cases, the high scattering or absorption properties of some pigments made it impossible to identify the spectral feature of interest for the application of the exponential fit. Specifically, the proposed method could not be applied to CR, CY, PBL, TW and ZW laid on a white background and PGL laid on a black background. In all the other cases, the quality of the fitting (coefficient of determination R^2) resulted in a value higher than 0.98.

Finally, the proposed method was tested on a sample of cobalt blue laid on canvas (Figure 7a). In the FORS spectrum (Figure 7b), obtained from the average of nine spectra, the maximum at 760 nm, previously identified on the mock-up for CB, resulted in R% = 82 ± 1. The corresponding thickness obtained using the exponential functions derived from both FORS and scanner analysis ranges from 145 to 178 μm (Figure 7c), which includes the actual thickness range measured with OCT, i.e., 150–165 μm (Figure 7d).

Table 3. Fit parameters of the exponential function (1) describing for each paint the dependence of the reflectance of the spectral feature identified with FORS and multi-spectral scanning on the layer thickness.

Paint	Background	Spectral Feature and Range [nm]	Technique	Fitting Parameters			Adjusted R^2
				y_0	A	R_0	
CB	w	max @808–811	FORS	75.29	23.78	-7×10^{-3}	0.98
			Scanner	71.46	23.13	-5×10^{-3}	
	b	max @710–760	FORS	70.1	−74.28	-9×10^{-3}	0.99
			Scanner	64.16	−79.75	-11×10^{-3}	
CR	w	n.a.	FORS	n.a.	n.a.	n.a.	n.a.
			Scanner	n.a.	n.a.	n.a.	
	b	flex @1400	FORS	94.19	−36.15	-9×10^{-3}	0.98
			Scanner	87.7	−36.6	-10×10^{-3}	
CY	w	n.a.	FORS	n.a.	n.a.	n.a.	n.a.
			Scanner	n.a.	n.a.	n.a.	
	b	flex @1705	FORS	65.93	154.23	-25×10^{-3}	0.99
			Scanner	69.01	−97.52	-19×10^{-3}	
PBC	w	max @950	FORS	33.03	66.86	-6×10^{-3}	0.99
			Scanner	36.52	53.01	-8×10^{-3}	
	b	max @925–950	FORS	38.46	−37.28	-8×10^{-3}	0.99
			Scanner	37.21	−34.98	-7×10^{-3}	
PBL	w	n.a.	FORS	n.a.	n.a.	n.a.	n.a.
			Scanner	n.a.	n.a.	n.a.	
	b	max @1230	FORS	91.73	−127.65	-24×10^{-3}	0.99
			Scanner	86.42	−78.84	-18×10^{-3}	
PGL	w	max @1050	FORS	69.17	22.57	-8×10^{-3}	0.99
			Scanner	64.72	22.90	-8×10^{-3}	
	b	n.a.	FORS	n.a.	n.a.	n.a.	n.a.
			Scanner	n.a.	n.a.	n.a.	
PRM	w	flex @610	FORS	45.97	−7.06	-4×10^{-3}	0.97
			Scanner	43.76	−3.72	-5×10^{-3}	
	b	flex @610	FORS	29.93	−20.94	-6×10^{-3}	0.99
			Scanner	29.34	−23.34	-10×10^{-3}	
PY	w	max @2010	FORS	32.97	29.38	-6×10^{-3}	0.99
			Scanner	33.13	24.45	-6×10^{-3}	
	b	max @550 570	FORS	91.12	−45.4	-6×10^{-3}	0.99
			Scanner	80.9	−42.05	-9×10^{-3}	
TW	w	n.a.	FORS	n.a.	n.a.	n.a.	n.a.
			Scanner	n.a.	n.a.	n.a.	
	b	max @1230	FORS	100.86	−72.65	-17×10^{-3}	0.99
			Scanner	89.51	−60.77	-19×10^{-3}	
ZW	w	n.a.	FORS	n.a.	n.a.	n.a.	n.a.
			Scanner	n.a.	n.a.	n.a.	
	b	max @420–520	FORS	101.20	−50.74	-14×10^{-3}	0.99
			Scanner	98.77	−46.81	-14×10^{-3}	

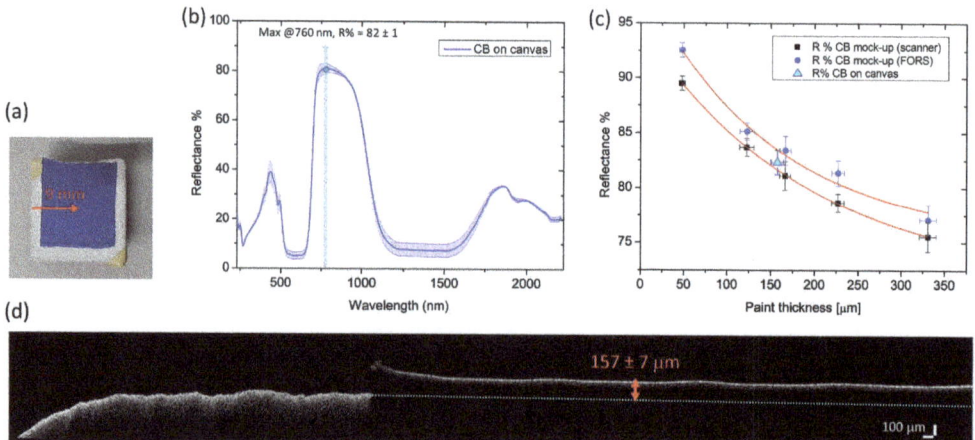

Figure 7. FORS and OCT results on CB laid on canvas: bright field image of the sample (**a**), with the arrow indicating the position and length of the OCT profile; averaged FORS spectrum (**b**) with a maximum at 760 nm considered for the analysis—the length of the error bars is the standard deviation; exponential functions previously obtained in the mock-up from FORS and scanner spectra, with R% measured on canvas sample reported in light blue (**c**); OCT tomogram of the paint layer with the measured thickness (**d**).

3. Materials and Methods

3.1. The Painting Mock-Up

The acrylic paints (extra-fine acrylic colours—Maimeri Brera™, Milan, Italy) selected for this study are reported in Table 2, with their commercial code and chemical composition (as declared by the manufacturer). The surface of the wooden support, a tablet with size $12.5 \times 26 \times 1$ cm^3, was prepared with a layer of acrylic primer (Lefranc Bourgeois™, Paris, France), which was accurately smoothed with fine-grit sandpaper. Half of the surface was covered with black acrylic paint (Carbon Black—PBk7—77266), as shown in Figure 8. Each acrylic colour was laid on five adjacent areas of 1 cm^2 (1–5 in Figure 8b) with increasing thicknesses. To this aim, a nearly 50 μm thick tape (Kapton Cypress—DuPont™, Wilmington, DE, USA) was used as a reference, i.e., depending on the desired thickness, 1 to 5 layers of tape were overlapped on both sides of the area to be painted. A glass spatula was used to spread the pigments in the various thicknesses in order to make their surface as even as possible.

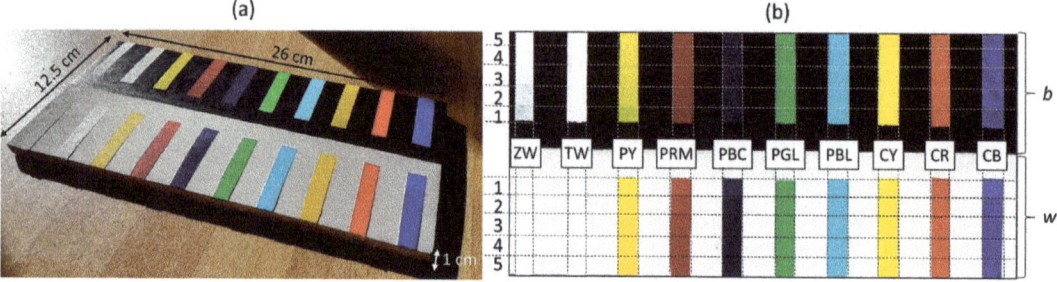

Figure 8. Acrylic painting mock-up: (**a**) photograph and (**b**) RGB image acquired with the multi-spectral scanner. For each acrylic, laid on the black (b) or white (w) backgrounds and labelled with the respective acronym (Table 1), the five areas with the increasing thickness (1→5) are delimited by the horizontal dotted lines.

The proposed method was finally tested on a sample made of a canvas industrially prepared with a gypsum-based primer and covered with a layer of Cobalt Blue acrylic paint.

3.2. Micro-Raman Spectroscopy

In order to enhance the detection of Raman signals emitted by high fluorescent paints such as Cadmium Yellow (CY), Cadmium Red (CR), and Primary Red Magenta (PRM), Raman analysis was carried out with two devices using different excitation wavelengths.

The first device is a Renishaw InVia 0310–02 System based on a continuous Nd:YAG laser excitation source at 532 nm. The diameter of the laser spot on the sample was diffraction limited to 1 µm by the objective lens (50×). The system is equipped with a Leica microscope (DMI 3000 M) and an electrically cooled CCD camera. The laser power was set between 0.15 and 0.30 mW.

The second device is a Renishaw Raman microscope RM1000 system coupled with an optical Leica DM LM microscope. The system is equipped with a refrigerated CCD camera and a CW He-Ne laser emitting at 632 nm, operating at a power of 3–30 mW, with a probing depth of 2 µm.

For all measurements, the spectral resolution was set at 4 cm^{-1} with an integration time in the range 10–60 s, the final spectra resulting from the accumulation of three individual ones.

3.3. Laser-Induced Breakdown Spectroscopy (LIBS)

The setup for LIBS analysis comprises a Q-switched Nd:YAG laser operating at its 4th harmonic at 266 nm, with a pulse duration of 15 ns and a repetition rate of 1 Hz. An f = 10 cm lens allows for fluences up to 6.6 J/cm^2. The luminous emission was collected and dispersed by a 0.30 m spectrograph with a 1200 lines/mm grating (TMc300, Bentham, London, UK) coupled to an intensified charged coupled device (ICCD, 2151 Andor Technologies, Belfast, UK). Spectra were recorded at a 0.2 nm resolution, with a temporal gate of 3 µs and a delay of 500 ns, in order to avoid the continuous Bremsstrahlung emission. The laser beam was directed toward the surface of the samples at an angle of 45° by using different mirrors. A cut-off filter of 300 nm was placed at the entrance of the spectrograph to reduce the scattered laser light and avoid second-order diffraction. The shot-to-shot fluctuation of laser pulse energy was less than 10%. The spectra resulted from summing the emissions of the ablation products after five successive laser pulses, a number that provided a good signal-to-noise ratio.

3.4. Spectral-Domain Optical Coherence Tomography (Sd-OCT)

The OCT device used in this study is a Thorlabs Telesto-II, working in the 1300 nm regime, with an axial resolution of 5.5 µm in air and a lateral resolution of 13 µm. The maximum field of view (FOV) is 10 × 10 mm^2, with a 3.5 mm imaging depth. The system is controlled via a 64-bit software running on a high-performance computer. The 3D scanning probe with an integrated video camera allows for high-speed imaging (76 kHz) for rapid volume acquisition and live display. The sample stage provides XY translation and rotation of the sample along with the axial travel of the probe. The 2D tomograms were acquired with a pixel size of 3.5 µm^2.

3.5. Fibre Optics Reflectance Spectroscopy (FORS)

FORS spectra were acquired with a Zeiss Multi-Channel Spectrometer, including two modules, the MCS601 UV-NIR C and the MCS611 NIR 2,2, with spectral sensitivity in the 190–1015 nm and 900–2200 nm range, and a spectral resolution of 0.8 nm/pixel and 5 nm/pixel, respectively. The size of the illumination spot was ⌀ = 3 mm, and the illumination/observation geometry was 45°/0°. The output signal was processed through dedicated software, providing also CIEL*a*b* coordinates with standard D65 illuminant and 2° observer. The calibration procedure was performed following CIE indications for non-contact spectrophotometric measurements by measuring a certified white 100% reflectance reference standard (Spectralon, Labsphere™, North Sutton, PA, USA) and background noise.

For this specific application, it was essential to obtain spectra with reproducible reflectance, as the absolute intensity value is the core physical quantity to prove our thesis. Therefore, we defined an optimized procedure to both stabilize the signal intensity at each measurement and minimize fluctuations between the two modules. For each area (each of the five thicknesses), five reflectance spectra were measured in five different points with three consecutive acquisitions interleaved with dark correction by automatic closure of the device shutter.

For each area, five averaged spectra (over three acquisitions to check the signal stability) were obtained, which were then averaged to obtain a single spectrum for each thickness, together with the standard deviation. The reflectance of the white and black backgrounds was also measured following the same procedure. Therefore, a total of 1530 spectra were acquired on the painting mock-up (510 spectra for each measured point). Spectral data were processed in OriginLab environment.

3.6. Reflectance Imaging Spectroscopy (RIS)

The multispectral scanner [36] developed at CNR-INO operates in the range 395–2550 nm providing simultaneously 32 narrow-band images (16 VIS + 16 NIR) and pointwise spectral information. The simultaneous movement of both the lighting system and collecting optics, placed in a 45°/0° illumination/detection geometry, allows for uniform illumination and minimal heating of the surface. During the scanning, an autofocus system ensures the optimal target-lens distance. The system has a 250 µm sampling step (4 points/mm) and 500 mm/s maximum speed. A proper calibration procedure was performed by measuring a certified white 100% reflectance reference standard (Spectralon™) and background noise, following CIE indications for non-contact spectrophotometric measurements. The spectra reported in this work were extracted from the spectral cube using an in-house developed software. Each spectrum is averaged over an area of $\varnothing$ = 8 mm selected on each painted region. Given that the system has a 250 µm sampling step (4 points/mm), each spectrum is averaged over ~805 pixels.

4. Conclusions

The chemical characterization of the considered ten acrylic paint materials was obtained by LIBS and micro-Raman spectroscopies, providing complementary information for the identification of elemental and molecular compounds in each paint and highlighting, in some cases, discrepancies with the chemical composition declared by the manufacturer.

The results obtained with the proposed method show that there is a clear correlation between the spectral response of the diffuse reflectance and the micrometric thickness of acrylic paint layers. Using two DRS modes (FORS and multispectral scanner), it was possible to identify significant spectral features (maxima and inflexion points) where the exponential decay of the reflectance on the layer thickness was evident. The resulting exponential fitting functions showed good agreement between the two techniques, both showing opposite trends depending on the substrate (white or black) for all paints. The fitting function defined for cobalt blue was used to estimate the thickness of a layer of paint spread on canvas, which was consistent with the thickness measured by OCT.

Measuring the thickness of a pictorial layer from its spectral characteristics can overcome some limitations of the OCT technique. Knowing a priori the refractive index of the material in order to correct optical distances in OCT tomograms is often not possible, especially for pigment mixtures typically found on paintings. Here, we have shown that the proposed method allows the measuring of the thickness of ten acrylic paints, including optically opaque pigments that cannot be measured by OCT, such as CY and CR (only in the presence of a black background), without the need to know their refractive index.

In this work, the feasibility of the method was demonstrated, assuming a single-layer homogeneous paint was present. For more complex situations, such as layers stratifications or pigment mixtures in painted artworks, more advanced techniques for feature extraction can be used [37]. Moreover, a systematic study based on specific spectral reflectance models simulating the painting composition and structure should be developed, considering the optical behaviour of each component. The presence of multilayers or pictorial mixtures, commonly found in paintings, requires the creation of a much larger database. In this view, approaches based on AI could be tested. An interesting development of the technique is the possibility of obtaining thickness maps on painted surfaces using hyperspectral reflectography, which offers spectral resolution comparable to FORS, but over large scanning areas.

Supplementary Materials: The following supporting information can be downloaded at: https://www.mdpi.com/article/10.3390/molecules28124683/s1, Figure S1: Chemical characterization of the acrylic paints with LIBS and micro-Raman spectroscopy.

Author Contributions: Conceptualization, A.D.F. and R.F.; methodology, A.D.F., R.F. and M.C.; validation, A.D.F., M.O., M.C. and R.F.; formal analysis, A.D.F., M.O., M.M.-W.; investigation, A.D.F., M.M.-W.; resources, R.F.; data curation, A.D.F. and M.O.; writing—original draft preparation, A.D.F.; writing—review and editing, A.D.F., M.C., M.O., M.M.-W. and R.F.; supervision, R.F.; funding acquisition, R.F. All authors have read and agreed to the published version of the manuscript.

Funding: This research was funded by the Spanish State Research Agency (AEI) through project PID2019-104124RB-I00/AEI/10.13039/501100011033, by project TOP Heritage-CM (S2018/NMT-4372) from Community of Madrid, and by the H2020 European project IPERION HS (Integrated Platform for the European Research Infrastructure ON Heritage Science, GA 871034).

Institutional Review Board Statement: Not applicable.

Informed Consent Statement: Not applicable.

Data Availability Statement: The data presented in this study are available on reasonable request from the corresponding author.

Acknowledgments: Support from Ana Crespo (Instituto de Estructura de la Materia, IEM-CSIC) for Raman measurements and by CSIC Interdisciplinary Platform "Open Heritage: Research and Society" (PTI-PAIS) are acknowledged.

Conflicts of Interest: The authors declare no conflict of interest.

Sample Availability: Samples of the compounds are available from the authors under reasonable request.

References

1. Iwanicka, M.; Moretti, P.; van Oudheusden, S.; Sylwestrzak, M.; Cartechini, L.; van den Berg, K.J.; Targowski, P.; Miliani, C. Complementary use of Optical Coherence Tomography (OCT) and Reflection FTIR spectroscopy for in situ non-invasive monitoring of varnish removal from easel paintings. *Microchem. J.* **2018**, *138*, 7–18. [CrossRef]
2. Targowski, P.; Iwanicka, M. Optical Coherence Tomography: Its role in the non-invasive structural examination and conservation of cultural heritage objects—A review. *Appl. Phys. A* **2012**, *106*, 265–277. [CrossRef]
3. Targowski, P.; Góra, M.; Wojtkowski, M. Optical Coherence Tomography for Artwork Diagnostics. *Laser Chem.* **2006**, *2006*, 35373–35383. [CrossRef]
4. Liang, H.; Cid, M.G.; Cucu, R.G.; Dobre, G.M.; Podoleanu, A.; Pedro, J.; Saunders, D. En-face optical coherence tomography—A novel application of non-invasive imaging to art conservation. *Opt. Express* **2005**, *13*, 6133–6144. [CrossRef] [PubMed]
5. Fontana, R.; Dal Fovo, A.; Striova, J.; Pezzati, L.; Pampaloni, E.; Raffaelli, M.; Barucci, M. Application of non-invasive optical monitoring methodologies to follow and record painting cleaning processes. *Appl. Phys. A* **2015**, *121*, 957–966. [CrossRef]
6. Abraham, E.; Fukunaga, K. Terahertz imaging applied to the examination of artistic objects. *Stud. Conserv.* **2015**, *60*, 343–352. [CrossRef]
7. Borg, B.; Dunn, M.; Ang, A.; Villis, C. The application of state-of-the-art technologies to support artwork conservation: Literature review. *J. Cult. Herit.* **2020**, *44*, 239–259. [CrossRef]
8. Rehorn, C.; Blümich, B. Cultural Heritage Studies with Mobile NMR. *Angew. Chem. Int. Ed.* **2018**, *57*, 7304–7312. [CrossRef]

9. Di Tullio, V.; Sciutto, G.; Proietti, N.; Prati, S.; Mazzeo, R.; Colombo, C.; Cantisani, E.; Romè, V.; Rigaglia, D.; Capitani, D. 1H NMR depth profiles combined with portable and micro-analytical techniques for evaluating cleaning methods and identifying original, non-original, and degraded materials of a 16th century Italian wall painting. *Microchem. J.* **2018**, *141*, 40–50. [CrossRef]
10. Brizi, L.; Bortolotti, V.; Marmotti, G.; Camaiti, M. Identification of complex structures of paintings on canvas by NMR: Correlation between NMR profile and stratigraphy. *Magn. Reson. Chem.* **2020**, *58*, 889–901. [CrossRef]
11. Dal Fovo, A.; Castillejo, M.; Fontana, R. Nonlinear optical microscopy for artworks physics. *La Riv. Del Nuovo Cim.* **2021**, *44*, 453–498. [CrossRef]
12. Dal Fovo, A.; Sanz, M.; Oujja, M.; Fontana, R.; Mattana, S.; Cicchi, R.; Targowski, P.; Sylwestrzak, M.; Romani, A.; Grazia, C.; et al. In-Depth Analysis of Egg-Tempera Paint Layers by Multiphoton Excitation Fluorescence Microscopy. *Sustainability* **2020**, *12*, 3831. [CrossRef]
13. Mari, M.; Filippidis, G. Non-Linear Microscopy: A Well-Established Technique for Biological Applications towards Serving as a Diagnostic Tool for in situ Cultural Heritage Studies. *Sustainability* **2020**, *12*, 1409. [CrossRef]
14. Dal Fovo, A.; Tserevelakis, G.J.; Papanikolaou, A.; Zacharakis, G.; Fontana, R. Combined photoacoustic imaging to delineate the internal structure of paintings. *Opt. Lett.* **2019**, *44*, 919–922. [CrossRef]
15. Tserevelakis, G.J.; Tsafas, V.; Melessanaki, K.; Zacharakis, G.; Filippidis, G. Combined multiphoton fluorescence microscopy and photoacoustic imaging for stratigraphic analysis of paintings. *Opt. Lett.* **2019**, *44*, 1154–1157. [CrossRef]
16. Dal Fovo, A.; Tserevelakis, G.J.; Klironomou, E.; Zacharakis, G.; Fontana, R. First combined application of photoacoustic and optical techniques to the study of an historical oil painting. *Eur. Phys. J. Plus* **2021**, *136*, 757. [CrossRef]
17. Striova, J.; Dal Fovo, A.; Fontana, R. Reflectance imaging spectroscopy in heritage science. *La Riv. Nuovo Cim.* **2020**, *43*, 515–566. [CrossRef]
18. Delaney, J.K.; Dooley, K.A. Visible and Infrared Reflectance Imaging Spectroscopy of Paintings and Works on Paper. In *Analytical Chemistry for the Study of Paintings and the Detection of Forgeries*; Springer: Cham, Switzerland, 2022; pp. 115–132.
19. Kubelka, P. Ein Beitrag zur Optik der Farbanstriche (Contribution to the optic of paint). *Z. Fur Tech. Phys.* **1931**, *12*, 593–601.
20. Cavaleri, T.; Giovagnoli, A.; Nervo, M. Pigments and Mixtures Identification by Visible Reflectance Spectroscopy. *Procedia Chem.* **2013**, *8*, 45–54. [CrossRef]
21. Bacci, M.; Fabbri, M.; Picollo, M.; Porcinai, S. Non-invasive fibre optic Fourier transform-infrared reflectance spectroscopy on painted layers: Identification of materials by means of principal component analysis and Mahalanobis distance. *Anal. Chim. Acta* **2001**, *446*, 15–21. [CrossRef]
22. Pouyet, E.; Miteva, T.; Rohani, N.; de Viguerie, L. Artificial Intelligence for Pigment Classification Task in the Short-Wave Infrared Range. *Sensors* **2021**, *21*, 6150. [CrossRef] [PubMed]
23. Pottier, F.; Gerardin, M.; Michelin, A.; Hébert, M.; Andraud, C. Simulating the composition and structuration of coloring layers in historical painting from non-invasive spectral reflectance measurements. *C. R. Phys.* **2018**, *19*, 599–611. [CrossRef]
24. Sessa, C.; Bagán, H.; García, J.F.; Navarro, H.B. Influence of composition and roughness on the pigment mapping of paintings using mid-infrared fiberoptics reflectance spectroscopy (mid-IR FORS) and multivariate calibration. *Anal. Bioanal. Chem.* **2014**, *406*, 6735–6747. [CrossRef] [PubMed]
25. Conti, C.; Botteon, A.; Colombo, C.; Pinna, D.; Realini, M.; Matousek, P. Advances in Raman spectroscopy for the non-destructive subsurface analysis of artworks: Micro-SORS. *J. Cult. Herit.* **2020**, *43*, 319–328. [CrossRef]
26. Kaszewska, E.A.; Sylwestrzak, M.; Marczak, J.; Skrzeczanowski, W.; Iwanicka, M.; Szmit-Naud, E.; Anglos, D.; Targowski, P. Depth-Resolved Multilayer Pigment Identification in Paintings: Combined Use of Laser-Induced Breakdown Spectroscopy (LIBS) and Optical Coherence Tomography (OCT). *Appl. Spectrosc.* **2013**, *67*, 960–972. [CrossRef]
27. Pagnin, L. *Characterization and Quantification of Modern Painting Materials by IR and Raman Spectroscopies*; Università Ca'Foscari Venezia: Venezia, Italy, 2017. Available online: http://hdl.handle.net/10579/11266 (accessed on 10 March 2023).
28. Aguayo, T.; Clavijo, E.; Villagrán, A.; Espinosa, F.; Sagüés, F.E.; Campos-Vallette, M. Raman vibrational study of pigments with patrimonial interest for the Chilean cultural heritage. *J. Chil. Chem. Soc.* **2010**, *55*, 347–351. [CrossRef]
29. Bell, I.M.; Clark, R.J.; Gibbs, P.J. Raman spectroscopic library of natural and synthetic pigments (pre- ≈ 1850 AD). *Spectrochim. Acta Part A Mol. Biomol. Spectrosc.* **1997**, *53*, 2159–2179. [CrossRef]
30. Barni, D.; Raimondo, L.; Galli, A.; Yivlialin, R.; Caglio, S.; Martini, M.; Sassella, A. Chemical separation of acrylic color components enabling the identification of the pigment spectroscopic response. *Eur. Phys. J. Plus* **2021**, *136*, 254. [CrossRef]
31. López-Ramírez, M.R.; Navas, N.; Rodríguez-Simón, L.R.; Otero, J.C.; Manzano, E. Study of modern artistic materials using combined spectroscopic and chromatographic techniques. Case study: Painting with the signature "Picasso". *Anal. Methods* **2015**, *7*, 1499–1508. [CrossRef]
32. Corden, C.; Matousek, P.; Conti, C.; Notingher, I. Sub-Surface Molecular Analysis and Imaging in Turbid Media Using Time-Gated Raman Spectral Multiplexing. *Appl. Spectrosc.* **2021**, *75*, 156–167. [CrossRef]
33. NIST. Atomic Spectra Database [Online]. Available online: http://physics.nist.gov/asd (accessed on 30 March 2023).
34. Infrared and Raman Users Group (IRUG). Spectral Database Index, (s. f.). Available online: http://www.irug.org/search-spectral-database/spectra-index?sortHeader=data_type_raman (accessed on 29 March 2023).
35. Dal Fovo, A.; Oujja, M.; Sanz, M.; Martínez-Hernández, A.; Cañamares, M.V.; Castillejo, M.; Fontana, R. Multianalytical non-invasive characterization of phthalocyanine acrylic paints through spectroscopic and non-linear optical techniques. *Spectrochim. Acta Part A Mol. Biomol. Spectrosc.* **2019**, *208*, 262–270. [CrossRef]

36. Striova, J.; Ruberto, C.; Barucci, M.; Blažek, J.; Kunzelman, D.; Dal Fovo, A.; Pampaloni, E.; Fontana, R. Spectral Imaging and Archival Data in Analysing *Madonna of the Rabbit* Paintings by Manet and Titian. *Angew. Chem.* **2018**, *130*, 7530–7534. [CrossRef]
37. Geldof, F.; Dashtbozorg, B.; Hendriks, B.H.; Sterenborg, H.J.; Ruers, T.J. Layer thickness prediction and tissue classification in two-layered tissue structures using diffuse reflectance spectroscopy. *Sci. Rep.* **2022**, *12*, 1698. [CrossRef]

Disclaimer/Publisher's Note: The statements, opinions and data contained in all publications are solely those of the individual author(s) and contributor(s) and not of MDPI and/or the editor(s). MDPI and/or the editor(s) disclaim responsibility for any injury to people or property resulting from any ideas, methods, instructions or products referred to in the content.

Article

Microanalytical Characterization of an Innovative Modern Mural Painting Technique by SEM-EDS, NMR and Micro-ATR-FTIR among Others

Pablo Aguilar-Rodríguez [1], Sandra Zetina [2], Adrián Mejía-González [1] and Nuria Esturau-Escofet [1,*]

[1] Instituto de Química, Universidad Nacional Autónoma de México, Mexico City 04510, Mexico
[2] Instituto de Investigaciones Estéticas, Universidad Nacional Autónoma de México, Mexico City 04510, Mexico
* Correspondence: nesturau@iquimica.unam.mx

Abstract: During the 20th century, modern painters experimented with different mediums and painting techniques, one of them was Rafael Coronel in his mural painting, *Paisaje Abstracto* (*Abstract landscape*). The painting was created with a peculiar pouring technique and an unknown binding medium; ageing produced fractures and severe conservation problems. Therefore, the characterization of the painting medium became an urgent matter in order to understand the current condition of the painting and to develop a proper treatment. The aim of this research was to characterize the chemical composition and painting technique of *Paisaje Abstracto*. To approach this goal two microsamples were taken and analyzed by optical microscopy (OM), scanning electron microscopy (SEM) with energy dispersive spectroscopy (EDS), nuclear magnetic resonance (NMR) spectroscopy, attenuated total reflection Fourier transform infrared spectroscopy (ATR-FTIR), micro attenuated total reflection Fourier transform infrared spectroscopy (micro-ATR-FTIR) and gas chromatography/mass spectrometry (GC/MS). The analysis allowed for the identification of cadmium sulfide (CdS) and titanium dioxide (TiO_2) as inorganic pigments; aluminosilicate fillers; poly(methyl methacrylate) (pMMA) as a binder; MMA monomer, red organic pigment PR181; benzoyl peroxide, dibutyl phthalate and 1-octadecanol as organic additives. This study presents an innovative painting technique with pMMA, a medium not commonly used by artists, which was probably polymerized onto the painting support.

Keywords: pMMA; multianalytical characterization; painting medium; mural painting art; NMR; micro-ATR-FTIR

1. Introduction

Paisaje Abstracto (*Abstract landscape*) is a transportable mural painted in 1964 by Rafael Coronel (1932–2019) to decorate the hall that leads to the Library at the Museo Nacional de Antropología (MNA, National Museum of Anthropology), in Mexico City. Rafael Coronel created an abstract composition that presents large areas of bright colors emerging from darker regions; a horizontal red haze at the center dominates the rectangular format. The painting technique is unusual, the composition lies over the texture qualities but there are no visible brushstrokes, as if the drying of the binding medium was slow. The resulting rough surface recalls a landscape, a coarse bubble-like crust with elevations and valleys. It seems to be created by pouring red, yellow and blue dense painting layers intermingled with black, gray and dark ocher thin layers, in a manner that recalls the experiments with automatist painting of American abstract expressionists or French tachistes.

The painting is not signed, and no documentation has been found at the museum archives about the author or the materials used by the artist. Between 1963 and 1964, at an early stage in his career, Rafael Coronel was experimenting with abstraction, but also with figurative painting, and it is evident that he was searching for new materials and techniques. Some sketches, painted with synthetic binding medium over canvas, were found at the artist collection; they bear the title Abstracción (Abstraction) and present a

red irregular area over a black rough surface. At the same time that the *Paisaje Abstracto* was painted, Coronel produced another well documented figurative painting, El mundo espiritual de los mayas peninsulares (The Spiritual World of Peninsular Mayas), for the same museum with a rather traditional brushwork technique using acrylic painting.

These paintings are part of the ambitious project developed by Architect Pedro Ramírez Vázquez for the creation of the acclaimed MNA in 1964. Developed under Adolfo López Mateos' government (1958–1964), it became a symbol of Mexico's cultural heritage [1]. The MNA project combined contemporary monumental architecture, innovative displays and Modern art, with the intention of giving a prominent place to Mesoamerican and indigenous art, but also to present Mexico as a cosmopolitan country through the inclusion of Modern tendencies [1].

The MNA Modern art project included sculptures, murals and assemblages, in which 23 artists associated with diverse and even contrasting artistic tendencies participated: social realists, advocates of pure painting or abstraction, and even surrealists, in an array of the proposals in conflict between the 1950s and 1960s in Mexico. Rafael Coronel pertained to the younger generation, sometimes called Ruptura, a term that inefficiently defines several divergent artistic groups and artists that rebelled against the Mexican School of Painting or muralism and their social realism [2]. Many of the younger painters, such as Rafael Coronel, preferred to be understood as individuals rather than as a group. Coronel was sometimes affiliated with the proposals of the New Presence or interiorista manifestos, an introspective current that evoked personal emotions through the violent application of painting, but was neither associated with abstraction or figuration [3].

The younger generation of Mexican painters rejected social realism associated with muralism, but were interested in David Alfaro Siqueiros's innovations with industrial binding mediums and techniques such as cellulose nitrate lacquers and spray guns, commonly used for the automotive industry. They appreciated the diversity of possibilities that offer synthetic mediums to create textures, volumes and color contrast, and in that sense, *Paisaje Abstracto* seems to be an experiment with industrial binding mediums to produce an abstract composition.

In this paper, we characterize the composition and painting technique of *Paisaje Abstracto*, which was painted with an unknown synthetic medium over a wood panel support; the painting layers are rigid and detached due to the expansion of the wood. During the last conservation process, two samples were provided by restorers in order to understand the materials and techniques used by the artist with the intention of designing the treatments. This paper presents the study of the binder; the additives, aggregates, and organic and inorganic pigments of these two samples that were analyzed by Optical Microscopy (OM), Scanning Electron Microscopy–Energy Dispersive X-ray Spectroscopy (SEM-EDS), Nuclear Magnetic Resonance (NMR) Spectroscopy, Attenuated Total Reflection Fourier Transform Infrared spectroscopy (ATR-FTIR), Micro Attenuated Total Reflection Fourier Transform Infrared spectroscopy (micro-ATR-FTIR) and Gas Chromatography/Mass Spectrometry (GC/MS).

The study is also aimed at examining the functionality, performance and time saving of scientific instruments suitable for the analysis of works of art of the period, many made with synthetic or innovative unknown materials. The selected methodology could be proposed as a protocol to be used for the art historical study and conservation of modern paintings created during the second half of the twentieth century.

Regarding the NMR spectroscopy technique, is important to mention that it is a powerful and frequently used technique in many fields, from basic to applied sciences, to characterize both molecular and supramolecular structures of organic and inorganic compounds. The main disadvantage of NMR spectroscopy is its low sensitivity compared to chromatographic separation coupled with any kind of mass spectrometry detection. This is an essential issue for cultural heritage analysis due to the limited amounts of samples available. However, the sensitivity of NMR spectroscopy has increased enormously because of recent improvements in hardware and instrumental advances. Therefore, in recent years, NMR spectroscopy as a tool for characterizing complex mixtures has increased, particularly in the field of cultural heritage, as it provides valuable insights in the identification of organic materials such as waxes, paint binders, pigments, additives and organic degradation products [4–9].

2. Results

2.1. Description of the Technique and Observations of the Conservation State

The painting structure of *Paisaje Abstracto* (Figure 1) is quite simple; the medium and pigments were applied directly onto a hardwood plywood panel, formed by two plywood sheets that were sustained by a reticular wood stretcher. Originally, a wooden frame reinforced the structure. These kinds of hardwood panels, of monumental dimensions (3.90 × 2.40 m) were produced by the museography staff and are similar in structure and dimensions to other paintings in the museum [10].

Figure 1. Photo of *Paisaje Abstracto* by Rafael Coronel after intervention. Photo: D.R. ©Digital Archive of the Collections of the National Museum of Anthropology, INAH, Canon, 2018.

The painting layer was possibly applied over the support placed in a horizontal position. It seems more like the materials were poured over the surface, creating textures and considerable elevations, a kind of topographic relief formed by the unevenness of the surface. When samples were observed under the magnifying lenses, the surface had a bubble texture of spheres. Under UV lighting, the surface of the painting presents strong fluorescence in red, dark brown and yellow areas [10]. The palette is restricted: blue, yellow, orange, red, dark ochre, black and white, were applied by areas. The intense colors emerge from a darker foreground; multiple textures increase the formation of a layered, volcano-like structure. The painting layer has an average of 1 cm of thickness with elevations that may protrude as high as 3 cm.

Only one side of the original frame was preserved; the stretcher was in good condition, but the plywood panels had deformations that detached two sections of the plywood from the stretcher and formed a concavity. The upper section to the right of the painting was separated by about 1.30 m and was detached almost 6 cm from the stretcher [10].

Possibly the loss of the frame, the changes in relative humidity and the particularly rigid, uneven and thick painting layer, produced a severe deformation of the plywood panels, and consequently a bulge detachment and the fracture of the painting layers in the heavier areas. The fractured and split areas from the support were veiled by conservators with non-woven fabric to prevent them from falling [10].

The rigidness of the painting layer and its lack of flexibility compromise its adhesion to the wood panel support. On the other hand, the wood support is reactive to the conditions of relative humidity and temperature, which alter its dimensions cyclically. The chemical composition of the painting layer forms a rigid solid that will not meet the dimensional changes of the wood panel.

2.2. Optical Microscopy Study

Figure 2a is a micrograph of the cross section of the sample with polarized light, where six layers can be distinguished: the first layer is yellow, followed by a greyish layer, then a dark red color and finally the red surface paint layer. In all the samples, an agglomeration of spherical structures is observed. Figure 2b shows the cross section of the sample with a UV filter from 460 to 490 nm, where the intermediate gray layer, identified in Figure 2a, presents different fluorescence caused by the overlay of rich binder layers and fillers, and pigments layers. Whitish regions with increased fluorescence that could be attributed to pigments or inorganic fillers are evident. The artist did not use a ground preparation for this mural; he applied, directly onto the support, the different layers of paint, where the surface paint layer was red (obverse side) (Figure 2c) and the first was a yellow layer (reverse side) (Figure 2d). The yellow painting layer has a flat surface because this side was attached to the panel.

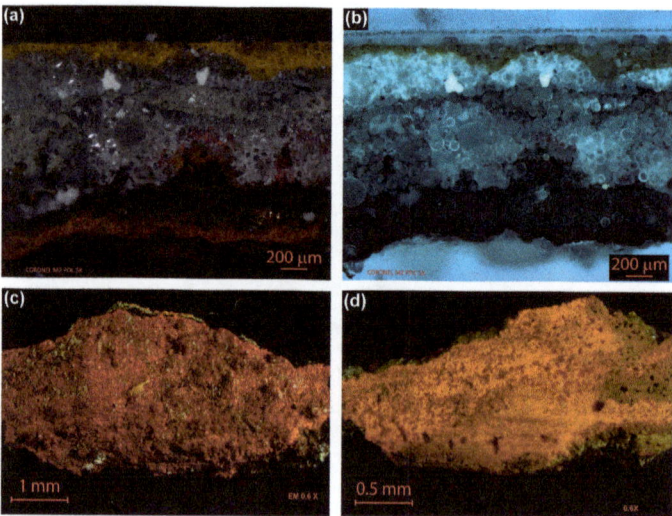

Figure 2. Cross section micrograph of the mural *Paisaje Abstracto* sample (area studied approximately 2100 × 1400 microns, magnification: 5×). (**a**) Polarized light; (**b**) UV filter, 460 to 490 nm (FITC); (**c**) sample without preparation, front surface area (approximately 3 × 6 mm); (**d**) back of the sample. Images (**a**,**b**), the cross section micrographs present the paint layers in reverse order (those closest to the panel are at the top).

2.3. Elemental Composition Analysis by SEM-EDS

Figure 3a shows the backscattered electrons (BSE) micrograph of the *Paisaje Abstracto* cross-section, in which low-contrast circular regions with a size between 20 and 150 m predominate, which are embedded in a heterogeneous matrix.

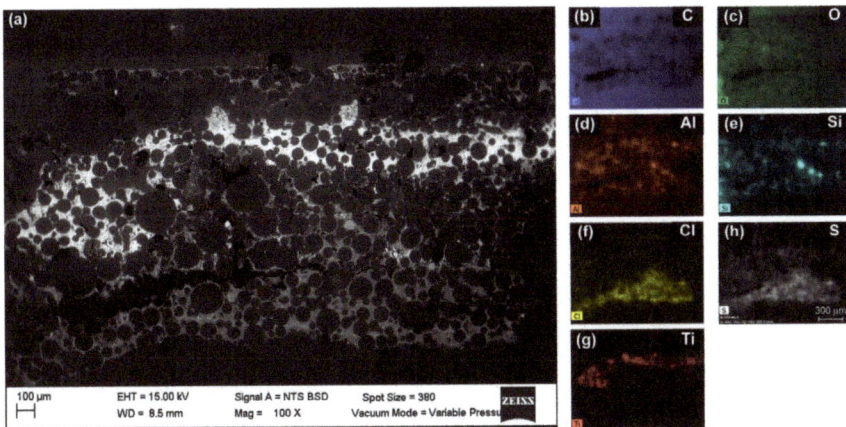

Figure 3. (a) SEM micrograph (BSE, 15.0 kV, 100×, area studied approximately 3000 × 1800 microns). Elemental mappings (EDS, 15.0 kV, 100×) of predominant elements: (b) C, (c) O, (d) Al, (e) Si, (f) Ti, (g) Cl and (h) S.

The matrix is composed of an organic material due to its high presence of C and O (Figure 3b,c). The presence of inorganic components with different geometries is observed. The yellow pigment was identified as cadmium yellow due to the presence of Cd and S in the same region (Figure S1). The presence of Al and Si (Figure 3d,e) in the same regions suggest the presence of an aluminosilicate filler. Ti (Figure 3f) was related to white titanium pigment TiO_2. In the red surface region of the sample, shown in Figure 2a, no elements that could be related to an inorganic pigment were detected. The predominant presences of Cl and S (Figure 3g,h) suggest the presence of an organic pigment.

2.4. Organic Compound Elucidation by NMR

The structure of the major organic compounds present in the sample of *Paisaje Abstracto* were elucidated by the interpretation of one-dimensional (^1H and ^{13}C) and two-dimensional (COSY, ed-HSQC and HMBC) NMR spectra. COSY spectra allowed us to identify homonuclear correlations between vicinal hydrogens separated by two or three bonds. The HSQC was used to determine correlations between directly bonded hydrogen to carbon atoms. Additionally, thanks to the multiplicity-edited HSQC, it was possible to differentiate between CH/CH_3 and CH_2 groups. The HMBC gives heteronuclear correlations between hydrogen and carbon atoms separated by two, three and, sometimes in conjugated systems, four bonds, giving connectivity information.

The spectra of ^1H, ^{13}C, ed-HSQC, HMBC and COSY are reported in Figures S2–S12. Based on the analysis of the NMR spectra, the compounds in the following subsections were elucidated (see Table 1 for NMR data and molecule structure). Atom numbering is indicated in each structure only for assignment purposes. The analytic process to reach them is presented below.

2.4.1. Binding Medium

The principal component in the sample is the binder which was identified as pMMA due to the HMBC correlation from the methylene protons broad signal H-2 (δ_{1H} 1.81 ppm) with three methyl carbon signals C-1 (δ_{13C} 16.66, 18.82 and 21.24 ppm), a methylene carbon C-2 (δ_{13C} 51.24–54.48 ppm), a quaternary carbon C-3 (δ_{13C} 44.74 ppm) and a carbon in a carbonyl group C-4 (δ_{13C} 176.26–178.44 ppm) (Figure S4). The carbonyl group was identified as an ester due to the HMBC correlation from the methyl protons signal H-5 (δ_{1H} 3.60 ppm) to C-4. All this indicates that the binder is exclusively composed of atactic pMMA due to the tacticity observed [11].

Table 1. Spectroscopic data (700 MHz, CDCl$_3$) of compounds identified in *Paisaje Abstracto* sample. Arrows in green indicate key HMBC correlations and in purple indicate key COSY correlations.

Compound/Structure/Assignment/Correlation	Label	δ_H/ppm (Multiplicity [1], J/Hz)	δ_C/ppm	HMBC (H→C)
poly(methyl methacrylate)	1	0.85 (bs)	16.66	C-2, 3, 4
	1	1.01 (bs)	18.82	C-2, 3, 4
	1	1.21 (bs)	21.24	C-2, 3, 4
	2	1.81 (bs)	51.24–54.48	C-1, 2, 3, 4
	3	—	44.74	-
	4	—	176.26–177.96 178.44	-
	5	3.60 (bs)	51.93	C-4
Methyl methacrylate	1	3.75 (s)	51.93	C-2
	2	—	167.86	-
	3	—	136.38	-
	4	1.94 (s)	18.49	C-2, 3, 5
	5′	5.56 (s)	125.61	-
	5″	6.10 (s)	125.61	C-2, 4
Benzoyl peroxide	1	7.60 (t, 7.49)	133.69	C-3
	2	7.47 (t, 7.81)	128.63	C-2, 3
	3	8.08 (d, 6.78)	130.28	C-1, 3, 5
	4	—	125.60	-
	5	—	167.79	-
Dibutyl phthalate	1	7.53 (dd, 5.8, 3.4)	131.03	C-2, 3
	2	7.71 (dd, 5.8, 3.4)	128.92	C-1, 3, 4
	3	—	132.63	-
	4	—	167.91	-
	5	4.31 (t, 6.7)	65.72	C-4, 6, 7
	6	1.72 (q, 6.9)	30.74	C-5, 7, 8
	7	1.44 (h, 7.4)	19.28	C-5, 6, 8
	8	0.96 (t, 7.4)	13.87	C-6, 7
Linear chain alcohol	1	0.88 (t, 7.0)	14.2	C-2, 3
	2	1.29 (bs)	22.80	C-1, 3
	3	1.25 (bs)	32.10	C-4
	4	1.25 (bs)	29.40	C-3
	5	1.34 (bs)	25.80	C-4
	6	1.56 (bs)	32.89	C-4, 5, 7
	7	3.64 (t, 6.7)	63.27	C-5, 6
PR181	1	—	NA [2]	-
	2	—	NA [2]	-
	3	—	125.09	-
	4	—	143.63	-
	5	7.09 (s)	129.05	C-3, 7, 9
	6	—	—	-
	7	7.36	121.98	C-3, 5
	8	—	NA [2]	-
	9	2.73	18.85	C-3, 4, 5

[1] Multiplicity: [s] singlet, [d] doublet, [t] triplet, [q] quintet, [h] hexaplet, [dd] doublet of doublets, [bs] broad signal. [2] NA: Not assigned.

2.4.2. Monomer

The MMA monomer was also identified due to the HMBC correlations from methyl protons H-4 (δ_{1_H} 1.94 ppm) to a carbonyl C-2 (δ_{13_C} 167.86 ppm), a quaternary carbon C-3 (δ_{13_C} 137.38 ppm) and a methylene carbon signal C-5 (δ_{13_C} 125.61 ppm), which indicate the presence of a methacrylate (Figure S5). The HMBC correlation, from methyl protons H-1 (δ_{1_H} 3.75 ppm) to C-2, confirmed the identification of the MMA monomer.

2.4.3. Catalyst

The catalyst benzoyl peroxide was elucidated from the COSY correlation between the aromatic protons H-1 (δ_{1_H} 7.60 ppm) and H-2 (δ_{1_H} 7.47 ppm) and between H-2 and H-3 (δ_{1_H} 8.08 ppm) (Figure S6). The HMBC correlations from H-3 to aromatic carbons C-1 (δ_{13_C} 133.69 ppm) and C-3 (δ_{13_C} 130.28 ppm) and a carbonyl signal C-5 (δ_{13_C} 167.79 ppm) confirmed the presence of benzoyl peroxide which is commonly used as a catalyst (Figure S7).

2.4.4. Plasticizer

The HMBC correlation from the aromatic protons H-2 to aromatic carbons C-1 (δ_{13_C} 131.03 ppm) and C-3 (132.63 ppm), and a carbonyl signal C-4 (δ_{13_C} 167.91 ppm), indicate the presence of a symmetric ortho disubstituted aromatic system (Figure S8). The aliphatic system of the molecule was identified with the COSY correlation between methylene protons H-5 (δ_{1_H} 4.31 ppm) and H-6 (δ_{1_H} 1.72 ppm), H-6 and H-7 (δ_{1_H} 1.44 ppm) and H-7 and methyl signal H-8 (δ_{1_H} 0.96 ppm) (Figure S9). The HMBC spectra confirmed the presence of dibutyl phthalate with the correlation from methylene protons H-5 to carbonyl signal C-4, and methylene carbons C-6 (δ_{13_C} 30.74 ppm) and C-7 (δ_{13_C} 19.28 ppm).

2.4.5. Additive

A linear chain alcohol structure was elucidated from COSY correlation between methyl protons H-1 (δ_{1_H} 0.88 ppm) and methylene H-2 (δ_{1_H} 1.29 ppm), H-2 and methylene H-3 (δ_{1_H} 1.25 ppm), methylene H-5 (δ_{1_H} 1.34 ppm) and methylene H-6 (δ_{1_H} 1.29 ppm), and H-6 and methylene H-7 (δ_{1_H} 3.64 ppm) (Figure S10). The HMBC correlation from H-4 (δ_{1_H} 1.25 ppm) to methylene carbon C-2 (δ_{13_C} 22.80 ppm) and from H-5 and H-6 to methylene carbon C-4 (δ_{13_C} 29.40 ppm) shows the linear connectivity from C-1 to C-7 (Figure S11).

2.4.6. Pigment

The HMBC correlation from methyl proton H-9 (δ_{1_H} 2.73 ppm) to aromatic carbons C-3 (δ_{13_C} 125.09 ppm), C-4 (δ_{13_C} 143.63 ppm) and C-5 (δ_{13_C} 129.05 ppm), with correlation from H-5 (δ_{1_H} 7.09 ppm) to C-3 and C-7 (δ_{13_C} 121.98 ppm) (Figure S12), indicate the presence of a tetra-substituted aromatic system which can be attributed to red pigment PR181 due to the identification of Cl and S in SEM-EDS (Figure 3g,h) in the red regions of the cross section (Figure 2a).

2.5. ATR-FTIR Study

ATR-FTIR analytical characterization was performed to identify the binder and pigments in the *Paisaje Abstracto* sample. The ATR-FTIR spectrum (Figure 4c) were assigned by relating them to the bands of the pMMA binder and these were compared with those of reference libraries. However, the bands of the characteristic groups of the additives, previously elucidated by NMR, were not observed due to their low concentration (see Table 2).

2.5.1. pMMA

The presence of pMMA was confirmed by comparing the profile of the ATR-FTIR spectra and the absorbance of diagnostic peaks within the literature. The signals attributed to the binder were the carbonyl stretching mode at 1722 cm^{-1}; the asymmetrical stretching mode of carbonyl group $_a$(C-CO-O) at 1240 and 1270 cm^{-1}; the asymmetrical stretching mode of ester group $_a$(C-O-C-) at 1144 and 1190 cm^{-1}; the C-H combination band of -CH$_3$ at 2847 and 2922 cm^{-1}; the band which included the symmetrical stretching mode $_s$(C-H) of O-CH$_3$ with $_s$(C-H) of -CH$_3$ and asymmetrical stretching mode $_a$(CH$_2$) at 2948 cm^{-1}; $_a$(C-H) of O-CH$_3$ with $_a$(C-H) of -CH$_3$ at 2994 cm^{-1}; the ones related to symmetrical bending modes $_s$(C-H) of -CH$_3$ and O-CH$_3$ at 1386 and 1434 cm^{-1}; the signal related to asymmetrical bending $_a$(C-H) of -CH$_3$ at 1446 cm^{-1}; the methylene scissoring mode signal (CH$_2$) at 1482 cm^{-1}; the signals that correspond to rocking mode (C-H) of CH$_2$, CH$_3$ and O-CH$_3$, respectively, at 842, 967, 988 cm^{-1}; and the skeletal stretching mode bands (C-C) at 749, 1063 cm^{-1} [12,13].

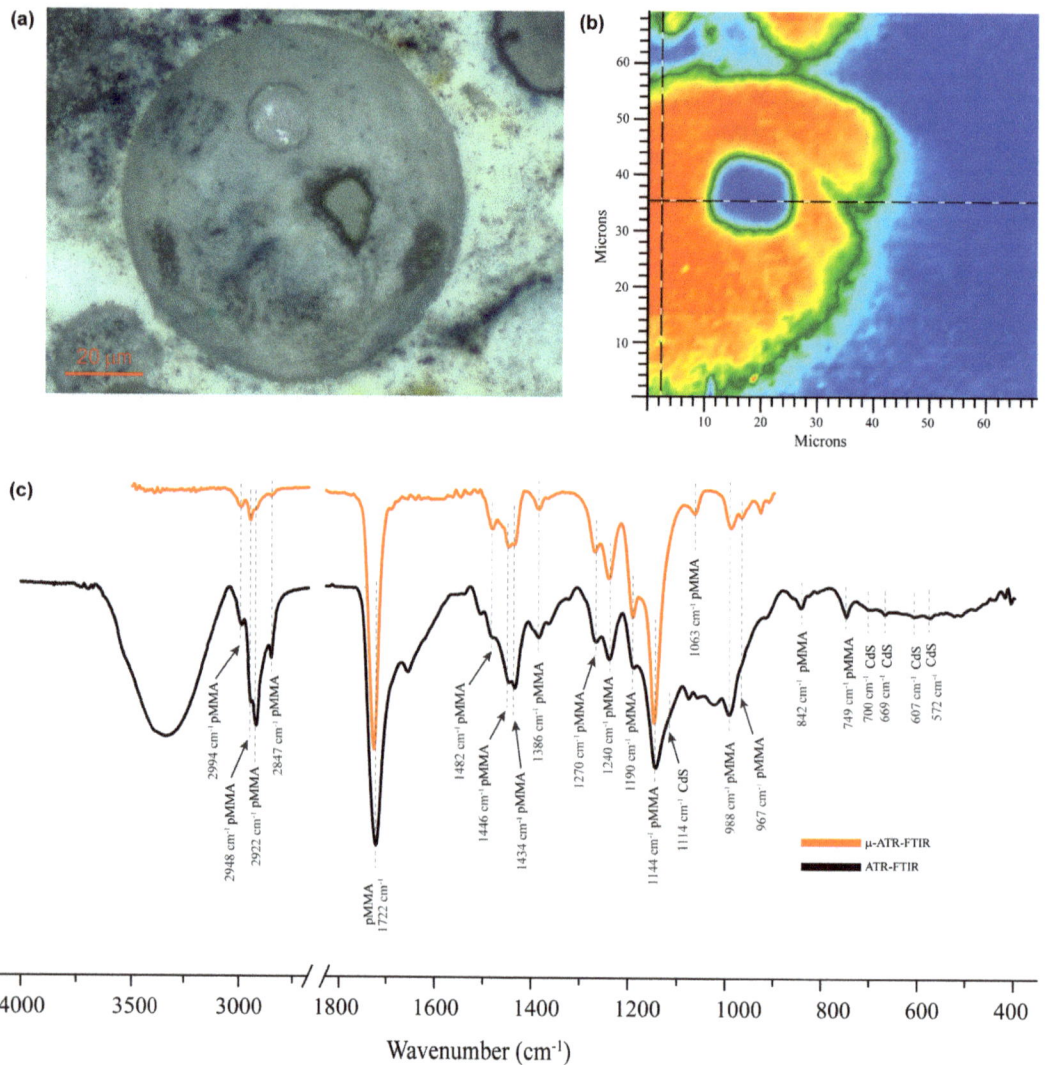

Figure 4. Cross sectional micrograph of the isolated layer yellow sample: (**a**) visible microscopic image, (**b**) chemical map displaying the intensity of the absorbance peaks at 2860 to 3040 cm^{-1}, (**c**) comparison between ATR-FTIR (dark line) and micro-ATR-FTIR spectra obtained from the highlighted region in false color image (orange line).

Table 2. ATR-FTIR absorption bands associated with the analyzed compounds in the *Paisaje Abstacto* sample. The anti-symmetric and symmetric stretching and blending modes are denoted with ν_a, ν_s, $_a$ and $_s$.

Component	Wavenumber (cm^{-1})/[Intensity [1]]/Assignment	References
poly(methyl methacrylate)	749 [m] ν(C-C) skeletal mode, 842 [m] γ(CH$_2$), 967 [sh] γ(α-CH$_3$), 988 [m] γ(O-CH$_3$), 1063 [m] ν(C-C) skeletal mode, 1144 [vs] and 1190 [vs] ν_a(C-O-C-), 1240 [s] and 1270 [s] ν(C-O), 1386 [m] δ_s(C-H) of α-CH$_3$, 1434 [s] δ_s(C-H) of O-CH$_3$, 1446 [s] δ_a(C-H) of α-CH$_3$, 1482 [m] δ(CH$_2$), 1722 [vs] ν(C=O), 2847 [vw] and 2922 [sh] combination band involving O-CH$_3$ and CH$_2$, 2948 [m] ν_s(C-H) of O-CH$_3$ with ν_s(C-H) of α-CH$_3$ and ν_a(C-H), 2994 [m] ν_a(C-H) of O-CH$_3$ and ν_a(C-H) of α-CH$_3$.	[11,12]
Red Pigment PR 181	1654 [vs], 1562 [m], 1434 [s], 1386 [w], 1294 [sh], 1240 [w], 1190 [sh], 1096 [sh], 1048 [sh], 842 [vw], 823 [vw], 781 [vw], 468 [vw].	[13,14]
Cadmium sulfide (CdS)	1114 [sh], 700 [vw], 669 [vw], 607 [vw], 572 [vw]	[15–18]

[1] Intensity: [vs] very strong, [s] strong, [m] medium, [w] weak, [vw] very weak, [sh] shoulder.

2.5.2. Organic Pigment

The ATR-FTIR spectra obtained from the red layer (Figure S13) shows bands associated with the red pigment PR181 [14,15]. Some bands that are present in the ATR-FTIR reference spectrum of the red pigment PR181 can also be associated with pMMA vibrations. Those bands are methyl group scissoring (H-C-H) at 1434 cm^{-1}; combination band at the 1386 cm^{-1} band of methyl group scissoring (H-C-H), benzene ring bending (C-C-H) and symmetrical stretching $_s$(C-C-C); combination band at 1240 cm^{-1} which includes the bending of benzene ring (C-C-H), asymmetrical stretching of thiophene ring $_a$(C-C-C), symmetrical stretching of benzene ring $_s$(C-C-C), and symmetrical bending of methyl groups $_s$(H-C-H); combination band at 1190 cm^{-1} that includes the bending of benzene ring (C-C-H), asymmetrical stretching of thiophene ring $_a$(C-C-C and C-C-S), symmetrical stretching of benzene ring $_s$(C-C-C); and a band identified at 842 cm^{-1}, of which its vibration was not assigned. However, and based on the compounds identified by NMR, the ATR-FTIR bands that can only be assigned to the red pigment and thus confirm its presence in the sample are band at 1654 cm^{-1} related to carbonyl in thiophene ring (C=O), stretching of benzene and thiophene ring (C-C-C) band at 1562 cm^{-1}; combination band at 1294 cm^{-1} that includes the asymmetrical stretching of benzene ring $_a$(C-C-C), symmetrical stretching of thiophene ring $_s$(C-C-C) and symmetrical bending of methyl groups $_s$(H-C-C-H); combination band at 1096 cm^{-1} scissoring of benzene ring (H-C-C-H), carbon-chloride stretching (C-Cl), symmetrical stretching of benzene ring $_a$(C-C-C), benzene ring and methyl group rocking (H-C-C-H) and asymmetrical stretching of thiophene ring $_s$(C-C-S), 1048 cm^{-1} thiophene ring stretching (C-C); combination band at 823 cm^{-1} of benzene ring rocking (H-C-C-H), symmetrical stretching of thiophene ring $_s$(C-S-C) and carbon-chloride stretching (C-Cl); the combination band at 468 cm^{-1} of benzene ring and methyl group symmetrical bending $_s$(H-C-C-H), thiophene ring bending (C-C-O) and carbon-chloride stretching (C-Cl); and a band at 781 cm^{-1}, of which its vibration was not assigned.

2.5.3. Inorganic Pigment

The presence of CdS pigment was confirmed with the bands related to the Cd-S bond at 700, 669 and 607 cm^{-1}. Additionally, bands related to cadmium yellow PY37 ATR-FTIR spectra (1114 and 572 cm^{-1}) were attributed to this pigment [16–19].

2.6. Micro-ATR-FTIR Analysis

In order to clarify the molecular composition of the spheres, a micro-ATR-FTIR mapping of the cross section of the sample was carried out. Figure 4a shows the optical microscope image of the mapped region and Figure 4b shows the chemical map obtained by integrating the bands at 2860 to 3040 cm^{-1}. Figure 4c shows the micro-ATR-FTIR spectra of the cross section of the sample, indicated in Figure 4b, and the stacked ATR-FTIR spectra obtained from the sample surface. The chemical map in Figure 4b confirms that the spheres are mainly composed of pMMA, both on their surface and inside.

2.7. GC/MS Analysis

The molecular weights of the volatile and thermostable compounds were confirmed by GC/MS (see chromatogram in Figure S14 and mass spectra in Figure S15a–d). The compounds identified were Benzoyl peroxide (RT = 11.38 min), dibutyl phthalate (RT = 23.38 min), 1-octadecanol also known as Stearyl alcohol (RT = 24.77 min) and the red pigment PR181 (RT = 40.84 min).

3. Discussion

Commercial artistic paintings are typically copolymerized MMA monomers with ethyl or *n*-butyl acrylate and methacrylate [19]. The finding of pMMA homopolymer as the binder of *Paisaje Abstracto* was unanticipated. The use of pMMA as a material for artistic purposes is uncommon and has not been reported before in artistic paintings.

The lack of organic and inorganic additives suggests that Rafael Coronel manufactured his own paint for this mural, mixing methyl methacrylate, poly(methyl methacrylate), benzoyl peroxide, aluminosilicate fillers, additives and organic and inorganic pigments. The layer sequence of polymer and pigment observed under the microscope support this statement. It seems that the artist poured mixtures of different colors with the see chromatogram in Figure S14 and mass spectra in Figure S15a–d support placed horizontally, producing elevations in the surface.

Based on the organic compounds identified by NMR, it could be proposed that Coronel used an application process similar to the Kulzer patented method to produce dental prosthetics in the mid-1930s [20]. Moreover, the alcohols most used in coating formulation are those with a maximum length of 5–8 carbons as biocides [21]. Therefore, the identification of 1-octadecanol could be more related for surgical or dental prostheses and dental repairs, as it was patented by Bayer Ag in 1952 [22].

Coronel may have used a higher proportion of pMMA/MMA, which could explain the formation of pMMA spheres [23]. The lack of miscibility of the polymer in the monomer can inhibit the chain transfer reaction of the monomer radical to the prepolymerized pMMA which did not allow optimal curing [24]. A modification in the pMMA/MMA proportion could affect the mechanical properties of the painting, especially the compressive strength, yield stress and compressive modulus [25,26].

4. Material and Methods

4.1. Sampling

During restoration, due to the dimensional changes (by expansion or contraction) of the wood panel, two microsamples were detached from the upper right corner. Conservators from the MNA provided the samples. Both samples had a red color on the outer surface and both had the same sequence of six layers: (i) yellow directly over the support, (ii) green, (iii) gray, (iv,v) two translucent ones, and (vi) red on top. In general, the painting layer was very prominent and thick; the stratigraphy of the samples measured approximately 2500 microns, but there were regions in the painting that were thicker than this section. The samples were inspected with an optical fiber microscope (Keyence, Osaka, Japan), then one sample section was reserved for chemical studies of isolated layers: yellow and red; the other sample section was embedded in Claro Cit©, a cold mounting acrylic resin produced by Struers (Roper Technologies, Sarasota, FL, USA), for the stratigraphic analysis.

4.2. Optical Microscopy

Samples were studied with an Axio Imager Z2 optical microscope (Carl Zeiss, Oberkochen, Germany) equipped with a Xenon arc lamp for UV fluorescence with 430–465 nm and 465–500 nm filters and a HAL100 light source in reflected light mode. The inspection under the optical microscope was performed over the cross sections.

4.3. Scanning Electron Microscopy—Energy Dispersive X-ray Spectroscopy

The electron micrographs were acquired with an EVOMA25 SEM (Carl Zeiss, Oberkochen, Germany) primarily with a backscattered electron detector. An accelerating voltage was applied between 15.0–17.0 keV, due to the organic composition of the sample and its sensitivity to be damaged by X-rays. The chemical elemental analysis of the samples was performed with an Energy Dispersive Spectroscopy microprobe, 30 mm (Bruker, Bremen, Germany) and PB/ZAF analysis. Samples were studied as cross sections. Double-stick carbon tape was used to attach and render the samples conductive. The images were taken using variable pressure (80 Pa) under nitrogen flow to avoid electrostatic charges on the samples' surfaces.

4.4. Nuclear Magnetic Resonance Spectroscopy

NMR spectra were acquired with a Bruker Avance III HD 700 spectrometer (Bruker, Billerica, MA, USA) operating at 16.4 T (700 and 175 MHz for ^{1}H and ^{13}C frequency, respectively) equipped with a 5 mm z-axis gradient TCI cryoprobe. The ^{1}H-NMR and 2D-NMR experiments were acquired at 298 K with standard pulse sequences from the Bruker library and processed with MestReNova software (v. 14.0, Mestrelab Research SL, Santiago de Compostela, Spain). The 2D-NMR experiments acquired were ^{1}H-^{1}H correlation spectroscopy (COSY), ^{1}H-^{13}C edited heteronuclear single quantum correlation spectroscopy (edited-HSQC) and ^{1}H-^{13}C heteronuclear multiple-bond correlation spectroscopy (HMBC). The chemical shifts (δ) are reported in ppm relative to the solvent resonance as the internal standard (CDCl$_3$: δ^1H = 7.26 ppm).

4.5. Attenuated Total Reflection Fourier-Transform Infrared Spectroscopy

ATR-FTIR was performed on samples without preparation with the Cary 600 spectrophotometer (Agilent Technologies, Santan Clara, CA, USA). The range used by the IR spectra was 4000–400 cm^{-1}. Spectra were acquired with 128 scans at 4 cm^{-1} resolution. All data were processed with Origin software. The signal identification was compared with the published literature.

4.6. Micro-Attenuated Total Reflection Fourier-Transform Infrared Spectroscopy

Micro-ATR-FTIR analyses were performed on cross sections with an Agilent Cary 620 FTIR microscope (Agilent Technologies, Santa Clara, CA, USA) with a 64 × 64 Focal Plane Array (FPA) detector coupled to an Agilent Cary 660 FTIR spectrometer. Data were collected at 1.1 µm per pixel resolution using a micro Germanium Attenuated Total Reflection crystal accessory and processed using Agilent Resolutions Pro software (Agilent Technologies, Santa Clara, CA, USA). The Field of View (FOV) was 70 × 70 µm. Spectra were acquired in the spectral range between 4000 to 900 cm^{-1}, performing 128 scans at 4 cm^{-1} resolution.

4.7. Gas Chromatography/Mass Spectrometry

Mass spectra were acquired using an Agilent 7890B gas chromatograph (Agilent Technologies, Santa Clara, CA, USA) equipped with capillary column HP-5ms cross-linked (5%-phenyl)-methylpolysiloxane (30 m × 0.25 mm × 0.25 µm) and an Agilent 5977A mass spectrometer (Agilent Technologies, Santa Clara, CA, USA) operated in positive mode with electron impact ionization (70 eV). The injector was set at 280 °C and the interface at 310 °C. The initial temperature for the GC column was 40 °C, held for 1 min, increased at 8 °C min^{-1} to 310 °C, held for 10.25 min. The helium flow rate was 1 mL min^{-1}. The mass range of the mass spectrometer was 30–600 m/z and mass fragments were identified by their respective spectra through NIST v. 14 mass library searches.

5. Conclusions

The combination of analytical techniques allowed for the precise identification of the major components present in *Paisaje Abstracto*. The pMMA identified as the binder is not a common polymer employed in artists' acrylic media alone; instead, the MMA is commonly copolymerized with other monomers and blended with inorganic additives, pigments and fillers. pMMA spherical micro-structures are not observed in other artist's acrylic formulations, but, nevertheless, these kinds of spheres are commonly observed in dental acrylic cement. Thus, it was possible to propose a hypothesis of the unique painting technique followed by Rafael Coronel to create the mural. The characterization of the polymer suggests a higher pMMA/MMA proportion which may have also affected the mechanical properties and conservation of the mural, causing a rigid painting layer that fractures easily.

Supplementary Materials: The following supporting information can be downloaded at: https://www.mdpi.com/article/10.3390/molecules28020564/s1, Figure S1: (**a**) SEM micrograph (BSE, 17.0 kV, 1500×). (**b**) Elemental mappings (EDS, 17.0 kV, 1500×) of predominant elements: (**c**) C, (**d**) O, (**e**) Cl, (**f**) Cd, (**g**) Ti, (**h**) S, (**i**) Al and (**j**) Si; Figure S2: ^{1}H-NMR spectra (700 MHz, CDCl$_3$) of isolated layers: (**a**) yellow and (**b**) red. Structures and colored signals of the binder (blue), MMA monomer (violet), catalyst (green), plasticizer (orange), additive (cyan) and red pigment PR181, are indicated; Figure S3: ^{13}C-NMR spectra (175 MHz, CDCl$_3$) of isolated yellow layer; Figure S4: HSQC (blue-orange) and HMBC (green) spectra (700 MHz, CDCl$_3$) of isolated yellow layer. The structure of the acrylic binder pMMA with the key correlation and the assignment of the signals in the spectra are shown; Figure S5: HSQC (blue-orange) and HMBC (green) spectra (700 MHz, CDCl$_3$) of isolated yellow layer. The structure of methyl methacrylate monomer with the key correlation and the assignment of the signals in the spectra are shown; Figure S6: COSY spectrum (700 MHz, CDCl$_3$) of isolated yellow layer. The structure of benzoyl peroxide (BPO) with the key correlation and the assignment of the signals in the spectrum are shown; Figure S7: HSQC (blue-orange) and HMBC (green) spectra (700 MHz, CDCl$_3$) of isolated yellow layer. The structure of benzoyl peroxide (BPO) with the key correlation and the assignment of the signals in the spectra are shown; Figure S8: HSQC (blue-orange) and HMBC (green) spectra (700 MHz, CDCl$_3$) of isolated yellow layer. The structure of dibutyl phthalate (DBP) with the key correlation and the assignment of the signals in the spectra are shown; Figure S9: COSY spectrum (700 MHz, CDCl$_3$) of yellow layer. The structure of dibutyl phthalate (DBP) with the key correlation and the assignment of the signals in the spectra are shown; Figure S10: COSY spectrum (700 MHz, CDCl$_3$) of isolated yellow layer. The structure of 1-octadecanol with the key correlation and the assignment of the signals in the spectra are shown; Figure S11: HSQC (blue-orange) and HMBC (green) spectra (700 MHz, CDCl$_3$) of isolated yellow layer. The structure of 1-octadecanol with the key correlation and the assignment of the signals in the spectra are shown; Figure S12: HSQC (blue) and HMBC (green) spectra (700 MHz, CDCl$_3$) of isolated red layer. The structure of red pigment PR181 with the key correlation and the assignment of the signals in the spectra are shown; Figure S13: ATR-FTIR spectra obtained from the red layer. The bands associated with the red pigment PR181 are indicated; Figure S14: Chromatogram of isolated red layer. The peaks with the identified compounds are presented at the respective retention times; Figure S15: Mass spectra of the identified compounds in the isolated red layer. (**a**) Benzoyl peroxide, (**b**) dibutyl phthalate, (**c**) 1-octadecanol and (**d**) red pigment PR181.

Author Contributions: P.A.-R.: investigation, methodology, formal analysis; data curation, visualization, writing—original draft. S.Z.: conceptualization, investigation, resources, project administration, funding acquisition, writing—review and editing. A.M.-G.: data curation, formal analysis, visualization, writing—review and editing. N.E.-E.: conceptualization, investigation, supervision, resources, project administration, funding acquisition, writing—review and editing. All authors have read and agreed to the published version of the manuscript.

Funding: This work was financially supported by Proyect IN402121 DGAPA-PAPIIT-UNAM.

Data Availability Statement: Not applicable.

Acknowledgments: P.A-R. and A.M-G. thank Consejo Nacional de Ciencia y Tecnología (CONACYT) for his PhD scholarship (grant 1022577 and 846597, respectively). This study made use of UNAM labs: LANCIC at IQ-UNAM and, IIE-UNAM which are funded by CONACYT [grant: LN232619 LN260779, LN271614, LN299076, LN314846 and LN315853], and LURMN which is funded by CONACYT [grant 0224747] and UNAM. We thank Marisol Reyes Lezama, Everardo Tapia Mendoza, Mayra León Santiago, Beatriz Quiroz García and Martha E. Garcia-Aguilera for technical assistance. Special thanks to Instituto Nacional de Antropología e Historia (INAH) who provided the microsamples and information about the mural for this study; to Antonio Saborit, director of the MNA, Patronato del Museo Nacional de Antropología, A. C.; Vanessa Fonseca and coworkers from Proyecto de Digitalización del MNA; to Laura Filloy Nadal and conservators from Laboratorio de Conservación del MNA; and to Gilda Elena Salgado Manzanares, coordinator of Proyecto de Conservación de Obra Moderna y Contemporánea- MNA. Additionally, we thank the conservators Sandra María Álvarez Jacinto, Levna Alejandra Caballero Acosta, Mitzi Vania García Toribio, Lourdes Ivette Navarrete Rodríguez, Astrid Sánchez Carrasco, students from Escuela Nacional de Conservación, Restauración y Museografía (ENCRyM-INAH), from the Seminario Taller de Restauración de Obra Moderna y Contemporánea (STROMC), under coordination of Ana Lizeth Mata Delgado.

Conflicts of Interest: The authors declare no conflict of interest.

Sample Availability: Samples from the mural prepared as cross sections are available from the LANCIC IIE UNAM sample archive for research purposes.

References

1. Coffey, M.K. *How a Revolutionary Art Became Official Culture*; Duke University Press: Durham, NC, USA, 2012; ISBN 9780822394273.
2. Rita Eder Variations. Painting in Mexico during the Fifties and Sixties. In *Defying Stability, Artistic Processes in Mexico, 1952–1967*; Eder, R., Ed.; UNAM: Ciudad de México, México, 2014; pp. 145–163, ISBN 978-84-15832-39-3.
3. Shifra, M. *Goldman Contemporary Mexican Painting in a Time of Change*; University of Texas Press: Austin, TX, USA, 1981; ISBN 0292710615.
4. Spyros, A.; Anglos, D. Study of Aging in Oil Paintings by 1D and 2D NMR Spectroscopy. *Anal. Chem.* **2004**, *76*, 4929–4936. [CrossRef] [PubMed]
5. Tortora, M.; Sfarra, S.; Chiarini, M.; Daniele, V.; Taglieri, G.; Cerichelli, G. Non-Destructive and Micro-Invasive Testing Techniques for Characterizing Materials, Structures and Restoration Problems in Mural Paintings. *Appl. Surf. Sci.* **2016**, *387*, 971–985. [CrossRef]
6. Mejía-González, A.; Jáidar, Y.; Zetina, S.; Aguilar-Rodríguez, P.; Ruvalcaba-Sil, J.L.; Esturau-Escofet, N. NMR and Other Molecular and Elemental Spectroscopies for the Characterization of Samples from an Outdoor Mural Painting by Siqueiros. *Spectrochim. Acta Mol. Biomol. Spectrosc.* **2022**, *274*, 121073. [CrossRef] [PubMed]
7. Saladino, M.L.; Ridolfi, S.; Carocci, I.; Martino, D.C.; Lombardo, R.; Spinella, A.; Traina, G.; Caponetti, E. A Multi-Analytical Non-Invasive and Micro-Invasive Approach to Canvas Oil Paintings. General Considerations from a Specific Case. *Microchem. J.* **2017**, *133*, 133–607. [CrossRef]
8. Tanasi, D.; Greco, E.; di Tullio, V.; Capitani, D.; Gullì, D.; Ciliberto, E. 1H-1H NMR 2D-TOCSY, ATR FT-IR and SEM-EDX for the Identification of Organic Residues on Sicilian Prehistoric Pottery. *Microchem. J.* **2017**, *135*, 140–147. [CrossRef]
9. Colantonio, C.; Baldassarri, P.; Avino, P.; Astolfi, M.L.; Visco, G. Visual and Physical Degradation of the Black and White Mosaic of a Roman Domus under Palazzo Valentini in Rome: A Preliminary Study. *Molecules* **2022**, *27*, 7765. [CrossRef] [PubMed]
10. Lizeth, M.-D.A.; Gilda, S.-M.; Laura, F.; Alejandro, M.; María, Á.-J.S.; Alejandra, C.-A.L.; Ivette, N.-R.L.; de Rafael Coronel, S.-C.A.P.A. *Informe de Los Trabajos de Conservación y Restauración Realizados en la Práctica Intersemestral del 17 Junio al 14 de Julio Del 2013*; Instituto Nacional de Antropología e Historia: Mexico City, Mexico, 2013. Available online: https://mediateca.inah.gob.mx/repositorio/islandora/object/informe%3A581 (accessed on 3 October 2022).
11. White, A.J.; Filisko, F.E. Tacticity Determination of Poly(Methyl Methacrylate) (PMMA) by High-Resolution NMR. *J. Polym. Sci. Polym. Lett. Ed.* **1982**, *20*, 525–529. [CrossRef]
12. Lipschitz, I. The Vibrational Spectrum of Poly(Methyl Methacrylate): A Review. *Polym. Plast. Technol. Eng.* **1982**, *19*, 53–106. [CrossRef]
13. Willis, H.A.; Zichy, V.J.I.; Hendra, P.J. The Laser-Raman and Infra-Red Spectra of Poly(Methyl Methacrylate). *Polymer* **1969**, *10*, 737–746. [CrossRef]
14. Price, B.A.; Pretzel, B. Suzanne Quillen Lomax Infrared and Raman Users Group Spectral Database. Available online: www.irug.org (accessed on 22 October 2022).
15. Sakthidharan, C.P.; Niewa, R.; Zherebtsov, D.A.; Podgornov, F.V.; Matveychuk, Y.V.; Bartashevich, E.V.; Nayfert, S.A.; Adonin, S.A.; Gavrilyak, M.V.; Boronin, V.A.; et al. Crystal Structures and Dielectric Properties of 4,4′-Dimethyl-6,6′-Dichlorothioindigo (Pigment Red 181). *Acta Crystallogr. B Struct. Sci. Cryst. Eng. Mater.* **2021**, *77*, 23–30. [CrossRef]
16. Zhang, K.; Yu, Y.; Sun, S. Influence of Eu Doping on the Microstructure and Photoluminescence of CdS Nanocrystals. *Appl. Surf. Sci.* **2012**, *258*, 7658–7663. [CrossRef]

17. Vahur, S.; Teearu, A.; Peets, P.; Joosu, L.; Leito, I. ATR-FT-IR Spectral Collection of Conservation Materials in the Extended Region of 4000–4080 cm^{-1}. *Anal. Bioanal. Chem.* **2016**, *408*, 3373–3379. [CrossRef] [PubMed]
18. Thambidurai, M.; Murugan, N.; Muthukumarasamy, N.; Agilan, S.; Vasantha, S.; Balasundaraprabhu, R. Influence of the Cd/S Molar Ratio on the Optical and Structural Properties of Nanocrystalline CdS Thin Films. *J. Mater. Sci. Technol.* **2010**, *26*, 193–199. [CrossRef]
19. Sobhana, S.S.L.; Vimala Devi, M.; Sastry, T.P.; Mandal, A.B. CdS Quantum Dots for Measurement of the Size-Dependent Optical Properties of Thiol Capping. *J. Nanoparticle Res.* **2011**, *13*, 1747–1757. [CrossRef]
20. Kulzer & Co., GmbH. Verfahren Zür Herstellung von Prothesen Für Zarnarztliche Order Andere Zweck Aus Polymerisierten Organischen Verbindungen. DE737058C. 1936. Available online: https://worldwide.espacenet.com/patent/search/family/025760434/publication/DE737058C?q=pn%3DDE737058C (accessed on 25 October 2022).
21. Bieleman, J. (Ed.) *Additives for Coatings*; Wiley: Hoboken, NJ, USA, 2000; ISBN 9783527297856.
22. BAYER AG—FARBENFABRIKEN BAYER. Improvements in or Relating to Material for Surgical or Dental Prostheses and for Dental Repairs. GB714652A. 1952. Available online: https://worldwide.espacenet.com/patent/search/family/006121592/publication/GB714652A?q=pn%3DGB714652A (accessed on 26 October 2022).
23. Samad, H.A.; Jaafar, M. Effect of Polymethyl Methacrylate (PMMA) Powder to Liquid Monomer (P/L) Ratio and Powder Molecular Weight on the Properties of PMMA Cement. *Polym. Plast Technol. Eng.* **2009**, *48*, 554–560. [CrossRef]
24. Silikas, N.; Al-Kheraif, A.; Watts, D.C. Influence of P/L Ratio and Peroxide/Amine Concentrations on Shrinkage-Strain Kinetics during Setting of PMMA/MMA Biomaterial Formulations. *Biomaterials* **2005**, *26*, 197–204. [CrossRef]
25. Jasper, L.E.; Deramond, H.; Mathis, J.M.; Belkoff, S.M. The Effect of Monomer-to-Powder Ratio on the Material Properties of Cranioplastic. *Bone* **1999**, *25*, 27S–29S. [CrossRef] [PubMed]
26. Belkoff, S.M.; Sanders, J.C.; Jasper, L.E. The Effect of the Monomer-to-Powder Ratio on the Material Properties of Acrylic Bone Cement. *J. Biomed. Mater. Res.* **2002**, *63*, 396–399. [CrossRef]

Disclaimer/Publisher's Note: The statements, opinions and data contained in all publications are solely those of the individual author(s) and contributor(s) and not of MDPI and/or the editor(s). MDPI and/or the editor(s) disclaim responsibility for any injury to people or property resulting from any ideas, methods, instructions or products referred to in the content.

Article

Analytical Investigations of XIX–XX Century Paints: The Study of Two Vehicles from the Museum for Communications of Frankfurt

Andrea Macchia [1,*], Lisa Maria Schuberthan [2], Daniela Ferro [1], Irene Angela Colasanti [1], Stefania Montorsi [1], Chiara Biribicchi [1], Francesca Irene Barbaccia [1] and Mauro Francesco La Russa [3]

1. YOCOCU (Youth in Conservation of Cultural Heritage), Via T. Tasso 108, 00185 Rome, Italy
2. Museum for Communication Frankfurt, Schaumainkai 53, 60596 Frankfurt am Main, Germany
3. Department of Biology, Ecology and Earth Sciences DIBEST, University of Calabria, Via Pietro Bucci, Arcavacata, 87036 Rende, Italy
* Correspondence: andrea.macchia@uniroma1.it

Abstract: Over the centuries, humans have developed different systems to protect surfaces from the influence of environmental factors. Protective paints are the most used ones. They have undergone considerable development over the years, especially at the turn of the 19th and 20th centuries. Indeed, between the two centuries, new binders and pigments have been introduced in the constituent materials of paints. The years in which these compounds have been introduced and spread in the paint market allow them to be defined as markers for the dating of paints and painted artifacts. The present work is focused on the study of the paints of two vehicles of the Frankfurt Museum of Communication, i.e., a carriage and a cart, that was designed for the German Postal and Telecommunications Service roughly between 1880 and 1920. The characterization of the paints was performed through in situ non-invasive techniques, i.e., portable optical microscopy and multispectral imaging, and laboratory non-destructive techniques, i.e., FT-IR ATR spectroscopy and SEM-EDS. The analytical investigation and the comparison with the data reported in the literature allowed us to determine the historicity of the paints, which are all dated before the 1950s.

Keywords: protective coating; paint; binder; pigment; characterization; non-invasive techniques; FT-IR ATR; SEM-EDS

1. Introduction

Over the centuries, several protective paints that could be applied to artifacts have been produced.

As to vehicles used for transportation, the function of the pictorial coating was protection from the outdoor environment and their consequent degradation [1–4]. Currently, protective paints are highly specific according to the substrate on which they have to be applied while, until the first half of the 20th century, the choice was based on the type of environment in which they would have been placed: namely indoor or outdoor environments [5–7].

Protective paints consist of three components: binders, dyes or pigments, and additives. Over the centuries, especially at the turn of the XIX and XX centuries, several developments in the production of paints were mainly linked to the introduction of new binders and pigments [4,6–11]. These developments are attributable to two main factors that have characterized the centuries: the Industrial Revolution and the automobile's mass production. These two events led to the growth of the first paint factories in the early 19th century, which started developing linseed oil paints, galvanizing processes, and TiO_2-based pigments in 1918. Linseed oil is a natural oil that is characterized by a concentration of linoleic acid-fatty acid whit formula $C_{18}H_{32}O_2$-around 70%, and consequently has a high number of double bonds. This characteristic makes linseed oil have a high drying power, which is one of the main reasons for its wide use in paintings and the production of paints until the first half of the 20th century [12,13]. With the introduction of synthetic resins, advancements in the coating industry quickly followed to meet market demand. Thus, in the middle of the XX century, natural oils that were traditionally used in paint formulations were being replaced by synthetic resins. Therefore, the identification of an oily medium in a coating may suggest that it was produced before 1950, while the presence of synthetic resins suggested a later dating. Furthermore, alkyd paints almost completely replaced oil paints in the second half of the XX century [4,6,7,11,13–15].

Pigments can be used as markers as well. Among them, the most important ones that have this role are lead-based pigments, zinc white, lithopone, and titanium white. Lead-based pigments, especially lead white (basic lead carbonate $2PbCO_3 \cdot Pb(OH)_2$), has been traditionally used to produce paint matrices. Due to the discovery of their toxicity, their use decreased between the XIX and XX centuries [4,6,7,16–18]. In the years leading up to World War II, consumers paid more attention to health and environmental risks, and paint manufacturers began to study safer alternatives to replace lead and other toxic elements.

Zinc white (ZnO) was discovered in the late 1700s; lithopone, a mixture of 70–72 wt% of barium sulfate ($BaSO_4$) and 30–28 wt% of zinc sulfide (ZnS), was discovered in the late 1800s. They were commonly used to replace lead white in the first half of the XX century [4,6,7,9,14–16,18]. In 1918, the first titanium white was produced as a mixture of 30% titanium dioxide (TiO_2) and barium sulphate ($BaSO_4$). Soon, titanium white started to be the most widely used white pigment in paint matrices of titanium white (TiO_2). Indeed, a study carried out by the International Agency for Research on Cancer (IARC) highlights an increase in the paints produced with titanium white (about 80%) and a decrease in paints produced with lead white or lithopone in these years [18]. Consequently, the presence of this pigment allows the paint to be dated after that year and with greater probability after 1940 when the use of white titanium spread.

Other markers can be defined through the evolution of colored pigments and dyes [6,7,16]. Until the first half of the 1900s, multiple pigments were used for the formulation of paints. In the XIX century, new pigments were introduced to the market, such as synthetic ultramarine blue ($Na_{8-10}Al_6Si_6O_{24}S_{2-4}$), zinc and lead chromates ($ZnCrO_4$ or $PbCrO_4$), cadmium sulfides (pigments containing CdS), and Paris green ($Cu(C_2H_3O_2)_2 \cdot 3Cu(AsO_2)_2$). Their presence suggests that the paints were produced before the 1950s. The use of these pigments decreased progressively due to the discovery of the toxicity of some of them, but especially due to the spread of synthetic dyes. Synthetic compounds such as aniline (an aromatic compound, C_6H_7N), azo dyes (organic compounds characterized by the functional group R-N=N-R′, where R and R′ are often aryl groups), and phthalocyanines (macrocyclic organic compounds whose formula is $(C_8H_4N_2)_4H_2$), became the most widely used compounds for paints. Therefore, their presence allows the paints to be dated after the 1950s [4,6,7,16].

The coatings' composition allows for the acquisition of useful information for the dating of objects and artifacts. However, paints are composed of different constituents, such as binders, pigments, and other fillers. Additionally, several chemical compounds have been used for their production over the centuries. Thus, their characterization is not simple and requires a multi-analytical approach [19].

A preliminary study is generally required, especially if several layers overlap. Optical microscopy (OM) is commonly used for this analysis and allows for the acquisition of information on the object's stratigraphy [19,20]. Additionally, multispectral imaging is often used in the cultural heritage sector since it is a non-invasive technique based on ultraviolet, visible, and near-infrared radiation. It allows more details of the surface to be obtained, such as alterations due to aging, and to discriminate any different constituent materials [21–23].

As for the characterization of the chemical composition of paints, numerous techniques can be used. Mass spectroscopy techniques, such as laser ablation-inductively coupled plasma-mass spectrometry (LA-ICP-MS), have been useful for obtaining trace-level information. Chromatographic techniques, such as pyrolysis–gas chromatography–mass spectrometry (Py-GC-MS), allow information to be obtained on the polymer fraction, namely the binder. Furthermore, X-ray fluorescence spectrophotometry (XRF) is used for elemental analysis, while X-ray diffractometry (XRD) allows for the determination of the inorganic fractions' composition and structure [19,20,24–33].

Despite the multiple possibilities, FT-IR spectroscopy, Raman spectroscopy, and SEM-EDS appear to be the most commonly used analytical techniques in the cultural heritage sector [19,20,24,25]. FT-IR and Raman spectroscopies are non-destructive techniques, often not requiring sample preparation. The first allows for the characterization of the organic fraction and only part of the inorganic one. The second allows for the identification of the inorganic component and, in particular, the metal compounds. However, the latter is often affected by the fluorescence phenomenon, which does not allow an interpretable spectrum to be obtained [19,24]. Furthermore, SEM-EDS allows information to be acquired on the morphological aspect of the samples, through secondary electrons, and on the elemental composition of the materials, by exploiting X-rays emitted by the sample [20,24].

This work is focused on the study of the protective paints of two vehicles belonging to the Museum für Kommunikation in Frankfurt and used by the German postal service at the turn of the two centuries. The first one is a cart for telecommunications services and is dated approximately 1920–1940 (Figure 1). This cart was constructed in the Reichspost era and was continuously used by the postal service also during the German Democratic Republic (GDR). The cart was found in the garden of a former postal employee in 2015. It was partially sunk into the ground, where it presumably stood for decades after having been taken out of service.

The second vehicle is a carriage of the German postal service, dated around 1880 (Figure 2). The carriage was found in 1999 when the King's family house was demolished. It is currently stored at the Museum für Kommunikation in Frankfurt. It has been probably used for the postal service between Schramberg and Rottwell until 1910.

Figure 1. Image of the cart for telecommunication service.

The main purpose of this work was the characterization of the inorganic and organic fractions of the vehicles' paints through the use of pigments as markers to allow the discrimination between original and non-original materials. A preliminary study of the surfaces has been carried out through the use of the multispectral imaging system and portable optical microscopy. Both these techniques have enabled the selection of more significant sampling areas to be analyzed with SEM-EDS and FT-IR ATR spectroscopy. Additionally, Raman spectroscopy analyses were performed. However, they were inconclusive due to fluorescence interference and are thus not reported in this paper. This study was thus aimed at identifying and dating the constituent materials by using analytical techniques that are available and easily implementable in most museums' laboratories for the identification of object materials. For this reason, specific techniques were used to assess the effectiveness of their combined application, defining a replicable operational methodology. This approach can have useful applications in many museums with relatively modern objects that can be characterized, such as vehicles of various natures, with a high historical value. Furthermore, the study aimed at detecting the areas requiring subsequent restoration intervention.

Figure 2. Image of the carriage of the family King.

2. Results

2.1. Multispectral Imaging

2.1.1. The Cart

The multispectral images of the cart's different areas show the different fluorescence that was induced by three layers of paint (Table 1).

Table 1. Summary of images obtained in different areas of the cart with multispectral analysis using visible light (VIS), ultraviolet fluorescence, and infrared reflectography (IR).

Table 1. Cont.

Area	VIS	UV FLUORESCENCE	IR 850
Right panel, plate			

The outermost layer, which is light grey in VIS light, is characterized by a low yellowish fluorescence that contrasts with the fluorescence induced on the surface of the intermediate grey-greenish layer. The dark grey layer in contact with the wooden support does not show any fluorescence (Figure 3).

Figure 3. (a) Details of the UV fluorescence image of the paints on the right panel; (b) Enlargement (10×) of the area.

Moreover, the investigation allowed the presence of some areas to be highlighted that appeared reddish in both visible and UV light while showing a black color in IR reflectography (Figure 4).

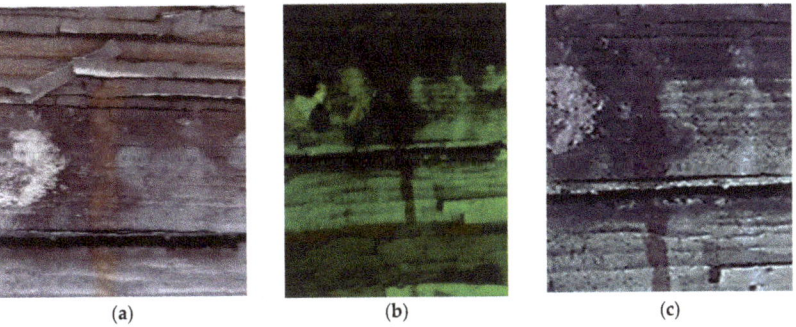

Figure 4. Details of the iron cart structure: leakage due to corrosion; (a) Image in visible light; (b) Image in UV fluorescence; (c) Image in IR reflectography.

These areas are related to the leakage that resulted from the degradation of the cart's iron structure, the solubilization of corrosion products, and the pouring of iron–ions-rich solutions.

2.1.2. The Carriage

The analysis performed by multispectral imaging in UV light showed a bluish-yellow ultraviolet UV fluorescence spread over the entire surface of the carriage (Table 2). Such fluorescence could be due to an oil-based paint that aged over time. The UV fluorescence also allowed the widespread presence of cracks in the paint layer to be highlighted, as well as localized corrosion in several areas of the metal panels that appeared darker.

Table 2. Multispectral images showing different areas of the carriage: visible light (VIS), ultraviolet fluorescence (UV), and infrared reflectography (IR).

Area	VIS	UV FLUORESCENCE	IR 850
Symbol on the right side			
Symbol on the left side			
Rear panel			
Left-side panel			
Rear-left wheel			
Left-side flap			

Table 2. *Cont.*

Area	VIS	UV FLUORESCENCE	IR 850
Rear wheel axle			

The multispectral investigation performed on the symbol present on the right and left side of the carriage underlined an uneven response under UV radiation (Figure 5a,b). Indeed, darker areas could be noticed, which appeared to be lighter in IR reflectography (Figure 5c). The areas in which the fluorescence is less marked (darker in the photo) present a greater reflection of the IR radiation, and these areas appear lighter. These differences can be presumably related to the loss of the finishing layer.

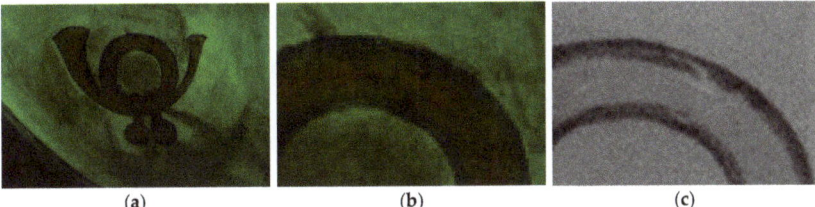

Figure 5. (**a**) Symbol panel of the right side of the carriage in UV fluorescence; (**b**) Enlargement (10×) of the area in UV fluorescence; (**c**) Enlargement (10×) of the area in IR reflectography.

On the rear panel (Figure 6), the IR reflectography allowed the presence of two yellow paints to be noticed that are characterized by different reflections of the IR radiation.

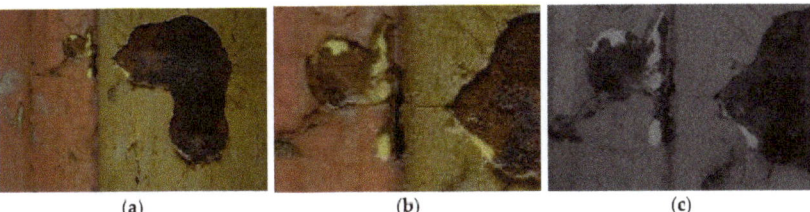

Figure 6. (**a**) Image of the rear panel of the carriage collected under visible light; (**b**) Enlargement (10×) of the area in VIS; (**c**) Enlargement (10×) of the area in IR reflectography. The chromatic differences between the two yellow paints, in visible light and their different reflection of the IR radiation are highlighted.

The more superficial layer appears darker in IR, while the underlying one shows a lighter tone due to the greater reflection of infrared radiation. These characteristics are attributable to the use of different paints. The same feature was also found in the rear axle of the wheels.

2.2. Portable Optical Microscopy

2.2.1. The Cart

The investigation was performed using the optical microscope and allowed the stratigraphy of the grey cart to be identified alongside the presence of a deposited particulate on the surface (Figure 7a).

(a)   (b)

Figure 7. Optical microscope images of the cart's stratification. (**a**) Image collected at 50× magnification in which the dark grey (1), greenish grey (2), and light grey (3) layers are visible; (**b**) Image collected at 200× magnification where the deposited particulate is present (4).

The layers, starting from the innermost one to the outermost one, can be described as follows:

1. Dark grey layer.
2. Greenish grey layer.
3. Light grey layer.

2.2.2. The Carriage

The analysis that was carried out with optical microscopy made it possible to observe the complex stratigraphy of the carriage's paints, which was mainly characterized by the superimposition of four paint layers (Figure 8).

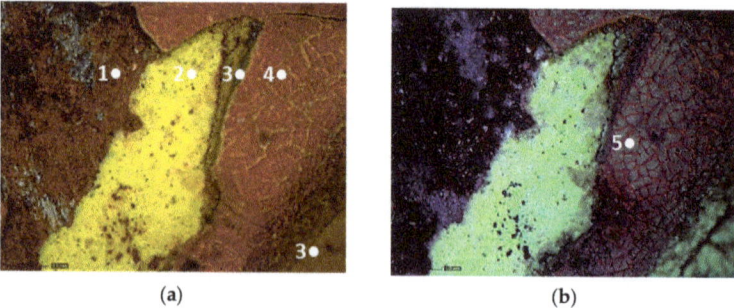

(a)   (b)

Figure 8. Optical microscope image of the layering on the carriage: (**a**) Image collected in VIS light, showing the substrate (1), the bright yellow layer (2), the yellow-brown layer (3), that is above or under the red layer, and the red layer (4); (**b**) Image collected in UV light, where the finishing layer-with a bluish fluorescence can be observed (5) above the colored layers.

Starting from the bottom, the stratigraphy can be described as follows:

4. The support: it is mostly metal-based, while some areas are made of wood.
5. The bright yellow layer: it is present in a few areas of the carriage, such as the back panel of the box letter, the rear axles of wheels, and the metal axes under the left panel of the box letter.
6. The yellow-brown layer.

7. The red layer.
8. The finishing layer.

The finishing layer is characterized by a bluish-yellow fluorescence. It shows widespread cracking and degradation signs which seem to be due to anomalous contractions of the material itself (Figure 9).

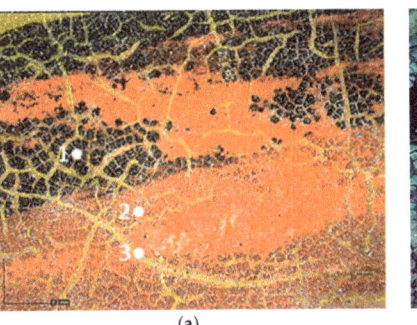
(a)
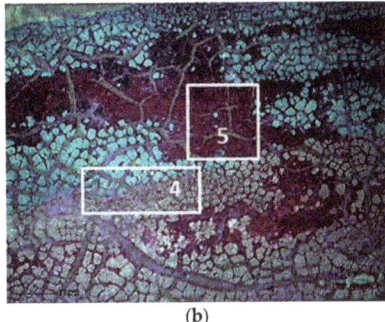
(b)

Figure 9. Optical microscope image (200× magnification) of the area corresponding to the postal service's symbol. (**a**) Image collected in VIS light showing the cracking process with thinner (1–2) and thicker cracks (3) that reach the yellow paint; (**b**) Image collected in UV light highlighting the islands of different sizes and shapes where the finishing layer is present (4), and an area with different fluorescence in which the layer is absent (5).

The craquelure was presumably induced by the imperfect adhesion between the finishing layer and the underlying ones but also by exposure to UV radiation and a high level of relative humidity [34]. In the overlapping areas of the red paint on the yellow one, such cracks are not limited to the surface only, but reach the underlying yellow layer. All types of branched cracking are according to variable, straight, or concentric crosslinks. These crosslinks intersect each other, forming islands of various sizes and shapes. Some areas of the carriage, such as the panels, show that the symbols are characterized by the absence of the finishing layer (Figure 10). Indeed, in these areas, it is possible to detect the underlying paints, which have a different fluorescence when observed in ultraviolet light (Figure 10b).

Moreover, the metal support of the carriage appears to be corroded, and some areas of the superimposed paint layers appear darker, presumably due to corrosion products of the underlying laminate (Figure 10).

(a)
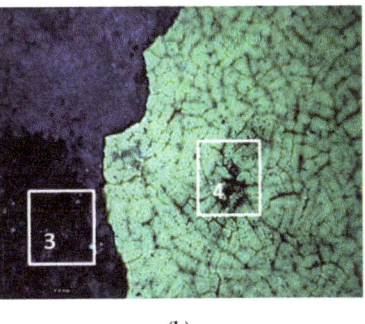
(b)

Figure 10. Optical microscope images (200× magnification) of one of the metal panels: corrosion of the laminate (1–3), causing the darkening of the yellow paint (2–4). (**a**) Image collected in VIS light; (**b**) Image collected in UV light.

Black paint was used exclusively for decorations (Figure 11).

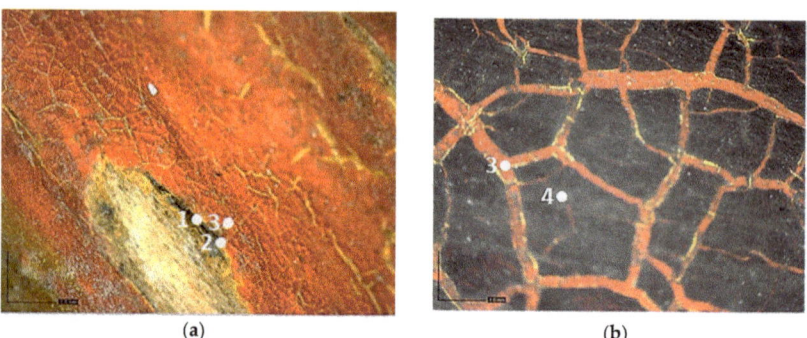

(a)   (b)

Figure 11. Optical microscope images of the areas showing the black paint. (**a**) Image (50× magnification) collected in VIS light showing the black paint (1) under the yellowish brown (2) and red (3) layers; (**b**) Image (50× magnification) collected in VIS light showing the black layer (4) above the red one (3).

2.3. Fourier Transform Infrared Coupled Attenuated Total Reflectance (FT-IR ATR)

The colored paints of both the carriage and the wagon are characterized by similar FT-IR ATR spectra (Figures 12 and 13).

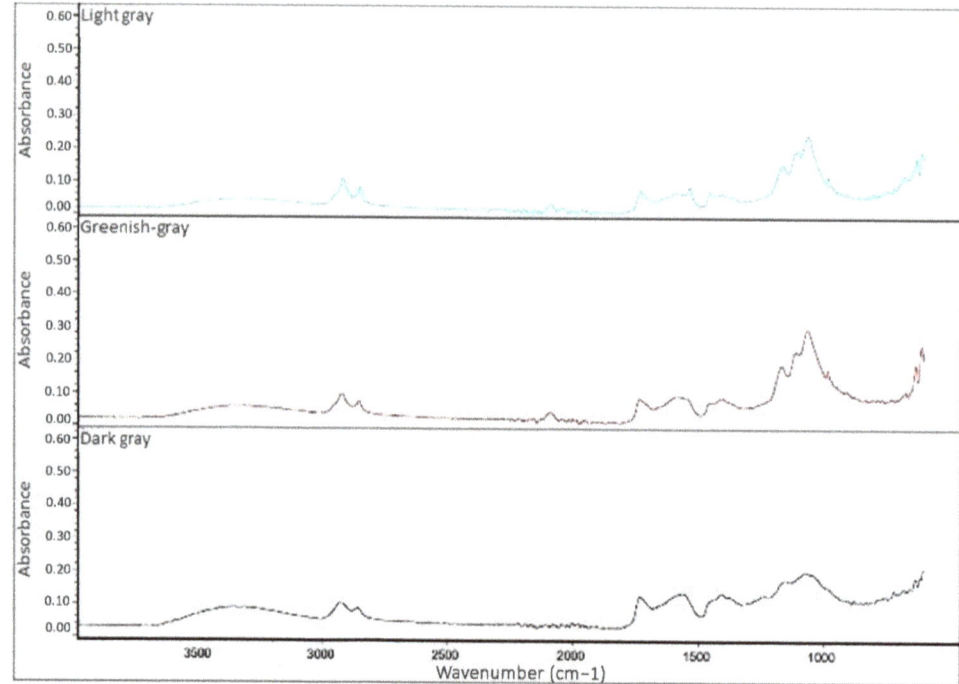

Figure 12. FT-IR ATR spectra of the three grey paints of the cart.

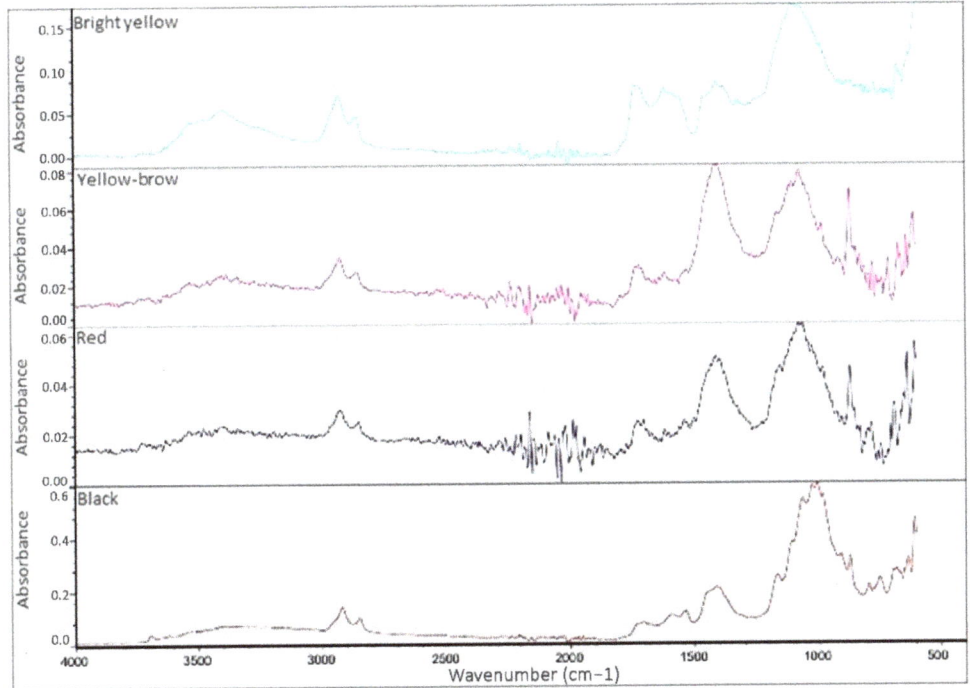

Figure 13. FT-IR ATR spectra of the colored paints of the carriage.

The main difference is in the bands at 1400 and 870 cm^{-1}, which are almost completely absent in the black paint of the carriage and the paints of the wagon. These bands can be linked to the presence of carbonates, which are therefore found in greater amounts in the yellow and red paints of the carriage compared to its total or almost absent in the grey-black paints. The interpretation of spectra allowed for the assignment of the following bands [35–43]:

- 3500–3400 cm^{-1}: stretching of the O-H bonds, present in linseed oil and also in some silicates;
- 2950–2800 cm^{-1}: stretching of the C-H bonds present in the aliphatic chains of the oil;
- 1730 cm^{-1}: stretching of the C=O bond of the esters present in the siccative oil;
- 1600–1400 cm^{-1}: stretching of the COO- group of carboxylates, the presence of which is due to the degradation of the drying oil in the presence of metal cations;
- 1400 and 870 cm^{-1}: asymmetric stretching of the carbonate group (CO$_3^{2-}$);
- 1163 cm^{-1}: asymmetric stretching of the C-C-O bonds of the esters present in the oils;
- 1080–1020 cm^{-1}: vibrational stretching of the Si-O and Si-O-Si bonds, which is typical of silicates.

The paints of the two vehicles are, therefore, characterized by the presence of aged drying oils, silicates, and sometimes carbonates, while the bands related to carboxylates (1600–1400 cm^{-1}) are due to the aging process and degradation of the drying oil.

The analysis of the finishing layer present on the surface of the carriage required the extraction of the finishing layer from the sample with methyl ethyl ketone (MEK). The spectrum acquired on the extracted materials showed the characteristic bands of an oil-based component and carboxylates, similar to the spectra collected on the colored paints (Figure 14).

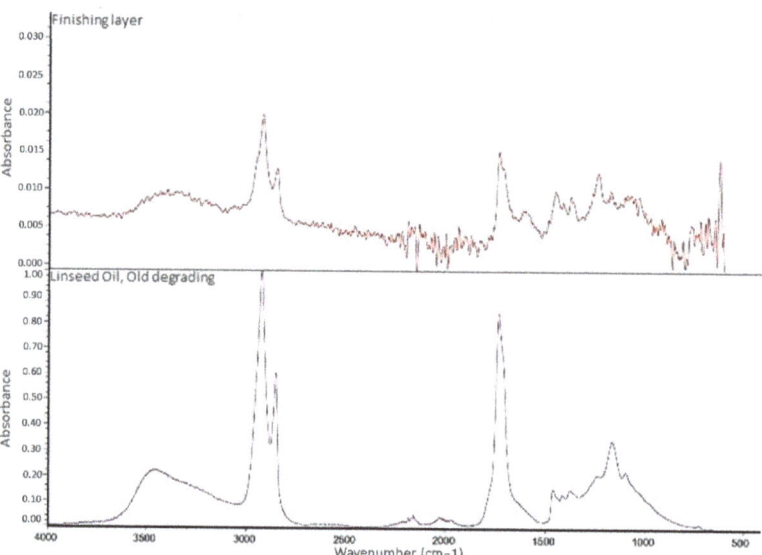

Figure 14. FT-IR ATR spectrum of the finishing layer extracted using MEK (**above**) and the spectrum of degraded linseed oil (**bottom**).

2.4. Scanning Electron Microscopy Coupled with Energy Dispersive Spectroscopy (SEM-EDS)

2.4.1. The Cart

SEM/EDS data of the cart's paints allows important differences between the light grey paint to be highlighted; hence, the most external and recent one compared with the innermost ones, with the dark grey paint directly applied on the wood.

The dark grey paint is characterized by the presence of zinc, barium, and carbon (Table 3). Arsenic is present in low percentages and cannot be related to the color of the paint. For this reason, it can be linked to the wooden support possibly having undergone a biocide treatment. Indeed, arsenic is a poisonous compound that is used for the disinfestation of insects, while copper is generally used for bacteria. The high percentages of carbon suggest that it could have been used with a pigment of organic origin (e.g., carbon black).

Table 3. SEM image collected on the grey inner paint of the cart. In the table, the data obtained by the EDS analysis are reported.

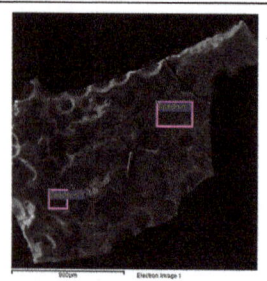

Spectrum	C	O	Mg	Si	S	Ca	Fe	Zn	As	Ba
Spectrum 1	45.7	36.2	0.2	0.3	4.8	0.2	0.8	4.2	0.3	7.3
Spectrum 2	44.8	32.6	0.4	0.3	5.4	0.3	1.1	5.5	0.4	9.3

Quantitative values in wt% (±0.2).

The SEM-EDS analysis carried out on the outermost paint layers (Table 4) made it necessary to exclude carbon from the elements' count in order to obtain more significant information on the elements present in smaller quantities, such as silicon, potassium, and calcium. It is, however, likely that the pigment used to give the grey color can be identified as carbon black, as in the previous paint's analysis.

The elemental analysis carried out on the intermediate paint layer (Table 4; Spectrum 3), i.e., the greenish-grey one, allowed a greater presence of zinc and barium to be noticed, which were probably used to make the dark grey paint lighter. The EDS analysis collected on the light grey layer (Table 4; Spectra 1–2) was characterized by the presence of zinc and titanium. The presence of titanium is of the utmost significance since it is absent in all other grey paints. This element can be linked to the use of titanium white together with zinc white to obtain the desired color gradation and opacity of the paint. Additionally, the presence of iron can be related to the chromatic alterations resulting from the solubilization and leaching of the iron corrosion products of the wagon's structure.

Table 4. SEM/EDS analysis of sample 16. Data obtained by the EDS analyses. The analyzed sample is characterized by the two outermost layers: the light gray layer (1) on which spectra 1 and 2 were collected; the intermediate layer (2) on which spectrum 3 was collected.

Spectrum	O	Al	Si	S	K	Ca	Ti	Fe	Zn	Ba
Spectrum 1						1.4	29.7	13.0	55.9	
Spectrum 2	66.3	1.8	3.9	2.7	0.3	1.1	13.6	3.4	6.9	

Qualitative values in wt% (±0.2).

In addition, the paints related to the iron axes (sample 18) showed a yellowish-brown color that was characterized by the presence of titanium and a high iron content (Table 5). Eventually, titanium was related to the use of titanium white, as already stated in the results obtained on the light-grey paint, while the iron was related to the corrosion of the metal axes containing superimposed paint layers.

Table 5. SEM image collected on sample 18. In the table, the EDS analysis is reported. Spectra 1–3 were collected on the brownish paint; Spectrum 2 was acquired on the underlying grey layer.

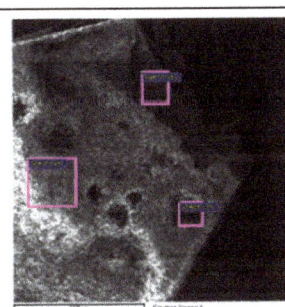

Spectrum	C	O	Al	Si	P	S	Cl	K	Ca	Ti	Fe	Zn	Ba
Spectrum 1	-	65.5	2.9	2.3	-	2.9	-	0.3	1.2	5.3	17.9	1.7	-
Spectrum 2	38.0	31.3	0.5	-	-	4.7	-	-	0.4	-	1.3	13.7	10.1
Spectrum 3	-	63.1	2.6	2.3	0.4	3.8	0.3	0.3	1.6	6.8	16.6	2.3	-

Quantitative values in wt% (±0.2).

2.4.2. The Carriage

The results of the SEM-EDS analysis allowed for the identification of the pigments based on the evaluation of detected chemical elements.

The bright yellow paint on the top of the metal support is characterized by a high amount of lead, which is not present in the other examined paints (Table 6). This paint represents the original layer of the carriage, whose yellow color is attributable to the presence of lead compounds.

Table 6. SEM image collected on the bright yellow paint of the carriage at high magnification. In the table, the data obtained from the EDS analysis are reported.

Spectrum	O	Al	Si	S	K	Ca	Fe	Zn	As	Br	Ba	Pb
Spectrum 1	62.6	1.4	-	-	-	1.4	-	13.0	-	-	-	20.8
Spectrum 2	62.2	1.5	1.9	4.1	0.5	1.6	9.7	11.3	0.0	-	4.0	3.2
Spectrum 3	44.8	-	-	-	-	2.4	-	13.4	-	0.0	5.4	34.0
Spectrum 4	74.7	2.6	2.5	3.5	-	1.6	1.8	6.0	-	-	2.0	5.3

Quantitative values in wt% (±0.2).

SEM-EDS analysis on the yellow-brown paint (second layer) allowed for the presence of silicates to be noticed. Due to the extensive corrosion of the metal panels (Figure 10), iron was excluded from the elemental analysis of the yellow paint, as it was attributable to both corrosion products and the use of iron-based pigments (Table 7). However, the lack of other chromophores indicates that the pigment of the paint is likely yellow earth.

Table 7. SEM image collected on the yellow-brown paint of the carriage. In the table the reported data were obtained from the EDS analysis (excluding iron).

Spectrum	C	O	Al	Si	S	K	Ca	Zn	Ba
Spectrum 1	52.4	46.3	0.3	0.2	0.2	-	0.5	0.3	-
Spectrum 2	-	77.0	-	1.3	9.4	0.6	10	1.7	-
Spectrum 3	24.8	51.5	-	3.8	4.8	-	9.2	1.3	4.6

Quantitative values in wt% (±0.2).

The SEM images of the red paint (Table 8) show the significant cracking forming regular blocks (third paint layer). It is a characteristic feature of the cracking induced by an imperfect adhesion between the red layer and the yellow paint layer underneath [34,44]. In the elemental analysis, carbon was omitted because its high value did not allow the semiquantitative determination of other elements, such as zinc, barium, sulfur, and calcium.

The results reported in Table 8, together with the elemental analysis acquired on a detail of the surface layer (Table 9), underline the presence of an organic compound. This information allows for the hypothesis that the paint was characterized by an organic dye that provided the red color. Moreover, higher magnification enabled the observation of uniform particles without a specific structure that could be attributed to materials not relevant to the paint. Indeed, the observation in the backscattered electrons (BSE) mode showed the presence of light particles on a dark matrix that was relevant to organic compounds. The presence of these inorganic particles was found in all paints and is attributable to earthy dust or plaster on the walls of the storage environment.

Table 8. SEM image of the red paint of the carriage. In the table are reported the data obtained from the EDS analysis.

Spectrum	O	Mg	Al	Si	S	K	Ca	Fe	Zn	Ba
Spectrum 1	78.0	-	2.5	4.0	5.3	0.5	4.3	1.2	4.2	-
Spectrum 2	70.6	0.6	1.3	2.0	11.2	-	11.2	0.8	2.3	-
Spectrum 3	52.0	-	2.5	5.2	11.0	0.4	3.4	2.8	6.2	16.5

Quantitative values in wt% (±0.2).

Table 9. SEM image collected on the red paint of the carriage at high magnification. In the table, the data obtained from the EDS analysis are reported.

Spectrum	C	O	Al	Si	S	Ca	Fe	Zn	Ba
Spectrum 1	47.6	44.0	-	-	0.4	7.0	0.4	-	0.6
Spectrum 2	-	78.3	1.5	1.8	-	18.4	-	-	-
Spectrum 3	43.4	31.0	-	-	3.8	0.8	0.9	1.0	19.0
Spectrum 4	35.0	29.0	-	-	6.3	0.7	0.8	1.0	27.0

Quantitative values in wt% (±0.2).

The black paint used for decorations on the red one is characterized by purely organic components, as shown by the SEM/EDS analysis (Table 10). No chromophore elements were detected. Thus, the black color was most likely given by the carbon present in the pigments of organic origin (e.g., carbon black). Traces of silicates, zinc, barium, sulfur, and calcium were also found, similar to the other paints.

Table 10. SEM image collected on sample 3. In the table, the data obtained from the EDS analysis of the black paint are reported.

Spectrum	C	O	Al	Si	S	Ca	Fe	Zn	Ba
Spectrum 1	39.0	51.4	0.7	1.2	1.1	4.4	0.2	0.5	1.6

Quantitative values in wt% (±0.2).

3. Discussion

The data provided by the multi-analytical investigation on the two vehicles, together with information found in the literature, have made it possible to obtain useful information regarding the composition of the matrices of paints, thus enabling their dating.

Multispectral imaging and optical portable microscopy allowed for the detection of the cart's stratigraphy, which appeared to have a simpler structure compared to the carriage. Indeed, it consists of three layers: the dark grey paint in direct contact with the wooden support, as well as the original paint; the greenish-grey paint, namely the intermediate one; the light grey paint, namely the outermost one. On the other hand, the carriage shows a stratigraphy consisting of the metal support, four paint layers (a yellow one, a yellow-brown one, a red one, and a black one corresponding to decorations only), and a finishing layer that is visible under UV radiation.

The chromophore element, which gives the grey color to the three paints, has been identified as carbon due to the use of a dye or an organic pigment.

SEM-EDS analysis allowed for the detection of high concentrations of lead in the innermost yellow paint layer, which could be related to the use of a yellow lead pigment. As previously mentioned, this pigment was used until the 1940s in Europe [4,6–9,14,15]. Zinc has also been found in the greenish-grey intermediate layer. It could be linked to the use of zinc white (zinc oxide) and/or lithopone, which have been used mainly since the late 19th century until the first half of the 20th century [5–8,14,16,18]. The presence of barium could be related to the use of lithopone (a mixture of barium sulfate and zinc sulfide) but also to the use of natural barite: a white pigment used as a filler. Further fillers that were detected in the vehicle's layer matrices consist of silicates and carbonates [4,6,7,14,18]. Furthermore, the chromophore element provided the grey color of the three paints, which were identified as carbon, while the presence of arsenic could be related to an insect biocide for the wooden support [30]. The outermost paint (the light-grey one) is mainly characterized by the presence of titanium, related to the use of titanium white. This pigment was discovered in the 1920s and was increasingly used from the 1940s [4,6–9,14,15]. The elemental analysis allowed for the identification of zinc in high concentrations in the outermost paint layer, suggesting the use of a composite titanium pigment made up of titanium and zinc whites. This information allowed for the dating of the paint presumably after the 1920s, when titanium white was discovered, and before the 1970s, when titanium oxide was no longer used as a composite pigment [16]. Eventually, SEM-EDS analysis also detected iron, especially in the outermost paint layers. The presence of this element is due to the corrosion of the metal axes of the cart structure, leading to the formation of chromatic alterations, which was also observed through multispectral imaging. The FT-IR ATR analysis highlighted the presence of silicates and carbonates while also detecting the characteristic bands of degraded oils in both the finishing layer, and the paint layer underneath. In all the paints, the presence of an oily binder could suggest that they might date before the 1940s when paints based on synthetic resins took over [4,6–9,14,15]. Whereas FT-IR spectroscopy detected the vibration bands associated with chemical bonds and oils of similar functional groups, it was not possible to define the exact type of oil. The most commonly used oily binder has been linseed oil, and it may have been used for the production of studied paints [4,6–9,14,15]. Despite these considerations, all oily binders were used up until the first half of the 20th century, when they were then completely replaced by synthetic resins [4,6–9,14,15].

The FT-IR ATR carried out on the carriage showed the presence of silicates and carbonates, as well as the characteristic bands of oil and carboxylates, highlighting once again the presence of degradation products of oily substances [43]. As in the cart, the presence of peaks at 1600–1400 cm^{-1} is related to the stretching of the COO- a group of carboxylates, which highlights the significant degradation of the drying oil.

While FT-IR ATR spectroscopy did not allow the exact identification of the pigments or dyes used for the red and black paints in the carriage, the SEM-EDS analysis allowed for the detection of a high concentration of lead in the innermost yellow paint layer, which could be related to the use of a yellow lead pigment. As previously mentioned, this pigment was used until the 1940s in Europe [14].

The bright yellow paint was lost in most areas of the carriage and was replaced by yellow-brown paint. Even though the SEM-EDS analysis of this paint's samples required the exclusion of iron since the high values could be related either to the corrosion of the metal sheets or the use of yellow ochre, this is one of the most used yellow pigments in history [45].

The two other paintings of the carriage, i.e., the red paint and the black one were used for decorations. They are characterized by a high carbon content, suggesting that both paints were related to the use of dyes or organic pigments.

The SEM-EDS analysis also highlighted the presence of calcium, silicon, barium, sulfur, and zinc in all the paint layers. The presence of calcium and silicon can be attributed, respectively, to the use of carbonates and silicates, while barium and sulfur enabled the hypothesis of barite. All these compounds were used in the paints' formulation as fillers to improve stability and decrease production costs. Zinc is attributable to the use of zinc oxide: a pigment that was discovered in the late 18th century and became common in the late 19th century. However, the presence of zinc and barite did not allow the exclusion of the use of lithopone: another opaque white pigment that has been used since 1880 [5–8,14,16,18].

In conclusion, the diagnostic investigations allowed the identification of the carriage's paints, which are oil-based paints containing pigments based on zinc, barium, and lead for the bright yellow paint only. The characterization suggested the dating of the paints before 1950 when the oil paints were replaced by paints based on synthetic resins, which were discovered in the first decades of the 20th century. The absence of titanium allows for estimating the dating of the paints before the 1940s.

4. Materials and Methods

4.1. The Vehicles

As previously mentioned, this article is focused on the study of protective paints used on two vehicles. They are conserved at the Museum für Kommunikation in Frankfurt and were used by the German postal service at the turn of the two centuries.

The first one is a cart for telecommunications services and is dated approximately 1920–1940 (Figure 1). This cart was constructed in the Reichspost era and was continuously used by the postal service throughout the German Democratic Republic (GDR). The cart was found in the garden of a former postal employee in 2015. It was partially sunk into the ground, where it presumably stood for decades after having been taken out of service.

The second vehicle is a carriage of the German postal service, dated around 1880 (Figure 2). The vehicle was found in 1999 when the King's family house was demolished. It is currently stored at the Museum für Kommunikation in Frankfurt. It is supposed to have been used for the postal service between Schramberg and Rottwell until 1910.

4.2. Portable Optical Microscopy

A portable optical microscope Dino-Lite AM411-FVW (Dino-Lite Europe, Almere, The Netherlands) was used to investigate the surface morphology. It was performed using different magnifications, from $40\times$ to $220\times$ (where X = enlargement), and three illumination modes: visible (VIS), ultraviolet (UV), and ranking (VIS-RAD) light.

4.3. Multispectral Imaging

Multispectral imaging was carried out through VIS, UV, and infrared (IR) spectrum bands. Fluorescence UV was used to identify and characterize the presence of film-forming substances on the surface, while mid- and near-infrared spectrum bands verified the presence of different materials based on their interaction with infrared radiation. The investigation was carried out with the Madatec multispectral system (Madatec, Pessano con Bornago, Italy) consisting of a Samsung NX500 28.2 MP BSI CMOS camera. Ultraviolet fluorescence was observed using CR230B-HP Madatec UV spotlights (365 nm), HOYA

UV-IR filter cut 52, and Yellow 495 52 mm F-PRO MRC 022. Infrared reflectography images were obtained using an 850 nm filter.

4.4. FT-IR ATR Spectroscopy

Spectroscopy FT-IR was performed to characterize the paint materials at the functional-group level. The spectra were gained using the Nicolet Summit FT-IR spectrometer (Thermo Fisher Scientific, Waltham, MA, USA) and equipped with the Everest™ Diamond ATR accessory, with an instrumental resolution of 8 cm^{-1}; 32 scans were performed on each sample, and the respective spectra were analyzed using the database library and the scientific literature. The analysis was carried out on 18 samples taken from the areas highlighted in Table 11. Samples (1 × 1 mm) were taken from the selected areas using a scalpel to sample the whole layer structure of the stratigraphy.

Table 11. Sampling areas.

Vehicle	Vehicle's Side	Sampling Areas
Carriage	Right side and rear axle (sample 6)	
Carriage	Rear panel of the box letter	

Table 11. *Cont.*

Vehicle	Vehicle's Side	Sampling Areas
Carriage	Left side, flap under the seat	
Carriage	Front panel	
Carriage	Left symbol (was detached; found inside the left flap)	

Table 11. *Cont.*

Vehicle	Vehicle's Side	Sampling Areas
Cart	Left panel	
Cart	Frontal panel	

4.5. SEM-EDS

The SEM-EDS technique was performed to carry out morphological, structural, and chemical analysis on the same samples used for FT-IR ATR analysis. The investigations were carried out with a VEGA3-Tescam instrument (Tescan, Brno, Czech Republic) coupled with an Inca 300 Energy Dispersive X-ray (Oxford Instruments Analytical, High Wycombe, United Kingdom) microanalysis system. Samples were analyzed in a high vacuum with a beam potential of 30 KeV: a suitable intensity to be able to perform the elemental composition in EDS acquisition. The SEM observations were performed in secondary electrons (SE) together with backscattered electrons (BSE). Furthermore, operating a high vacuum allowed for reliable quantitative values and even light elements.

5. Conclusions

The exposure of objects to the environment led to changes in their properties, functionality, and integrity. For this reason, protective paints have been traditionally used to protect surfaces from the external environment. They have undergone great evolutions over the years. Specifically, the transformations increased at the turn of the 19th and 20th centuries, thanks to the discovery and development of new compounds, together with the withdrawal from the market of toxic elements.

Oily binders were used until the first half of the 20th century, while synthetic binders were discovered in the 1920s, completely replacing oily binders in the second half of the century. Between the 19th and 20th centuries, significant developments occurred in the pigments' manufacturing process as well. These changes were mainly represented by the progressive disposal of lead-based pigments and the discovery of new pigments, such as zinc white, lithopone, and titanium white.

Due to the potential of such binders and pigments to be used as markers for the dating of objects and artworks, the present study aimed at characterizing the paints of the cart and the carriage belonging to the Museum für Kommunikation in Frankfurt while estimating their dating and identifying the authentic materials to be preserved. The paints of the two vehicles might have been applied before 1950, as they are oil-based. The paints of the carriage and the innermost paints of the cart might be dated before 1940, before the widespread use of titanium white. In addition, the outermost paints of the cart are characterized by the presence of titanium and are thus dated between the 1920s and the 1950s.

In conclusion, the analytical investigations allowed the carriage and the cart to be dated, confirming their authenticity and also gaining significant information on the paints used in Germany for the decoration of vehicles in the postal service. Furthermore, obtaining information on the compositional nature and historical importance of the paints is essential for directing any restoration project which aims to preserve the historical and testimonial importance of these vehicles.

Author Contributions: Conceptualization, A.M.; methodology, A.M.; validation, A.M.; formal analysis, A.M., L.M.S., I.A.C. and S.M.; investigation, A.M., D.F. and I.A.C.; resources, L.M.S.; data curation, A.M., D.F., I.A.C. and F.I.B.; writing—original draft preparation, A.M., I.A.C. and D.F.; writing—review and editing, A.M., I.A.C. and C.B.; visualization, A.M., I.A.C. and C.B.; supervision, A.M. and M.F.L.R.; project administration, A.M. All authors have read and agreed to the published version of the manuscript.

Funding: This research received no external funding.

Institutional Review Board Statement: Not applicable.

Informed Consent Statement: Not applicable.

Data Availability Statement: Publicly available datasets were analysed in this study to compare the acquired FT-IR spectra with already available spectra of standard materials. Data can be found here: https://spectra.chem.ut.ee/ (accessed on 5 November 2022).

Conflicts of Interest: The authors declare no conflict of interest.

References

1. D'Agostino, D.; Macchia, A.; Cataldo, R.; Campanella, L.; Campbell, A. Microclimate and salt crystallization in the crypt of lecces duomo. *Int. J. Archit. Herit.* **2014**, *9*, 290–299. [CrossRef]
2. Hughes, J. An artifact is to use: An introduction to instrumental functions. *Synthese* **2009**, *168*, 179–199. [CrossRef]
3. Fotovvati, B.; Namdari, N.; Dehghanghadikolaei, A. On coating techniques for surface protection: A review. *J. Manuf. Mater.* **2019**, *3*, 28. [CrossRef]
4. Bierwagen, G.P. *Surface Coating*; Encyclopedia Britannica: Chicago, IL, USA, 2016.
5. Mathiazhagan, A.; Joseph, R. Nanotechnology-A New Prospective in Organic Coating-Review. *Int. J. Chem. Eng. Appl.* **2011**, *2*, 225–237. [CrossRef]
6. Standeven, H.A.L. *House Paints, 1900–1960: History and Use*; Getty Publications: Los Angeles, CA, USA, 2011.
7. Standeven, H.A.L. Oil-based house paints from 1900 to 1960: An examination of their history and development, with particular reference to ripolin enamels. *J. Am. Inst. Conserv.* **2013**, *52*, 127–139. [CrossRef]
8. Todd, J. Cars, paint, and chemicals: Industry linkages and the capture of overseas technology between the wars. *Aust Econ. Hist Rev.* **1998**, *38*, 176–193. [CrossRef]
9. Gooch, J.W. History of Paint and Coatings Materials. In *Lead-Based Paint Handbook*; Springer: New York, NY, USA, 2002; pp. 13–35.
10. Myers, R.R. History of Coatings Science and Technology. *J. Macromol. Sci.* **1981**, *15*, 1133–1149. [CrossRef]
11. Croll, S. Overview of developments in the paint industry since 1930. In *Modern Paints Uncovered*; Getty Publications: Los Angeles, CA, USA, 2006; pp. 17–29.
12. Islam, M.N.; Rahman, F. Production and modification of nanofibrillated cellulose composites and potential applications. In *Green Composites for Automotive Applications*; Woodhead Publishing: Sawstone, UK, 2018; pp. 115–141.
13. Eastlake, C. *Materials for a History of Oil Painting*; Longmans, Green & Company: London, UK, 1847.
14. Källbom, A. The Concept of Historical Aluminium-Pigmented Anticorrosive Armour Paints, for Sustainable Maintenance of Ferrous Heritage. *Int. J. Archit. Herit.* **2022**, *16*, 1112–1129. [CrossRef]

15. Källbom, A.; Almevik, G. Maintenance of Painted Steel-sheet Roofs on Historical Buildings in Sweden. *Int. J. Archit. Herit.* **2022**, *16*, 538–552. [CrossRef]
16. Van Driel, B.A.; Berg, K.J.V.D.; Gerretzen, J.; Dik, J. The white of the 20th century: An explorative survey into Dutch modern art collections. *Herit. Sci.* **2018**, *6*, 16. [CrossRef]
17. Macchia, A.; Biribicchi, C.; Carnazza, P.; Montorsi, S.; Sangiorgi, N.; Demasi, G.; Prestileo, F.; Cerafogli, E.; Colasanti, I.A.; Aureli, H.; et al. Multi-Analytical Investigation of the Oil Painting "Il Venditore di Cerini" by Antonio Mancini and Definition of the Best Green Cleaning Treatment. *Sustainability* **2022**, *14*, 3972. [CrossRef]
18. IARC Working Group on the Evaluation of Carcinogenic Risks to Humans. and International Agency for Research on Cancer. Some organic solvents, resin monomers and related compounds, pigments and occupational exposures in paint manufacture and painting. *IARC Monogr. Eval. Carcinog. Risks Hum.* **1989**, *47*, 1.
19. McIntee, E. Forensic Analysis of Automobile Paints by Atomic and Molecular Spectroscopic Methods and Statistical Data Analyses. *ETD* **2008**, 2004–2019.
20. Mahmoud, H.H.M. Investigations by Raman microscopy, ESEM and FTIR-ATR of wall paintings from Qasr el-Ghuieta temple, Kharga Oasis, Egypt. *Herit. Sci.* **2014**, 2. [CrossRef]
21. Zhao, Y.; Berns, R.S.; Taplin, L.A.; Coddington, J. An investigation of multispectral imaging for the mapping of pigments in paintings. In *Computer Image Analysis in the Study of Art*; SPIE: Cergy Pontoise, France, 2008; Volume 6810, p. 681007. [CrossRef]
22. Aldrovandi, A.; Bertani, D.; Cetica, M.; Matteini, M.; Moles, A.; Poggi, P.; Tiano, P. Multispectral image processing of paintings. *Stud. Conserv.* **1988**, *33*, 154–159.
23. Triolo, P.A.M. *Manuale Pratico di Documentazione e Diagnostica per Immagine per i BB.CC*; Il Prato: Saonara, Italy, 2019.
24. Malek, M.A.; Nakazawa, T.; Kang, H.W.; Tsuji, K.; Ro, C.-U. Multi-modal compositional analysis of layered paint chips of automobiles by the combined application of ATR-FTIR imaging, raman microspectrometry, and SEM/EDX. *Molecules* **2019**, *24*, 1381. [CrossRef] [PubMed]
25. Ploeger, R.; Scalarone, D.; Chiantore, O. The characterization of commercial artists' alkyd paints. *J. Cult. Herit.* **2008**, *9*, 412–419. [CrossRef]
26. De Nolf, W.; Janssens, K. Micro X-ray diffraction and fluorescence tomography for the study of multilayered automotive paints. *Surf. Interface Anal.* **2010**, *42*, 411–418. [CrossRef]
27. Franquelo, M.L.; Duran, A.; Castaing, J.; Arquillo, D.; Perez-Rodriguez, J.L. XRF, μ-XRD and μ-spectroscopic techniques for revealing the composition and structure of paint layers on polychrome sculptures after multiple restorations. *Talanta* **2012**, *89*, 462–469. [CrossRef] [PubMed]
28. Kanngießer, B.; Malzer, W.; Rodriguez, A.F.; Reiche, I. Three-dimensional micro-XRF investigations of paint layers with a tabletop setup. *Spectrochim. Acta Part B At. Spectrosc.* **2005**, *60*, 41–47. [CrossRef]
29. Sawczak, M.; Kamińska, A.; Rabczuk, G.; Ferretti, M.; Jendrzejewski, R.; Śliwiński, G. Complementary use of the Raman and XRF techniques for non-destructive analysis of historical paint layers. *Appl. Surf. Sci.* **2009**, *255*, 5542–5545. [CrossRef]
30. Colombini, M.P.; Modugno, F. *Organic Mass Spectrometry in Art and Archaeology*; Wiley: Hoboken, NY, USA, 2009.
31. Germinario, G.; van der Werf, I.D.; Sabbatini, L. Chemical characterisation of spray paints by a multi-analytical (Py/GC–MS, FTIR, μ-Raman) approach. *Microchem. J.* **2016**, *124*, 929–939. [CrossRef]
32. Hobbs, A.L.; Almirall, J.R. Trace elemental analysis of automotive paints by laser ablation-inductively coupled plasma-mass spectrometry (LA-ICP-MS). *Anal. Bioanal. Chem.* **2003**, *376*, 1265–1271. [CrossRef]
33. Smith, K.; Horton, K.; Watling, R.J.; Scoullar, N. Detecting art forgeries using LA-ICP-MS incorporating the in situ application of laser-based collection technology. *Talanta* **2005**, *67*, 402–413. [CrossRef]
34. Pauchard, L.; Giorgiutti-Dauphiné, F. Craquelures and pictorial matter. *J. Cult. Herit.* **2020**, *46*, 361–373. [CrossRef]
35. Madejová, J. FTIR techniques in clay mineral studies. *Vib. Spectrosc.* **2003**, *31*, 1–10. [CrossRef]
36. Meilunas, R.J.; Bentsen, J.G.; Steinberg, A. Analysis of Aged Paint Binders by FTIR Spectroscopy. *Stud. Conserv.* **1990**, *35*, 33.
37. Nayak, P.S.; Singh, B.K. Instrumental characterization of clay by XRF, XRD and FTIR. *Bull. Mater. Sci.* **2007**, *30*, 235–238. [CrossRef]
38. Van der Weerd, J.; Van Loon, A.; Boon, J.J. FTIR studies of the effects of pigments on the aging of oil. *Stud. Conserv.* **2005**, *50*, 3–22. [CrossRef]
39. Mazzeo, R.; Roda, A.; Prati, S. Analytical chemistry for cultural heritage: A key discipline in conservation research. *Anal. Bioanal. Chem.* **2011**, *399*, 2885–2887. [CrossRef] [PubMed]
40. Li, K.-M.; Jiang, J.-G.; Tian, S.-C.; Chen, X.-J.; Yan, F. Influence of silica types on synthesis and performance of amine-silica hybrid materials used for CO_2 capture. *J. Phys. Chem. C* **2014**, *118*, 2454–2462. [CrossRef]
41. Teresa, O.H.; Choi, C.K. Comparison between SiOC thin films fabricated by using plasma enhance chemical vapor deposition and SiO2 thin films using fourier transform infrared spectroscopy. *J. Korean Phys. Soc.* **2010**, *56*, 1150–1155.
42. Pantoja-Castro, M.A.; González-Rodrìguez, H. Study by infrared spectroscopy and thermogravimetric analysis of Tannins and Tannic acid. *Rev. Latinoam Quim.* **2011**, *39*, 107–112.
43. Filopoulou, A.; Vlachou, S.; Boyatzis, S.C. Fatty acids and their metal salts: A review of their infrared spectra in light of their presence in cultural heritage. *Molecules* **2021**, *26*, 6005. [CrossRef] [PubMed]

44. Kim, S.; Park, S.M.; Bak, S. Investigation of craquelure patterns in oil paintings using precise 3D morphological analysis for art authentication. *PLoS ONE* **2022**, 17. [CrossRef] [PubMed]
45. Hradil, D.; Grygar, T.; Hradilová, J.; Bezdička, P. Clay and iron oxide pigments in the history of painting. *Appl. Clay Sci.* **2003**, 22, 223–236. [CrossRef]

Disclaimer/Publisher's Note: The statements, opinions and data contained in all publications are solely those of the individual author(s) and contributor(s) and not of MDPI and/or the editor(s). MDPI and/or the editor(s) disclaim responsibility for any injury to people or property resulting from any ideas, methods, instructions or products referred to in the content.

Article

Analytical Techniques Applied to the Study of Industrial Archaeology Heritage: The Case of *Plaiko Zubixe* Footbridge

Ilaria Costantini *, Kepa Castro, Juan Manuel Madariaga and Gorka Arana

Department of Analytical Chemistry, Faculty of Science and Technology, University of the Basque Country UPV/EHU, P.O. Box 644, 48080 Bilbao, Spain; kepa.castro@ehu.eus (K.C.); juanmanuel.madariaga@ehu.eus (J.M.M.); gorka.arana@ehu.eus (G.A.)
* Correspondence: ilaria.costantini@ehu.eus

Abstract: In this work, micro-Raman spectroscopy and micro-energy-dispersive X-ray fluorescence spectroscopy (μ-EDXRF) were applied on microsamples taken from the *Plaiko Zubixe* footbridge (1927) located in Ondarroa (Basque Country, Spain) in order to investigate the original paint coating and make an evaluation of the conservation state before its restoration. Elemental and molecular images were acquired for the study of the compounds distribution. Some modern pigments such as phthalocyanine blue and green pigments, minium, calcium carbonate, Prussian blue, and hematite were identified. Barium sulfate and titanium dioxide were recognized as opacifier agents. Thanks to the study of the stratigraphies, it has been possible to determine the original paint layer, which includes lead white, ultramarine blue, carbon black, and barium sulfate. In addition, colorimetric analyses made it possible to know the CIELab values of the original layer in order to reproduce the original colour during the planned restoration work. The massive presence of chlorine detected by μ-EDXRF and the corrosion products of the rust layer, in particular akaganeite and hematite, highlighted the atmospheric impact in the conservation of the bridge because they were due to the effect of both marine aerosol and to the presence of acidic components in the environment coming from anthropogenic activity. This work demonstrated the usefulness of a scientific approach for the study of industrial archaeology heritage with the aim to contribute to its conservation and restoration.

Keywords: industrial heritage; μ-Raman spectroscopy; μ-EDXRF; pigments; conservation state

Citation: Costantini, I.; Castro, K.; Madariaga, J.M.; Arana, G. Analytical Techniques Applied to the Study of Industrial Archaeology Heritage: The Case of *Plaiko Zubixe* Footbridge. *Molecules* **2022**, *27*, 3609. https://doi.org/10.3390/molecules27113609

Academic Editors: Maria Luisa Astolfi, Maria Pia Sammartino and Emanuele Dell'Aglio

Received: 8 May 2022
Accepted: 1 June 2022
Published: 4 June 2022

Publisher's Note: MDPI stays neutral with regard to jurisdictional claims in published maps and institutional affiliations.

Copyright: © 2022 by the authors. Licensee MDPI, Basel, Switzerland. This article is an open access article distributed under the terms and conditions of the Creative Commons Attribution (CC BY) license (https://creativecommons.org/licenses/by/4.0/).

1. Introduction

The archaeological industrial heritage is a relatively new concept, born in the 1970s, when the need to preserve the proofs of industrialization process after they had been fallen into disuse or abandoned was declared [1]. Indeed, the industrial heritage concerns a particular type of heritage that includes objects, infrastructures, and works created during the industrial revolution, mainly for practical rather than decorative purposes that have had a strong impact on the territory and some of which have now been recently declared historical heritage.

On the one hand, the conservation of the industrial heritage concerns individual objects that can represent a symbol for the city and its inhabitants. On the other hand, the rehabilitation of the industrial heritage can involve entire urban areas, producing an evident increase in tourism and promoting a social, environmental, and economic development of the cities [2,3]. Therefore, regardless of the social impact, the industrial heritage conservation is unquestionably a topical issue.

Its restoration could be carried out with scientific standards, respecting the original appearance and the economic and technological environment of the time, or by creating a new work, a replica of the existing. In other situations, although the original appearance of a work was different, the authorities can decide to restore the appearance that the work had in the last few years. These latter cases are mainly linked to social or cultural reasons.

However, a scientific diagnostic approach, through the use of diagnostic techniques, capable of gaining knowledge of the technologies and materials used, has not been widely adopted in this field of research as has been the case for other types of works of art [4–6]. Currently, there are not many examples of scientific research in the literature concerning the study of industrial heritage. One of the most recent is the work by Tissot et al. [7] on the paint coatings of three energy generators from the early 20th-century power plant at Levada de Tomar (Portugal) that shows the importance of applying a scientific-diagnostic method even for the study of objects belonging to the industrial revolution [5].

The conservation of the industrial cultural heritage is strongly influenced by the environment in which the object is located. Both metal and steel, of which the industrial heritage is mainly composed, are considered among the most resistant materials, and for this reason they have been used for the construction of bridges and infrastructures. Despite this, if they are located near sources of humidity or environments with high relative humidity values, they can suffer from faster oxidation processes over the years. In particular, a marine atmosphere is one of the most corrosive environments for metallic structures due to the influence of marine aerosol. It is composed by organic and inorganic matter dissolved in water and includes primary (PMA) and secondary aerosol (SMA) particles. The primary aerosol is composed by suspended sea water drops rich in chloride-ions, generated by the interaction between wind and waves on the surface of the sea, which are deposited on the terrestrial surfaces according to a dry or wet deposition process [8]. The high content of airborne chlorides, mainly in the form of NaCl or KCl, react extensively with iron materials [9].

In addition, it is well known that the presence of SO_2 as well as the action of other acid gases, such as NO_x and CO_2, can cause the increase of the corrosion rate in metals through wet or dry deposition mechanisms [10,11]. In wet deposition, the atmospheric acid gases react with the humidity and/or rainwater, giving rise to their acidic aerosols (H_2CO_3, H_2SO_4, and HNO_3). The acidic nature of the moisture film deposited on the surface generates first the oxidation and then the dissolution of the metal, accelerating the corrosion mechanisms and the consequent formation of nitrate, sulfate, and carbonate salts. In dry deposition, the atmospheric gases can react directly with solid particles deposited on the surfaces [12]. Moreover, in an urban site close to the coast, the marine aerosol is rich in airborne particulate matter including metals such as Pb, Cd, Cr, Mn, Cu, Mo, Rh, Ni, As, Ti, V, and Hg coming from combustion processes, traffic, and industrial activity [13]. In addition, factors such as turbulence of the air, chemical affinity between pollutants, and the material and reactivity of the pollutants can accelerate the deposition phenomena [10]. Although in some cases the corrosion products have a protective function [14–16], in coastal atmospheres the presence of certain corrosion products, such as akaganeite ($FeO_{0.833}(OH)_{1.167}Cl_{0.167}$) [17], can accelerate the corrosion rate in metal works of historical interest.

Thus, the rehabilitation of the iron-building heritage has been necessary because of the deterioration produced by natural and anthropogenic factors that endangered their survival and their usefulness to the society for which they were designed. This is the case with the Ondarroa footbridge. The footbridge belongs to the tradition of mobile iron bridges built in numerous navigable channels. Its particularity of being one of the few remaining rotating bridges preserved today, the only one in Spain, makes it a unique architectural element worthy of being preserved and a symbol of the country. Due to its precarious state of conservation, a restoration intervention was planned, which also had the aim of restoring its original colour.

Thus, the present work aims at identifying the original colour of the footbridge of Ondarroa with a view to its future restoration so that it would be possible to recover its original appearance, since the metallic structure has been subjected, as a whole, to various chromatic changes from its construction to the last interventions. In addition, the impact of marine aerosol and the harbour environment in the bridge will be documented by the characterization of different corrosive compound and biomarkers.

For this purpose, a scientific diagnostic study was necessary. The study was carried out on six micro samples, five of them as cross-sections, by means of elemental (micro-energy-dispersive X-ray fluorescence spectroscopy) and molecular analysis (Raman spectroscopy) after a careful observation under an optical microscope. Colorimetric analyses were also performed to know the colour values of the different pigments used.

2. Results and Discussion

2.1. Characterization of the Paint Layers

Two samples collected in different areas in the largest piece, which belongs to the low part of the railing (Figure S1a, Supplementary Materials) received in the laboratory, presented the same stratigraphic composition; therefore, only the results of one sample are shown below. Specifically, three different homogeneous layers, two outermost green ones (Figure S2a,b) and one inner red/orange (Figure S2c), were recognized by observing the samples with a stereomicroscope.

Thanks to Raman analyses, it was possible to identify the compounds that characterize the different paint layers. The outermost layer (layer a in Figure S2) of dark green colour was composed of phthalocyanine green ($C_{32}H_3Cl_{13}CuN_8$, Raman bands at 688, 740, 744, 815, 977, 1079, and 1208 cm^{-1}, Figure 1a) [18], while the intermediate layer (layer b in Figure S2) of lighter green colour was made mainly with the phthalocyanine blue pigment ($C_{32}H_{16}N_8Cu$, Raman bands at 236, 257, 483, 594, 680, 747, 779, 832, 953, 1007, 1108, 1143, and 1193 cm^{-1}, Figure 1b) [18]. In addition, in many Raman spectra a very weak peak at 1038 cm^{-1} was visible (Raman band marked in the red circle in Figure 1b). This band could belong to the yellow azo pigment PY100 used in small amounts, mixed with phthalocyanine blue, to obtain a green colour. Contrary to the green coloured layers, the innermost orange layer (layer c in Figure S2) was made by mixing two compounds, minium (Pb_3O_4 Raman bands: 122, 152, 224, 313, 390, and 548 cm^{-1}) [19] and a smaller amount of barium sulfate ($BaSO_4$, Raman bands: 453, 460, 618, and 987 cm^{-1}, Figure 1c) [20] since all the Raman spectra recorded in this area showed the main features of both compounds.

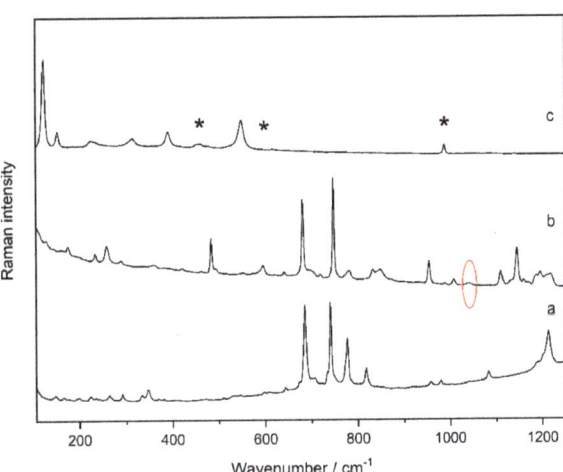

Figure 1. Raman spectra of the compounds identified in the subsample (SUBS-1a) in cross-section: phthalocyanine green (**a**), phthalocyanine blue with traces of azo pigment PY100 (in the red circle) (**b**), minium plus barium sulfate (*) (**c**).

As can be seen in the Raman spectroscopy images (Figure 2) carried out considering the band with the highest intensity of each compound, the green layers consist of phthalocyanine green (Figure 2b) and phthalocyanine blue (Figure 2c), respectively. On the other hand, the presence of barium sulfate used in mixture with minium was confirmed due to the presence of the two compounds in the same pictorial layer, as is evident from the overlapping of the Raman images in Figure 2e,f. The presence of the yellow azo pigment was identified on two pictorial layers in mixture with phthalocyanine blue and even with minium and barium sulfate like in the Raman image shown in Figure 2d. The Raman map was obtained considering the band at 1038 cm^{-1}, assigned to SO_3^- symmetric stretch [21], in order to avoid the overlapping with barium sulfate since both compounds have a Raman bands in the same position (989 cm^{-1}).

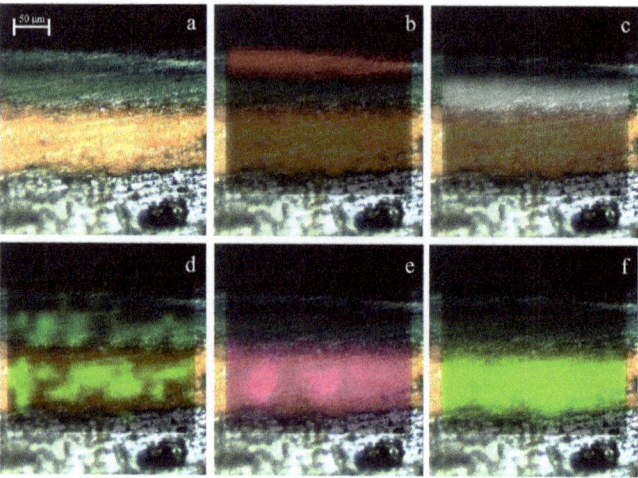

Figure 2. Raman image of a cross-sectional sample (sample SUBS-1a) (50×) collected from the railing piece (**a**) and its molecular composition: phthalocyanine green (**b**), phthalocyanine blue (**c**), yellow azo pigment (**d**), barium sulfate (**e**) and minium (**f**).

Thus, in this part of the bridge, after applying red lead, barium sulfate, plus a yellow azo pigment, a green paint (phthalocyanine blue and yellow azo pigment) was applied, and over it a dark-green one (phthalocyanine green),which was the colour that is visible currently. The use of red lead, currently banned due to its toxicity, was probably employed as antioxidant paint, following the rules in the second half of 20th century [22].

Unlike the previous case, the subsamples that were collected from the piece that permit the movement of the bridge (Figure S1b) showed a different stratigraphic composition from a first observation with the stereoscopic microscope. The first analysed subsample (SUBS-1b), collected from a green area, shows four homogeneous and well-defined layers (Figure S3). The outermost pictorial layer was entirely composed of phthalocyanine green (Figure S4a). In the second one, white in colour, which probably represents the primer layer, rutile (α-TiO$_2$, Raman bands a: 442 and 608 cm^{-1}) [23] and calcium carbonate (CaCO$_3$, Raman bands at: 282 and 1086 cm^{-1}) [23] were detected, as seen in Figure S4b. Additionally, the innermost green layer of a lighter shade was composed of a mixture of Prussian blue (Fe$_4$ [Fe (CN)$_6$]$_3$, Raman bands a: 277, 364, and 530 cm^{-1}) [24] and barium sulfate (Figure S4c).

Another pictorial layer was recognized, previously applied, thanks to the Raman spectroscopy images showed in Figure 3. At first glance, this layer had the same hue as the previous one and only by observation with a stereoscopic microscope it was not possible to recognize the two different layers. However, the Raman images made it possible to distinguish one more pictorial layer, entirely composed of phthalocyanine blue (Figure 3g). Thus, in this sample four layers were found, respectively, from the oldest to the most recent: phthalocyanine green, rutile mixed with calcium carbonate, Prussian blue with barium sulfate, and phthalocyanine blue.

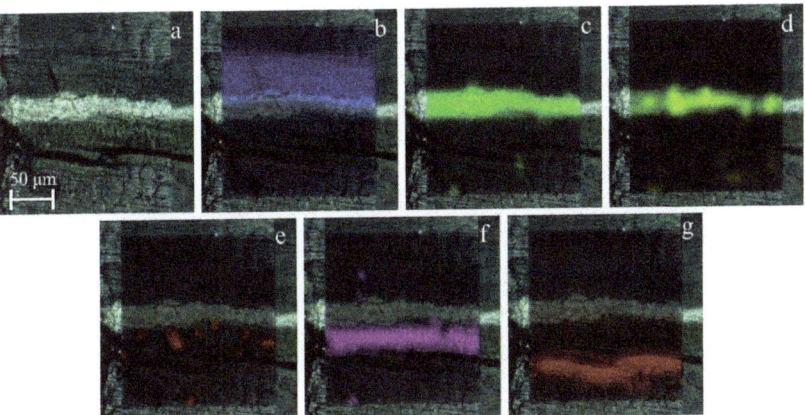

Figure 3. Optical image (**a**) and Raman spectroscopy image of the sample in section showing its molecular composition: phthalocyanine green (**b**), rutile (**c**) and calcium carbonate (**d**), barium sulfate (**e**), Prussian blue (**f**), and phthalocyanine blue (**g**). The black line through the sample is a crack in the sample.

Consequently, the stratigraphy of this subsample showed that on a black layer (its stratigraphy is described in the following subsample), a phthalocyanine blue primer was applied (layer g in Figure 3), on which another layer was composed of Prussian blue and barium sulfate (layer e + f in Figure 3). Next, another white primer composed of rutile and calcium carbonate was applied (layer c + d in Figure 3), and finally, a green layer of phthalocyanine green (layer b in Figure 3), which was the one visible nowadays.

Raman analysis on the surfaces of the subsample (SUBS-2b) taken from a black area showed the presence of carbon black, homogeneously distributed throughout the entire surface. Underneath, there was a heterogeneous layer consisting mainly of hematite found in grains of different sizes (Fe_2O_3, Raman bands a: 222, 240, 290, 405, 490, and 608 cm^{-1}, Figure S5a) [25]. In addition to iron oxide, the analyses on this layer have made it possible to recognize at various points of the Raman spectrum characteristic of a material composed of silicon and carbon (similar to silicon carbide wire, SiC, with Raman bands at: 150, 763, 786, 795, and 964 cm^{-1}, Figure S5b) [26]. Raman bands belonging to barium sulfate were also identified in the same layer. The distribution of the three compounds is indicated in the Raman images in Figure S6.

The observation of one of the subsamples (SUBS-3b) taken in the piece shown in Figure 1b, which is part of the system that allowed the movement of the bridge, allowed us to recognize a more complex stratigraphy. This sample was collected in the lower area of the dark piece where probably the original painting could remain. The first outer layer was dark in colour applied over a heterogeneous red and white layer that presented grains of different sizes as in the piece previously described. Additionally, a thicker layer of black colour and another of green colour could be clearly recognized. According to the observation with the stereomicroscope, the first layer applied in the sample was a heterogeneous layer of grey colour with dark and blue grains as shown in Figure 4. The detail of the blue-greyish layer, whose thickness was around 1.5 mm, is showed in Figure 4c. In the lower part of the mentioned layer of blue-grey paint, remains of iron oxide flakes (reddish colour) detached from the metallic surface were also observed.

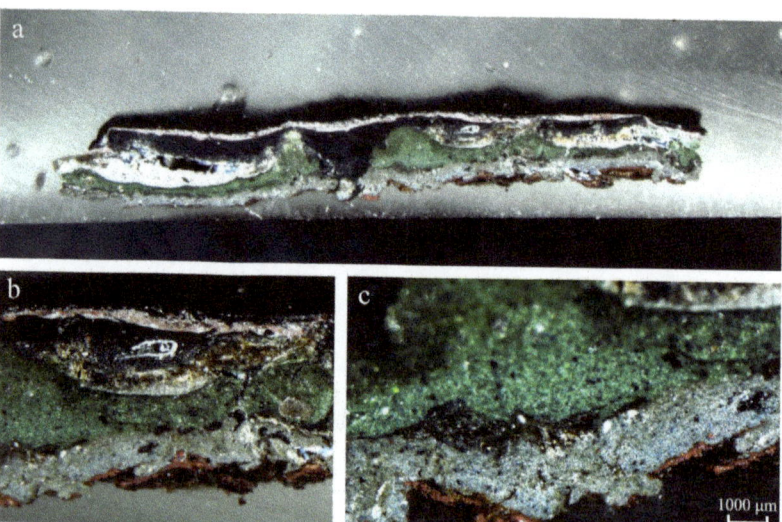

Figure 4. Cross-sections at diferent magnification (**a**–**c**) of the painting sample SUBS-3b from a black area where the original paint layer is visible.

At first, the subsample was analysed by micro-energy dispersive X-ray fluorescence (μ-EDXRF) to study its elemental composition. The elemental maps of each element on the analysed sample are shown in Figure 5. The μ-XRF analysis indicated the presence of chlorine, iron, silicon, barium, and zinc, uniformly distributed in the outer part of the subsample. Surprisingly, chlorine was not detected in the green layer indicating the absence of phtalocyanine green in this case as the green pigment. Iron and silicon belong to the first interior layer where hematite and the silicon carbide compound were detected. On the other hand, some of these elements such as iron, chlorine, silicon, barium, and zinc did not belong to the exterior green paint layer, but rather they were elements trapped by the atmospheric particles of the marine aerosol that arrived at the bridge continuously. Although the marine aerosol is mainly composed of chlorides, it can also carry fine particles in suspension composed of beach sand (in this site, there is a beach close to the bridge) that mainly contributes to increase Si and Fe concentration on the surface. The evidence of iron in the lower area of the sample, on the other hand, refers to remains of the metal support that belongs to the bridge structure and this contributes to the mass fraction of Fe in the sample. The presence of barium inside the subsample was evident occupying much of its surface and it was the element with the highest concentration in the sample. Among the main elements of this interior, sulfur, lead, calcium, and zinc stand out. The semi-quantitative values of each element in the sample SUBS-3b are shown in Table 1.

Table 1. EDXRF elemental data (wt.%) of sub sample SUBS-3b.

Sample ID	Mg	Al	Si	S	Cl	K	Ca	Fe	Zn	Sr	Ba	Pb	Cr
SUBS-3b	1.63	0.78	3.4	7.2	16.4	0.3	10.8	12.8	3.44	0.98	28.2	13.9	0.2

In addition, in urban and harbour areas, other ions are also present in a suspended way such as Ba^{2+}, Zn^{2+}, Ca^{2+}, K^+, Mg^{2+}, Fe^{3+}, Al^{3+}, Sr^{2+}, NH_4^+, HCO^{3-}, and Br^-. The source of these anions and cations can reside in the influence of maritime traffic, port activities, and also industry or road traffic [27].

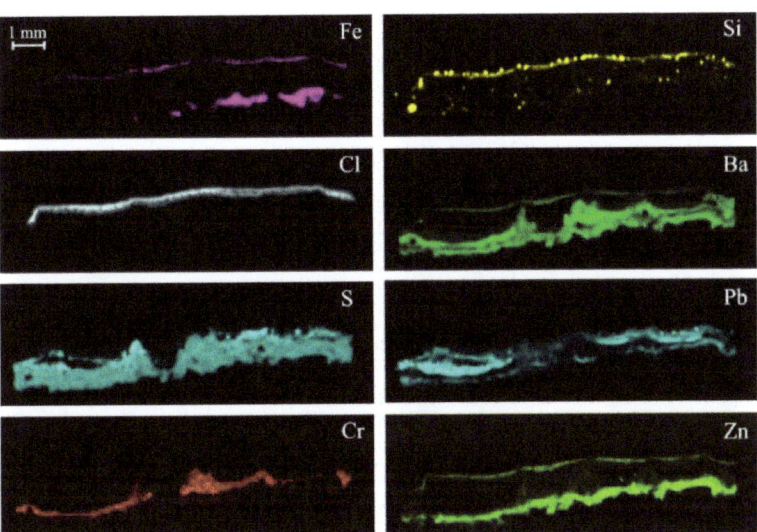

Figure 5. Micro-energy-dispersive X-ray fluorescence spectroscopy (μ-EDXRF) maps of the sample SUBS-3b from a black area.

The molecular composition of the most layers was the same as in the previous subsample (SUBS-2b), since carbon black was identified on the outside and, underneath, there was a heterogeneous layer with hematite, barium sulfate, and the silicon carbide. This pigment composition accounts for the presence of the elements Si, S, Ba, and Fe identified by μ-EDXRF analysis. In this sample, the green colour was not related with the use of phthalocyanine pigments. Indeed, Prussian blue, the yellow pigment lead chromate (Raman bands at 337, 360, 376, 402, and 840 cm^{-1} Figure S7a) and barium sulfate were the major components of the green colouration identified in the interior of the subsample. In addition, lead white (2PbCO$_3$·Pb(OH)$_2$) was found in some points of analysis in this area. Thus, the presence of lead was very irregular, and it represents the second main element, of the total sample after Ba. However, a major content of lead appears in the lower part of the sample (the one in contact with metallic iron) that could be associated with the use of lead white, identified mostly in the blue and grey layers (as will be seen later as well) and only in traces in the green layer. The low values of chromium in the XRF semi-quantification, belonging to lead chromate, was justified with the high absorption coefficient both of lead of the same pigment and of lead white used in the mixture. Additionally, by Raman spectroscopy other minor compounds were detected in that blue-green layer, such as gypsum (Raman bands at 412, 492, 617, 668 1008, and 1134 cm^{-1}, Figure S7b) [28] and anatase (β-TiO$_2$, Raman bands at 140, 192, 393, 512, and 635 cm^{-1}, Figure S7c) [29].

In the oldest grey paint layer, in most of the analysed points, Raman spectra of barium sulfate were recorded, with all the characteristic bands (Raman bands: 453, 460, 618, 648, 987, and 1140 cm^{-1}, Figure 6a) both in the matrix and in loose grains of different sizes. Additionally, in the black and blue grains, Raman spectra of carbon black (Raman bands: 1347, and 1602 cm^{-1}, Figure 6b) and ultramarine blue (Al$_6$Na$_8$O$_{24}$S$_3$Si$_6$, Raman bands: 260, 546, 583, 805, 1095, and 1644 cm^{-1}, Figure 6c) [30] were recorded, respectively. In addition, lead white (Raman band at 1050 cm^{-1}) was identified in the original paint layer, although in lesser quantity compared to the other compounds (Figure 6d). All the pigments found in the original paint layer of the footbridge are shown in the Raman spectra collected in Figure 6.

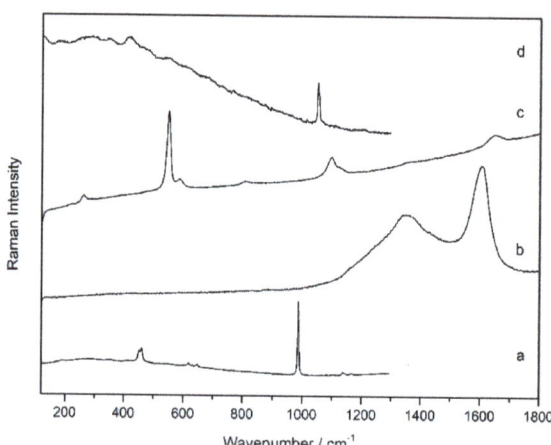

Figure 6. Raman spectra recorded in the original grey layer (sample SUBS-3b): barium sulfate (**a**), carbon (**b**), ultramarine blue (**c**), and lead white (**d**).

No chlorine compounds were found by Raman spectroscopy in this sample. The presence of this element, identified by µ-EDXRF only in the outermost part of the sample, is possibly due to the characteristics of the natural marine environment in which the bridge is located, as discussed later in detail.

The results allow reconstructing the execution of painting on the railing structure. Its original colour was greyish blue (see blue layer of Figure 4) composed of the mixture of ultramarine blue and carbon black. These pigments were mixed with lead white and anatase, with these latter ones probably being used as opacifiers [31]. On top of this greyish blue layer, another paint of a more greenish blue hue was applied (see greenish layer above the blue in Figure 4) composed of Prussian blue mixed with lead chromate and dispersed in barium sulfate (with traces of gypsum) and anatase. Over time, a hematite antioxidant primer was applied and over it, a black to cover possibly all rust formation and colour flakes.

2.2. Colorimetric Studies

Colour measurements in CIELab colour space were collected from the pieces of the bridge (Figure S1) delivered to the laboratory (Table 2) that permit the replication of colour. Therefore, the values of the green colour and of the original greyish blue colour were recorded, obtained by carefully scraping the surface of the piece that allows the rotation of the bridge.

Table 2. Lab values of the colours measured in the samples in Figure S1.

Colour	L*	a*	b*
Green (Figure S1a)	34.96	−11.22	1.68
Greyish blue (Figure S1b)	44.47	−0.14	2.07

* The CIELAB color space is referred with asterisks to prevent confusion with Hunter Lab.

2.3. Evaluation of the State of Conservation for the Iron Structure of the Bridge

The Raman analyses were also applied for the study of corrosion products in the iron structure generated by the exposure of the bridge to the marine environment that favoured its disintegration. The state of conservation of the pieces delivered to the laboratory was different from each other; therefore, an oxidized chip (SUBS-4b) without a paint layer from the piece in Figure S1b and a sample (SUBS-2a) with paint layer from the base of the railings (Figure S1a) were selected.

The Raman maps (Figure 7) carried out in the oxidized sample SUBS-2a, treated as cross-section, highlighted the presence of goethite (α-FeOOH, Raman bands: 250, 330, 390, 478, and 550 cm^{-1} as shown in Figure 8a) [32] as the main compound in the internal area of the oxidized chip, as seen in the Raman image shown in Figure 7a. The presence of magnetite (Fe$_3$O$_4$, Raman bands: 550 and 663 cm^{-1} shown in Figure 8c) [33] was also identified by point-by-point Raman spectroscopy, and its presence is important in the rust area as shown in the Raman image of Figure 7c. On the other hand, the presence of lepidocrocite (γ-FeO(OH), (Raman bands: 215, 249, 305, 345, 376, 523, 645, and 1300 cm^{-1} as depicted in Figure 8b) in the edges of the sample and in correspondence to microfractures of the subsample was identified as seen from the Raman image shown in Figure 7b. This observation is in good agreement with other investigations that documented the formation of lepidocrocite associated to local more aerated conditions [34,35]. This species of iron oxyhydroxide is known to be one of the most unstable forms of the corrosion compounds, which can transform into the more stable goethite with the succession of wet–dry cycles during the passivation of the corrosion processes [36]. This means that in the sample, the decay process has not been yet completed and is still going on.

Figure 7. Raman images show the distribution of goethite (**a**), lepidocrocite (**b**), and magnetite (**c**) in the sample SUBS-2a.

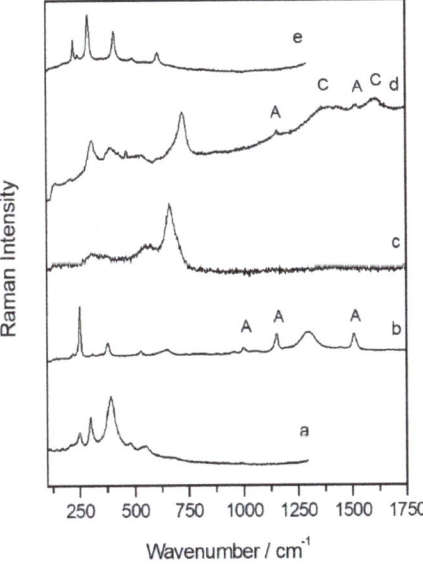

Figure 8. Raman spectra of goethite (**a**), lepidocrocite plus astaxanthin (A) (**b**), magnetite (**c**), akaganeite with traces of astaxanthin (A) and carbon (C) (**d**), and hematite (**e**) from the sample SUBS-4b.

In agreement with our results, previous studies on rust surfaces on mild steel demonstrated that in marine environments, lepidocrocite develops preferentially on the outermost surface, irrespective of the chloride ion deposition rate, while magnetite and akaganeite (not found in this sample), an oxyhydroxide formed in chlorine rich atmospheres, mainly form near the base steel. In addition, according to Diaz et al. [37] with the increase of the exposition time, rust layers become thicker, and the lepidocrocite is partially transformed into goethite generating a stratified bilayer structure of rust consisting of a porous outer layer of lepidocrocite and an inner layer of compact goethite.

As mentioned, even magnetite was identified in the subsample as a corrosion product. Its formation is commonly detected as a decay compound in rust developed in marine atmospheres, and it is usually detected in the inner zone closest to the base steel, where the lower oxygen availability favours its development [38,39].

Raman results revealed that the structure was also affected by biological colonization. The carotenoid pigment astaxanthin, recognized by the three main bands at 1509 cm^{-1} (v_1 C=C), 1152 cm^{-1} (v_2 C-C and 1001 cm^{-1} (v_3 C-H) and even by the overtones at 957, 1191, 1448, 2150, 2296, and 2650 cm^{-1}, was detected in extended areas of the sample [40]. The spectrum in Figure 8b shows astaxanthin and lepidochrocite in the same spot area, showing how the colonization process extends also to the oxidized rusts. Among the carotenoid pigments, astaxanthin is the most oxidized species and is known to be synthesized by photosynthetic organisms such as cyanobacteria, fungi, and algae as a defense mechanism against atmospheric pollution. For this reason, it was proposed as bioindicators of high concentration of SO_2 in the atmosphere [41]. In our study, the presence of astaxanthin in corrosion patina was probably related to the acidic environment in which the bridge was located. The industrial activity present in the outskirts of the city and the close proximity to the port are the responsible for the high concentration of this compound in the oxidized microsample.

Furthermore, surface analyses were carried out on a microsample (SUBS-4b) collected from an oxidized area of the sample that allowed the movement of the bridge (Figure S1b). Only the presence of lepidocrocite, with a globular morphology, was identified in the inner surface of the metal fragment. On the other hand, the exterior side showed a more heterogeneous composition. The presence of akaganeite (FeO$_{0.833}$(OH)$_{1.167}$Cl$_{0.167}$), a highly unstable Cl$^-$ bearing corrosion phase, was detected in some black areas of the sample. This compound was characterized by its Raman bands at 310, 390, 535, and 724 cm^{-1} (Figure 8d), and its identification in the sample suggested the high impact of the chloride rich marine aerosol in the rotating bridge structure. As the akaganeite structure is characterized by tunnels partially occupied by chloride anions parallel to the c-axis of the tetragonal lattice, it tends to form low density, compared with other corrosion products, and fragile rust layers promoting cracking and exfoliation phenomena [42]. According to the investigation of Li et al. [43], once the akaganeite is saturated with Cl$^-$ and it cannot take up any more Cl$^-$, free Cl$^-$ can be available to accelerate corrosion at anodic sites.

In the same spectrum showing akaganeite, the presence of black carbon (Raman bands: 1360 and 1600 cm^{-1} as seen in Figure 8d) was evident. This suggests an important amount of carbon in the small spot of the Raman observation (approximately 10 µm in diameter) that can only be explained by deposition of atmospheric particulate matter containing soot.

The presence of hematite (Figure 8e) was also identified in some points on the rust layer. This formation of hematite was not largely documented in the literature between the corrosion products. In the research of De la Fuente et al., it was identified from the ferrihydrite formation only in a rust layer exposed to an industrial and marine atmosphere. The presence of hematite could be related to the influence of acid rainwater, and an SO_2-rich atmosphere could have favoured the transformation.

Thus, the identification of compounds such as hematite, carbon or the most oxidized form of carotenoid pigments reflect the environmental conditions that affected the iron of the bridge during the years. In fact, it is located in an area where SO_2 emissions are generated by the burning of fuel oils from boats, or by gases produced by industries located nearby and transported by winds.

3. Materials and Methods

Microsampling and Laboratory Instrumental Set Up

The Ondarroa footbridge, known as *Plaiako Zubixe* (the Bridge of the Beach in the local Basque language) was inaugurated in 1927 with the name *Pasarela de Alfonso XIII* and until the 1980s, due to its peculiarity as a revolving bridge, it allowed boats to pass through the Artibai river. Since 2016, the walkway has been closed to pedestrians for safety reasons, due to its serious state of deterioration due to corrosion caused by the marine environment in which it is located. In 2008, it was declared a Cultural Asset by the Basque Government since, due to its rotating system, it represents an exceptional example that remains today of the solutions adopted, in previous decades, to allow the transits of boats and vehicles in the estuaries.

Two metal pieces were received in the laboratory and collected from different parts of the bridge. The samples have been taken in some points (Figure S1) where different paint layers appeared using a scalpel. Two subsamples were taken for their stratigraphic study from the piece that belongs to the lower part of the bridge railing, which is green in colour on the outside and has a circular geometric decoration (Figure S1a). Four subsamples were taken from the second piece that belongs to the base of the railings and that permitted the movement of the bridge (Figure S1b). In this case, the piece was partially covered with a layer of green paint while other areas show a green/black tone, with extensive areas where no pictorial layer appears and only the highly degraded metallic material could be seen. All samples have been treated as cross-sections, but since two samples of the railing piece were identical from a compositional point of view, only the results of one sample were showed. All the samples collected on each piece and the techniques employed are summarized in Table S1.

Six samples taken from the two pieces (Figure S1a,b) were prepared as cross-sections to acquire complete information on their stratigraphy. For the cold encapsulation of the collected samples, an acrylic polymer based on methyl methacrylate was used. After encapsulating the samples, they were polished to obtain a completely smooth surface. The polishing of the cross-sections was carried out with the Metkon Forcipol 1 polisher (Barcelona Quálites, s.l.) using WS FLEX 18C and 16 waterproof sandpaper (P320-2000) and cloths for the last finish with diamond paste with a granulometry of 1 µm.

First, high resolution images of the fragments were obtained using an SMZ-U stereomicroscope (Nikon, Japan) coupled with a Nikon DigitalSight DS-L1 camera.

Then, Raman analyses were performed using an InVia micro-Raman confocal spectrometer (Renishaw, UK) coupled to a Leica DMLM microscope equipped with 5×, 20×, 50×, 50× (long distance), and 100× objectives and using 532 and 785 nm as the excitation laser source. The lasers were set at low power (no more than 1 mW) to avoid thermal transformation of the samples. Data acquisition was carried out using Wire 4.2 software (Renishaw). Spectra were acquired between 150 and 3200 cm^{-1} (mean spectral resolution 1 cm^{-1}) and several scans were recorded for each spectrum in order to improve the signal-to-noise ratio (3-20 s, accumulations of 5-200). Around 40 Raman spectra were recorded for each sample. For samples with a more complex stratigraphy, nearly 50 spectra were collected.

The elemental distribution maps were acquired using an M4 TORNADO EDXRF spectrometer (Bruker Nano GmbH, Germany). The lateral resolution used for spectral acquisitions was 20 microns. For most of the elements, the selected line was the $K_{\alpha 1}$, except for Ba for which its corresponding $L_{\alpha 1}$ line was employed. In addition, to improve the detection of the lightest elements (Z > 11), the measurements were acquired under a vacuum (20 mbar) by means of a diaphragm pump MV 10 N VARIO-B. Data collection and interpretation were carried out by means of the software M4 TORNADO (Bruker Nano GmbH), whereas the quantitative analyses were carried out using the deconvolution M-Quant software package based on the application of fundamental parameters quantitative methods.

Colour measurements were made with a PCE-CSM 5 Trisstimulus colorimeter with a measurement spot size of 4 mm, measurement geometry in the integrating sphere with an incidence angle of 8° and the reflectance measured diffusely, CIELab colour space with illuminant D65 and standard 10-degree observer.

4. Conclusions

In this work, micro-Energy Dispersive X-ray fluorescence (µ-EDXRF) and Raman spectroscopy have been successfully used at first in the characterization of the pictorial layers of Ondarroa's rotating walkway. Considering the lack of knowledge of the paints used for metal structures at the beginning of the twentieth century, a scientific diagnostic study was necessary. The samples collected from the two green pieces have made it possible to identify the original polychrome of the Ondarroa footbridge, *Plaiako Zubixe*, as well as the following applications of different pictorial layers.

The original colour of the current green parts of the bridge was not green. It was a greyish blue colour that was obtained by mixing ultramarine blue, in very fine grains (~30/90 µm), with larger grains (~300 µm) of carbon black and lead white, dispersed in a greater amount of sulfate of barium, used as opacifier. The discovery of the original blue-grey layer is fully compatible with the black and white images of the catwalk, taken at the time of its inauguration.

However, despite the fact that the diagnostic analyses had shown that the current green colour did not correspond to the original, the inhabitants decided to restore the same colour that the bridge had in recent years. This work is a clear example of how the restoration of a work of art does not only concern the object itself. When it is part of a community or it is considered as a symbol of a city, the rules that are generally applied to the restoration do not apply when the opinion of an entire community prevails.

The bad oxidation state of the bridge has been the reason for its closure to the pedestrian and boat passage. Undoubtedly, the port marine environment in which it was constructed accelerated the oxidation state of the iron structure. This assumption was confirmed by the presence of iron degradation compounds, mainly akaganeite, which causes a cyclic alteration phenomenon that often ends with the total consumption of the iron core. In the same way, the identification of hematite and carbon demonstrated the SO_2 impact from ships, road traffic, and industrial activity. Thus, the different iron oxides identified as corrosion products of the metal structure as well as the organic carotenoid (astaxanthin) acting as a bioindicator of the bad air quality, to which the bridge has been exposed since its construction, denotes the influence of both natural and anthropogenic factors in the state of conservation of the rotating bridge.

Supplementary Materials: The following supporting information can be downloaded at: https://www.mdpi.com/article/10.3390/molecules27113609/s1, Figure S1: Parts of the rotating bridge from which the samples analysed were extracted, Figure S2: Image of the cross-section (sample SUBS-1a) from the piece that belongs to the lower part of the railing showing the three paintings layers and the iron support, Figure S3: Image with a stereoscopic microscope of the sample collected in a green area, Figure S4: Raman spectra of the compounds found in the subsample collected from a green area: phthalocyanine green (a), rutile plus calcium carbonate (C) (b), Prussian blue plus barium sulfate (SB) (c) and phthalocyanine blue (d), Figure S5: Raman spectra of hematite (a) and of the silicon-carbon compound (b), Figure S6: Raman images show the distribution of hematite (b), a silicon and carbon material (c) and barium sulfate (d), Figure S7: Raman spectra of barium lead chromate with Prusian blue (PB) and lead white (LW) sulfate plus gypsum (G) (b) and anatase (c) from the sample SUBS-3b, Table S1: Subsamples collected from each piece of the rotating bridge and techniques employed for their characterization.

Author Contributions: Conceptualization, Data collection, formal analysis, writing—review and editing, I.C.; formal analysis, writing—review and editing, K.C.; methodology, project administration, funding acquisition, G.A.; supervision, J.M.M. All authors have read and agreed to the published version of the manuscript.

Funding: This research was funded by DEMORA project (Grant No. PID2020-113391GB-I00), funded by the Spanish Agency for Research (through the Spanish Ministry of Science and Innovation, MICINN, and the European Regional Development Fund, FEDER).

Institutional Review Board Statement: Not applicable.

Informed Consent Statement: Not applicable.

Data Availability Statement: The data supporting the findings of this study are available within the article.

Acknowledgments: The DEMORA project (Grant No. PID2020-113391GB-I00) supported the development of this work. Costantini, I. thanks her post-doctoral contract from the University of the Basque Country (UPV/EHU).

Conflicts of Interest: The authors declare no conflict of interest.

Sample Availability: Samples of the compounds are not available from the authors.

References

1. Douet, J. *Industrial Heritage Re-Tooled: The TICCIH Guide to Industrial Heritage Conservation*; Routledge: Oxfordshire, UK, 2016.
2. Ćopić, S.; Đorđevića, J.; Lukić, T.; Stojanović, V.; Đukičin, S.; Besermenji, S.; Stamenković, I.; Tumarić, A. Transformation of Industrial Heritage: An Example of Tourism Industry Development in the Ruhr Area (Germany). *Geogr. Pannonica* **2014**, *18*, 43–50. [CrossRef]
3. Jones, C.; Munday, M. Blaenavon and United Nations World Heritage Site Status: Is Conservation of Industrial Heritage a Road to Local Economic Development? *Reg. Stud.* **2001**, *35*, 585–590. [CrossRef]
4. Coccato, A.; Moens, L.; Vandenabeele, P. On the Stability of Mediaeval Inorganic Pigments: A Literature Review of the Effect of Climate, Material Selection, Biological Activity, Analysis and Conservation Treatments. *Herit. Sci.* **2017**, *5*, 12. [CrossRef]
5. Ziemann, M.A.; Madariaga, J.M. Applications of Raman Spectroscopy in Art and Archaeology. *J. Raman Spectrosc.* **2021**, *52*, 8–14. [CrossRef]
6. van den Berg, K.J.; Bonaduce, I.; Burnstock, A.; Ormsby, B.; Scharff, M.; Carlyle, L.; Heydenreich, G.; Keune, K. *Conservation of Modern Oil Paintings*; Springer Nature: Berlin, Germany, 2020.
7. Tissot, I.; Fonseca, J.F.; Tissot, M.; Lemos, M.; Carvalho, M.L.; Manso, M. Discovering the Colours of Industrial Heritage Characterisation of Paint Coatings from the Powerplant at the Levada de Tomar. *J. Raman Spectrosc.* **2021**, *52*, 208–216. [CrossRef]
8. Morillas, H.; Maguregui, M.; García-Florentino, C.; Marcaida, I.; Madariaga, J.M. Study of particulate matter from Primary/Secondary Marine Aerosol and anthropogenic sources collected by a self-made passive sampler for the evaluation of the dry deposition impact on built heritage. *Sci. Total Environ.* **2016**, *550*, 285–296. [CrossRef]
9. De la Fuente, D.; Díaz, I.; Simancas, J.; Chico, B.; Morcillo, M. Long-term atmospheric corrosion of mild steel | Elsevier Enhanced Reader, (n.d.). *Corros. Sci.* **2011**, *53*, 604–617. [CrossRef]
10. Aramendia, J.; Gomez-Nubla, L.; Arrizabalaga, I.; Prieto-Taboada, N.; Castro, K.; Madariaga, J.M. Multianalytical Approach to Study the Dissolution Process of Weathering Steel: The Role of Urban Pollution. *Corros. Sci.* **2013**, *76*, 154–162. [CrossRef]
11. Martínez-Arkarazo, I.; Angulo, M.; Bartolomé, L.; Etxebarria, N.; Olazabal, M.A.; Madariaga, J.M. An integrated analytical approach to diagnose the conservation state of building materials of a palace house in the metropolitan Bilbao (Basque Country, North of Spain). *Anal. Chim. Acta* **2007**, *584*, 350–359. [CrossRef]
12. Sarmiento, A.; Maguregui, M.; Martinez-Arkarazo, I.; Angulo, M.; Castro, K.; Olazábal, M.A.; Fernández, L.A.; Rodríguez-Laso, M.D.; Mujika, A.M.; Gómez, J.; et al. Raman spectroscopy as a tool to diagnose the impacts of combustion and greenhouse acid gases on properties of Built Heritage. *J. Raman Spectrosc.* **2008**, *39*, 1042–1049. [CrossRef]
13. Arruti, A.; Fernández Olmo, I.; Irabien, A. Regional Evaluation of Particulate Matter Composition in an Atlantic Coastal Area (Cantabria Region, Northern Spain): Spatial Variations in Different Urban and Rural Environments. *Atmos. Res.* **2011**, *101*, 280–293. [CrossRef]
14. Kamimura, T.; Hara, S.; Miyuki, H.; Yamashita, M.; Uchida, H. Composition and Protective Ability of Rust Layer Formed on Weathering Steel Exposed to Various Environments. *Corros. Sci.* **2006**, *48*, 2799–2812. [CrossRef]
15. Aramendia, J.; Gomez-Nubla, L.; Bellot-Gurlet, L.; Castro, K.; Paris, C.; Colomban, P.; Madariaga, J.M. Protective Ability Index Measurement through Raman Quantification Imaging to Diagnose the Conservation State of Weathering Steel Structures. *J. Raman Spectrosc.* **2014**, *45*, 1076–1084. [CrossRef]

16. Aramendia, J.; Gomez-Nubla, L.; Castro, K.; Martinez-Arkarazo, I.; Vega, D.; Sanz López de Heredia, A.; García Ibáñez de Opakua, A.; Madariaga, J.M. Portable Raman Study on the Conservation State of Four CorTen Steel-Based Sculptures by Eduardo Chillida Impacted by Urban Atmospheres. *J. Raman Spectrosc.* **2012**, *43*, 1111–1117. [CrossRef]
17. Ståhl, K.; Nielsen, K.; Jiang, J.; Lebech, B.; Hanson, J.C.; Norby, P.; van Lanschot, J. On the Akaganéite Crystal Structure, Phase Transformations and Possible Role in Post-Excavational Corrosion of Iron Artifacts. *Corros. Sci.* **2003**, *45*, 2563–2575. [CrossRef]
18. Pozzi, F.; Basso, E.; Rizzo, A.; Cesaratto, A.; Tagu, T.J., Jr. Evaluation and optimization of the potential of a handheld Raman spectrometer: In situ, noninvasive materials characterization in artworks. *J. Raman Spectrosc.* **2019**, *50*, 861–872. [CrossRef]
19. Zhao, Y.; Wang, J.; Pan, A.; He, L.; Simon, S. Degradation of red lead pigment in the oil painting during UV aging. *Color Res. Appl.* **2019**, *44*, 790–797. [CrossRef]
20. Singer, B.W.; Gardiner, D.J.; Derow, J.P. Analysis of White and Blue Pigments from Watercolours by Raman Microscopy. *Pap. Conserv.* **1993**, *17*, 13–19. [CrossRef]
21. Ropret, P.; Centeno, S.A.; Bukovec, P. Raman identification of yellow synthetic organic pigments in modern and contemporary paintings: Reference spectra and case studies. *Spectrochim. Acta Part A Mol. Biomol. Spectrosc.* **2008**, *69*, 486–497. [CrossRef]
22. Lindqvist, S.-Å.; Vannerberg, N.-G. Corrosion-inhibiting properties of red lead—I. Pigment suspensions in aqueous solutions. *Mater. Corros.* **1974**, *25*, 740–748. [CrossRef]
23. De Gelder, J.; Vandenabeele, P.; Govaert, F.; Moens, L. Forensic analysis of automotive paints by Raman spectroscopy. *J. Raman Spectrosc.* **2005**, *36*, 1059–1067. [CrossRef]
24. Coupry, C.; Lautié, A.; Revault, M.; Dufilho, J. Contribution of Raman spectroscopy to art and history. *J. Raman Spectrosc.* **1994**, *25*, 89–94. [CrossRef]
25. Hernanz, A.; Gavira-Vallejo, J.M.; Ruiz-López, J.F. Introduction to Raman microscopy of prehistoric rock paintings from the Sierra de las Cuerdas, Cuenca, Spain. *J. Raman Spectrosc.* **2006**, *37*, 1054–1062. [CrossRef]
26. Bechelany, M.; Brioude, A.; Cornu, D.; Ferro, G.; Miele, P. A Raman Spectroscopy Study of Individual SiC Nanowires. *Adv. Funct. Mater.* **2007**, *17*, 939–943. [CrossRef]
27. Gómez-Laserna, O.; Arrizabalaga, I.; Prieto-Taboada, N.; Olazabal, M.Á.; Arana, G.; Madariaga, J.M. In Situ DRIFT, Raman, and XRF Implementation in a Multianalytical Methodology to Diagnose the Impact Suffered by Built Heritage in Urban Atmospheres. *Anal. Bioanal. Chem.* **2015**, *407*, 5635–5647. [CrossRef]
28. Jehlička, J.; Vítek, P.; Edwards, H.g.m.; Hargreaves, M.D.; Čapoun, T. Fast detection of sulphate minerals (gypsum, anglesite, baryte) by a portable Raman spectrometer. *J. Raman Spectrosc.* **2009**, *40*, 1082–1086. [CrossRef]
29. Zhang, W.F.; He, Y.L.; Zhang, M.S.; Yin, Z.; Chen, Q. Raman scattering study on anatase TiO_2 nanocrystals. *J. Phys. Appl. Phys.* **2000**, *33*, 912–916. [CrossRef]
30. Osticioli, I.; Mendes, N.F.C.; Nevin, A.; Gil, F.P.S.C.; Becucci, M.; Castellucci, E. Analysis of natural and artificial ultramarine blue pigments using laser induced breakdown and pulsed Raman spectroscopy, statistical analysis and light microscopy. *Spectrochim. Acta Part A Mol. Biomol. Spectrosc.* **2009**, *73*, 525–531. [CrossRef]
31. Titanium Dioxide Pigments. In *Surface Coatings: Vol I-Raw Materials and Their Usage*; Springer: Dordrecht, The Netherlands, 1983; pp. 305–312. [CrossRef]
32. de la Fuente, D.; Alcántara, J.; Chico, B.; Díaz, I.; Jiménez, J.A.; Morcillo, M. Characterisation of rust surfaces formed on mild steel exposed to marine atmospheres using XRD and SEM/Micro-Raman techniques. *Corros. Sci.* **2016**, *110*, 253–264. [CrossRef]
33. Morcillo, M.; Chico, B.; Alcántara, J.; Díaz, I.; Wolthuis, R.; de la Fuente, D. SEM/Micro-Raman Characterization of the Morphologies of Marine Atmospheric Corrosion Products Formed on Mild Steel. *J. Electrochem. Soc.* **2016**, *163*, C426. [CrossRef]
34. Monnier, J.; Neff, D.; Réguer, S.; Dillmann, P.; Bellot-Gurlet, L.; Leroy, E.; Foy, E.; Legrand, L.; Guillot, I. A corrosion study of the ferrous medieval reinforcement of the Amiens cathedral. Phase characterisation and localisation by various microprobes techniques. *Corros. Sci.* **2010**, *52*, 695–710. [CrossRef]
35. Misawa, T.; Hashimoto, K.; Shimodaira, S. The mechanism of formation of iron oxide and oxyhydroxides in aqueous solutions at room temperature. *Corros. Sci.* **1974**, *14*, 131–149. [CrossRef]
36. Cook, D.C.; Oh, S.J.; Balasubramanian, R.; Yamashita, M. The Role of Goethite in the Formation of the Protective Corrosion Layer on Steels. *Hyperfine Interact.* **1999**, *122*, 59–70. [CrossRef]
37. Diaz, I.; Cano, H.; de la Fuente, D.; Chico, B.; Vega, J.M.; Morcillo, M. Atmospheric Corrosion of Ni-Advanced Weathering Steels in Marine Atmospheres of Moderate Salinity. *Corros. Sci.* **2013**, *76*, 348–360. [CrossRef]
38. Misawa, T.; Asami, K.; Hashimoto, K.; Shimodaira, S. The Mechanism of Atmospheric Rusting and the Protective Amorphous Rust on Low Alloy Steel. *Corros. Sci.* **1974**, *14*, 279–289. [CrossRef]
39. Alcántara, J.; Chico, B.; Díaz, I.; de la Fuente, D.; Morcillo, M. Airborne Chloride Deposit and Its Effect on Marine Atmospheric Corrosion of Mild Steel. *Corros. Sci.* **2015**, *97*, 74–88. [CrossRef]
40. Ibarrondo, I.; Prieto-Taboada, N.; Martínez-Arkarazo, I.; Madariaga, J.M. Resonance Raman imaging as a tool to assess the atmospheric pollution level: Carotenoids in Lecanoraceae lichens as bioindicators. *Environ. Sci. Pollut. Res. Int.* **2016**, *23*, 6390–6399. [CrossRef]
41. Morcillo, M.; Chico, B.; de la Fuente, D.; Alcántara, J.; Wallinder, I.O.; Leygraf, C. On the Mechanism of Rust Exfoliation in Marine Environments. *J. Electrochem. Soc.* **2016**, *164*, C8. [CrossRef]

42. Post, J.E.; Buchwald, V.F. Crystal Structure Refinement of Akaganéite. *Am. Mineral.* **1991**, *76*, 272–277.
43. Li, S.; Hihara, L.H. A Micro-Raman Spectroscopic Study of Marine Atmospheric Corrosion of Carbon Steel: The Effect of Akaganeite. *J. Electrochem. Soc.* **2015**, *162*, C495. [CrossRef]

Article

Photons for Photography: A First Diagnostic Approach to Polaroid Emulsion Transfer on Paper in Paolo Gioli's Artworks

Zeynep Alp [1], Alessandro Ciccola [2,*], Ilaria Serafini [2], Alessandro Nucara [3], Paolo Postorino [3], Alessandra Gentili [2], Roberta Curini [2] and Gabriele Favero [1]

[1] Department of Environmental Biology, Sapienza University of Rome, 00185 Rome, Italy
[2] Department of Chemistry, Sapienza University of Rome, 00185 Rome, Italy
[3] Department of Physics, Sapienza University of Rome, 00185 Rome, Italy
* Correspondence: alessandro.ciccola@uniroma1.it

Abstract: The aim of this research is to study and diagnose for the first time the Polaroid emulsion transfer in the contemporary artist Paolo Gioli's artworks to provide preliminary knowledge about the materials of his artworks and the appropriate protocols which can be applied for future studies. The spectral analysis performed followed a multi-technical approach first on the mock-up samples created following Gioli's technique and on one original artwork of Gioli, composed by: FORS (Fiber Optics Reflectance), Raman, and FTIR (Fourier-Transform InfraRed) spectroscopies. These techniques were chosen according to their completely non-invasiveness and no requirement for sample collection. The obtained spectra from FTIR were not sufficient to assign the dyes found in the transferred Polaroid emulsion. However, they provided significant information about the cellulose-based materials. The most diagnostic results were obtained from FORS for the determination of the dye developers present in the mock-up sample which was obtained from Polacolor Type 88 and from Paolo Gioli's original artwork created with Polacolor type 89.

Keywords: polaroid chemistry; polacolor; fibre optics reflectance spectroscopy; Paolo Gioli; chromium (III) complex of azomethine dye; chromium complex of an azo pyrazolone; blue copper phthalocyanine

1. Introduction

Conservation of contemporary artworks is a new thrilling field that needs continuous research due to the utilization of the newest and ever-changing materials, mediums, and ideologies in the making of a work of art. Contemporary art considers the use of all materials, along with different forms of exposition, ranging from the absence of material to its accumulation. A restorer must pay good attention to the composition of the artwork and consecutively to the decision-making process regarding the conservation steps which will be applied to the artwork. At this point, the detailed knowledge about the artwork materials on a scientific level is crucial. Therefore, what conservation science provides to the literature through analytical techniques is of great importance for art preservation. From this perspective, this study presents research diagnosing Paolo Gioli's Polaroid emulsion transfers by following a multi-spectral approach. Gioli was an Italian contemporary artist born in Sarzano (Rovigo) on 12 October 1942. The above-mentioned significance of the artist's material plays even a bigger role for Gioli's art because he used the Polaroid for its material plasticity and versatility. He transferred the emulsion to different receptor layers (paper, silk, wood, etc.), while denying having used fully what the Polaroid Corporation has to offer, from the cameras to the instructions. The artworks he created are of great complexity on an aesthetic level and, also, on a chemical one. Since knowledge about the chemistry and the conservation of Polaroid materials is not broad enough, analysing the Polaroid materials transferred to a different support is an important contribution and a discovery of a different aspect of the photographic materials from the scientific point of view. It is fundamental to mention that the original photographic material on which the analysis will be applied is aesthetically complete and in very good condition, and it is a perfect representation of the artists' intention: every single trace contributes to the wholeness of the artwork. Therefore, also material-wise, the sampling from the material surface was impossible. For this reason, a completely non-invasive spectroscopic protocol was chosen both for the mock-up sample, produced by using the film Polacolor type 88, and the original artwork of Paolo Gioli, in which he used Polacolor type 89. Photographic material is expected to give complications while performing non-invasive diagnostics, due to its organic components, and this is a limitation of this research. Moreover, since no prior diagnostics has been carried out on Polaroid materials transferred on different receptor layers, in our case paper, it is hard to foresee the information potential achievable from single techniques. However, for the same reason, this represents the first ever research to provide primary information about the protocols to be followed for the analysis of such materials in the future. From the conservation science point of view, Polaroid film represents an undiscovered field and needs further research to fully determine the substances used. Paolo Gioli's works being constructed of complex experimental layers of materials add a further dimension to future research.

Chemistry of photographic materials has been studied extensively by past scholars [1–27] through several analytical techniques [4,5,26,28–44]. The colour photography mechanism is based on silver halides (AgX) just as black and white photography. The silver halides in colour photography act as mediators for transforming light into organic dye images [15]. The use of colour filters for the colour photography dates to the production of colour images by James Clerk Maxwell in 1861, which relies on the additive colour system whereby the three primary colours (red, green, and blue) produce the full gamut of colours on the composite final image. On the other hand, the subtractive colour process, subtracts red, green, or blue light from the visible spectrum. Therefore, the original image produces a positive, through the intermediate step of a negative image. Cyan, magenta, and yellow are the colours used for the subtractive colour system [23].

The mechanism of instant photography was first published in 1947 [45]. The Polaroid Corporation published the early instant photography concepts in 1948 of black and white [46]. The beginning of the coloured peel-apart system was put on the market in 1963 by Polaroid Corporation, Cambridge, NA, USA Afterwards, integrated (mono-sheet) systems called SX-70 (1972, Polaroid), PR-10 (1976, Kodak, New York, NY, USA), and FI-10 (1981, Fuji Photo Film, Tokyo, Japan) appeared as more elaborate systems [15]. Polaroid prints film has the same subtractive colour principle as regular colour negative films. The main difference is that all the chemicals, including the developer and the dyes (referred as the developer-dye) are enclosed within a thin sheet of film. In the so-called "peel-apart" films, the negative and the positive is stripped apart to reveal the image [21].

A reagent system is an essential part of each instant film. Dry processing is realised by using a highly viscous gel reagent and therefore restricting the amount enough to complete a single photo. The viscosity of the reagent is provided by water-soluble polymeric thickeners including hydroxyethyl cellulose, the alkali metal salts of carboxymethyl cellulose, and carboxymethyl hydroxyethyl cellulose. The high viscosity enables the accurate metering to form a thin film and serves as an adhesive for the two sheets during processing. The reagent is highly alkaline, and this property can remain stable due to the sealed pod that contains the reagent. Sealing also protects the reagent from oxygen until the development of the film. The reagent contains reactive components that participate in image formation, deposition, and stabilisation [47]. The function of the developing agent consists in reducing exposed silver halide grains to metallic silver and thus facilitating the formation of the oxidized species of the developing agent. This species reacts with a colour coupler in the emulsion to form a dye [48].

In the Polacolor instant film, a single reagent is used to obtain a negative and a positive image. The quantity of reagent that is available to produce the positive image is controlled as a function of the development of the silver halide latent image as an oxidation product of the reagent, which is a substance that is immobile in a photosensitive element while depositing the unreached reagent on a print-receiving material to provide the positive image. The reagent, when oxidised, provides a reaction product which has a low solubility in a processing liquid than the unreached reagent [23]. In particular, the multilayer negative contains a set of blue-, green-, and red-sensitive emulsions. Each layer contains respectively a dye developer complementary in colour to the emulsion's spectral sensitivity. A negative section comprises the lower part of the film, and a positive section the upper part. These sections are physically one unit until they become separated by the viscous alkaline reagent after exposure. The alkaline reagent is enclosed in pods at the edge of the film. It is spread between the two sections by automatic rollers [48]. The dye developers get oxidated and immobilised due to the development of the exposed grains during processing. The non-immobilised dye developer then migrates through the layers of the negative to the image-receiving surface to form the positive image. Dye developers must have good diffusion properties and must stay inert within the negative before that it is processed. The diffusion must also remain stable to light and have suitable spectral absorption characteristics after the processing [10].

When Polaroid Corporation first introduced Polacolor film in 1963, the dye developers were a set of two azo dyes for yellow and magenta and anthraquinone for cyan [23]. However, it was seen that over time these dyes needed further light fastness and stability. More recent developments in dye-developer chemistry have primarily focused on achieving further light fastness and stability and have resulted in the introduction of premetallized dyes in the Polacolor 2 and SX-70 material. The technique used for metallized dye images used the image wise transfer of dye developers which included chelating dye systems [49], especially metal complex azo dyes, renowned for their great stability to light [16,50–53].

2. Results and Discussion

2.1. Mock-Up

2.1.1. FORS

Out of 20 points where acquisions were made, points T2, T5, T8, T14, T10, and T20 show the most intense bands (Figure 1), which result useful for the identification of the dyes. Bands centered around 490–505, 562–565, and 681–684 nm were the most recurring, as shown in Table 1.

The paper substrate does not affect or distort the obtained results because the colourant is abundant enough to absorb the light and, therefore, does not make it possible for the light to arrive to the paper surface. This has been proved by obtaining the apparent absorbance spectrum of the paper and comparing it with the absorbance spectra of the analogues of the coloured areas. Therefore, the reflectance of the paper has been ignored for the interpretation of the results obtained.

For the identification of the dyes, we compared the experimental data with the literature characteristic absorption ranges and maxima for several dyes used in Polaroid films. In general, yellow, magenta, and cyan dye couplers ideally absorb the wavelength ranges specifically: 600–700 nm is the typical range for a cyan dye, 500–600 nm for a magenta dye, and 400–500 nm for a yellow dye.

In our case, as shown in Table 1, the absorption density around 682–685 nm is present for every analysed point, and it is associated with the absorption of the red component of visible light by the cyan dye. From research made on copper phthalocyanine dyes in solution [52–55], the absorption maxima of such dyes result at 678–688 nm, which are compatible to our results obtained from every sample point [15,24,48,49] (Figure 2).

The maxima around 560–565 nm wavelength range are also present in all the spectra; this is indicative of the magenta dye. For what is reported about Polaroid films, magenta dyes have azo-pyrazolone structure in coordination with Chromium, with absorption maxima varying also on the substituent groups of the dye [49] (Figure 3). In particular, magenta dye developer used for Polacor 2 films—with the X group being $SNO_2(CH_2CH_2OH)_2$, the Y group being C_6H_5, and the Z group being CH_3—presents absorption maximum in the mentioned range, according to [49]. Therefore, this is likely the structure of the dye developer present in the mock-up (Figure 4).

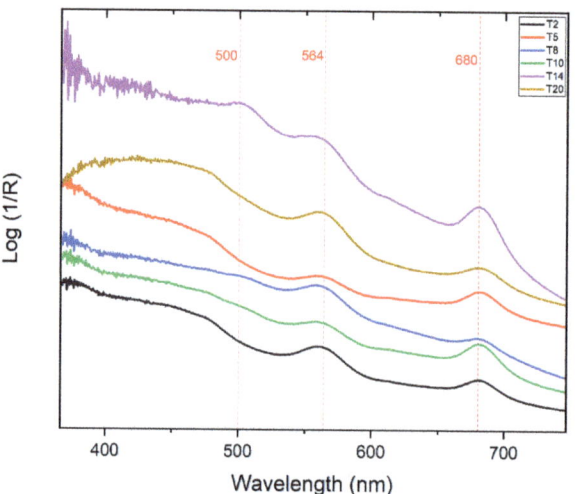

Figure 1. Absorption bands of the sample Polaroid 88 transferred on paper obtained by FORS spectrophotometer.

Table 1. Absorption maxima of the mock-up Polaroid 88 transferred on paper obtained by FORS spectrophotometer.

Points	Absorbance Max (nm)
T1	560, 682
T2	564, 685
T3	564, 683
T4	500, 561, 682
T5	443, 564, 682
T6	501, 564, 682
T7	500, 564, 685
T8	562, 682
T9	505, 562, 682
T10	560, 682
T11	501, 565, 682
T12	502, 563, 682
T13	501, 560, 682
T14	503, 564, 682
T15	505, 564, 684
T16	563, 684
T17	564, 683
T18	505, 564, 683
T19	562, 682
T20	456, 564, 684

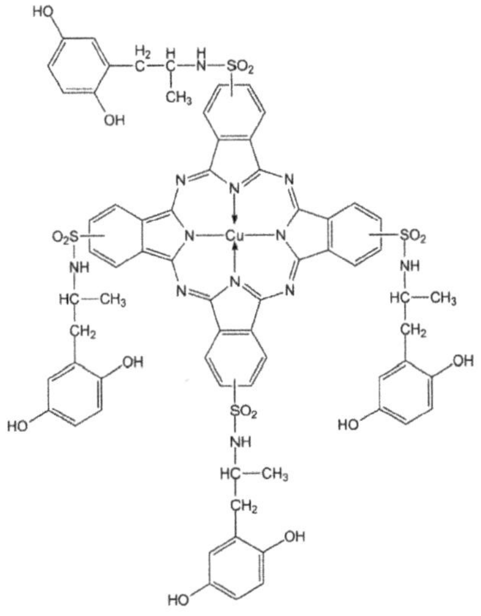

Figure 2. Blue copper phthalocyanine compound.

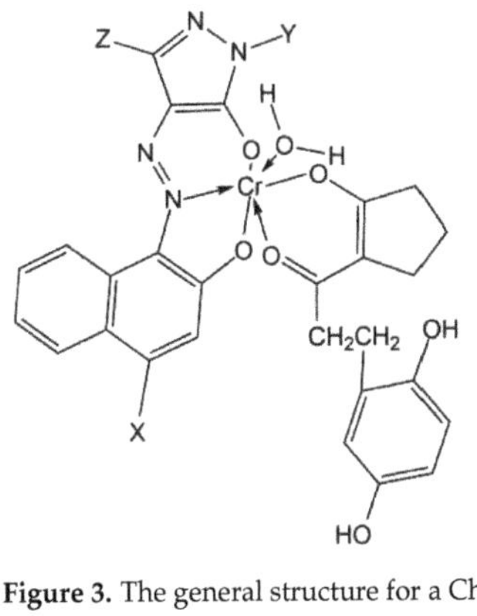

Figure 3. The general structure for a Chromium complexed magenta dye developer in Polaroid films; the X, Y, and Z substituents could vary in different dyes of the same typology.

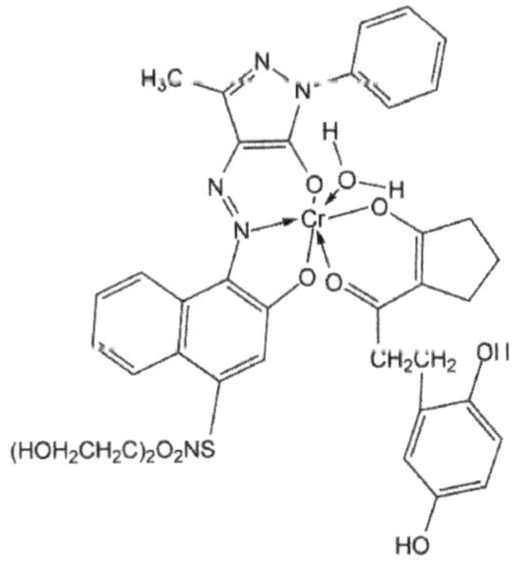

Figure 4. Structure of the magenta dye developer present in Polacolor 2 films and identified in the studied mock-up.

Regarding the yellow dye, it is important to notice that only the point T5 presents an evident maximum at 443 nm, while points T4, T6, T7, T9, T11, T12, T13, T14, T15, and T18 have maxima shifted around 500–505 nm. The other points do not show evident maxima in this range (Table 1), even if for all the spectra broad absorptions are observable in the extended 430–500 nm range. The research of [32] demonstrates that, due to degradation, the absorption maxima at around 440 nm of the yellow dye shift towards lower wavelengths, which in our case is the least detectable and least stable zone. Moreover, the bands tend to become wider and lower in intensity. Finally, yellow dyes could have some unwanted green absorption due to degradation phenomena [48]. In order to highlight which were the points which could be indicative of the original yellow dye and which could be more affected by degradation processes, colorimetric data (L* a* b* coordinates) were also obtained to understand the correlation of the absorbance results and the state of the yellow dye (Table 2).

Table 2. The L* a* b* data of the acquired spectra of sample points T1–T20.

Point	L*	a*	b*
T1	78.37	7.243	23.830
T2	68.49	4.805	19.961
T3	65.46	6.279	18.444
T4	59.29	6.329	14.921
T5	65.13	3.359	23.101
T6	49.90	8.333	14.516
T7	45.78	7.725	13.677
T8	54.82	10.898	13.000
T9	39.94	9.091	12.545
T10	63.73	6.769	18.838
T11	41.12	13.288	15.784
T12	47.02	11.160	15.140
T13	46.17	11.040	15.779
T14	30.58	13.827	13.844
T15	40.39	11.700	14.025
T16	58.35	7.935	17.138
T17	54.93	7.852	15.279
T18	49.30	15.432	14.817
T19	52.27	10.981	16.194
T20	56.07	12.205	21.171

In the spectra corresponding to points T2, T5, and T20, a wide curve is observable between 409 and 480 nm (Figure 1), even if its remarkably broader in comparison to the band of other dyes (for instance, in the case of T2, the maximum cannot be identified with certainty). Moreover, no evident band is observable around 500 nm. The colorimetric data for these points presented the highest value for the b* coordinate, confirming a higher concentration for the yellow dye, while no particular evidence of degradation products involving a greenish shade is observable. On the other side, the point T9 corresponds to the lowest value of b* coordinate, while in the apparent absorbance spectrum a maximum of T9 at 500 nm is remarkable (see Supplementary Figure S6). At this point, the degradation of the yellow dye to green degradation products can be hypothesized. A similar trend is observable for the other points. From these data, it is possible to affirm that, for Polaroid emulsions, bands between 400 and 490 nm are indicative of original yellow dyes, but their eventual broadness could not allow a specifical attribution, while signals at 500 nm are likely indicative of green degradation products, with similarities to phenomena observed for other typologies of photographic films [48].

2.1.2. FTIR

Reflectance FTIR spectra present some drawbacks for the spectral interpretation; for example, since the sent light is being reflected, depending on the physical state of the material of interest, the results may show shifted, noisy, or negative peaks where there is supposed to be a maximum. This factor was considered while interpreting the results.

The FTIR spectra corresponding to the area where the emulsion transfer was made and the paper itself showed a similar pattern due to the gel state and the thinness of the measured film (Figure 5). The signals at 1088, 1129, 1370, and 1641 cm^{-1} and the broad absorption band between 3443 and 3715 cm^{-1} are characteristics of cellulose found in the paper [56], while the peaks at 1377 and 1433 cm^{-1} may suggest the presence of Arabic gum, used as binder of the cellulosic paper. Point F3, the lightest area in hue, shows a more similar spectrum to the paper than point F1 and F2.

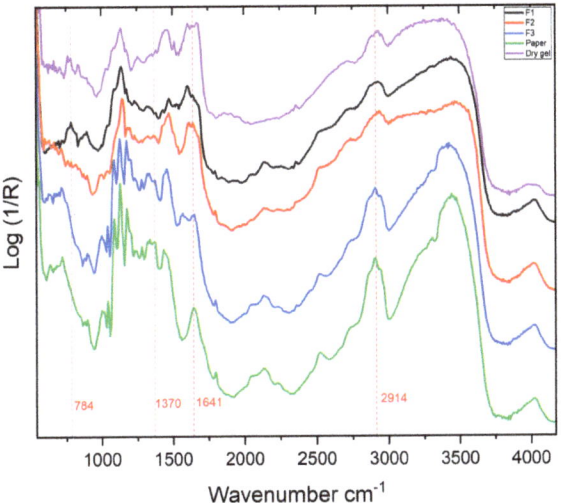

Figure 5. FTIR spectra of points F1, F2, F3, paper, and the dry obtained from the mock-up sample.

F1 differentiates from F3 and the paper from the fact that the peaks at 1088 and 1177 cm^{-1} in F1 are way less intense; meanwhile, the peak, found in F3 and paper at 1129 cm^{-1}, is shifted to 1145 cm^{-1} in F1 and F2. The peak at 1281 cm^{-1} was found in F3, and the paper does not exist in F1 and F2, while the signals at 1473 and 1609 cm^{-1}, observable in F1 and F2, cannot be seen in F3 and the paper. The stretching band between 3400–3300 cm^{-1} region gets lower in intensity in F1, F2, as it occurs in the spectrum of the dry gel. The spectrum of point F1, which has the highest dye density, is the most similar to the spectrum of the dry gel, for instance, the peak at 784 cm^{-1} is only observable in these spectra. Finally, peaks at 840 and 896 cm^{-1}, which are not present in point F3 and the paper, are observed in the fingerprint region of the spectra acquired in correspondence to both the point F1 and the dry gel area. These differences in spectra of F1 and F2 are attributed to the photographic emulsion transferred on paper. In particular, the signals at 830–840 and 890–897 cm^{-1} are reported in the transmission spectra of hydroxyethyl cellulose, with reference to [57]. Moreover, the single firm peak at around 1140–1145 cm^{-1} found in the spectra of the dry gel, F1, and F2 is also present in the hydroxyethyl cellulose. The comparison with the FTIR reflection spectra of hydroxyethyl cellulose (directly on the powder and on the film obtained from the drying of its dispersion in water) confirmed this attribution (see Supplementary Figures S7 and S8), even if it is fundamental to highlight that some artifacts are present in the reference spectra due to several factors. In particular, if the peak at 893 cm^{-1} is observable in all the spectra, it is important to mention a probable overlapping with a close band at 899 cm^{-1} of the paper. The band at 835 cm^{-1} could be affected from deformations for the aggregation state of the material: it results barely visible in the solid-state spectrum, while it has a maximum in the spectrum of the film from the water dispersion. The band at 1140 cm^{-1}, instead, presents a derivative-like shape in both the reference spectra, while it could overlap with paper bands in the mock-up spectrum. Considering the materials used for producing Polaroid gels [47], where the viscosity of the reagent was provided by water-soluble polymeric thickeners including hydroxyethyl cellulose, the attribution of the above-mentioned peaks to hydroxyethyl cellulose should be cited taking into account these spectral deformations and the eventual overlapping with cellulose signals.

2.2. Paolo Gioli's Original Artwork

2.2.1. FORS

The paper of Gioli's artwork has an absorption maximum at 374 nm, which is found to be identical to the paper which is used for the mock-up.

All points turn out to have maxima in similar wavelengths to those of yellow, magenta, and cyan dyes used in the Polacolor 2 system, which also correspond to our FORS results from the mock-up emulsion transfer (Table 3). However, in the original artwork of Paolo Gioli, the colours are much better preserved simply because the artist used recently expired or non-expired films. Therefore, the maxima of the bands obtained from the FORS analysis result in different intensities based on the prevailing dye found in the point of interest, and the band intensities are significantly higher compared to the mock-up (Figure 6).

Table 3. Absorption maxima of the points chosen from Paolo Gioli's original artwork obtained by FORS spectrophotometer.

Points	Absorbance Max (nm)
G1	450, 535, 569, 681
G2	570, 620, 680
G3	528, 570, 626, 680
G4	450, 536, 573, 684
G5	536, 573, 615, 684
G6	536, 575, 683
G7	536, 575, 683
G8	580, 615, 683
G9	577, 617, 677

With reference to the maxima of the bands reported in Table 3, hypotheses about the molecular structure of the dyes can be formulated. The wide maximum at 450 nm, evident in G1 and G4, can be attributed to the yellow dye used in Polacolor 89, a chromium (III) complex of azomethine type dye showing a maximum at this wavelength as expected from a Polacor 2 dye complex [48,49]. The absorption density is higher in the zone of yellow and red (400–600 nm) and lower in in the zone of blue (600–700 nm) which explains the reddish colour of the point.

The maxima around 535 and 575 nm, which change in relative intensity according to the overlapping with the bands of other dyes, are indicative of the magenta dye. These maxima differ from the analogue observed at 567 nm for the mock-up, so a different dye can be hypothesized. The difference is probably caused by the X substituent group of the magenta dye complex, which is fundamental for the differentiation of the dye complex.

In the original artwork of Gioli, Y group is C_6H_5, the Z group is CH_3, and the X group is found to be a cyanide group (CN) with reference to [35] (Figure 7).

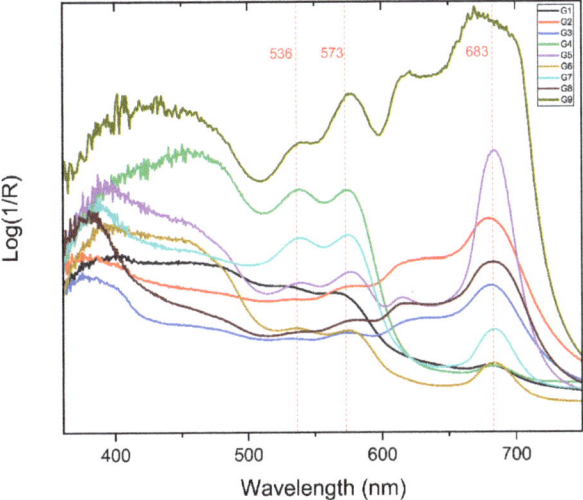

Figure 6. Absorption bands obtained by FORS spectrophotometer from Paolo Gioli's original artwork.

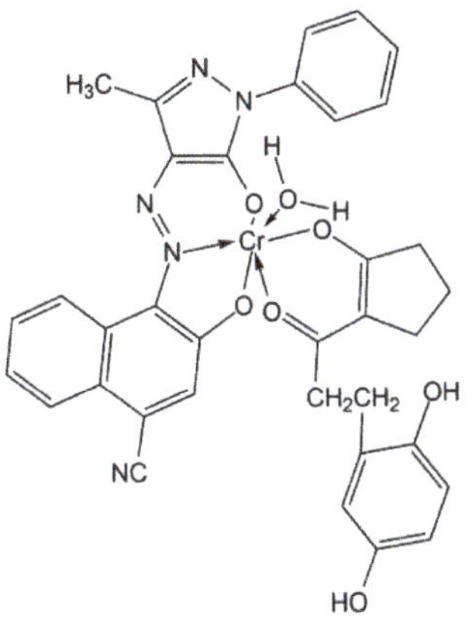

Figure 7. Magenta a chromium complex of an azo pyrazolone of type.

A main band at 680 nm, along with another one at about 615–620 nm, is observable in all the points, even if its intensity is dominant for G5. These signals, as observed for the mock-up, are compatible with the characteristic ones of copper phthalocyanine compounds [15,24,48,49].

2.2.2. FTIR

The spectra of the points GF3 and GF4 had interference fringes that made it difficult to acquire accurate information. Therefore, the results obtained from points GF3 and GF4 were not taken into consideration. The main signals for the spectra of points GF1, GF2, GF5, and paper are shown in Table 4.

It can be stated that all three spectra are significantly similar to each other even though they represent different areas of the artwork having different hues (Figure 8). The characteristic signals of cellulose, found in the paper and reported in Section 2.1.2, can be noted [56].

Table 4. Bands of the FTIR spectra of points GF1, GF2, G5, and paper.

Acquisition Point	Wavenumber
GF1	724, 820, 893, 976, 1078, 1099, 1139, 1206, 1313, 1335, 1386, 1403, 1459, 1481, 1604, 1655, 2109, 2725, 2911, 3482, 4004, 5141 cm^{-1}
GF2	820, 904, 945, 1066, 1088, 1145, 1161, 1206, 1268, 1285, 1313, 1335, 1352, 1380, 1436, 1464, 1492, 1604, 1655, 1671, 2103, 2137, 2523, 2725, 2882, 3409, 4026, 4777, 5153 cm^{-1}
G5	713, 825, 903, 1083, 1127, 1162, 1240, 1318, 1364, 1448, 1644, 2103, 2143, 2513, 2720, 2866, 2916, 3522, 4021, 4766, 5203 cm^{-1}
Paper	720, 877, 899, 1003, 1016, 1043, 1090, 1130, 1176, 1205, 1236, 1282, 1321, 1336, 1357, 1373, 1433, 1452, 1643, 1795, 2069, 2129, 2518, 2723, 2904, 3443, 4021, 4749, 5177 cm^{-1}

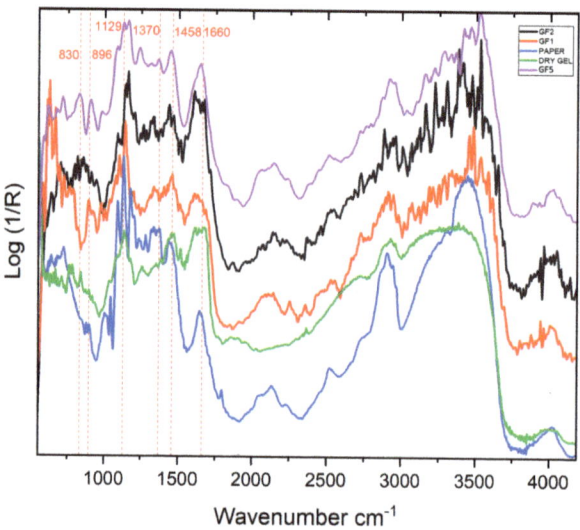

Figure 8. FTIR spectra of points GF1, GF2, and GF5; of the paper obtained from the original artwork of Gioli; and of the reference dry gel. Signals of the dry gel are highlighted.

Compared with FTIR spectra of the the mock-up emulsion transfer, the spectra of Paolo Gioli's original artwork shows major similarities. In the fingerprint region of the three spectra, all points present peak around 825–830, 893–904, and 1139–1145 cm^{-1} which show their analogues in the spectrum obtained for the dry gel. As observed in the mock-up spectra, these peaks could be indicative of hydroxyethyl cellulose, even if a certain overlapping with the signals of cellulose cannot be excluded, as discussed above, which would explain a certain shift (Supplementary Figures S9 and S10). With reference to the literature, in a study that covers the identification of copper complexes of pyrazolone dyes by FTIR in transmission mode [54], the signals in the range at 1600–1680 cm^{-1} are due to the aromatic functionality of the pyrazolone dyes. The absorption bands around 1445–1465 cm^{-1} may indicate the presence of N=N stretching vibrations. It is fundamental to highlight that these signals overlap with those of cellulose at 1641 and 1433 cm^{-1}, respectively, but the higher similarity between GF1 and GF2 spectra and the dry gel one in comparison to the paper one is evident: in particular for GF1, the spectrum shows the same pattern in the 1580–1680 cm^{-1}, with two maxima at 1604 and 1655 cm^{-1}, observable for the dry gel and attributable to aromatic groups of pyrazolone dye, while no maximum for the cellulose is evident at 1641 cm^{-1} [58]. These signals would confirm the presence of pyrazolone complexes, constituting the magenta and yellow dyes, previously considered based on FORS data.

3. Materials and Methods

3.1. Sample Preparation

In order to test the diagnostic protocol, a Polaroid transfer mock-up was prepared following a technique similar to Paolo Gioli's. The following equipment has been used: Polaroid Polacolor Type 88; receptor paper: Fabriano water colour papers (as a smoother surface will capture the image better, hot pressed watercolour paper was preferred); Polaroid EE 100 land camera; a rolling pin; a hard smooth working surface; scissors; timer; surgical gloves. The dry transfer method consisted of the steps explained in detail by [59].

The chosen artwork of Paolo Gioli to be analysed takes part of the "Cameron Obscura" series that he created in 1981.

3.2. The Spectroscopic Analysis

For the analysis of the mock-up created by transferring the Polacolor 88 film to paper and the original artwork of Paolo Gioli, three different spectroscopic methods are chosen to be applied: FORS (Fiber Optics Reflectance Spectroscopy, Plainsboro, NJ, USA), Raman Spectroscopy and FTIR (Fourier Transform InfraRed, Ettlingen, Germany) spectroscopy. The reason to apply more than one analytical method is from the simple fact that each method is complementary to each other, and therefore, their results are comparable and more informative [60–62].

3.2.1. FORS

Analysis of the Mock-Up Polaroid Emulsion Transfer

In this research, FORS was used to have a preliminary and complementary method to determine the dyes and the dye developers found in the transferred Polaroid 88 emulsion transfer on paper.

Since the film was expired and out of the developing gel sacs, of the eight films placed in the camera, only one was in a liquid phase, and only one mock-up could be obtained. The developing gel contained in the other seven films was dry and therefore could not provide an image. The film containing the liquid gel had two sacks, and only one has broken to release the gel. Therefore, the image obtained had an irregular form, though it presented different tonalities of a similar hue. For statistical purposes, 20 points were analysed in correspondence with the overall image on the mock-up, while an area was analysed on the gel contained in the sac (Supplementary Materials: Figure S1). For every analysed point, 10 spectra were acquired.

The analysis was conducted using Exemplar LS from B&W Tek, Inc., Plainsboro, NJ, USA, with a wavelength Range of 200 to 850 nm, spectral Resolution 1.5 nm together with the BPS101 Tungsten Halogen Light Source with a spectral range of e 350 nm to > 2600 nm. Data were acquired via a handheld probe, with an analysis area of 2 mm^2, placed at 45° in close contact with the sample. This equipment was calibrated using a certified reflectance standard provided by B&W Tek, Inc., Plainsboro, NJ, USA. Spectra were recorded in 10 scans with an integration time of 80 µs par scan. In order to compare the obtained spectra with absorbance ones available in literature for dyes, we calculated the apparent absorbance spectra from the reflectance ones using Origin 2021 Software, according to procedures reported in literature [52,54,55] and using the conversion in Equation (1):

$$\text{Abs} \propto \log(\frac{1}{R}) \qquad (1)$$

where Abs is the absorbance and R is the reflection spectrum. The experimental apparent absorption spectra can be used for qualitative identification of the dyes based on the apparent absorption maxima, assuming that transmission of the light through the sample can be considered neglectable. It has been preferred to analyse mainly the absorption spectra of the dyes because spectral data of the dyes in the literature are mainly found in solutions and, therefore, in absorbance. Comparing dyes in a solid state with the dyes in solution involves some limitations to their identification, since differences among spectra of the same molecule in different phases must be considerd (different wavelenght maxima, presence of multiple bands, different band broadness, etc.) However, taking into account the scarce availability of standards and reference spectra of solids for this typology of analytes and with reference to the fact that this is a preliminary study for Polaroid matrices, this approach was chosen.

Analysis of Paolo Gioli's Original Artwork

The analysis was conducted with the same instrumentation as the mock-up sample. Spectra were recorded with 70 µs integration time with 10 acquisitions. Inflection points in the reflectance mode were determined using the first derivative spectrum, and the pseudo absorbance spectrum was generated by calculating the negative logarithm of the reflectance of the sample of each spectrum using Origin 2021 Software. Ten points, including the paper, were chosen to be analysed from Gioli's original artwork (see Supplementary Figure S2). Ten acquisitions were made from each point.

3.2.2. Raman

Three different points were chosen from the transferred emulsion on paper sample, R1 being the darkest, R2 being the medium, and R3 being the lightest in hue (see Supplementary Figure S3). Paper and the dry gel contained in the sac of the Polacolor 88 were also analysed.

The instrument used was LabRAM HR Evolution Confocal Raman Microscope (laser wavelength: 633 nm), by HORIBA, Kyoto, Japan. The used objective was 100× for each point. For the paper 100% laser intensity was set, for the dry gel 2.5%, and for the transferred emulsion on paper 5%. For all points, the acquisition ranges were 100–1000, 1000–1900, and 2400–3300 cm^{-1}. The acquisition time for paper was 20 s with 30 acquisitions, for the dry gel 3 s with 500 acquisitions, and for the transferred emulsion on paper, it was 5 s with 60 acquisitions.

However, the results obtained from Raman Spectroscopy did not provide information about the analysed sample. In fact, strong interference fringes were observed. For this reason, the method was not used for the Paolo Gioli's artwork. Indeed, with reference to the great potential of Raman spectroscopy in providing molecular information, this analytical issue should be deepened. This will be the object of further studies, when other experimental conditions will be tested: for instance, variation of the spot size. However, we preferred highlighting this aspect in order to make the reader aware of possible setbacks for other studies on similar matrices.

3.2.3. FTIR

Analysis of the Mock-Up Sample

The instrumentation used for the FTIR analysis was Bruker Optics-IFS 66 v/s Vacuum FT-IR, Ettlingen, Germany. The instrument was combined with the Hyperion 2000 IR microscope with 15× objective and used in the Reflection mode, since the sample was not transparent enough to transmit the IR radiation. Combining microscopy with FTIR allowed us to perform a completely non-invasive analysis.

Four different points were chosen from the transferred emulsion on paper sample, F1 being the darkest, F2 being the medium, and F3 being the lightest in hue and P0 being the paper (see Supplementary Figure S4). In addition, the powdered dry developing gel was analysed too. As reference, a small gold plate as a reflective surface was used. Three measurements were taken from every sample point using the following parameters: 4 cm^{-1} resolution, 256 scans, wavenumber covering from 400 to 6000 cm^{-1} using OPUS software. To obtain a pseudo-absorbance spectrum, the following Equation (2) is used (R = reflectance, R_1 = reference:

$$\text{Abs} \propto -\log\left(\frac{R}{R_1}\right) \quad (2)$$

Analysis of Paolo Gioli's Original Artwork

The instrumentation used for the FTIR was the same as the mock-up. Five different points were chosen from the transferred emulsion on paper sample (see Supplementary Figure S5). As reference, a small gold plate as a reflective surface was used. Three measurements were taken from every sample point using the following parameters: 4 cm^{-1} resolution, 256 scans, wavenumber covering 400 to 6000 cm^{-1} using OPUS software. A pseudo absorbance spectrum was obtained with the above-mentioned equation.

4. Conclusions

The aim of this research was the characterization of the Polaroid emulsion transfer in Paolo Gioli's artworks, contributing preliminary knowledge to the literature about the materials of the artworks and the appropriate protocols which can be applied for future studies. We faced several limitations during the making of the mock-up sample due to the aging of the Polaroid 88 film. The film expired in November 1985. Therefore, the alkaline gel, which was supposed to reach the film layers to provide a positive image, was dry. Moreover, since production of these films was discontinued by the Polaroid Corporation, it was significantly hard to find the Polacolor films that Gioli used.

When performing FTIR, combination of the microscope with the FTIR in reflectance mode was used, since our sample will not be transparent enough to transmit IR radiation and since it was crucial to perform a technique that was non-invasive. The obtained spectra from FTIR gave significant results about the cellulose found in the paper, and the presence of Arabic gum as a binder of the paper was detected. FTIR results also showed the presence of hydroxyethyl cellulose as the thickening agent of the gel contained in the sac (in our case in a dry state), differing from the fingerprint region of only the paper itself. However, this assignation is only hypothetical because working in reflectance mode brought limitations to the assignment of the peaks due to spectral deformation and overlapping with cellulose signals. The FTIR spectra acquired from the original artwork of Paolo Gioli showed very similar results to our mock-up sample, but the presence of pyrazolone dyes could be confirmed from the spectra of the points analysed. Furthermore, the presence of hydroxyethyl cellulose as the thickening agent of the gel was evaluated also in this case. However, further research is necessary to define the efficacy of FTIR for the analysis of Polaroid emulsions transferred on paper by Paolo Gioli. In particular, extended studies on the materials present in Polacolor film, aimed to precise individuation of spectral artifacts and determination of single component contribution to the spectrum, are required, and they will be the object of further studies. About the determination of the dye developers present in both the mock-up sample, obtained from Polacolor 88, and Paolo Gioli's Original artwork created with Polacolor type 89, the most diagnostic results were obtained from FORS. With reference to the literature, the main dyes used in the Polacolor Types 88 and 89 were determined from the apparent absorbance maxima, which allowed hypothesizing their structures and molecular substituents. However, other studies involving further characterization of original materials are necessary, in order to confirm the data reported in literature and also to support these results. The characterization of the dye developers with complementary techniques (e.g., chromatography, mass spectrometry [63,64]) is foreseen and desirable. In conclusion, new perspectives are open: from the conservation science point of view, Polaroid film characterization represents a new frontier to be investigated, whose results could be fundamental for future preservation, as it could be the case with Paolo Gioli's works. Further research, involving complementary analytical techniques and ageing studies, is required in order to deepen our knowledge about instant photography materials, conservation, and occurrences.

Supplementary Materials: The following supporting information can be downloaded at: https://www.mdpi.com/article/10.3390/molecules27207023/s1, Supplementary Figures Figure S1: Points of the mock-up sample analysed by FORS spectroscopy; Figure S2: Points of Paolo Gioli's artwork analysed by FORS spectroscopy;.Figure S3: Points of the mock-up sample analysed by Raman spectroscopy; Figure S4: Points of the mock-up sample analysed by FTIR spectroscopy; Figure S5: Points of Paolo Gioli's artwork analysed by FTIR spectroscopy; Figure S6: A comparison of the absorption spectra of points T1 and T9; Figure S7: A comparison of FTIR spectra obtained for the mock-up (F1 point), paper, dry-gel and hydroxyethyl cellulose (solid and film from water dispersion); Figure S8: A comparison of FTIR spectra obtained for the mock-up (F1 point), paper, dry-gel and hydroxyethyl cellulose (solid and film from water dispersion) in the range 750–1800 cm^{-1}; Figure S9: A comparison of FTIR spectra obtained for the artwork (GF5 point), dry-gel and hydroxyethyl cellulose (solid and film from water dispersion); Figure S10: A comparison of FTIR spectra obtained for the artwork (GF5 point), dry-gel and hydroxyethyl cellulose (solid and film from water dispersion) in the range 750–1800 cm^{-1}.

Author Contributions: Conceptualization, A.C. and I.S.; methodology, A.N. and P.P.; validation, Z.A. and A.C.; formal analysis, Z.A. and A.C.; investigation, Z.A. and A.C.; resources, P.P., A.G. and G.F.; data curation, Z.A.; writing—original draft preparation, Z.A. and A.C.; writing—review and editing, Z.A. and A.C.; visualization, Z.A. and A.C.; supervision, G.F.; project administration, R.C. All authors have read and agreed to the published version of the manuscript.

Funding: This research received no external funding.

Institutional Review Board Statement: Not applicable.

Informed Consent Statement: Not applicable.

Acknowledgments: This study was supported by the Sapienza University, and all analyses in this study were performed at the Sapienza University. We would like to express our deepest appreciation to Paolo Vampa and Maria Francesca Bonetti for their support.

Conflicts of Interest: The authors declare no conflict of interest.

References

1. Blacklow, L. *New Dimensions in Photo Processes: A Step-by-Step Manual in Alternative Photography*; Focal: Oxford, UK, 2007; ISBN 978-0-240-80789-8.
2. Greenwald, R.B. 4-Hydropyrazole Developing Agents. US4102684, 20 September 1977.
3. Bloom, S.M. Wayland, Boris Levy Image Receiving Elements. US43048US430483535, 8 December 1981.
4. Fischer, M. Short Guide to Film Base Photographic Materials: Identification, Care, and Duplication. *Northeast Doc. Conserv. Cent.* **2007**, *5*, 112–117.
5. Wilker, A. *The Composition and Preservation of Instant Films*; University of Texas at Austin: Austin, TX, USA, 2004.
6. Attridge, G.G.; Pointer, M.R. Colour Science in Photography. *J. Photogr. Sci.* **1994**, *42*, 197–209. [CrossRef]
7. Yeoung-Chan, K. The Synthesis of Green-Sensitizing Dye for Photographic Emulsion. *J. Korean Appl. Sci. Technol.* **1997**, *14*, 57–64.
8. Yeoung-Chan, K.; Tai-Sung, K. The Synthesis of Sensitizing Dye for Photographic Emulsion. *J. Korean Appl. Sci. Technol.* **1995**, *12*, 5–11.
9. Yeoung-Chan, K. The Synthesis of Red-Sensitizing Dye for Color Photography. *J. Korean Appl. Sci. Technol.* **2001**, *18*, 136–141.
10. Witten, N.M. The Chemistry of Photography. Senior Theses, University of South Carolina-Columbia, Columbia, SC, USA, 2016.
11. Van de Sande, C.C. Dye Diffusion Systems in Color Photography. *Angew. Chem. Int. Ed. Engl.* **1983**, *22*, 191–209. [CrossRef]
12. Chalmers, A.N. A Colorimetric Comparison of Colour Reproductions. *Color Res. Appl.* **1979**, *4*, 217–224. [CrossRef]
13. Sofen, S. Instant Imaging. In Proceedings of the IS&T's 50th Annual Conference, Cambridge, MA, USA, 18–23 May 1997; Polaroid Corporation: Minnetonka, MN, USA, 1997.
14. Bergthaller, P. Organic Sulfur Compounds in Silver Halide Photography. *Sulfur Ceports* **2006**, *23*, 1–45. [CrossRef]
15. Fujita, S. *Organic Chemistry of Photography*; Springer Berlin Heidelberg: Berlin, Heidelberg, 2004; ISBN 978-3-642-05902-5.
16. Venkataraman, K. *The Chemistry of Synthetic Dyes*; Academic Press: Cambridgem, MA, USA, 1971; Volume 4, p. 714.
17. Simon, M.S. New Developments in Instant Photography. *J. Chem. Educ.* **1994**, *71*, 132. [CrossRef]
18. Land, E.H. Photographic Product Comprising a Ruptuble Container Carrying a Photographic Processing Liquid. US2543181, 27 February 1951.
19. Mervis, S.H.; Walworth, V.K. Photography, Instant. *Kirk-Orthmer Encycl. Chem. Technol.* **2001**, *50*. [CrossRef]
20. Renner, E. *Pinhole Photography: From Historic Technique to Digital Application*, 4th ed.; Routledge: London, UK, 2012; ISBN 978-0-08-092789-3.

21. Dupré, H.F. Online Book: Polaroid Image Transfers Tools and Techniques. Available online: https://www.21gradi.it/repository/POLAROID%20TRANSFER%20TECHNIC.pdf (accessed on 6 July 2022).
22. Berg, W. Polaroid One-Step Color Photography. *Naturwissenschaften* 1977, *64*, 1–7. [CrossRef]
23. Rogers, H.G. Processes and Products for Forming Photographic Images in Color. US2983606, 9 May 1961.
24. Idelson, M. SX-70: A Unique System of Dyes. *Res. Lab. Polaroid Corp. Camb. Massachus* 1982, *3*, 191–201. [CrossRef]
25. Guida, W.C.; Raber, D.J. The Chemistry of Color Photography. *J. Chem. Educ.* 1975, *52*, 622. [CrossRef]
26. Orraca, J. The Conservation of Photographic Materials. *Int. Inst. Conserv. Hist. Artist. Works* 1973, *13*, 32–38. [CrossRef]
27. Bloom, S.M.; Green, M.; Idelson, M.; Simon, M.S. The Dye Developer in the Polaroid Color Photographic Process. In The Chemistry of Synthetic Dyes. Elsevier: Amsterdam, The Netherlands, 1978; pp. 331–387, ISBN 978-0-12-717008-4.
28. Haines, A. Conservation of Photographic Chemicals. *J. Soc. Motion Pict. Eng.* 1943, *41*, 409–411. [CrossRef]
29. Rempel, S. A Conservation Method for Nitrate Based Photographic Materials. *Pap. Conserv.* 1977, *2*, 44–46. [CrossRef]
30. Casoli, A.; Fornaciari, S. An Analytical Study on an Early Twentieth-Century Italian Photographs Collection by Means of Microscopic and Spectroscopic Techniques. *Microchem. J.* 2014, *116*, 24–30. [CrossRef]
31. Maillette de Buy Wenniger, T. The Quest towards a Non-Destructive Identification Method for Polaroid Integral Film Types. Master Thesis, University of Amsterdam, Amsterdam, The Netherlands, 2020.
32. Silva, J.; Ferreira, J.L.; Laia, A.T.; Parola, A.J.; Lavédrine, B.; Ramos, A.M. New Approaches for Monitoring Dye Fading in Chromogenic Reversal Films: UV-Vis Spectrophotometry and Digitisation. In Proceedings of the ICOM-CC 18th Triennial Conference, Copenhagen, Denmark, 4–8 September 2017; p. 11.
33. Pietro, G.D. Chapter 4 Examples of Using Advanced Analytical Techniques to Investigate the Degradation of Photographic Materials. In *Physical Techniques in the Study of Art, Archaeology and Cultural Heritage*; Elsevier: Amsterdam, The Netherlands, 2007; Volume 2, pp. 155–198, ISBN 978-0-444-52856-8.
34. Ricci, C.; Bloxham, S.; Kazarian, S.G. ATR-FTIR Imaging of Albumen Photographic Prints. *J. Cult. Herit.* 2007, *8*, 387–395. [CrossRef]
35. Sclocchi, M.C.; Damiano, E.; Matè, D.; Colaizzi, P.; Pinzari, F. Fungal Biosorption of Silver Particles on 20th-Century Photographic Documents. *Int. Biodeterior. Biodegrad.* 2013, *84*, 367–371. [CrossRef]
36. Rampazzi, L.; Brunello, V.; Campione, F.P.; Corti, C.; Geminiani, L.; Recchia, S.; Luraschi, M. Non-Invasive Identification of Pigments in Japanese Coloured Photographs. *Microchem. J.* 2020, *157*, 105017. [CrossRef]
37. Simon, M.S. Opacification: Solving the Fundamental Problems. *Dyes Pigment.* 1989, *11*, 1–12. [CrossRef]
38. Trumpy, G. Photographic film and its interaction with light: Detection of Dust and Scratches for Image Restoration. Ph. D. Thesis, Universität Basel, Basel, Switzerland, 2013.
39. Lavedrine, B. *Photographs of the Past: Process and Preservation, Getty Publications*, 2nd ed.; The Getty Conservation Institute: Los Angeles, CA, USA, 2007; ISBN 978-0-89236-957-7.
40. Cattaneo, B.; Chelazzi, D.; Giorgi, R.; Serena, T.; Merlo, C.; Baglioni, P. Physico-Chemical Characterization and Conservation Issues of Photographs Dated between 1890 and 1910. *J. Cult. Herit.* 2008, *9*, 277–284. [CrossRef]
41. Van Roosmalen, P.M.B.; Biemond, J.; Lagendijk, R.L. Restoration and Storage of Film and Video Archive Material. *Signal Process. Multimed.* 1999, *1*, 167–191.
42. Bryan, F.R.; Runge, E.F. Sensitometric Properties of a Polaroid Emulsion as Applied to Spectroscopic Analysis. *Appl. Spectrosc.* 1961, *15*, 16–18. [CrossRef]
43. Ann Daffner, L.; McGlinchey, C. The Big Picture: Conservation Research Program for Contemporary Color Photographs. *Stud. Conserv.* 2004, *49*, 109–113. [CrossRef]
44. Fenech, A.; Strlič, M.; Cassar, M. The Past and the Future of Chromogenic Colour Photographs: Lifetime Modelling Using near-Infrared Spectroscopy & Enhancement Using Hypoxia. *Appl. Phys. A* 2012, *106*, 411–417. [CrossRef]
45. Land, E.H. A New One-Step Photographic Process. *J. Opt. Soc. Am.* 1947, *37*, 61. [CrossRef]
46. Rogers, D. The Chemistry of Photography, From Classical to Digital Technologies. In *The Royal Society of Chemistry*; RSC publishing: London, UK, 2007; ISBN 9780854042739.
47. Mervis, S.H. Photography, Instant. In *Kirk-Othmer Encyclopedia of Chemical Technology*; Wiley-Interscience: Hoboken, NJ, USA, 2000.
48. Theys, R.D.; Sosnovsky, G. Chemistry and Processes of Color Photography. *Chem. Rev.* 1997, *97*, 83–132. [CrossRef]
49. Idelson, Martin Chemistry of Metallized Dyes. *J. Soc. Photogr. Sci. Technol* 1984, *47*, 421–431.
50. Luísa, A.; Baddini, D.Q.; Luiz, J.; Paula, V.D.; Reiner, R.; Silva, L.; Paula, D.; Araújo, C.; Freitas, R.P. PLS-DA and Data Fusion of Visible Reflectance, XRF and FTIR Spectroscopy in the Classification of Mixed Historical Pigments. *Spectrochim. Acta Part A Mol. Biomol. Spectrosc.* 2022, *265*, 120384. [CrossRef]
51. Defeyt, C.; Pevenage, J.V.; Moens, L.; Strivay, D.; Vandenabeele, P. Spectrochimica Acta Part A: Molecular and Biomolecular Spectroscopy Micro-Raman Spectroscopy and Chemometrical Analysis for the Distinction of Copper Phthalocyanine Polymorphs in Paint Layers. *Spectrochim. Acta. A. Mol. Biomol. Spectrosc.* 2013, *115*, 636–640. [CrossRef]
52. Beaulieu-Houle, G.; Gilson, D.F.R.; Butler, I.S. Pressure-Tuning Micro-Raman Spectra of Artists' Pigments: α- and β-Copper Phthalocyanine Polymorphs. *Spectrochim. Acta. A. Mol. Biomol. Spectrosc.* 2014, *117*, 61–64. [CrossRef] [PubMed]
53. Anghelone, M.; Jembrih-simbürger, D.; Schreiner, M. Identification of Copper Phthalocyanine Blue Polymorphs in Unaged and Aged Paint Systems by Means of Micro-Raman Spectroscopy and Random Forest. *Spectrochim. Acta. A. Mol. Biomol. Spectrosc.* 2015. [CrossRef] [PubMed]

54. Aceto, M.; Agostino, A.; Fenoglio, G.; Idone, A.; Gulmini, M.; Picollo, M.; Ricciardi, P.; Delaney, J.K. Characterisation of Colourants on Illuminated Manuscripts by Portable Fibre Optic UV-Visible-NIR Reflectance Spectrophotometry. *Anal. Methods* **2014**, *6*, 1488–1500. [CrossRef]
55. Albay, C.; Koç, M.; Altin, I.; Bayrak, R.; Değirmencioğlu, I.; Sökmen, M. New Dye Sensitized Photocatalysts: Copper(II)-Phthalocyanine/TiO2 Nanocomposite for Water Remediation. *J. Photochem. Photobiol. Chem.* **2016**, *324*, 117–125. [CrossRef]
56. Cheng, S.; Huang, A.; Wang, S.; Zhang, Q. Effect of Different Heat Treatment Temperatures on the Chemical Composition and Structure of Chinese Fir Wood. *BioResources* **2016**, *11*, 4006–4016. [CrossRef]
57. Ostrovskii, D.; Kjøniksen, A.L.; Nyström, B.; Torell, L.M. Association and Thermal Gelation in Aqueous Mixtures of Ethyl(Hydroxyethyl)Cellulose and Ionic Surfactant: FTIR and Raman Study. *Macromolecules* **1999**, *32*, 1534–1540. [CrossRef]
58. Hussain, G.; Ather, M.; Khan, M.U.A.; Saeed, A.; Saleem, R.; Shabir, G.; Channar, P.A. Synthesis and Characterization of Chromium (III), Iron (II), Copper (II) Complexes of 4-Amino-1-(p-Sulphophenyl)-3-Methyl-5-Pyrazolone Based Acid Dyes and Their Applications on Leather. *Dyes Pigment.* **2016**, *130*, 90–98. [CrossRef]
59. Dupre, H.F. *Polaroid Image Transfers: Tools and Techniques*; Polaroid publishing: San Francisco, CA, USA, 2000.
60. Socrates, G.; Socrates, G. *Infrared and Raman Characteristic Group Frequencies: Tables and Charts*, 3rd ed.; Wiley: Chichester, UK; New York, NY, USA, 2001; ISBN 978-0-471-85298-8.
61. Larkin, P. *Infrared and Raman Spectroscopy: Principles and Spectral Interpretation*; Elsevier: Amsterdam, The Netherlands, 2011; ISBN 978-0-12-386984-5.
62. Serafini, I.; Ciccola, A. Nanotechnologies and Nanomaterials. In Nanotechnologies and Nanomaterials for Diagnostic, Conservation and Restoration of Cultural Heritage. Elsevier: Amsterdam, The Netherlands, 2019; pp. 325–380, ISBN 978-0-12-813910-3.
63. Calà, E.; Benzi, M.; Gosetti, F.; Zanin, A.; Gulmini, M.; Idone, A.; Serafini, I.; Ciccola, A.; Curini, R.; Whitworth, I.; et al. Towards the Identification of the Lichen Species in Historical Orchil Dyes by HPLC-MS/MS. *Microchem. J.* **2019**, *150*, 104140. [CrossRef]
64. Persechino, S.; Toniolo, C.; Ciccola, A.; Serafini, I.; Tammaro, A.; Postorino, P.; Persechino, F.; Serafini, M. A New High-Throughput Method to Make a Quality Control on Tattoo Inks. *Spectrochim. Acta. A Mol. Biomol. Spectrosc.* **2019**, *206*, 547–551. [CrossRef]

Article

A Silver Monochrome "Concetto spaziale" by Lucio Fontana: A Spectroscopic Non- and Micro-Invasive Investigation of Materials

Margherita Longoni [1], Carlotta Beccaria [2], Letizia Bonizzoni [3,*] and Silvia Bruni [1,*]

[1] Department of Chemistry, University of Milan, 20133 Milan, Italy; margherita.longoni@unimi.it
[2] Beccaria Carlotta & C. Studio Di Restauro, 20122 Milan, Italy; beccaria.restauro@gmail.com
[3] Department of Physics, University of Milan, 20133 Milan, Italy
* Correspondence: letizia.bonizzoni@mi.infn.it (L.B.); silvia.bruni@unimi.it (S.B.)

Abstract: In several of his artworks, for instance the *Venezie* cycle, Fontana employed metallic paints; previous investigations on such materials highlighted the use of different synthetic binders and of thick paint layers below the metal one, having different colours to change the visual perception of the metallic surface. In the present work, a monochrome silver "Concetto spaziale" by the Italo–Argentine artist belonging to a private collection recently gifted to the museum of the Church of San Fedele in Milano, Italy, was investigated to deepen the knowledge of this particular group of Fontana's paintings. The artwork was initially visually inspected in visible and ultraviolet (UV) light. Subsequently, a non-invasive spectroscopic investigation was performed by X-ray fluorescence (XRF), reflection Fourier-transform infrared (FTIR) and Raman spectroscopy. A minute fragment of silver-coloured paint was taken from the reverse of the painting, near the cut edge, and examined by scanning electron microscopy coupled with energy dispersive X-ray analysis (SEM-EDX) and micro-Fourier-transform Raman (FT-Raman) spectroscopy. The analytical data made it possible to identify the composition of the metallic paint layer and of the underlying dark one, both from the point of view of the pigments and of the binders used, also highlighting the potential of the non-invasive and micro-invasive methods adopted in the investigation.

Keywords: XRF; FTIR; Raman spectroscopy; SEM-EDX; FT-Raman; modern painting; metallic paint

Citation: Longoni, M.; Beccaria, C.; Bonizzoni, L.; Bruni, S. A Silver Monochrome "Concetto spaziale" by Lucio Fontana: A Spectroscopic Non- and Micro-Invasive Investigation of Materials. *Molecules* **2022**, *27*, 4442. https://doi.org/10.3390/molecules27144442

Academic Editors: Maria Luisa Astolfi, Maria Pia Sammartino, Emanuele Dell'Aglio and Gavino Sanna

Received: 31 May 2022
Accepted: 8 July 2022
Published: 11 July 2022

Publisher's Note: MDPI stays neutral with regard to jurisdictional claims in published maps and institutional affiliations.

Copyright: © 2022 by the authors. Licensee MDPI, Basel, Switzerland. This article is an open access article distributed under the terms and conditions of the Creative Commons Attribution (CC BY) license (https://creativecommons.org/licenses/by/4.0/).

1. Introduction

The worldwide recognized importance of the pictorial works by Lucio Fontana, the well-known Italian–Argentine artist (1899–1968), derives both from the continuous research and innovation of his figurative and abstract way of expression, and from his pioneering use of innovative materials. These include, for example, fluorescent colours and many different synthetic paints, such as nitrocellulose, alkyd, polyvinyl acetate and acrylics, which became commercially available as he created his artworks and which he often used in multiple layers and together [1–4].

In the present work, a multi-technique approach was applied to investigate the materials in a painting by Lucio Fontana, preserved at the exhibition hall of the Church of San Fedele in Milano and belonging to Fontana's *Tagli* (Cuts) series. The artwork, a "Concetto spaziale-Attesa" (Spatial Concept-Wait) dated 1961, is particularly interesting as it is a silver monochromatic painting and therefore belongs to those works in which the artist employed a metallic paint.

A few previous investigations have been published about the materials of Fontana's works based on metallic paints. In the paintings of the *Venezie* cycle, even if considered to be part of the *Olii* series, the use of a synthetic alkyd resin as a binder has been identified [1,2,5]. The reason for this choice was traced to the considerable thickness of the paint layer which, if obtained with a drying oil, would probably show significant drying craqueleure. More precisely, in Fontana's silver paintings, the thick paint layer usually lies under a very thin layer of silver-coloured paint and has different colours, ranging from ochre to red or black, to change the perception of the metal surface above [2]. Furthermore, pyrolysis–gas chromatography–mass spectrometry analyses showed, at least in one case (painting 61 O 41), the presence of low quantities of polyvinyl acetate (PVA) in addition to alkyd resin [1], while in another *Venezia* painting (61 O 53), a silver acrylic coating was detected on a hard polyester layer [6].

In the present work, characterization of painting materials was carried out to answer specific questions raised by conservation issues and restoration needs. It is not necessary to point out that the precise knowledge of materials is mandatory to perform correct and long-lasting restoration. A sequence from general to specific was adopted in the investigation. It is worth noting that scientific analyses, and in particular non-invasive procedures, are becoming more and more applied also in cases of modern and contemporary paintings [7–9] and to get information about degradation phenomena [10]. After a visual inspection in both visible (also with an optical digital microscope, OM) and ultraviolet light, we approached the problem using non-invasive techniques, in detail X-ray fluorescence (XRF), reflection Fourier-transform infrared (FTIR) and Raman spectroscopy. These techniques were chosen for their proven synergy on pictorial materials [11,12] and due to their different penetration depths, which can help reconstruct stratigraphic sequences without sampling, opening up interesting applications to off-limits masterpieces, such as the present case. Furthermore, all these techniques together present the ability to investigate different types of materials (organic, or inorganic such as vitreous or metallic). XRF is useful to get a rapid and reliable determination of medium-heavy elements, even when present in light matrix [13,14], without any sample preparation and regardless of the substrate material. This allows individuation of the pigments used in a wide variety of artefacts in a non-invasive way [15,16], even on modern materials [17]. XRF penetration depth depends on the matrix and the atomic number of the investigated element, and it can range from tens of micrometres to a few millimetres; the synergy with techniques with different penetration depths can give hints about the stratigraphy of the different layers of material [18]. Reflection FTIR spectroscopy can provide information on binders, even synthetic ones [19], adhesives [20] and inorganic materials such as carbonates or sulphates [21] with a penetration depth of about ten microns. Raman spectroscopy is suitable for the identification of pigments, both inorganic [22] and synthetic organic [23,24]. The penetration depth of the Raman technique depends on the excitation wavelength used and its absorption by the material examined, and can in principle be greater when using NIR exciting radiation [25]. In the present work, due to the physical properties and chemical composition of the metallic paint, more complete characterization of this material also required micro-analyses in addition to the non-invasive ones, carried out on a small sample taken from the back of the painting near the edge of the cut. To this end, scanning electron microscopy coupled with energy dispersive X-ray analysis (SEM-EDX) was used for elemental analysis, while micro-Fourier-transform Raman (FT-Raman) spectroscopy was exploited to identify the binder.

2. Materials and Methods

2.1. The Painting

The work, preserved in the exhibition hall of the Church of San Fedele in Milano (as indicated above), was created on a canvas measuring 99.8 × 79.8 cm. It bears the artist's signature and a numerical sequence (1 + 1-7741F) on the back.

Belonging to the *Nanda Vigo* collection, this work was created by Fontana using silver-coloured paint and is characterized by a diagonal texture of considerable thickness, probably made with the aid of a paintbrush.

The canvas showed deformations caused by the considerable layer of colour spread on the front, as well as an extension of the original cut in the upper part (about 10 cm) where the threads of the original canvas had broken. The cut was "closed" on the back by the "teletta"; this had been placed by the artist, but the painting suffered structural damage in the area of the tear, where it was detached (Figure 1).

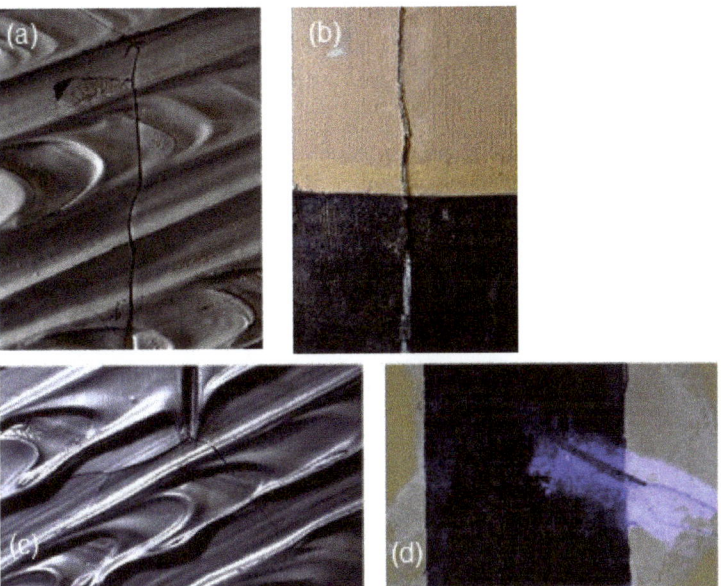

Figure 1. Some details of alterations observed in the monochrome silver "Concetto spaziale" by Lucio Fontana: (**a**) crack developed from the upper end of the cut; (**b**) tear in the original canvas and "teletta"; (**c**) degradation of the pictorial material; (**d**) old restoration in the lower area of the cut, with presence of a textile patch, observed under UV light.

An old restoration intervention was visible on the lower side of the cut; in fact a patch, also black, had been placed over the "teletta".

A previous restoration intervention was also visible due to the presence of altered retouchings carried out in the upper left and right corners of the painting.

The front of the work was affected by an aesthetic deterioration caused during the drying of the material (Figure 1); yellowed areas could be seen in the thickest and most depressed spots due to accumulation of binder.

Some small micro-holes of the silver layer suggested the presence of an underlying black layer. This led to the supposition that the silver layer was only a surface finish, as reported by some of the texts examined [?]

The artist's cut, placed in the central part of the painting, showed slight deformation, and its edges fell toward the "teletta".

The work was placed inside a wooden frame with plexiglass protection.

Restoration was also performed on the basis of the scientific analyses. After the surface was dusted with soft brushes, the "teletta" was detached and repaired, the interruption of the support canvas was restored, the structure was reinforced from the back by strips of synthetic veil with Beva® Gel as an adhesive, and then the "teletta" was repositioned. Aesthetic intervention was limited to lightening the visual interruption points of the monochrome drafting in order to restore the continuity and correct reading of the work. Finally, the work was placed back in the frame, and on the back, a cardboard plume panel was added to protect the work from weather variations and atmospheric dust.

2.2. Reference Materials

Alkyd and poly(vinyl)acetate binders were purchased from Kremer Pigmente (Aichstetten, Germany) to acquire reference spectra.

2.3. Measurement Areas

The different spectroscopic techniques listed above were applied to several areas of the silver-coloured surface, including some points where surface inhomogeneity was present, such as areas with pictorial layer leaks or binder stains, and an area with evident restoration. The reverse of the work was also considered, focusing in particular on the canvas and on the silver-coloured dripping near the edges of the cut.

The measurement areas on which XRF, reflectance FTIR and Raman analyses were performed are listed in Table 1. OM was also performed on some measurement areas, as indicated in the same table.

Table 1. Areas of Lucio Fontana's monochrome silver painting on which in situ spectroscopic measurements were performed.

Measurement Area	Techniques	Description
1 *	reflectance FTIR, XRF, Raman	front side, silver-coloured paint
2 *	reflectance FTIR, XRF	front side, black spot (gap in the silver-coloured paint)
3 *	reflectance FTIR, XRF	front side, yellow stain
4 *	reflectance FTIR, XRF	front side, upper left corner, possible retouching
5 *	reflectance FTIR, XRF, Raman	back side, canvas
6	reflectance FTIR	back side, adhesive residue
7	reflectance FTIR	back side, adhesive residue
8 *	XRF	back side, silver paint dripping

* These measurement areas were also subjected to optical digital microscope (OM).

As mentioned in the Introduction, a small sample of silver-coloured paint taken from the back side of the painting near the edge of the cut (near Point 8) was also examined with FT-Raman and SEM-EDX.

2.4. Optical Microscopy

A portable digital optical microscope (DinoLite, 5Mpx) was used to get images with $50\times$ or $200\times$ magnification to support chemical analyses. The microscope was equipped with a polarising filter to reduce the gloss of the shiny surface.

2.5. X-ray Fluorescence (XRF)

XRF investigation was performed using an Assing LITHOS 3000 portable spectrometer, with quasi-monochromatic excitation at 17.4 keV (100 µm transmission Zr filter on a Mo target X-ray tube); the analysed area on the sample was about 4 mm radius. Working conditions were 25 kV and 300 µA, and the measuring time was 100 s (live time); the energy efficiency of the spectrometer is low for energies below 2.5 KeV. Sensitivity of the spectrometer is low for energy below 2.5 keV, not allowing the detection of light elements (from approximately $Z = 17$).

2.6. Reflection FTIR Spectroscopy

A Bruker Alpha FTIR spectrophotometer in reflection mode was used for non-invasive analyses. The instrument is equipped with a reflection module for contactless measurements and a deuterated triglycine sulphate (DTGS) detector, which operates at room temperature and guarantees a linear response in the spectral range between 7500 and 375 cm^{-1}. An integrated camera of approximately 6 mm diameter allows the operator to select the area to be measured. FTIR spectra were acquired with a resolution of 4 cm^{-1} as a sum of 100 scans after the acquisition of the background spectrum on a gold mirror. The reflection spectra in the mid-IR (MIR) region were processed by Kramers–Kronig transform using Bruker OPUS software.

2.7. Raman Spectroscopy

A Bruker BRAVO handheld spectrometer was used for Raman measurements. This instrument is based on patented SSE™ technology, which provides the excitation of spectra by means of two diode lasers operating at different temperatures and emitting, respectively, at 785 and 850 nm. An appropriate algorithm allows extraction of the final Raman spectral data. The spectra are collected in two sequential steps, from 300 cm^{-1} to 2000 cm^{-1} and from 2000 cm^{-1} to 3200 cm^{-1}. The average spectral resolution is approximately 11 cm^{-1}. The applied laser power is less than 100 mW for both lasers, and the beam is focused on an area of approximately 500 μm × 100 μm, so the power density is really limited. The acquisition time and the number of accumulations were automatically set by the instrument. In particular, the Raman spectrum on Area 1 was recorded with an exposure time of 13 s and 2 accumulations, while the spectrum on Area 5 was recorded with an exposure time of 5.4 s and 8 accumulations.

2.8. FT-Raman Spectroscopy

FT-Raman spectra were recorded between 4000 and 200 cm^{-1} directly on the sample taken from the back of the painting, without any preparation. A Jasco FT-Raman RFT-600 spectrometer was employed, using the 1064 nm emission of a Nd:YAG laser for excitation. The laser output power was about 180 mW. Spectra were acquired as the sum of 300 scans. The resolution was 4 cm^{-1}.

2.9. SEM-EDX Analysis

SEM-EDX analyses were performed on the same sample from the back of the painting using a Hitachi TM 1000 microscope with a resolution of 1 nm and equipped with an energy dispersion X-ray (EDX) spectrometer. The accelerating voltage was 15 kV.

3. Results and Discussion

Due to the synergy of the applied techniques, which exploited different penetration depths, and due to the presence of small gaps in the paint layers, it was possible to investigate the whole stratigraphy, composed of the preparation layer of the canvas, a dark layer and the silver-coloured paint.

3.1. Front of the Painting

On the front side of the artwork, the silver-coloured paint (Area 1) returned FTIR reflectance spectra with a characteristic slope towards lower wavenumbers due to the optical properties of the metallic surface (Figure 2a). In the same spectra, the only bands attributable to the binder were found around 2975, 2940 and 1740 cm^{-1} and did not allow its precise identification. At the same time, no relevant signals could be observed in the Raman spectrum.

In the small black areas observed in some empty spaces of the silver-coloured paint (Area 2), the presence of an alkyd component could be recognized, as demonstrated by the characteristic FTIR bands at 1280, 1123, 1068, 744 and 703 cm^{-1} (Figure 2b) [26], which are clearly evident in the reflection spectrum of an alkyd binder spread on canvas acquired as reference (Figure 2c). As its bands were detected in the gaps of the metallic paint, the alkyd binder is more reasonably associated with the underlying black paint layer.

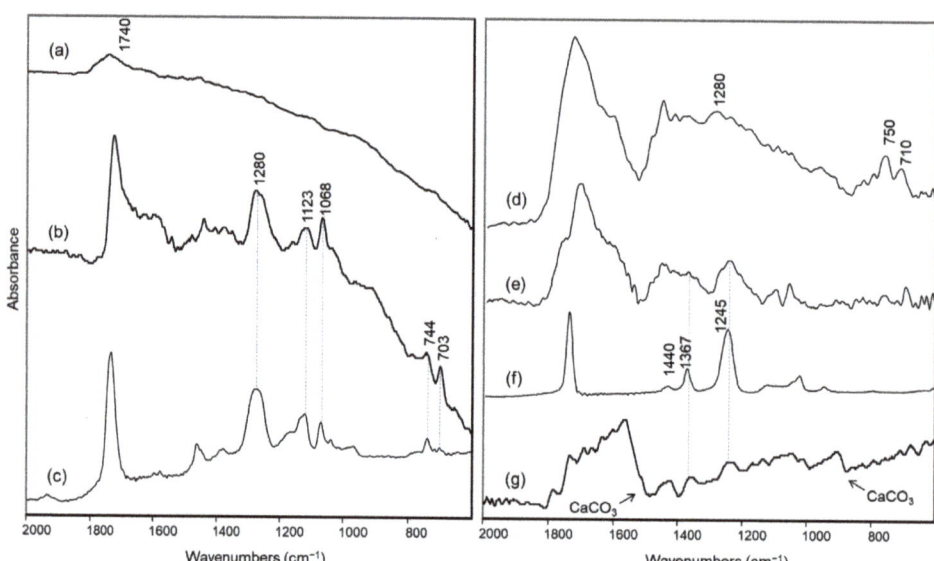

Figure 2. FTIR specular reflection spectra of areas on the front of the "Concetto spaziale": (**a**) silver paint (Area 1); (**b**) black spot in an empty space of the silver paint (Area 2); (**c**) reference alkyd resin spread on canvas; (**d**) yellowish spot corresponding to an accumulation of binder (Area 3); (**e**) differential spectrum between spectra (**b**,**d**); (**f**) reference polyvinyl acetate spread on canvas; (**g**) possible retouching in the upper left corner of the painting (Area 4). In all cases the spectra shown were obtained from the experimental reflectance data using the Kramers–Kronig transform, with the exception of spectrum (**d**), which was obtained from a transparent material superimposed on the metallic surface, and is thus shown as pseudo-absorbance, log(1/R). In spectrum (**g**), the bands due to calcium carbonate are distorted due to specular reflectance of a material with a higher particle size.

For this reason, to acquire possible information about the binder in the silver-coloured paint, a yellowish stain present on the front of the painting (Area 3) and presumably due to accumulation of binder was examined by reflection FTIR spectroscopy. Even if the spectrum obtained was not of excellent quality, it still allowed us to observe once again the main bands due to an alkyd resin (Figure 2d), with the doublet around 700 cm^{-1} slightly displaced, probably due to distortion associated with the reflection conditions. However, in the region of the spectrum from 1250 to 1000 cm^{-1}, different components seem to overlap. In order to possibly highlight these components, we decided to subtract the FTIR spectrum of the alkyd resin from that of the yellow stain. In the difference spectrum (Figure 2e), a band around 1245 cm^{-1} can now be clearly detected and, together with the signal around 1370 cm^{-1}, can be possibly assigned to polyvinyl acetate (PVA) [20,27], as also confirmed by the comparison with the reference spectrum acquired for this binder (Figure 2f).

If, as suggested by Gottschaller for other paintings with silver-coloured paint by the same artist [2], the silver-coloured layer is very thin and superimposed on a sort of underlying "bole", it can be hypothesized that the alkyd binder is due to the latter, while the vinyl binder is associated with the metallic paint. On the portion of silver-coloured paint in the micro-sample taken from the cut on the back of the work, it was in fact possible to acquire an FT-Raman spectrum with bands at 2939, 1734, 1438 and 632 cm^{-1} (Figure S1), attributable to polyvinyl acetate [27], but in that case it could also be due to the adhesive used to fix the "teletta" (see Section 3.2). Nevertheless, it should be emphasized that neither in the IR spectra nor in the Raman spectra obtained on the silver-coloured paint have ever been observed characteristic signals of an acrylic binder for this painting. In particular, the typical band of such a binder around 1170 cm^{-1} [19,20] is lacking in the IR spectra

(Figure 2a,d), and the Raman bands at about 800–840 cm^{-1} reported in the literature for an acrylic polymer [28,29] were also not observed in the FT-Raman spectrum of the silver paint in the micro-sample (Figure S1).

As for the pigments, considering the XRF results summarised in Table 2, the comparison between the spectra obtained on the silver-coloured area (Point 1) and the gap of paint do not show any difference in the detected elements, but the change in the Compton scattering peak intensity indicates a higher presence of light elements in the silver-coloured paint. The same applies to Point 3, the yellow binder stain. A difference is instead evident if comparing the spectra obtained on the silver-coloured area on the front with those acquired on the back of the painting: in the front, the presence of Ba and Co is highlighted. The former, typical of modern painting, is in the form of barium sulphate or lithopone (if mixed with zinc sulphide), often used as filler or extender due to its stability. Possibly the painter used it to disperse silver-coloured material (see also MO image in Figure 3). Co is instead usually linked to the use of blue pigments; we can thus speculate (also on the basis of the MO images in Figure 3) the presence of small quantities of cobalt blue, which could also be the matter of the dark priming used as a ground. Unfortunately, no further evidence could be obtained for this pigment from the non-invasively acquired vibrational spectra, and no samples from the front of the painting were available for microanalysis in the laboratory. Anyway, even if the presence of Co alone could indicate a possible use of smalt blue, with the painting being a contemporary one, this pigment can be correctly excluded. Indeed, modern cobalt acrylic paints have been recognized in other Fontana works by Gottschaller et al. [1] and Ferriani et al. [3]. In particular, the latter paper reports the use of cobalt blue (acrylic).

Table 2. Elements detected by XRF on the areas described in Table 1, with the exception of Area 8, which is extensively discussed in the text.

Measurement Area	Principal Elements	Trace Elements
1 *	Pb	Cu, Co, Ba (Zn)
2 *	Pb	Cu, Co, Ba
3 *	Pb	Cu, Co, Ba
4 *	Ca, Pb	Ti, Fe, Cu
5 *	Pb	

* These measurement areas were also subjected to optical digital microscope (OM).

Figure 3. (left) OM (200×) of the front of the painting: small metallic grains and blue pigment grain are visible; (right) OM (200×) of the restored area on the front of the painting: the different aspect of the silver area is evident, together with the green painting.

The material responsible for the metallic appearance of the paint could definitely be recognised as aluminium thanks to SEM-EDX analysis of the small fragment taken from the edge of the cut on the back of the painting (Figure 4). It is interesting to note that in the part of the fragment not covered by the silver paint, only lead could be observed, in agreement with the results of non-invasive XRF analysis.

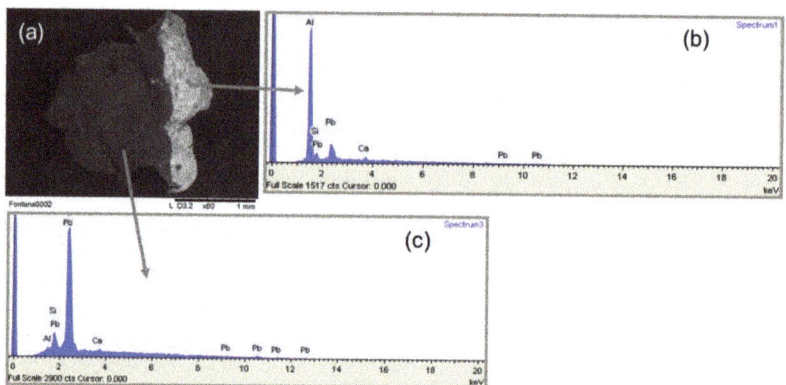

Figure 4. (**a**) SEM image of the sample taken from the silver dripping near the cut on the back of the "Concetto spaziale"; (**b**) EDX spectrum of the sample area with the highest mean Z value, corresponding to the silver paint; (**c**) EDX spectrum of the sample area with the lowest mean Z value, corresponding to preparation layer of the canvas.

Finally, possible retouching has been identified in the upper left corner of the painting. The FTIR spectrum obtained there clearly shows the presence of calcium carbonate, again with a vinyl binder (Figure 2g). XRF as well reveals a high presence of Ca, together with Ti and Zn, while the Ba signal completely disappears (Figure 5).

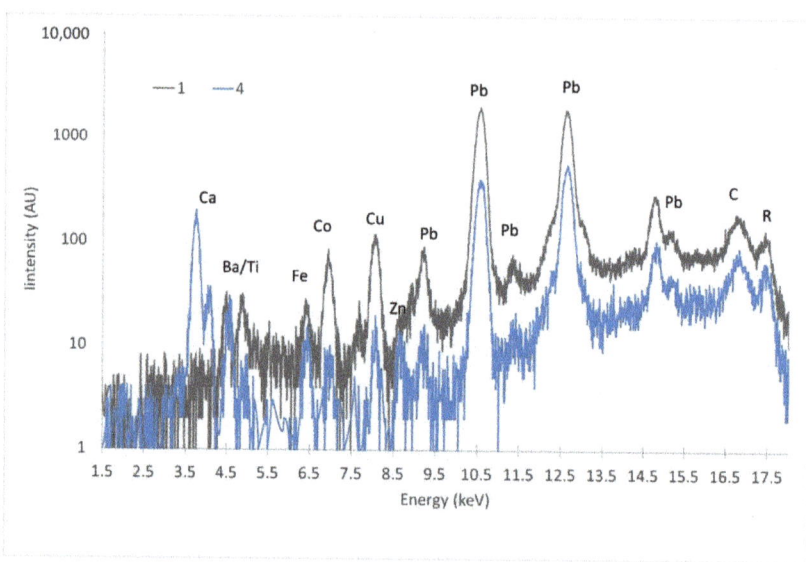

Figure 5. Comparison of XRF spectra in logarithmic scale for the original (1) and restored (4) areas on the front of the painting.

3.2. Back of the Painting

The back of the canvas, uniformly beige in colour, merely shows the presence of lead (from white lead, as identified by reflectance FTIR, see below), for which both L series (9–15 keV) and M series (2.3–2.4 keV) are detected in XRF spectra. We can thus claim that the light brown/yellow shade is not given by the presence of heart/ochre. The detection of M lines indicates that lead is also present in the surface layer of the back of the canvas; these signals disappear when investigating the silver-coloured dripping, as even its small thickness can absorb such low-energy signals. On the other hand, in the dripping spectrum, the Compton scattering peak is more intense, indicating a lower average Z of the material, and thus the presence of non-detectable elements, such as Al, S and/or Cl, besides organic compounds. In this same spectrum, low signals of iron and copper are also present, correlated to the silver-coloured pigment, mainly composed of light elements, as shown by SEM-EDX (see Section 3.1). Indeed, very low signals due to traces of barium and cobalt are also detectable in this spectrum.

Reflectance FTIR measurements were also performed on the back of the canvas (Area 5) and showed bands at 1420, 1044 and 683 cm^{-1} typical of lead white (Figure 6a). The Raman spectrum obtained on a corresponding area was also characterized by a sharp peak at 1050 cm^{-1} due to the same pigment (Figure S2). No traces of the components of so-called cementite, used in the preparation of the reverse of many of Fontana's paintings [1] and usually containing calcium carbonate, could be detected by such techniques. In the literature, the use of lead white is mentioned for the preparation layer of another "Concetto spaziale" (60 O 81) [3] belonging to the *Olii* series, but has not been reported yet as primer for the reverse of Fontana's painting. The only similarity with cementite-based preparations is the high density of the pigment [1], which is demonstrated by the fact that the bands due to lead white dominate the FTIR spectrum (also in the near IR region, data not shown), with the exception of a weak signal around 1730 cm^{-1}, making it difficult to identify the binder used. Anyway, the creamy colour observed for the layer is most likely due to the ageing of the binder itself.

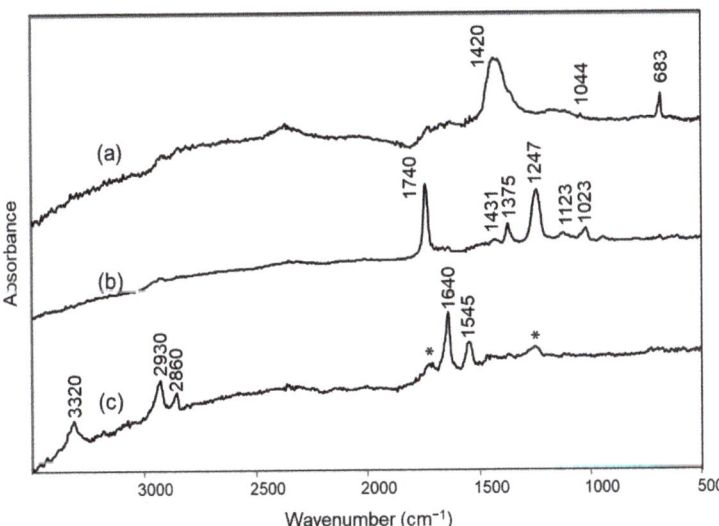

Figure 6. FTIR specular reflection spectra (after Kramers–Kronig transformation) of areas on the back of the "Concetto spaziale": (**a**) beige canvas (Area 5); (**b**) adhesive residue that gives yellowish fluorescence in UV light (Area 6); (**c**) adhesive residue which gives white fluorescence in UV light (Area 7). In spectrum (**c**), the bands marked with an asterisk (*) are probably due to residues of the PVA adhesive.

Still on the reverse of the painting, near the black gauze "teletta", two different adhesives could be identified based on FTIR reflectance spectra (Areas 6 and 7). The first one, which produced a yellow fluorescence under UV light, was, as expected, a PVA-based material, most likely Vinavil®, commonly used by the artist to glue the "teletta" on the reverse of the cuts [2]. It was easily identified thanks to the bands at 1740, 1431, 1375, 1247, 1123 and 1023 cm^{-1} [27] (Figure 6b). The second, which produced white emission when irradiated with UV light (Figure 1d), was instead recognized as a polyamide adhesive based on its bands at 3320, 2930, 2860, 1640 and 1545 cm^{-1} [30] (Figure 6c), and was presumably used in the previous restoration on the underside of the cut.

4. Conclusions

Lucio Fontana was an indefatigable experimenter who borrowed the materials he used in his masterpieces from other fields, even from industrial experimentation. This attitude results, nowadays, in restoration challenges that cannot be faced without the help of scientific investigations to give information about materials and techniques. The present work takes a key piece in the bare literature about the subject and confirms the fundamental importance of scientific investigations prior to restoration for modern and contemporary art where artists experimented with new materials to express a new function of art.

Indeed, the analytical data obtained made it possible to identify the composition of the metallic paint and of the underlying dark layer, both from the point of view of the pigments and of the binders used, also highlighting the potential of the non-invasive and micro-invasive methods. The coupling of techniques with different penetration depths and the presence of some small micro-lacunae of the silver-coloured layer let us understand the stratigraphy of the painting, where the silvered layer is only a surface finish, as reported in other cases [2]. Moreover, two different types of binder for the silver and black colours were detected, confirming this hypothesis. In particular, in the silver-coloured paint, aluminium could be recognized as a metal, and poly(vinyl)acetate was hypothesized as a binder, while for the underlying black paint layer, an alkyd resin could be identified as a binder, and the presence of cobalt suggested the use of cobalt blue as a pigment. Finally, on the back of the painting, two different adhesives were recognized, namely one based on PVA, commonly used by the artist to fix the "teletta", and a polyamide possibly employed in a previous restoration.

Results were thus used for the critical restoration performed, in particular for detachment of the "teletta" and for mending the tear. Assays and solubility tests were also performed based on the analytical results to decide which solvent to apply for the removal of the black canvas and the subsequent suturing of the interrupted threads. Acetone in a 10% and 20% thickened form in Nevek was used, with progressively longer setting times, in line with the vinyl nature of the adhesive used in laying the black cloth, as shown by the results of the scientific analyses. Thanks to this solution, it was then possible to detach the upper part of the canvas from the back of the cut by making the minimal amounts of solvent necessary for the reactivation of the glue, proceeding in parallel with a dry method for the mechanical removal of the adhesion at the points of greatest fragility of the fabric.

Supplementary Materials: The following supporting information can be downloaded at https://www.mdpi.com/article/10.3390/molecules27144442/s1. Figure S1: FT-Raman spectrum of the silver paint in the sample taken from the dripping on the back of the canvas; Figure S2: Raman spectrum from the back of the canvas.

Author Contributions: Conceptualization, S.B., L.B., C.B. and M.L.; methodology, S.B., L.B., C.B. and M.L.; formal analysis, L.B., S.B., C.B. and M.L.; investigation, L.B., S.B., M.L. and C.B.; resources, C.B., L.B. and S.B.; data curation, S.B., L.B., C.B. and M.L.; writing—original draft preparation, L.B., S.B. and C.B.; writing—review and editing, S.B., L.B., C.B. and M.L.; visualization, L.B., S.B. and C.B.; supervision, L.B., S.B. and C.B. All authors have read and agreed to the published version of the manuscript.

Funding: The acquisition of the portable Bruker Bravo Raman spectrometer was funded by the Fondazione Banca del Monte di Lombardia with granting of 22 October 2019.

Institutional Review Board Statement: Not applicable.

Informed Consent Statement: Not applicable.

Data Availability Statement: Not applicable.

Acknowledgments: The authors acknowledge Fondazione San Fedele for cooperation with this project.

Conflicts of Interest: The authors declare no conflict of interest.

Sample Availability: Not available.

References

1. Gottschaller, P.; Khandekar, N.; Lee, L.F.; Kirby, D.P. The evolution of Lucio Fontana's painting materials. *Stud. Conserv.* **2012**, *57*, 76–91. [CrossRef]
2. Gottschaller, P. *Lucio Fontana: The Artist's Materials*; The Getty Conservation Institute: Los Angeles, CA, USA, 2012.
3. Ferriani, B.; Barbero, L.M.; Izzo, F.C. 1949–1968. Concetti spaziali by Lucio Fontana: A historical-artistic and technical study. In *Science and Art: The Contemporary Painted Surface*; Sgamellotti, A., Ed.; The Royal Society of Chemistry: London, UK, 2020; pp. 39–66.
4. Ciccola, A.; Tozzi, L.; Romani, M.; Serafini, I.; Ripanti, F.; Curini, R.; Vitucci, F.; Cestelli Guidi, M.; Postorino, P. Lucio Fontana and the light: Spectroscopic analysis of the artist's collection at the National Gallery of Modern and Contemporary Art. *Spectrochim. Acta Part A Mol. Biomol. Spectrosc.* **2020**, *236*, 118519. [CrossRef] [PubMed]
5. Izzo, F.; Ferriani, B.; van den Berge, K.; van Keulen, H.; Zendria, E. 20th century artists' oil paints: The case of the Olii by Lucio Fontana. *J. Cult. Herit.* **2014**, *15*, 557–563. [CrossRef]
6. Chiantore, O.; Ploeger, R.; Poli, T.; Ferriani, B. Materials and techniques in the pictorial oeuvre of Lucio Fontana. *Stud. Conserv.* **2012**, *57*, 92–105. [CrossRef]
7. Bacci, M.; Casini, A.; Cucci, C.; Picollo, M.; Radicati, B.; Vervat, M. Non-invasive spectroscopic measurements on the Il ritratto della figliastra by Giovanni Fattori: Identification of pigments and colourimetric analysis. *J. Cult. Herit.* **2003**, *4*, 329–336. [CrossRef]
8. Anselmi, C.; Vagnini, M.; Cartechini, L.; Grazia, C.; Vivani, R.; Romani, A.; Rosi, F.; Sgamellotti, A.; Miliani, C. Molecular and structural characterization of some violet phosphate pigments for their non-invasive identification in modern paintings. *Spectrochim. Acta Part A Mol. Biomol. Spectrosc.* **2017**, *173*, 439–444. [CrossRef]
9. Rosi, F.; Miliani, C.; Clementi, C.; Kahrim, K.; Presciutti, F.; Vagnini, M.; Manuali, V.; Daveri, A.; Cartechini, L.; Brunetti, B.G.; et al. An integrated spectroscopic approach for the non-invasive study of modern art materials and techniques. *Appl. Phys. A* **2010**, *100*, 613–624. [CrossRef]
10. Ciccola, A.; Serafini, I.; Guiso, M.; Ripanti, F.; Domenici, F.; Sciubba, F.; Postorino, P.; Bianco, A. Spectroscopy for contemporary art: Discovering the effect of synthetic organic pigments on UVB degradation of acrylic binder. *Polym. Degrad. Stab.* **2019**, *159*, 224–228. [CrossRef]
11. Bonizzoni, L.; Bruni, S.; Gargano, M.; Guglielmi, V.; Zaffino, C.; Pezzotta, A.; Pilato, A.; Auricchio, T.; Delvaux, L.; Ludwig, N. Use of integrated non-invasive analyses for pigment characterization and indirect dating of old restorations on one Egyptian coffin of the XXI dynasty. *Microchem. J.* **2018**, *138*, 122–131. [CrossRef]
12. Bonizzoni, L.; Bruni, S.; Galli, A.; Gargano, M.; Guglielmi, V.; Ludwig, N.; Lodi, L.; Martini, M. Non-invasive in situ analytical techniques working in synergy: The application on graduals held in the Certosa di Pavia. *Microchem. J.* **2016**, *126*, 172–180. [CrossRef]
13. Pessanha, S.; Guilherme, A.; Carvalho, M.L. Comparison of matrix effects on portable and stationary XRF spectrometers for cultural heritage samples. *Appl. Phys. A* **2009**, *97*, 497–505. [CrossRef]
14. Fermo, P.; Andreoli, M.; Bonizzoni, L.; Fantauzzi, M.; Giubertoni, G.; Ludwig, N.; Rossi, A. Characterisation of Roman and Byzantine glasses from the surroundings of Thugga (Tunisia): Raw materials and colours. *Microchem. J.* **2016**, *129*, 5–15. [CrossRef]
15. Galli, A.; Caccia, M.; Alberti, R.; Bonizzoni, L.; Aresi, N.; Frizzi, T.; Bombelli, L.; Gironda, M.; Martini, M. Discovering the material palette of the artist: A µ-XRF stratigraphic study of the Giotto panel 'God the Father with Angels'. *X-Ray Spectrom.* **2017**, *46*, 435–441. [CrossRef]
16. Bonizzoni, L.; Canevari, C.; Galli, A.; Gargano, M.; Ludwig, N.; Malagodi, M.; Rovetta, T. A multidisciplinary materials characterization of a Joannes Marcus viol (16th century). *Herit. Sci.* **2014**, *2*, 15. [CrossRef]
17. Martins, A.; Coddington, J.; van der Snickt, G.; van Drie, B.; McGlinchey, C.; Dahlberg, D.; Janssens, K.; Dik, J. Jackson Pollock's Number 1A, 1948: A non-invasive study using macro-x-ray fluorescence mapping (MA-XRF) and multivariate curve resolution-alternating least squares (MCR-ALS) analysis. *Herit. Sci.* **2016**, *4*, 33–56. [CrossRef]
18. Bonizzoni, L.; Colombo, C.; Ferrati, S.; Gargano, M.; Greco, M.; Ludwig, N.; Realini, M. A critical analysis of the application of EDXRF spectrometry on complex stratigraphies. *X-Ray Spectrom.* **2011**, *40*, 247–253. [CrossRef]

19. Rosi, F.; Daveri, A.; Moretti, P.; Brunetti, B.G.; Miliani, C. Interpretation of mid and near-infrared reflection properties of synthetic polymer paints for the non-invasive assessment of binding media in twentieth-century pictorial artworks. *Microchem. J.* **2016**, *124*, 898–908. [CrossRef]
20. Doménech-Carbó, M.T.; Doménech-Carbó, A.; Gimeno-Adelantado, J.V.; Bosch-Reig, F. Identification of Synthetic Resins Used in Works of Art by Fourier Transform Infrared Spectroscopy. *Appl. Spectrosc.* **2001**, *55*, 1590–1602. [CrossRef]
21. Zaffino, C.; Guglielmi, V.; Faraone, S.; Vinaccia, A.; Bruni, S. Exploiting external reflection FTIR spectroscopy for the in-situ identification of pigments and binders in illuminated manuscripts. Brochantite and posnjakite as a case study. *Spectrochim. Acta Part A Mol. Biomol. Spectrosc.* **2015**, *136B*, 1076–1085. [CrossRef]
22. Bell, I.M.; Clark, R.J.H.; Gibbs, P.J. Raman spectroscopic library of natural and synthetic pigments (pre- ≈1850 AD). *Spectrochim. Acta Part A Mol. Biomol. Spectrosc.* **1997**, *53*, 2159–2179. [CrossRef]
23. Scherrer, N.C.; Zumbuehl, S.; Delavy, F.; Fritsch, A.; Kuehnen, R. Synthetic organic pigments of the 20th and 21st century relevant to artist's paints: Raman spectra reference collection. *Spectrochim. Acta A* **2009**, *73*, 505–524. [CrossRef]
24. Fremout, W.; Saverwyns, S. Identification of synthetic organic pigments: The role of a comprehensive digital Raman spectral library. *J. Raman Spectrosc.* **2012**, *43*, 1536–1544. [CrossRef]
25. Pätzold, R.; Keuntje, M.; Anders-von Ahlften, A. A new approach to non-destructive analysis of biofilms by confocal Raman microscopy. *Anal. Bioanal. Chem.* **2006**, *386*, 286–292. [CrossRef]
26. Dredge, P. Sidney Nolan's adventure in paint—An analytical study of the artist's use of commercial paints in the 1940s and '50s. *AICCM Bull.* **2014**, *34*, 15–23. [CrossRef]
27. Renuka Devi, K.B.; Madivanane, R. Normal coordinate analysis of poly vinyl acetate. *Eng. Sci. Technol. Int. J.* **2012**, *2*, 795–799.
28. Ellis, G.; Claybourn, M.; Richards, S.E. The application of Fourier Transform Raman spectroscopy to the study of paint systems. *Spectrochim. Acta Part A Mol. Biomol. Spectrosc.* **1990**, *46*, 221–241. [CrossRef]
29. Pagnin, L. Characterization and Quantification of Modern Painting Materials by IR and Raman Spectroscopies. Master's Thesis, Università Ca' Foscari, Venezia, Italy, 2017.
30. Chércoles Asensio, R.; San Andrés Moya, M.; de la Roja, J.M.; Gómez, M. Analytical characterization of polymers used in conservation and restoration by ATR-FTIR spectroscopy. *Anal. Bioanal. Chem.* **2009**, *395*, 2081–2096. [CrossRef]

Review

Artefact Profiling: Panomics Approaches for Understanding the Materiality of Written Artefacts

Marina Creydt [1,2,*] and Markus Fischer [1,2]

[1] Institute of Food Chemistry, Hamburg School of Food Science, University of Hamburg, Grindelallee 117, 20146 Hamburg, Germany; markus.fischer@uni-hamburg.de

[2] Cluster of Excellence, Understanding Written Artefacts, University of Hamburg, Warburgstraße 26, 20354 Hamburg, Germany

[*] Correspondence: marina.creydt@uni-hamburg.de; Tel.: +49-40-42838-8803; Fax: +49-40-42838-4342

Abstract: This review explains the strategies behind genomics, proteomics, metabolomics, metallomics and isotopolomics approaches and their applicability to written artefacts. The respective sub-chapters give an insight into the analytical procedure and the conclusions drawn from such analyses. A distinction is made between information that can be obtained from the materials used in the respective manuscript and meta-information that cannot be obtained from the manuscript itself, but from residues of organisms such as bacteria or the authors and readers. In addition, various sampling techniques are discussed in particular, which pose a special challenge in manuscripts. The focus is on high-resolution, non-targeted strategies that can be used to extract the maximum amount of information about ancient objects. The combination of the various omics disciplines (panomics) especially offers potential added value in terms of the best possible interpretations of the data received. The information obtained can be used to understand the production of ancient artefacts, to gain impressions of former living conditions, to prove their authenticity, to assess whether there is a toxic hazard in handling the manuscripts, and to be able to determine appropriate measures for their conservation and restoration.

Keywords: ancient artefacts; written manuscripts; omics strategies; panomics; artefact profiling; archaeometry

Citation: Creydt, M.; Fischer, M. Artefact Profiling: Panomics Approaches for Understanding the Materiality of Written Artefacts. *Molecules* **2023**, *28*, 4872. https://doi.org/10.3390/molecules28124872

Academic Editor: Maria Luisa Astolfi

Received: 30 May 2023
Revised: 15 June 2023
Accepted: 18 June 2023
Published: 20 June 2023

Copyright: © 2023 by the authors. Licensee MDPI, Basel, Switzerland. This article is an open access article distributed under the terms and conditions of the Creative Commons Attribution (CC BY) license (https:// creativecommons.org/licenses/by/ 4.0/).

1. Introduction

In recent years, it has increasingly become apparent that cooperation between the humanities and natural sciences represents a particularly valuable symbiosis for the preservation of the common cultural heritage [1]. In this context, the so-called omics disciplines for the chemical analysis of historical artefacts have received enormous attention. However, its application to ancient objects is still in its infancy. Nevertheless, in manuscript research, the term "biocodicology" has taken on special significance, introduced only a few years ago by Fiddyment et al., referring to the study of biological information in a manuscript using genomics and proteomics strategies [2]. We propose to extend this approach to include further omics disciplines such as metabolomics and metallomics as well as isotopolomics, as these are also suitable for chemical analyses of ancient manuscripts and provide additional information.

While omics disciplines have been applied for several years in medicine, plant breeding, or food quality control, the application to ancient objects is relatively new [3,4]. Occasionally, the term paleo-omics is also used in this context [5]. Paleo-omics strategies have so far been applied mainly to understand biodeterioration processes and, if necessary, to adopt appropriate strategies for conservation. Therefore, the focus has been predominantly on the analysis of microorganisms. Thus, these are actually meta-omics methods since it is not the objects themselves that are detected, but residues from other organisms [6–11]. Nevertheless, the analysis of ancient materials using omics strategies has also increased in recent years. In the case of manuscripts, the focus is in particular on the writing surfaces and the inks, dyes, and pigments [2,12,13].

In general, the suffix -omics refers to comparative chemical analyses with which a complete or almost complete cellular level, tissue, organ, or organism can be recorded. However, in this regard, it should be noted that not all elements, molecules, or sequences can usually be fully detected. The classic omics disciplines include genomics (DNA), transcriptomics (RNA), proteomics (peptides and proteins), and metabolomics (metabolites), which together describe the flow of information from genotype to phenotype (Figure 1). For this reason, this context is often referred to as the omics cascade or as the central dogma of biology. However, RNA is mostly not very stable to environmental influences, which is why it is not usually used for the analysis of ancient artefacts. Nevertheless, sometimes ancient RNA can be detected if the right environment conditions are present [14]. In addition, it should be noted in metabolomics analyses that not only are metabolites often detected, but also other small organic compounds (<1500 Da) that are not naturally formed but synthesised by humans, the so-called xenobiotics. These can be, for example, dyes, pesticides, drugs, or cosmetics [15].

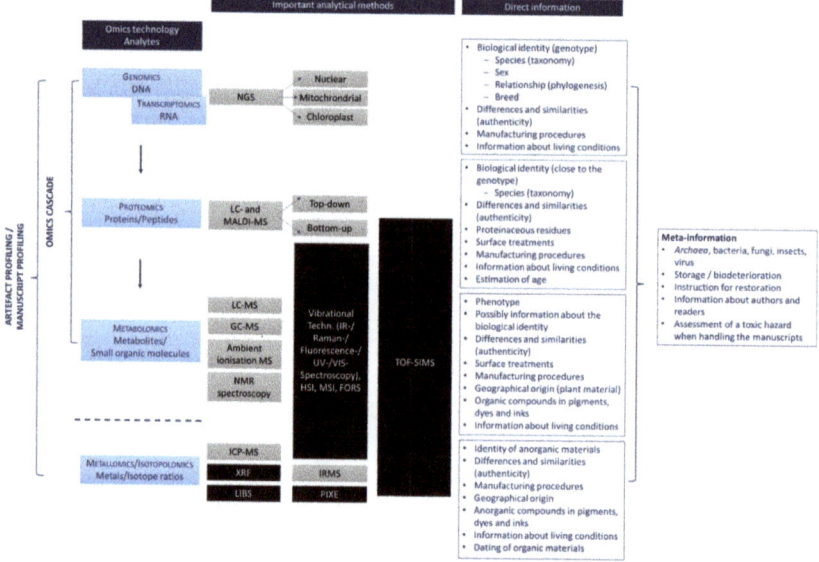

Figure 1. Omics disciplines that can be applied to the analysis of ancient artifacts including the analysis methods commonly used. High-resolution platforms typically applied for omics analysis are colored grey. Low-resolution methods are coloured black. They are also commonly applied for analysing ancient artefacts, but they do not offer the resolution and sensitivity required for omics approaches. In addition, it is shown for which scientific questions the different strategies can be used. While the direct information can be taken from the writing surfaces, inks, dyes, or pigments, meta-information results from residues from other organisms. Abbreviations: GC-MS, gas chromatography mass spectrometry; FORS, Fiber Optics Reflectance Spectroscopy; HIS, hyperspectral imaging; ICP-MS, inductively coupled plasma mass spectrometry; IR, infrared; IRMS, isotope ratio mass spectrometry; LC-MS, liquid chromatography mass spectrometry; LIBS, laser-induced breakdown spectroscopy; MALDI-MS, matrix-assisted laser desorption/ionisation mass spectrometry; MSI, multispectral imaging; NGS, next generation sequencing; NMR, nuclear magnetic resonance; PIXE, particle-induced X-ray emission; TOF-SIMS; time-of-flight secondary ion mass spectrometry; UV, ultraviolet; VIS, visible; XRF, X-ray fluorescence.

The terms genome, transcriptome, proteome, and metabolome refer to the entirety of the analytes with regard to the respective omics discipline. Based on these terminologies, other omics approaches have emerged such as metallomics (metals), isotopolomics (isotope ratios), or microbiomics (community of microorganisms. Furthermore, omics strategies are carried out without prior identification of the detected compounds, which is why they are also called non-targeted and are necessarily based on comparison with reference data. The use of non-targeted omics strategies requires high-resolution and sensitive analytical technologies to capture as many analytes as possible and to increase the information content accordingly. Since very large data sets are inevitably generated using omics approaches, bioinformatic and chemometric methods must be used for data evaluation. Bioinformatic approaches primarily include sequence analysis and are used in genomics and proteomics datasets. While chemometric methods are mainly used for pattern recognition using multivariate methods. They are suitable for the evaluation of metabolomics and metallomics data sets, but can also be used for the evaluation of genomics and proteomics data in order to visualize relationships [16].

Data evaluation should not be underestimated and, depending on the selected omics approach and the data structure, can involve a great deal of time and effort. These requirements are accompanied by an appropriate infrastructure and trained staff, which can often only be guaranteed in specially equipped laboratories. The non-targeted strategies are in contrast to the so-called targeted methods, in which analytes are known a priori and are usually also absolutely quantified. This achieves maximum comparability of the data sets but is accompanied by a loss of information. The advantage is that targeted strategies usually require less complex technological equipment. However, since no comprehensive analysis is carried out, these are not classic omics disciplines [3,13,17,18].

The analysis of the genome, proteome and metabolome is particularly suitable for organic materials such as writing surfaces made of papyrus, paper, parchment, leather, plant leaves, wood and wood bark, bamboo, silk, tapa, amate, textiles, and wax as well as inks, pigments, or dyes containing organic components. In addition to biological identity and taxonomy, further research questions may concern the production process or conservation methods, e.g., palm leaves manuscripts are protected with various plant extracts, which can be easily identified by mass spectrometric metabolomics methods [19]. In addition, such studies can be used to record similarities and differences, for example, to determine whether certain fragments or pages belong together, or to assign geographical origins [20–22]. In particular, the meta-information obtained with these techniques can provide indications of bacterial or fungal infestation of manuscripts, which can be helpful to stop biodeterioration processes and also to protect the people who handle the analysis material [23–25]. In addition, they are also suitable for obtaining information about the authors or the readers [26–32]. The analysis of certain stable isotopes using isotopolomics approaches can provide further information. The ratio of ^{14}C and ^{12}C, which can be applied to date artefacts, is of particular importance. It should be noted, however, that this method can only date the material used, not when a manuscript was actually written by an author [33,34]. Metallomics strategies including certain isotope ratios of metals are mainly suitable for analysing inorganic writing materials such as metals, stones, ceramics, clay tablets, glass, or bones [35]. They are also of great relevance for the analysis of inorganic components in pigments, dyes, and inks [18,36].

By combining the various omics levels, a comprehensive elementary and molecular profile can be created about an object and a maximum depth of information can be generated. This procedure is also called panomics or multi-omics profiling and is used, for example, in medicine or in the analysis of food (food profiling). In the context of manuscript analyses, we propose the terms "manuscript profiling" and "artefact profiling" to cover not only manuscripts but also other objects that play an important role in the context of cultural heritage such as images, sculptures and other ancient manmade objects items [3,37,38]. The analysis of ancient artefacts and manuscripts using omics strategies is still a relatively young field of research, and the terminology is evolving accordingly. Table 1 lists some of the most important terms.

Table 1. Glossary of key terms.

Term	Explanation
Archaeometry	Application of physics, chemistry, geology, and engineering sciences to analyse various archaeological materials [39].
Biocodicology	Application of genomics and proteomics approaches to ancient manuscripts [2].
Codicology	Analysis of the technical craft aspects and physical properties of a manuscript [40,41].
Omics/Paleo-omics	Comparative analysis of different elemental or molecular entireties [3]. In the context of studies on ancient manuscripts or artefacts, the focus is on the analysis of the actual object, e.g., the writing surfaces or the inks used. The prefix "paleo" means old or ancient [5,42].
Meta-omics	The focus of meta-omics studies is not the object itself, but the residues of other organism, e.g., microorganisms or the authors and readers [11]. Depending on the omics strategy chosen, the terms meta-genomics, meta-proteomics or meta-metabolomics are used, for example.
Paleography	The term refers to the study of ancient writings to be able to make spatial or temporal classifications. The focus is, for example, on the forms of the letters, spelling, or the use of typical abbreviations [40,41].
Panomics	The prefix "pan" comes from the Greek and means all, every, or whole [43]. The term panomics is used for personalised medical questions or when analyzing food. It is a networked, symbiotic approach based on genomics, transcriptomics, proteomics, and metabolomics-based data [38,44].

In addition to the analytical methods explained in this publication, there are numerous archaeometric approaches to deciphering information about ancient manuscripts. This also includes, e.g., physical methods such as computed tomography as well as various microscopic methods, which, however, do not fall within the scope of panomics applications and are therefore not considered in this review. For the same reason, we have not gone into detail about low-resolution technologies either, as they typically capture only a fraction of the information compared to the high-resolution technological platforms used for typically omics investigations, although the boundaries can certainly not be drawn clearly.

2. Sampling Strategies

A challenge with all higher resolution archaeometric approaches is the sampling. Measurements directly on the object (in situ) offer an alternative, but these usually operate in a lower resolution range. In this context, a total of three approaches are being pursued as follows: non-invasive, minimally invasive, and invasive. (i) The gentlest methods are non-invasive approaches that do not require sampling or contact between the instrument and the object. However, even with such procedures it must be considered that long-term damage to the artefacts can occur; for example, if the objects are exposed to ultraviolet (UV) light for a long time, the ageing processes can be accelerated. (ii) With minimally invasive methods, sampling or contact with the measuring instrument is not macroscopically visible. In contrast, invasive approaches require either close contact with the object or sampling that is macroscopically visible. In addition, these procedures can be categorised into non-destructive and destructive methods. Non-destructive methods require that part of the object is removed but not destroyed and made available for further analysis. This is the case, for example, with fibre analyses. Using destructive approaches, on the other hand, the sample material is completely destroyed when, for example, the mass spectrometric measurements are carried out [18]. In general, the greatest information density can be generated with invasive methods precisely because they can be used not only to detect analytes on the surface, but also those elements, molecules, or sequences in the deeper layers. In addition, there is a larger quantity of analytes obtained in this way, so that the sensitivity of the analytical methods increases accordingly. However, since the greatest damage can also be caused with invasive approaches, a careful cost-benefit analysis is required to determine which extraction method is best suited. An alternative may be to use small pieces that naturally fall off old objects for analysis. In most cases, about 2–5 mm^2, respectively a few milligrams, are already sufficient to perform the corresponding analyses [19,45–49].

2.1. Sampling Strategies for High Resolution Omics Approaches That Can Be Performed on Site

Most of the high-resolution analytical methods listed in Figure 1 typically require minimally invasive or invasive sampling. Furthermore, in most cases the sample is also destroyed or must be prepared specifically for the measurement, so that no further measurements are possible with this part. Since sampling is one of the most challenging aspects of working with ancient objects, much research has been performed in this area in recent years and alternative sampling techniques have been developed. In addition to the classic use of adhesive tapes, scalpels to scrape the surface, sponges, and swabs, a micro-aspiration technique has recently been introduced that allows molecules to be picked up from the surface with a simple vacuum [45,50–54]. A more destructive alternative is the simple and inexpensive polyvinyl chloride (PVC) rubber method described by Fiddyment et al. [2,55]. Promising results were also obtained using a film of ethyl vinyl acetate (EVA) in which C8 and/or C18 resins as well as cation/anion exchangers or metal chelators were fused. Before use, the film is moistened and placed on the manuscript so that the corresponding analytes are adsorbed. About 1–2 cm^2 of the surface of the samples are covered with the film [27,56,57]. Some research groups have also had good experience with hydrophilic nanogels, which can be enriched with a variety of solvents, as well as enzymes such as trypsin (see Section 4). The area that the authors of this study covered with the gel was only 3 mm^2 [58]. Another approach is based on the use of the fungal protein hydrophobin Vmh2, which can be obtained from the edible fungi *Pleurotus ostreatus*. The protein is placed on a cellulose loose acetate surface covering an area of 2 cm^2. Hydrophobins are characterised by their surface-active and amphiphilic properties. However, they are currently still expensive as they are not produced on a large scale [59,60]. Further experiments were carried out with a polishing film that was applied to polystyrene rods with a diameter of a few millimeters. The film was carefully rubbed over the samples. During this process, the analytes were absorbed by the film and could then be analysed [61]. Moreover, it may already be sufficient to take up analytes from the surface of a manuscript with small volumes of solvent and then to analyse the solution obtained [62].

For the analysis of small organic molecules and metabolites by mass spectrometry (MS), the atmospheric solids analysis probe (ASAP) procedure can be used. For this approach, a glass capillary probe is rubbed over the surface of a manuscript and the adhering analytes are then introduced into a mass spectrometer via an atmospheric pressure chemical ionisation (APCI) source. The glass capillary probes can also be shipped, which is why this procedure can be carried out on site, e.g., in libraries or private collections. By using a APCI source, analytes <2000 Da can be detected [63,64]. In probe electrospray ionisation (PESI) technology, a similar approach is taken, but it can be assumed that larger molecules are still detectable since an electrospray is generated. The difference is that instead of a glass capillary, a metal needle (similar to an acupuncture needle) is used for sampling. By moistening and by applying a high voltage to the needle, an electrospray is generated, which leads to the ionisation of the analytes, and which can then be analysed using a mass spectrometer. According to the best of our knowledge, this approach has not yet been carried out on ancient objects, but on various biological samples [64–66]. However, it could still be a suitable alternative in the future. Further possibilities for on-site sampling are provided by special laser systems, which are discussed in Section 2.2, due to the technological requirements.

Not all the options listed here have been tested on manuscripts at this point, though some have also been tried out on paintings, for example, which is why it is currently not possible to make a conclusive assessment as to which area sizes need to be sampled. In addition, such considerations must also take into account the material used, the state of preservation of the manuscript, and the aim of the study in order to decide which sampling method is the most appropriate in each case.

Although all these methods are nearly non-invasive or minimally invasive, curators often have reservations about allowing sampling because some of these methods require the surface of the manuscript to be moistened. However, these methods also have the advantage that sampling can be carried out quickly and easily as well as on site so that the historical documents or measuring equipment do not have to be transported at great expense. In addition, immovable objects such as gravestones or inscriptions on walls can also be sampled. Although this type of sampling is normally invisible to the naked eye, slight colour changes can still occur, which should be carefully checked in advance.

2.2. Sampling Strategies for High Resolution Omics Approaches That Require No Sample Preparation but Must Be Performed in a Laboratory Environment

In contrast to the on-site strategies described, there are micro-invasive alternatives that require a laboratory infrastructure. The advantage of these techniques is that they enable spatial resolution and require no sample preparation. These include, in particular, ambient ion sources for coupling to mass spectrometers such as direct analysis in real time (DART), desorption electrospray ionisation (DESI) and nanospray desorption electrospray ionisation (nano-DESI). Those designs are particularly suitable for the analysis of metabolites and small organic molecules in a mass range <2000 Da since the ionisation energy is not sufficient for larger molecules [64].

For example, minimally invasive DART-MS analyses were performed on parchment to identify potential conservation treatments. In the context of old manuscripts, it became clear that it is important not to set the temperature of the DART carrier gas too high in order to avoid damage such as burn marks on the manuscripts [67]. Further, analyses using DART-MS were also carried out to differentiate between various types of paper, although the paper was not sampled directly, but rather small pieces that had previously been separated with tweezers [68]. In addition, there is a study on different inks using DART-MS. It turned out that inks change very significantly after the first application since volatile compounds evaporate before a stable state is established [69]. However, all these studies were not carried out on very old material, which is why a conclusive evaluation of this strategy in the context of ancient manuscripts is not possible at this stage. Further research in this area is needed, although DART sources are currently not very widespread.

While DART-MS designs use plasma for ionisation, the ionisation of DESI-MS or nano-DESI-MS couplings is based on an electrospray to which the surface of a manuscript is exposed. In contrast to DESI, a second capillary is used with nano-DESI to transport the analytes to the entrance of the MS. In this way, the efficiency is improved and more stable mass spectrometric signals are obtained, which is particularly helpful when analysing old artefacts since the analytes are present in comparatively low concentrations [64]. DESI-MS approaches have been applied to ancient artefacts to detect peptides in historical flint plates and pottery shards, and to analyse inks [70–73]. However, as with the DART-MS studies, the focus has so far been on more recent samples. Nevertheless, such studies can provide indications of the extent to which a transfer to historical manuscripts is possible and what damage can occur.

In secondary ion mass spectrometry (SIMS) technology, the surface of an object is bombarded with primary ions, and the actual analytes are released as secondary ions for further analysis by an MS analyser. SIMS must be performed in ultra-high vacuum, i.e., the sample must be vacuum stable and the samples must not be too large and thick, depending on the design of the equipment used. The advantage is that both organic and inorganic components can be detected and that a depth profile can be created. A further development of SIMS technology are nano-SIMS applications, which are characterised by a particularly high spatial resolution. SIMS platforms have been widely used for the study of ancient paintings [74]. In addition, other materials such as glass or bronze were also analysed [75]. Studies carried out on ancient manuscripts using SIMS do not yet exist. However, there have been numerous studies on the analysis of inks and also one on parchment, which have so far focused mainly on recent samples, but which can serve as a starting point for further developments [76–81].

In addition, the detection of inorganic elements by means of laser ablation inductively coupled plasma mass spectrometry (LA-ICP-MS) is relatively well established for the analysis of ancient artefacts. The ablation site is not visible to the naked eye. Small spots or grooves can only be seen under a microscope. An LA-ICP-MS instrument usually consists of an ablation cell in which the object is located and can be moved in x-, y-, and z- axes. Using a focused laser beam, the analytes on the surface of the sample are transferred into the gas phase, which is transported to an ICP-MS system by means of a carrier gas (Ar or He). Different laser types (solid state, e.g., Nd:Yag or gas eximer, e.g., ArF) with different wavelengths and pulse durations can be used for the ablation process, which on the one hand have an impact on the damage to the object and on the other hand on the signal intensities. Especially when analysing precious ancient artefacts, lasers with shorter wavelengths are preferable since the damage is the least (no frayed spots appear) and the ablation is efficient and fast [82–84]. However, the size of the artefact is limited by the ablation chamber in the currently commercially available designs. Nevertheless, there are alternative options to overcome this disadvantage. One possibility is open and moveable ablation cells or the complete omission of a cell, i.e., the laser is used under ambient conditions. In order to ensure the exchange of air for argon, diaphragm pumps or special designs of gas exchange instruments are applied in such constructions [82,85,86]. LA-ICP-MS instruments for the analysis of historical manuscripts have in particular been used thus far to require the chemical composition of iron gall inks since the metals obtained in these colours can be catalytically active and can accelerate the degradation processes of the historical documents (see Section 6.2) [36,87].

Based on the laser systems described, portable ablation systems were also developed with which sampling can be carried out directly on site, e.g., for sampling wall inscriptions or gravestones that cannot be transported to a laboratory. Consequently, such systems are more likely to be assigned to the methods from Section 2.1. The technological similarities have been mentioned in this section. The challenge with portable systems is on the one hand the size as well as the weight, and on the other hand the technical requirements such as efficient cooling of the laser unit. Diode-pumped solid-state (DPSS) lasers with a wavelength of 532 nm and a pulse rate of <1 nm, which can be operated with air cooling, are suitable for meeting these requirements. Since these lasers can only generate comparatively long wavelengths above the UV range, highly transparent materials cannot be analysed. Nevertheless, these designs can be used to generate appropriate aerosols from suitable objects, which can be collected on filters using a membrane pump and then analysed using, for example, ICP-MS instruments. However, compared to laser-induced breakdown spectroscopy (LIBS), which is quite similar in terms of sampling technique, and X-ray fluorescence (XRF) measurements, the limits of detection achieved using this approach are in the lower ranges (see Section 6.1) [88–90]. In the meantime, there are some successful application examples of this portable laser system which have been carried out on antique objects such as archaeological silver, antique glass beads, or various gold objects [91–93].

So far, sampling techniques from historical artefacts based on laser ablation methods are mainly used for the analysis of inorganic analytes. Nevertheless, there are also ways of removing organic molecules from surfaces gently and as non-destructively as possible using laser processes. Both UV and infrared (IR) lasers are suitable for this demand whereby the water content can be relevant and with which method better results can be obtained since the IR laser initially triggers a sudden evaporation of water molecules, taking the analytes with them. The analytes can be ionised directly from the resulting aerosol, e.g., by means of an electrospray, and then passed into an MS instrument. Alternatively, the ablation cloud can be collected on site with the help of membrane pumps or as condensate on slides and later analysed in a laboratory. Depending on the type of laser used, these methods are called laser ablation electrospray ionisation (LAESI) if an IR laser is used, or electrospray laser desorption ionisation (ELDI) if a UV laser is used. In the past, both methods could be applied to detect both small polar molecules and larger peptides, as well as proteins in biological materials [94–96]. Intact DNA molecules could also be removed from organic materials using an IR laser [97]. At the present time, it is difficult to estimate whether a transfer of such a technical strategy to old materials is possible and how high the information content is in comparison to the other methods presented since there is currently no experience in this regard. Furthermore, it is currently not known how destructive such processes are for writing materials. However, in recent years, lasers have been increasingly used to clean paintings. It was also shown that the same type of laser (UV excimer KrF laser) applied for cleaning processes can also be applied for laser-induced fluorescence (LIF) measurements. Possibly, such developments can also be suitable for omics investigation of ancient manuscripts [98].

3. Paleogenomics and Metagenomics Analyses

Genome analyses, which focus on examining DNA sequences, are particularly suitable for determining the biological identity (genotyping) of plant and animal writing surfaces by sequencing the extracted DNA. In addition to the taxonomic classification, this also includes the determination of sex, breed, and degree of relationship. Furthermore, the analysis of exogenous DNA sequences allows gaining meta-information about the microbiome. Not only is nuclear DNA suitable for such investigations, but also DNA from mitochondria and chloroplasts. The latter two DNA types have the advantage of being available in higher copy numbers, which can facilitate the analysis. In addition, they are in most cases only inherited from the mother's side, which can often make data interpretation easier. The investigation of ancient DNA (aDNA) is particularly challenging since the sequences are highly degraded (<100 base pairs) depending on the environment conditions and their age. Both endogenous and exogenous nucleases are responsible for the degradation processes. There are also environmental influences such as temperature, humidity, pH value, and UV radiation. In addition, the DNA of the writing surface can already be largely destroyed during the production process. This is the case, for example, with paper, which undergoes various chemical and physical processing steps during its manufacture, which is why the DNA can no longer be detected in most cases. Conversely, certain conditions such as low temperatures and a dry environment can have a strong preservative effect [2,99–102].

First, DNA fragmentation begins with the hydrolysis of the bases, especially the purine bases, whereby the bases are cleaved from the sugar-phosphate backbone of the DNA double strand (depurination). As a result, single-strand breaks occur in the backbone of DNA sequence and the DNA double helix breaks into smaller pieces. In addition, deamination reactions take place in which cytosine is converted into uracil and 5-methylcytosine into thymine. Furthermore, guanine can be deaminated to xanthine and adenine to hypoxanthine. Due to the conversion of cytosine to uracil, thymine is incorporated instead of guanine during the generation of the complementary counterstrand when amplifying DNA by PCR prior to sequence analysis. Thus, such reactions have the consequence that in sequence analyses, erroneous information can be generated [101,103]. However, since this reaction occurs particularly frequently in the single-stranded overhangs of the fragmented aDNA pieces, the relatively high thymine concentration can also be used to check the authenticity of aDNA and distinguish recent DNA from aDNA [102,104,105]. Further, aDNA is characterised by intra-molecular and inter-molecular cross-links. These are formed by bonds of the sugars of the DNA backbone with the amino groups of DNA bases or with amino acids of proteins (Maillard reaction). In addition, thymine dimers can form. Such modifications mean that the DNA cannot be amplified by PCR and cannot be sequenced. However, they occur comparatively little, and each DNA molecule occurs in many copies. This redundancy ultimately helps in the detection of the correct sequence. Usually, the conversion of cytosine into uracil predominates [101,103].

In addition to the high degree of fragmentation and the small amount of sufficiently long DNA sequences, the extraction of aDNA is also challenging because many other DNA sequences from various organisms (i.e., from microorganisms and animal pests) are often also extracted, which are usually present in abundant concentrations [99,100]. In order to avoid additional contamination, it is advisable to carry out laboratory work in an appropriate clean room under special protective conditions [102,106,107]. Furthermore, several software tools and strategies have been developed that can help distinguish contaminations of aDNA with modern DNA [108–110].

3.1. Analytical Procedure for Genomics Analyses

The amount of samples required for DNA analysis can vary greatly and depends primarily on the material of the manuscript and its state of preservation. In addition, it remains to be considered whether, for example, the microbiome should only be examined superficially, or whether a comprehensive characterization of the writing surface or ink should be carried out. Furthermore, the method used for the detection of the DNA sequences must be taken into account. In the case of parchment manuscripts, which have so far been researched most frequently with DNA analyses, fragments with a size of approx. 5 mm^2 were used by most working groups. A visual impression of the required size of the fragments can be found in the publication by Piñar et al. [48,111]. From historical papyrus manuscripts, it was possible to extract DNA from pieces with a size of approx. 1 cm^2 [112]. In the case of non-destructive work, e.g., with the rubber method, an assessment is much more difficult since only superficial DNA can be obtained, which may be heavily contaminated with microbial DNA. In such cases, we recommend using the relevant literature and comparable methods and materials as a guide.

Various protocols exist for the extraction of aDNA from historical artefacts, which are used depending on the material and the scientific question. It is difficult to estimate how much DNA and in what condition the DNA is actually still present in old materials. Furthermore, the success of the extraction depends strongly on the method. In this context, there are various studies dealing in particular with the optimization of the extraction process of aDNA [51,113–115].

In parts, the extraction procedures are similar to those used for the extraction of modern DNA. However, there are also some special features to consider. In general, a lysis buffer is added after DNA sampling, which can be completed using the rubber method, for example (see Section 2.1). The buffer contains detergents such as cetyltrimethylammonium bromide (CTAB) or sodium dodecyl sulfate (SDS) to destroy the secondary and tertiary structure of membrane proteins; mercaptoethanol or dithiothreitol (DTT) to cleave disulfide bridges in proteins; polyvinylpyrrolidone (PVP) to bind polyphenols and polysaccharides; and if necessary, enzymes to degrade proteins. After incubation for about 24 h at temperatures between 37 and 55 °C, depending on the temperature optimum of the enzymes used, the DNA is purified, for example, with silica columns or by two-phase extraction by means of phenol and chloroform. Subsequently, the DNA is precipitated with ethanol or isopropanol and stored in a buffer of tris-(hydroxymethyl)-aminomethanhydrochlorid (TRIS-HCl) and ethylenediaminetetraacetic acid (EDTA) as well as Tween-20 for long-term storage [51,100,102,111,116].

Since, as already described, the deamination of cytosine to uracil leads to errors in sequencing, it can be helpful to treat the DNA extract with certain enzymes to reduce error rates. Uracil DNA glycosylase (UDG) is suitable for this purpose, as it removes uracil from the DNA sequence. Abasic sites remain, which are cut off with endonuclease VIII

(Endo VIII). This procedure is also referred to as uracil-specific excision reagent (USER) treatment and is sold as an enzyme mixture by New England Biolabs. However, the aDNA sequence is fragmented even more by this reaction, and the higher thymine content is lost as an authentication parameter. For this reason, it is appropriate to treat one part of the DNA extract with and one part of the DNA extract without these enzymes [117]. Additional enzymatic methods have been developed to correct other aDNA modifications. An overview can be found in the reference cited [103].

In the past, DNA sequencing was commonly performed using Sanger's chain termination method. However, this procedure is both very time consuming and expensive. With the development of next generation sequencing (NGS) instruments, ultra-high throughput parallel sequencing have become possible. Consequently, both the costs and the time required could be reduced. A whole range of different instrument types are now available, which have significant differences in read lengths, capacities, error rates, runtimes, acquisition, and maintenance costs. Currently, instruments from the manufacturer Illumina dominate the sequencing market, with about 80% of the instruments sold [118,119]. While second-generation devices first amplify the target DNA sequence using polymerase chain reaction (PCR), which can be very error-prone, third- and fourth-generation platforms do not need this step. However, relatively long DNA sequences are required for the third-generation and fourth-generation designs [118]. Since aDNA is usually highly fragmented, platforms of the second-generation are often preferred. Nevertheless, there are efforts to concatenate short DNA molecules in such a way that they become accessible for third- and fourth-generation instruments [102,120]. However, at the present time, the newer instruments are currently still used primarily in the context of cultural heritage issues in order to obtain meta-information from recent DNA sequences, especially to record the microbiome. In this context, the small MinION platform by Oxford Nanopore Technologies plays a special role (see Section 3.3) [121,122].

The second-generation instruments are based on the shotgun and sequencing by synthesis (SBS) approach, in which the genome is first fragmented enzymatically or mechanically, since only short reads can be taken into account [118]. Due to the high degree of fragmentation of aDNA, the first step can be omitted, and the so-called library preparation can take place directly after the DNA extraction. However, the manufacturers' standard protocols for preparing libraries are often not optimal for the analysis of aDNA, since they are associated with a high loss of the template DNA sequences [99,123]. For this reason, different strategies have been developed specifically for fragmented and damaged aDNA sequences. Classically, a double-stranded library preparation is performed in which double-stranded adapters are bound to the ends of the double-stranded DNA sequences. The DNA sequences are then denatured and bound single-stranded to a carrier material by means of the adapters, and an amplification is carried out. Depending on the NGS system, a bridge or emulsion PCR is used to be able to detect sufficiently strong signals in the subsequent sequencing. During sequencing, a complementary DNA strand is synthesised and a corresponding signal in the form of a light beam or a change in pH value, e.g., is generated when the respective complementary base is incorporated [124,125]. However, the disadvantage of this procedure is that only double-stranded DNA sequences can be detected, but aDNA sequences are often present as single strands. For this reason, a method for the production of a single-stranded library was developed specifically for severely damaged DNA sequences. This approach considers both single-stranded and double-stranded DNA sequences. For this purpose, the double-stranded DNA molecules are first denatured with heat and then a single-stranded adapter is ligated to the respective ends. The adapter contains biotin, which is bound to streptavidin-coated beads. After adding an oligonucleotide that binds to the adapter, a double strand can be generated from the single strand by means of PCR. A double-stranded adapter is tied to the opposite free end of the double strand. The biotin-streptavidin bond is broken with heat and the DNA double strand is converted into two single strands, with which amplification and sequencing can be carried out analogously to double-stranded library preparation [126].

The single-stranded library approach is both more time consuming and expensive. For this reason, various modifications were made [127–129]. This includes in particular the so-called single-tube library preparation procedures, in which the many individual steps otherwise required for library preparation are combined in a few steps, so that the hands-on time is reduced. In addition, DNA loss when transferring the extract from a used tube to a new tube can be minimised because fewer reaction tubes are needed overall. A prerequisite for this simplification is an optimised composition of the chemicals and enzymes required for the individual steps [129,130].

The analysis of NGS data from aDNA sequences can be very challenging and time-consuming, as only short reads are obtained, which often contain errors and artefacts. For this reason, assignments are made using reference genomes from organisms that may be closely related. De novo assembly, as can be performed using NGS data of modern DNA sequences using overlapping reads, is mostly not possible due to the short read lengths. An overview of suitable software tools that can be used for the analysis of aDNA expressions was recently published by Orlando et al. [102].

As an alternative to the described non-targeted NGS methods, target enrichment sequencing approaches are also suitable, in which previously known DNA sequences are first bound to a target-specific single-stranded DNA bait, amplified, and then detected using NGS technologies. Such procedures are cheaper and can simplify the analysis especially when only a small amount of aDNA is present, but since this is a targeted approach and no omics technique, there is a risk that potentially relevant DNA sequences will not be detected, and information will be lost [131].

3.2. Analysis of Endogenous DNA Sequences

In the past, endogenous genomic analysis of aDNA was used specifically to study parchment manuscripts [48,55,132–134]. Such analyses can help, for example, to reassemble fragments of manuscripts or archives into a whole since it can be assumed that fragments originating from the same animal belong to each other. In addition, they provide information about the way of life of people in certain regions, for example, whether cow or sheep breeding was more common or whether manuscripts were produced in other places and only later reached a distant region. Examples include the study of the Dead Sea Scrolls as well as the analysis of a New Zealand founding document to which an additional blank parchment page could be added [20,135]. Further studies in this regard were carried out on a gospel book produced using various animal species (calves and sheep) and different sexes. It was particularly noticeable that four female animals and one male animal were used for the production of the parchment. Since female animals have a higher value as breeding animals, one would not initially expect this ratio. A possible explanation could be the occurrence of a cattle disease at the time of writing the manuscript, or that female animals were deliberately chosen as sacrificial offerings for the text. Conversely, it is also possible that there was an oversupply of females, as the males were mainly used as work animals. [133]. Another example is the analysis of endogenous DNA of Slavonic codices made from parchment of different animal species [48,136].

An older study from 2002 describes the decay of endogenous DNA in papyri and suggests a half-life of DNA of 19–24 years. According to this study, DNA in papyrus manuscripts is not detectable up to an age of 500–700 years. The detection was performed on chloroplast DNA, which was first amplified by PCR as well as specific primers and then analysed by electrophoresis [112]. Since new possibilities have emerged in the meantime due to the introduction of NGS strategies, it would be useful to check this assumption with the more recent developments. Further studies on plant materials such as wood, which was used both as writing material and for book bindings, and palm leaves suggest that the extraction and analysis of endogenous DNA from degenerate manuscripts should be successful and may provide information about their origin and use [51,137,138]. To the best of our knowledge, however, written artefacts have not yet been sufficiently well researched in this context [139].

3.3. Analysis of Exogenous DNA Sequences

In addition to the analysis of endogenous DNA from historical manuscripts, metagenomic analyses of ancient manuscripts are performed to detect the microbiome of insects, fungi, bacteria, archaea, as well as viruses, in particular to track and, if necessary, to stop biodeterioration processes. Such analyses may also be important to protect people handling the ancient materials from potentially dangerous organism. The majority of published studies in this area are based on culture-based methods and the use of various genetic fingerprinting methods such as polymerase chain reaction denaturing gradient gel electrophoresis (PCR-DGGE). Nevertheless, NGS developments have also found their way into this research field, especially since the frequently used culture-dependent methods can only detect a fraction of the total microorganisms present in environmental samples [140–143]. NGS approaches are distinguished between metabarcoding methods, also known as amplicon sequencing, and whole metagenome shotgun sequencing (WMS) strategies. In the former, only specific marker genes are sequenced by NGS strategies and matched with a database. For the taxonomic determination of bacteria and archaea, the highly conserved 16S rRNA sequence, and for fungi, the internal transcribed spacer (ITS) regions 1 and 2 as well as the 5.8S rRNA, 18S, and 28S sequences, are particularly suitable. In contrast, in WMS, the total DNA that can be extracted from an organism or sample is used for sequencing. In this way, WMS allows not only taxonomic classification, but also identification at the species or strain level. However, due to the enormous amount of data obtained in this approach, data analysis remains a challenge [144,145]. In addition to analysing the microbiome, examining exogenous DNA sequences can also be used to obtain information about the authors and readers. For example, it is possible to deduce the frequency of use of the manuscripts based on the amount of human DNA, although special care must be taken to determine whether it is modern DNA or aDNA in order not to draw false conclusions [133].

Numerous microbiomes analyses have been carried out on manuscripts made of parchment. A compilation recently published can be found in Piñar et al. [146]. One of the first studies performed with NGS on parchment manuscripts concerned the appearance of purple spots. According to Migliore et al., halophilic archaea are responsible for the purple spots that appear on some parchment manuscripts over time. It is assumed that parchment comes into contact with numerous microorganisms during the manufacturing process. These include halophilic and halotolerant microorganisms that enter into the parchment through treatment with sea salt. Under the right conditions, the archaea can reproduce and gain energy from bacteriorhodopsin and light. In the process, they also attack the collagen structures. With a decrease in salinity and an increase in humidity, they collapse, and purple stains remain. At the same time, more marine bacteria begin to grow, feeding on the remains of archaea and the attacked collagen matrix. Gradually, more and more bacteria and fungi can colonise, causing the collagen to break down so that the manuscripts begin to disintegrate [24,147]. Another research group held Actinobacteria and *Aspergillus* species responsible for the appearance of the purple spots. The different findings are not necessarily mutually exclusive since presumably various microorganisms colonise manuscripts at different times [46].

Other microorganisms detected are mainly associated with the human microbiome. These include, for example, propionibacteria as well as *Staphylococci* and *Streptococci*. It is suspected that the bacteria infected the manuscripts through handling and use, but also through kisses. Such bacteria can also be involved in the decomposition of the manuscripts [48,133,146].

In addition, viruses were also detected in parchment samples, which only became really possible with the use of NGS technologies [146]. Some of the viruses (*Siphoviridae*) could be identified as bacteriophages, the occurrence of which was associated with colonisation of the parchments by certain bacteria [48,136,146]. Other viruses such as Merkel cell polyomavirus are associated with the human "virobiota" and can therefore be traced back to contact with human skin [48,136,146].

Paper manuscripts are particularly prone to fouling by fungi, which release enzymes such as cellulases, amylases, gelatinases, proteases, and lipases that lead to degradation of the historical materials by attacking the cellulosic structures. Furthermore, fungi often form acidic metabolites that can further degrade the chemical structures of the paper through acid hydrolysis. In addition, there are stains and discolorations due to fungal infestation, as chromophoric molecules are released. The stain-causing fungi that colonise paper include, in particular, *Aspergillus* and *Penicillium* [148]. Since paper such as parchment is a hygroscopic material, it is therefore particularly important that the documents are stored in suitable conditions, avoiding temperatures above 23 °C and humidity above 65% [140,141,149]. In most published studies on paper manuscripts in recent years, culture-dependent methods or fingerprinting approaches have been used, too [141,142,148,150,151]. However, in a recently published study on a paper manuscript from the 11th century, it could be shown that the combination of culture-dependent methods and NGS methods in particular has significant advantages since the two approaches complement each other and provide complementary information [25]. A similar result was obtained from the analysis of a book, in which a higher level of biodiversity could be detected using a culture-independent NGS approach that targeted the 16S and 28S rRNA genes [152].

Microbiological examinations of papyrus manuscripts have so far hardly been carried out, even if there are numerous attempts to stop the microbiological deterioration of them by applying various substances [153]. To our knowledge, there are no studies available in connection with NGS for the detection of microorganisms. There is only one study that was recently published to identify fungi using a culture-dependent approach. The authors describe this study as the first of its kind and show how little is actually known in this context. [154].

Furthermore, some studies were conducted on wax seals, also to better understand biodeterioration processes. Bacteria as well as various fungal species were detected. NGS methods were applied in the studies so as to be able to detect nob-cultivable microorganisms as well. In addition to a second-generation NGS system (MiSeq Illumina), the MinION platform by Oxford Nanopore Technologies was used to analyse bacterial 16S rRNA and fungal ITS as well as 28S rRNA genes [155–157]. The Minion platform is a small and portable sequencer, the size of a small glasses case, which is based on the so-called nanopore technology. Sequencing with this design is comparatively inexpensive, and can be performed directly on site. By measuring current changes, the nucleobases can be detected, and the sequence can be read out. Comparatively, long reads can be sequenced with this technology, but the error rate is quite high, which can be reduced by increasing the coverage or, in the case of double-stranded DNA molecules, by using a duplex approach [118]. Currently, this technology is more suitable for the sequencing of modern and exogenous non-fragmented sequences and is therefore well suited for microbiome analysis, but not for aDNA [121,122].

4. Paleoproteomics and Metaproteomics Analyses

The first investigations of amino acids in ancient fossils date back about 70 years. However, the big breakthrough came only with the use of soft-ionisation mass spectrometer instruments at the beginning of 2000 by Ostrom et al. [158]. Until then, peptides or proteins had to be analysed using gel-based techniques and protein sequences determined using Edman sequencing, which is time-consuming, expensive, and limited to a length of around 20 amino acids. Compared to the investigations that are carried out on aDNA, paleoproteomics analyses are still a comparatively young field of research [159].

The results obtained from proteomics studies show some overlap with those from genomic analysis (Figure 1). Thus, proteomics approaches can also be used to investigate the biological identity of manuscripts. In contrast to DNA analysis, the taxonomic resolution is not as high, e.g., it is not possible to determine gender or the degree of relationship. However, proteomics analysis can be applied to identify different types of tissue, e.g., if egg yolk or egg white was used. In addition, proteomics analyses are suitable for the detection of proteinaceous residues, e.g., from paint binders, glues, and certain surface treatments [13,159,160]. The major advantage of proteomics compared to genomic analysis, especially for historical artefacts, is that ancient peptide sequences are more stable and may still yield results when aDNA sequences are too fragmented to analyse. The stability of proteins depends on various parameters. On the one hand, this includes environmental factors because, similar to DNA molecules, high temperatures, extreme pH values, moisture, or enzymes can lead to the degradation of the protein sequences. On the other hand, the primary and thus the secondary, tertiary, and possibly quaternary structure also have a major influence on stability, which can lead to some proteins being broken down more quickly than others. Compared to DNA molecules, proteins are also present in larger quantities. However, unlike genomics techniques, it is not possible to carry out an amplification of the analytes. The qualitative and quantitative occurrence of proteins depends on various endogenous and exogenous factors. This means that compared to DNA sequences, greater diversity must be expected, which can have both advantages and disadvantages: on the one hand, this diversity makes data analysis more difficult, and on the other hand, in the best case scenario, additional information about an object is obtained [2,13,159,160]. The performance of proteomics is faster and cheaper compared to DNA analysis. However, since genomics and proteomics studies often provide complementary information, both techniques are used at best [2,13,159]. A procedure that is already being carried out by some working groups [55,133,152].

During the ageing of proteins, fragmentation occurs through hydrolysis of the backbone of the protein sequences, in which the peptide bonds are cleaved. Smaller peptides or free amino acids are also obtained as reaction products [161]. Further, ageing reactions can affect the side chains of the amino acids. The focus here is on the non-enzymatic deamidation reactions of glutamine to glutamic acid and asparagine to aspartic acid, which can be used to distinguish between ancient and modern proteins and sometimes even as a tool for estimating ageing ("molecular clock"). However, it must be noted that environmental factors and the protein structure can have a strong influence on such reactions. In addition, deamidation reactions can also occur in vivo, which is why the results are sometimes not very meaningful [161–165]. Nevertheless, it has recently been proposed to use the deamidation grade of glutamine for the calculation of a Parchment Glutamine Index (PQI). However, the focus is not on the time-dependent deamidation, but on the production process, in particular the liming with $Ca(OH)_2$, the animal type used, and the thickness, since these factors also have an influence on the deamidation grade [166].

Except for glycines, all proteinogenic amino acids have at least one chiral carbon atom and can occur in the L- or D- conformation. In most cases, amino acids occur in biological systems as L-isomers. The conversion of L-amino acids to the corresponding D-enantiomers is called amino acid racemisation (AAR) and can also be used as a marker of ageing. However, these reactions are also influenced by many exogenous factors and the protein structures, which must be taken into account for a reliable assessment [161]. In addition to the reactions described, numerous other degradation and conversion processes can occur during the ageing of proteins. These include chemical processes such as Maillard reactions, dehydration, decarboxylation, lactamisation, aldol cleavage, oxidation, phosphorylation, and dephosphorylation as well as hydroxylation and dehydroxylation. Additionally, there are enzymatic degradation reactions by the microbiome. Such reactions make the data more complicated, but can also serve as important authenticity markers for distinguishing between modern and ancient sequences [161].

4.1. Analytical Procedure for Proteomics Analyses

Proteins or peptides can be extracted from ancient manuscripts either directly from smaller fragments or using the methods described in Section 2 such as the rubber method, the use of EVA films, hydrophobin, or gels. Researchers that prefer destructive sampling for proteomics analyses also use small fragments with a size of approx. 5 mm^2 or about 5 mg, similar to DNA analyses [134,167]. The proteins and peptides can then be dissolved with different buffers. For further analysis, two different approaches are followed, referred to as bottom-up and top-down strategies. As in many other research areas, proteomics analyses of ancient manuscripts are usually carried out using the bottom-up approach, in which the proteins and peptides are first enzymatically digested (Figure 2). The enzymatic digestion is usually performed with the enzyme trypsin. Trypsin is an endopeptidase and cleaves peptide bonds behind the amino acids lysine and arginine at the C-terminus. The smaller peptides obtained can then be analysed using mass spectrometers and assigned to the corresponding proteins with the help of matching to databases. The disadvantage of this approach is that complete sequence coverage can often not be achieved and that fragments with post-translational modifications (PTMs), protein truncations, as well as alternative splicing events cannot be identified [13,147,168]. In the top-down approach, intact proteins are analysed so that proteoforms can also be characterised.

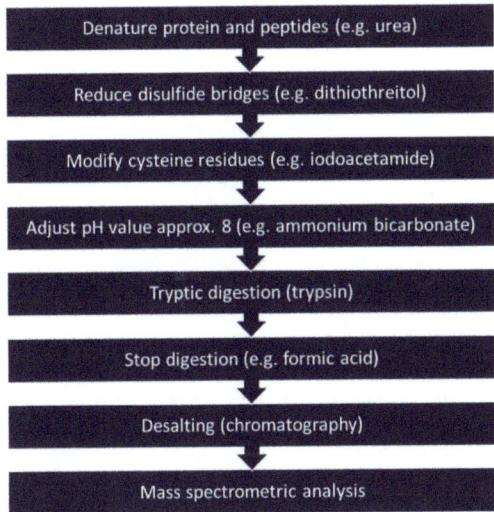

Figure 2. Typical sample preparation steps for performing bottom-up analysis.

However, there are some physical and technological limitations with this approach. On the one hand, these result from the fact that the extraction and purification of intact proteins is significantly more difficult because there are in many cases limits caused by their solubility, which is why some of the sampling methods described in Section 2 also suggest in situ digestion with trypsin. On the other hand, there are limitations set by the mass spectrometric detectors since the sensitivity of most detectors is inversely proportional to the mass of the proteins. However, some efforts currently exist to overcome this disadvantage [169]. The extent to which there are still really intact proteins in ancient manuscripts is questionable and, as already mentioned above, depends on the material used and on the exogenous factors during storage. Nevertheless, such an approach could certainly provide information about the composition of very stable molecules such as collagen. Furthermore, information about the degree of degradation of the ancient materials could be obtained [13,170]. In addition to these two classic approaches, there is also the middle-down proteomics approach, with which the low sequence coverage of the bottom-up approach can be improved, and simultaneously it is ensured that the limitations of the top-down approach are reduced. In the middle-down approach, the enzymatic process is carried out with proteases, which do not work quite as efficiently as trypsin, so that larger peptides are obtained. This strategy is relatively new and to our knowledge, has not been used to research ancient manuscripts [13,171].

Mass spectrometric analysis of proteomics experiments is performed in most cases with matrix-assisted laser desorption/ionisation time-of-flight (MALDI-TOF), liquid chromatography electrospray ionisation (LC-ESI)-Orbitrap, or liquid chromatography electrospray ionisation-quadrupole time-of-flight (LC-ESI-QTOF) instruments. Both MALDI and ESI are soft ionisation techniques that largely avoid initial fragmentation of the analytes. MALDI and ESI sources are complementary to each other. In the best case, both techniques are used to obtain as much information as possible about a sample, especially since both technologies can cause ion suppression effects, which can lead to analytes not being detected [13,168]. To reduce ion suppression effects in MALDI and to achieve the highest possible ion yield, so-called MALDI-2 methods have been developed in recent years. In this procedure, normal desorption and ionisation of the matrix as well as analytes first take place using a UV laser. Immediately afterwards, when the MALDI plume has formed, post-ionisation is carried out with a second laser in order to achieve a charge transfer from post-ionised matrix molecules to neutral analyte molecules [172,173]. This technique should be particularly useful for examining extracts from ancient artefacts in which only low concentrations of the analytes are present, either because a large part has already been degraded or because, for example, the sampling was non-invasive and only a few analyte molecules are present. However, although this strategy was developed a few years ago, it has only recently become commercially available and, to our knowledge, no studies of ancient artefacts or manuscripts have yet been published.

To minimise ion suppression in ESI-based instruments, the analytes are usually pre-separated using capillary electrophoresis- (CE) or LC-systems. NanoLC systems are particularly suitable for this purpose. In addition to improving analyte coverage due to the increase in sensitivity, these systems have the advantage that the amounts of solvent required can be reduced, which has a positive effect on costs and sustainability [13,174]. The coupling of a nanoLC-ESI system with an orbitrap analyser enables platforms with a particularly high resolution and sensitivity. Orbitrap analysers, currently only available for sale by Thermo Fisher, are often used as hybrid instruments by implementing a quadrupole or an ion trap. The newer designs, known as Orbitrap tribrid mass spectrometer, have three mass analysers. These are a quadrupole, an ion trap, and the orbitrap. Together, they enable various multiple analyte fragmentation options, which simplifies structure elucidation by recording MS/MS or MS^n spectra [175]. Such designs are particularly suitable for bottom-up approaches in which a peptide fragment fingerprinting (PFF) analysis is performed. In this way, sequence information can be obtained and, if necessary, de novo sequencing can also be carried out. Different strategies are pursued for the recording of MS/MS spectra, such as data-dependent analysis (DDA also known as information dependent acquisition (IDA)) or data independent analysis (DIA). Typically, in non-targeted measurements, DDA experiments are performed, and MS/MS are recorded from the most abundant ions. In order to ensure the best possible coverage of the various precursor ions by means of MS/MS spectra, decision or exclusion algorithms are often used. However, it may still be the case that only the most intense signals are captured, and important smaller signals are not fragmented. This risk can be reduced by means of DIA experiments and the application of strategies such as elevated-energy mass spectrometry (MS^E, supported by Waters, Milford, CT, USA), all-ion fragmentation mode (AIF, from Thermo Fischer Scientific, Waltham, MA, USA and Agilent Technologies, Santa Clara, CA, USA), parallel accumulation serial fragmentation (PASEF, supported by Bruker, Billerica, MA, USA) or sequential window acquisition of all theoretical mass spectra (SWATH, introduced by Sciex, Framingham, MA, USA). However, the evaluation of the data is more complex [44].

In contrast, in peptide mass fingerprinting (PMF) experiments, full scan spectra are recorded, and the peak lists obtained are compared to appropriate databases to carry out identification. This procedure is relatively simple and fast, but has the disadvantage that the corresponding databases must be available. For the latter approach, MALDI-TOF designs are often used, too [13,159,160].

TOF analysers can be coupled using both ESI and MALDI sources. In the meantime, dual platforms are also available that can be operated with both sources and can therefore be used more flexibly. When coupling a TOF analyser with an ESI interface, an LC or nanoLC unit for separating the sample extracts and a quadruple are usually also installed in order to be able to carry out fragmentation experiments. One of the most disruptive developments in recent years regarding the construction of such LC-ESI-QTOF-MS platforms is the integration of ion mobility (IM) cells, with which the ions can be separated according to their size, shape, and charge using electric fields and a drift gas. In this way, isomers and isobaric molecules can also be distinguished from each other. In addition, the collision cross section (CCS) value is received as a further identification parameter and the MS/MS spectra rate can increased which can be helpful when performing PFF experiments [44,64]. Carrying out proteomics studies, the use of IM cells is suitable for both top-down and bottom-up approaches to improve sensitivity and the number of detectable features. Although LC-ESI-QTOF-MS instrumentation currently plays a minor role in proteomics analysis because the mass resolution is not comparable to orbitrap designs, the implementation of IM cells offers a high added value that makes these designs interesting for proteomics experiments [176]. There is also the possibility to upgrade Orbitrap analysers with field asymmetric ion mobility spectrometry (FAIMS) cells, but these IM cells have a different physical principle than the usual IM cells of TOF analysers, which is why, for example, no CCS values can be generated [44,64]. Other suitable mass analysers are fourier-transform ion cyclotron resonance (FT-ICR) designs and, under certain circumstances, triple quadrupole (QqQ) or QTrap couplings. The latter two are used when analytes that are already known are to be detected and, in most cases, also to be quantified, since they are particularly sensitive. However, this is a targeted approach and not a non-targeted omics procedure. FT-ICR analysers have the best mass resolution and mass accuracy, but scan rates are comparatively slow, making coupling with LC techniques cumbersome. They are also relatively expensive to purchase and maintain, which is why they are not very widespread. An overview of the various advantages and disadvantages of the various analysers can be found in the references given [175,177,178].

The introduction of mass spectrometric methods for the analysis of proteins and proteins has brought many advantages, but it also has some disadvantages. This includes the fact that many signals cannot be correctly identified because the signal-to-noise ratio is too low, databases are incomplete or unexpected PTMs are present. In addition, fragments of sequences can be detected, but not the complete sequence itself. All these factors together mean that only a fraction of the information can be obtained. Especially when analysing ancient proteins or peptides, it can be assumed that these disadvantages have a particularly negative effect. To overcome these drawbacks, efforts are currently being made to make the method described in Section 3.3 for sequencing DNA and RNA using nanopores accessible for protein sequence analyses. At the moment, there are still a few challenges in development that need to be solved. Nevertheless, such a strategy could take proteomics investigations to a new level in the future [179].

4.2. Analysis of Endogenous Proteomics Sequences

Endogenous proteomics analyses are particularly suitable for the rapid taxonomic determination of the animal species used in parchment manuscripts. The first studies in this regard were carried out on a pocket Bible, also known as the "Marco Polo Bible", from the 13th century, as well as on a Qur'an folio from the 9th century. In both studies, sampling was performed destructively by removing small pieces of parchment [167,180]. Just a few years earlier, in 2010, Buckley and Collins coined the term "zooarchaeology by mass spectrometry" (ZooMS) and described the identification of animal species from proteinaceous material using characteristic peptide sequences of collagen I by means of an PMF approach [181,182]. In 2015, this method was further developed and minimally invasive sampling with the PVC rubber mentioned in Section 2.1 was suggested. In this method, collagen molecules or fragments are detached from the surface by careful erasing and bound to the rubber residues using the triboelectric effect. Subsequently, analytes can in turn be extracted from the erase residues, digested using trypsin and analysed by a MALDI-TOF instrument. The authors of this study refer to this procedure as electrostatic zooarchaeology by mass spectrometry (eZooMS) [55]. Recently, this approach has been extended by an automated data evaluation procedure [183]. Since the first application, several studies have been carried out on manuscripts from parchment using the eZooMS approach. In the majority of publications, genomics- and proteomics-based methods were used to take advantage of both omics strategies. The results obtained allow conclusions to be drawn about the animal species used and thus in turn interpretations of geographical origin, production processes, and livestock [55,133,134,184–188]. In addition, certain animal skins were probably used deliberately, either because of price differences or because of their practical utility. For example, legal deeds from thirteenth to twentieth century in the United Kingdom were probably mostly made of sheepskin because fraudulent changes can be more easily traced on it [189]. Furthermore, proteomics analyses are suitable for the investigation of illuminated manuscripts and inks since these were often produced with proteinaceous binders and glues, e.g., from the egg yolk and/or egg white of various birds, as well as from gelatin, collagen, or milk of various animals and certain plants, e.g., gum arabic [12,190].

The examples listed show that proteomics analyses are particularly suitable for the analysis of protein-rich materials such as parchment or leather. However, it was recently possible to extract proteins from Tibetan paper fragments and thus draw conclusions about the manufacturing process. Proteins from *Stellera chamaejasme*, which often served as a plant basis for Tibetan paper, could be detected, as well as proteins from milk and wheat, which probably served as binders and fillers [191]. However, it must not be forgotten that paper is often very highly processed, and it can therefore be difficult to identify peptide sequences still present in the paper, which is why proteomic analyses are not always promising depending on the type of paper production [2].

4.3. Analysis of Exogenous Proteomics Sequences

Analysis of the exogenous proteome of writing artefacts can be used to characterise the microbiota, as in genomics-based studies. In a direct comparison between genomics- and proteomics-based studies, almost the same results were obtained. However, a better characterisation of the microorganism was possible with the NGS technologies, which is partly due to the currently available databases [152,192]. To date, comparatively few proteomics studies were carried out to identify microorganisms for the investigation of ancient artefacts. In other areas such as food analysis and medical issues, proteome-based approaches have already established themselves as a standard method for detecting microorganisms [193].

In addition to examining bacteria that settle on ancient manuscripts or that normally occur on the human skin flora but are relatively unspecific, bacterial residues were also identified in some cases, which allow some conclusions to be drawn about the authors of the manuscripts. This includes, for example, a study conducted on a typewritten letter by George Orwell. A bottom-up analysis was used to identify peptide markers that confirmed that Orwell was a carrier of *Mycobacterium tuberculosis* [30]. On the basis of an original manuscript by the Russian satirist Mikhail Bulgakov, it could also be proven that the author died of nephrotic syndrome. This study was preceded by other investigations in which morphine was detected. However, it was not initially clear whether the morphine entered the manuscript through consumption by the author or by other people. With the detection of protein markers specific to kidney disease, the authors of the two studies concluded that morphine was used by Bulgakov himself to alleviate his suffering [27,28,56]. Further proteomic investigations were carried out on the death registers from Milano of 1630. In addition to peptides that could be used to identify *Yersinia pestis* pathogens, anthrax proteins were also unexpectedly detected. In addition, numerous other peptide markers could be identified that originate from humans, mice, and rats and indicate vegetable protein residues from potato, corn, rice, carrot, and chickpeas, which presumably came from the authors' meals [26]. Protein residues from foodstuffs such as honey, eggs, cereals, milk, and legumes have been detected on parchment manuscripts used as birth girdles. Presumably, these foods were used for treatment during pregnancy and childbirth. In addition, the scientists also found numerous human proteins, probably from vaginal secretions, among other things, which indicates a practical use of the parchment rolls [188].

5. Paleometabolomics and Metametabolomics Analyses

As the end product of genetic and enzymatic processes, metabolites are closest to the phenotype and at the last stage of the omics cascade. In addition to endogenous factors, the presence and absence of metabolites as well as their various concentrations within an organism are particularly influenced by exogenous factors. Compared to the proteome, the metabolome shows stronger changes to exogenous factors because metabolites often serve as inhibitors and activators for enzymes and changes in gene expression or enzyme transcription are slower, so that modifications of the proteome do not always correlate with changes in the phenotype. In animals, relevant exogenous factors can be, for example, different food sources, while in plants, the climate and the availability of nutrients play a role. In addition, metabolome analyses are also suitable for detecting xenobiotics, i.e., analytes that are foreign to the organism such as drugs, cosmetics, or organic dyes, as already mentioned in Section 1. Metabolomics analyses are relatively widely used in the analysis of foods to distinguish geographical origins, different varieties or cultivation, storage and processing conditions. In addition, they are often applied for medical issues, e.g., to detect certain diseases and to understand the course of the disease [3,17,44,64]. However, the use of truly non-targeted metabolomics approaches for the analysis of ancient manuscripts is currently the least common. This circumstance is probably also related to the fact that metabolites are often not very stable to exogenous factors and can be broken down and transformed quickly, which can make interpretation of the data more difficult.

Nevertheless, some studies have recently been published in this context, as is made clear in Sections 5.1 and 5.2. However, so far, the part of the studies of small organic molecules in which single selected organic molecules were in the foreground (targeted studies) predominates.

5.1. Analytical Procedure for Metabolomics Analyses

Metabolites can be destructively obtained directly from small pieces of the manuscript, but also with less destructive sampling strategies, e.g., EVA films or ASAP and PESI approaches, as well as with special ionisation techniques such as DART and DESI sources or with the help of laser systems (see Section 2). The amount of sample required for metabolomics analysis can vary greatly when working destructively with fragments. While a few µg of sample material are sufficient for pyrolysis–gas chromatography (Py-GC-MS) analyses, 5–50 mg of the samples are required in most cases for gas chromatography–mass spectrometry (GC-MS), liquid chromatography–mass spectrometry (LC-MS), or nuclear magnetic resonance (NMR) spectroscopy analyses [19,190,191,194–196]. It may be possible to further reduce the amount of sample required by using new technologies such as nanoLC-MS couplings [64]. An alternative could also be to perform different analyses from the same extraction approach. Such a procedure has been performed, for example, at a mural for the analysis of lipids and proteins by exploiting the different solubilities of the analytes [197].

Solvents such as water, acetonitrile, methanol, isopropanol, or chloroform are suitable for extracting metabolites, for example, from the EVA film or from small pieces. The selection of the appropriate solvent depends on the physical and chemical properties that are the focus of the analysis and the device platforms applied. In order to bring as many metabolites as possible into solution, it can also be helpful to use methods for cell disruption, e.g., ball mills or ultrasonic waves [21].

Usually, MS and NMR spectroscopy platforms are used for non-targeted studies of metabolites (Table 2). For non-targeted investigations of metabolites using MS, the high-resolution mass analysers already mentioned in Section 4 such as QTOF, Orbitrap, or FT-ICR are generally applied. These are often also coupled with LC or LCxLC, nanoLC, or CE units via ESI sources. Alternatively, MALDI-TOF instruments can be used to analyse metabolites, or surface-assisted laser desorption/ionisation (SALDI)-TOF experiments can be performed [3,44]. SALDI is a modification of MALDI. The organic matrices typically used in MALDI experiments often lead to interferences in low mass ranges, which makes it difficult to detect the actual analyte molecules. With SALDI-based approaches, inorganic matrices are usually used, with which these interferences can be avoided, which is why SALDI is more suitable for the analysis of metabolites [151]. In addition, the chromatographic separation of metabolites can be carried out using GC or GCxGC couplings if the analytes are correspondingly volatile or have been chemically derivatised beforehand. GC-MS approaches have the advantage that they allow headspace analysis to investigate volatile compounds without having to destroy the samples [198]. Another option for rapid sample preparation is to use Py-GC-MS platforms. With these instruments, small sample quantities of a few µg are heated to decomposition and then the thermal degradation products are characterised using GC-MS [194,195].

The GC units are usually coupled to a mass spectrometer via electron ionisation (EI) or chemical ionisation (CI) sources. However, a higher number of analytes can usually be detected with LC-MS couplings compared to GC-MS-based platforms. To our knowledge, no comparative studies have been carried out on ancient artefacts in this context, but on other matrices so that the results can be transferred accordingly [44,64].

Mass spectrometers are very sensitive and allow the detection of analytes with a wide variety of chemical and physical properties, as long as they can be ionised. However, mass spectrometric measurements that are carried out at longer time intervals are not very reproducible, which can make it difficult to compare the acquired data directly. There are various strategies how to deal with the low reproducibility, especially in the case of high-resolution non-targeted measurements [44]. Alternatively, NMR spectroscopy instruments are also suitable for use in metabolomics studies, which can deliver repeatable measurement results over long periods of time. NMR spectroscopy devices, however, are less sensitive and usually only small numbers of different analytes can be recorded with them, which can be crucial when studying old manuscripts if only a small amount of sample is available anyway. Nevertheless, it can be helpful to use MS and NMR spectroscopy platforms in combination since they can be used to detect different metabolite classes [199–201].

In addition to the relatively widespread high-resolution NMR spectroscopy instruments with field strengths of several hundred MHz, small and compact tabletop low-resolution NMR spectroscopy devices equipped with permanent magnets are also used in manuscript research. These have the advantage that they are mobile and only require a power connection with no cooling with liquid nitrogen or helium since no superconducting magnets are used [200,202]. In addition, special designs such as the NMR-MOUSE instrument enable non-destructive surface analyses and the creation of depth profiles of objects of any size as long as they are hydrogen-containing materials. However, the information content obtained with these smaller instruments is significantly lower since no non-targeted analyses are carried out to detect individual metabolites; instead, structural and moisture analyses are the priority to obtain information about the composition, condition, and age of manuscripts. Similar information can also be obtained by solid-state NMR spectroscopy (^{13}C CP MAS (Cross Polarisation/Magic Angle Spinning) NMR spectroscopy) [203,204].

Metabolomics analysis data are usually evaluated using multivariate methods to extract differences between various sample groups. However, it is often quite challenging to identify the most relevant metabolites. In the case of LC-MS data, this is completed by acquiring additional MS/MS spectra, the interpretation of which requires some experience. The CCS values mentioned in Section 4.1, which are recorded using an additional IM cell, can significantly simplify this step. GC-MS data are typically acquired at 70 eV with an EI source. Since these are standard parameters, which, unlike LC-MS applications, also deliver very reproducible results, identification using database comparisons is well established.

In the case of high-resolution NMR spectroscopy data, the substances are assigned either on the basis of the chemical shift, the integration of the signals and the scalar coupling, or with the aid of reference substances [178].

Table 2. Advantages and disadvantages of the different technology platforms suitable for metabolite detection [199,205–207].

Technology	Metabolite Coverage	Advantages	Disadvantages
GC-MS	- mainly polar metabolites, e.g., sugars, amino acids, organic acids as well as fatty acids after hydrolysis also from non-polar lipids - depending on the sample, several thousand metabolites can be detected	- depending on the analysis material, up to several thousand metabolites can be detected - very sensitive (approx. nmol/L-µmol/L) - relatively robust and cheap - good reproducibility	- analytes must be easily ionisable - thermally labile analytes cannot be measured - the injected sample volume cannot be measured again - the analytes often have to be derivatised
LC-MS	- polar metabolites e.g., sugars, amino acids, organic acids or fatty acids and lipids e.g., such as glycerols, phospholipids as well as sterols - depending on the sample, several thousand metabolites can be detected	- depending on the analysis material, up to several thousand metabolites can be detected - very sensitive (approx. pmol/L-nmol/L-nmol/L) - numerous ionisation sources are available to detect analytes with a wide variety of chemical and physical properties - it is possible to carry out very fast direct infusion experiments	- analytes must be easily ionisable - the injected sample volume cannot be measured again - relatively expensive to purchase and maintain - low reproducibility - limited robustness
NMR spectroscopy	- mainly polar metabolites e.g., sugars, amino acids, organic acids - depending on sample, only hundreds of metabolites can be detected	- poorly ionisable metabolites can also be detected - the sample extract is still available after the measurement and can be measured again - very reproducible results	- the detectable number of metabolites is significantly lower compared to GC-MS or LC-MS platforms - less sensitive compared to MS instruments (approx. µmol/L-nmol/L) - very expensive to purchase and maintain depending on the field strength

5.2. Analysis of Endogenous Metabolites

So far, metabolomics studies have been used in particular to understand the degradation and ageing processes of writing surfaces. In most of these studies, not only were antique samples analysed, but also artificially aged samples that were previously exposed to higher temperatures or humidity as well as treated with acids and alkalis in order to be able to use corresponding reference standards for comparative analyses. In addition to CE-MS and NMR spectroscopy instruments, Py-GC-MS platforms were used particularly frequently in order to be able to detect specific degradation metabolites. Such studies have been carried out for all relevant organic writing surfaces such as paper [196,198,208–212], papyrus [213], parchment, and leather [214,215].

In addition, metabolomics studies can be applied to gain information about the materials used for the production of the writings surfaces, e.g., what plant materials were used [191,194,195,216] or whether animal ingredients were added to a predominantly cellulose-based paper. The latter can be demonstrated, for example, by analysing the amino acid profile [196].

Furthermore, metabolomics studies have been applied to the investigation of preservation methods of manuscripts. Thus, it was possible to detect the surface treatment of parchments with castor oil and glycerol by means of a DART-MS based approach using multivariate methods for data evaluation [67]. Additionally, for the study of conservation methods, but on Indian palm leaf manuscripts, a GC-MS approach was used to identify diverse plant oils acting as insecticides and fungicides [19].

Moreover, metabolomics studies have been used to study inks, e.g., to identify the natural dyestuffs or binders [12,217]. Further investigations in this context were carried out on medieval illuminated scrolls of parchment in order to elucidate the paint recipes [190], to characterise the production of orchil, a purple dye that can be obtained from various lichens [218] and on the writings of the Italian physicist Alessandro Volta. In the latter study, around 1800 metabolites were recorded using a GC-MS platform, which indicate, among other things, that the ink was produced from *Rubia tinctorum* [219].

By carrying out metabolomics analyses of different sample groups, it is also possible to determine the geographical origin of materials or to prove whether artefacts have certain similarities and differences. We were able to show this recently with wood samples, for example [21,22]. However, it can be assumed that this approach can also be transferred to other organic plant materials such as palm leaf manuscripts or papyrus manuscripts and that classification studies of ancient documents should also be possible if corresponding reference samples are available. An example of this is a study performed on ancient Tibetan paper to characterise the plant material used. The analysis was based on a comparison of an ancient paper manuscript with a modern reference sample of handmade paper [191].

In addition, metabolomics approaches can also be used for questions about authentication, as was illustrated, for example, by the works of the Scottish poet Robert Burns. The focus of this study was the examination of inks and paper in order to distinguish the author's original works from fakes [62].

5.3. Analysis of Exogenous Metabolites

Like genomics and proteomics approaches, metabolomics studies of exogenous metabolites are also suitable for detecting the infestation of written artefacts with microorganisms. For example, the volatile metabolome of moulds growing on parchments was analysed using headspace GC-MS analysis to determine whether active mould growth could be detected in this way. The results should help conservators to take appropriate measures or not in order to avoid unnecessarily burdening the sensitive documents with disinfection measures [220]. Furthermore, the foxing of paper samples was analysed using a SALDI-TOF-MS approach to obtain information about the microbial mechanisms involved in foxing formation and to identify the substances that lead to staining of the paper [151]. Another study, which was also carried out using SALDI-TOF-MS, focused on bee wax seals. Some of the detected metabolites could be associated with microbiological degradation processes, too [156].

In addition, information about the authors or the readers of a written artefact can also be obtained by means of metabolomics studies. For example, a recent analysis of a booklet, a certificate for a thermometer, an article, and other personal objects belonging to the American writer Jack London, could not prove the consumption of opium, morphine and heroin. This result contradicts the relatively widespread assumption that Jack London consumed these substances regularly. However, the authors of this study were instead able to identify twelve other drugs believed to have been used by Jack London [221].

6. Metallomics and Isotopolomics Analyses

Metallomics analyses are particularly useful for examining inorganic materials such as metals, stones, ceramics, clay tablets, glass or bones. They are also of importance when analysing inks, as these often contain inorganic components [18,36,87]. In addition, metallomics analyses also provide important information about the toxicity of ancient objects and thus about how to handle them safely [35,36].

6.1. Analytical Procedure for Metallomics and Isotopolomics Analyses

Various destructive, non-destructive, and micro-destructive analytical methods are suitable for the analysis of metals from written artefacts. The non-destructive methods include XRF and PIXE, while LIBS is a micro-destructive method [18]. In addition, ICP-MS, in particular LA-ICP-MS (see Section 2.2) instruments, can also be used for metallomics approaches, which are the most sensitive. Usually, samples that are to be measured using ICP-MS are digested with acids (e.g., nitric acid) and oxidizing substances (e.g., hydrogen peroxide) to remove any organic residues before analysis. However, this step requires comparatively large sample volumes, which is why laser ablation methods have been developed. These are directly coupled with ICP-MS instruments, which means that wet-chemical digestion can be omitted [44,82].

The application of XRF instruments for the analysis of manuscripts is one of the most widely used research methods, and its usefulness has now been documented in numerous publications [18]. Using XRF instruments, electrons are excited from the inner atomic shells and ejected so that electrons fall back from the outer shells to the inner shells. During this process, characteristic fluorescent X-rays are released, which can be used to draw conclusions about the respective element. When working with X-rays, it is important to remember that the rays can cause long-term damage to the objects to be examined. Since the devices can often be easily transported, they can be used directly on site [222].

Alternatively, elemental analysis can also be performed using ion beams such as those generated by proton-induced X-ray emission (PIXE) instruments. In this technique, protons or alpha particles are directed at the ancient object, which also cause electrons to be ejected off the element's inner shells, which in turn leads to the release of characteristic X-rays. For PIXE analyses, the samples must not be too large since they are usually placed in a vacuum chamber. However, there are now also designs that do not require a vacuum, but the lateral resolution is then lower and the detection limits increase [223]. Unlike the XRF devices, PIXE instruments cannot be transported. XRF platforms have reduced sensitivity compared to PIXE instruments when analysing light elements, which is why XRF devices are more suitable for detecting elements with higher masses [222].

In LIBS analyses, parts of the sample are ablated with a laser and transferred to a plasma at temperatures of several 1000 K. The excited atoms and ions emit characteristic emission spectra when returning to a lower-energy state, from which the composition of the sample can be deduced. LIBS instruments are also transportable [222].

In a comparative study on egg tempera pigments, oil based pigments, decorated glazed ceramics, and coins, it was shown that all three platforms (XRF, PIXE, and LIBS) have different advantages and disadvantages, which is why the authors recommend a combined analysis or to base their choice on the sample matrix, the most appropriate technology [222].

While the detection limit for XRF, LIBS, and PIXE is in the ppm range, LA-ICP-MS instruments can measure down to the ppb range. However, it must be noted that it depends heavily on the devices used and the different matrices. In addition, the instruments differ in the analytical coverage with regard to the detectable elements and in the resolving power, which is relevant, for example, if not only the individual elements are to be distinguished from one another, but also their isotope ratios. Further differences result from the lateral resolution and the penetration depth [224]. Accurate quantitation is a challenge with all of the techniques listed. In general, certified reference materials are used that have a matrix similar to that of the objects to be examined. However, such reference materials for ancient manuscripts are not actually available, so in most cases only qualitative analyses are performed [54,225,226].

Isotopolomics data are usually evaluated using multivariate analysis methods, as is the case with metabolomics analyses. However, it is not necessary to identify the variables here, which is why the data analysis can be carried out much more easily and quickly, especially since the number of detected elements is also kept within a manageable range [3].

6.2. Analysis of Endogenous Metals and Isotope Ratios

Metallomics analyses are part of numerous publications that have been carried out on written artefacts, as already mentioned, the analysis of the inks and illuminated manuscripts is in the foreground. More detailed information on this can be found in the recently published review by Burgio [18]. The questions of ink analyses relate in particular to their composition in order to gain information about their manufacture, but also to prevent ink corrosion, as occurs with iron gall inks, and which can lead to damage to the historical documents due to acid catalysis and/or redox reactions [36,87,227]. However, Cappa et al. recently pointed out that metal analysis should not be used alone to discriminate inks, as only the elemental compositional information is not sufficient to provide an adequate characterisation. Therefore, the authors recommend combining inorganic analyses with other methods that also record organic compounds [18,228].

Furthermore, metallomics approaches can also be used to date inks, to prove their authenticity, or to determine the geographical origin [18]. For the latter, the isotope ratios of lead, strontium, and iron have proven to be particularly suitable [229,230].

In addition, information about the writing surfaces themselves can also be obtained. For example, a high proportion of calcium in parchments indicates certain production methods and can provide information about the microbial flora [48,136]. Moreover, it is possible to distinguish newer and older documents from parchment by means of element analyses and also to identify the type of animal that was used for the production of the parchment [231]. Metallomics analyses are also suitable for analysing paper samples, e.g., to differentiate between different types of paper [232].

6.3. Analysis of Exogenous Metals and Isotope Ratios

Metallomics studies can also be used to obtain information about the authors and readers of the written artefacts. For example, elevated levels of lithium could be found in books that Joseph Stalin read and that also contained comments by him. Lithium is used to treat various mental disorders. Accordingly, the authors hypothesised that the lithium residues detected were related to Stalin's treatment of a mental disorder [32]. The same working group also analysed some pages of a book written by Johannes Kepler. Comparatively, high amounts of gold, silver, mercury and lead were detected, which is why the authors of the study suggested that Kepler may have been involved with alchemy [29]. Another study in this regard was carried out on the memoirs of Giacomo Casanova. Increased levels of mercury sulphide (cinnabar) could be detected. Mercury sulphide is highly toxic, but was previously used to treat gonorrhoea, a disease from which Casanova also suffered, which is probably why the increased cinnabar levels were detectable on the historical documents [31]. In addition, information about the writing process itself can be obtained. Metal analyses on Greek papyrus rolls were able to detect lines drawn, which the authors of the manuscripts probably used as layout templates [233].

7. Application of Panomics Strategies to Manuscript Research

Most of the published studies on ancient manuscripts are based on the application of a single omics strategy. However, there are also examples where different omics approaches were combined for the analysis of old manuscripts. Some examples in this context can be found in Table 3. Nevertheless, with such analyses it must always be noted that the palaeographic interpretation must not be disregarded, as was evident, for example, in the relatively well-known case of the "Gospel of Jesus' Wife", where the chemical analyses indicated that it was an ancient manuscript. In the meantime, however, it is assumed that it is a forgery, since the text and writing style do not correspond to the style of the time [49].

Table 3. Exemplary studies for the analysis of ancient manuscripts employing more than one omics technology.

Scientific Issue	Omics Technology	Analytical Method	Reference
Characterization of Parchment	Genomics Proteomics	NGS LC-MS and MALDI-TOF	[55]
Characterization of Parchment	Genomics Proteomics	NGS MALDI-TOF	[133]
Characterization of Parchment	Genomics Proteomics	NGS MALDI-TOF	[134]
Characterization of Parchment	Genomics Metallomics	NGS FTIR, XRF, MSI	[136]
Characterization of Parchment	Proteomics Metabolomics Metallomics	MALDI-TOF, FTIR Amino acid analysis, FTIR, GC-MS FTIR, Raman, Energy dispersive X-ray spectroscopy	[185]
Organic composition of parchment and paint binders	Proteomicse Metabolomics	MADLI-TOF GC-MS	[190]
Studying the manufacturing process of Tibetan paper	Proteomicse Metabolomics Isotopolomics	LC-MS GC-MS Radiocarbon Dating	[191]
Composition of ink binders	Proteomics Metabolomics	MALDI-TOF GC-MS and LC-MS	[12]
Analysis of paper foxing	Genomics Metabolomics	NGS SALDI-TOF	[151]
Identification of microbial communities in book collections	Genomics Proteomics	Culture-dependent analysis and NGS MALDI-TOF	[152]
Disease and treatment of nephrotic syndrome by the author Mikhail Bulgakov	Proteomics Metabolomics	LC-MS GC-MS	[27,28,56]
Infection of Casanova with gonorrhea bacteria	Proteomics Metallomics	LC-MS Special mini-Hg-sensor	[31]
Investigating the microbiological and metabolic diversity of beeswax seals	Genomics Metabolomics	NGS SALDI-TOF	[156]
Characterization and study of the biodeterioration as well as the associated microbiome of a wax seal	Genomics Metabolomics Metallomics	Culture-dependent analysis and NGS Raman, FTIR Energy dispersive X-ray spectroscopy	[155]

8. Conclusions

The use of omics technologies offers numerous possibilities for the analysis of written artefacts. Each omics strategy contributes different information, which ensures a significant increase in knowledge about the written heritage, especially through the combination of several omics disciplines. In the meantime, there are already numerous examples in which the various strategies are combined with one another, so it is possible to draw more comprehensive conclusions about the individual artefacts [12,49,151,156,190,191]. It is to be expected that this trend will continue to increase in the coming years and will also be further strengthened by the introduction of new analytical technologies and databases.

Author Contributions: Conceptualization, M.C. and M.F.; writing—original draft preparation, M.C.; writing—review and editing, M.C. and M.F.; visualization, M.C. and M.F.; supervision, M.F.; project administration, M.C.; funding acquisition, M.C. All authors have read and agreed to the published version of the manuscript.

Funding: This research was funded by the Deutsche Forschungsgemeinschaft (DFG, German Research Foundation) under Germany's Excellence Strategy–EXC 2176 'Understanding Written Artefacts: Material, Interaction and Transmission in Manuscript Cultures', project no. 390893796. The research was conducted within the scope of the Centre for the Study of Manuscript Cultures (CSMC) at Universität Hamburg. Furthermore, this project was supported by the Akademie der Wissenschaften in Hamburg and financed by the Freie und Hansestadt Hamburg.

Institutional Review Board Statement: Not applicable.

Informed Consent Statement: Not applicable.

Data Availability Statement: Not applicable.

Conflicts of Interest: The authors declare no conflict of interest.

Sample Availability: Not applicable.

References

1. Piñar, G.; Sterflinger, K. Natural sciences at the service of art and cultural heritage: An interdisciplinary area in development and important challenges. *Microb. Biotechnol.* **2021**, *14*, 806–809. [CrossRef] [PubMed]
2. Fiddyment, S.; Teasdale, M.D.; Vnouček, J.; Lévêque, É.; Binois, A.; Collins, M.J. So you want to do biocodicology? A field guide to the biological analysis of parchment. *Herit. Sci.* **2019**, *7*, 35. [CrossRef]
3. Creydt, M.; Fischer, M. Omics approaches for food authentication. *Electrophoresis* **2018**, *39*, 1569–1581. [CrossRef] [PubMed]
4. D'Adamo, G.L.; Widdop, J.T.; Giles, E.M. The future is now? Clinical and translational aspects of "Omics" technologies. *Immunol. Cell Biol.* **2021**, *99*, 168–176. [CrossRef] [PubMed]
5. Amann, R.; Braus, G.; Gemeinholzer, B.; Häuser, C.; Jahn, R.; Lohrmann, V.; Lüter, C.; Meyer, A.; Misof, B.; Raupach, M.; et al. *Herausforderungen und Chancen der Integrativen Taxonomie für Forschung und Gesellschaft—Taxonomische Forschung im Zeitalter der OMICS-Technologien*; Misof, B., Wägele, J.W., Eds.; Deutsche Akademie der Naturforscher Leopoldina e. V.: Halle, Germany, 2014, ISBN 978-3-8047-3281-0.
6. Marvasi, M.; Cavalieri, D.; Mastromei, G.; Casaccia, A.; Perito, B. Omics technologies for an in-depth investigation of biodeterioration of cultural heritage. *Int. Biodeterior. Biodegrad.* **2019**, *144*, 104736. [CrossRef]
7. Pyzik, A.; Ciuchcinski, K.; Dziurzynski, M.; Dziewit, L. The bad and the good—Microorganisms in cultural heritage environments—An update on biodeterioration and biotreatment approaches. *Materials* **2021**, *14*, 177. [CrossRef]
8. Beata, G. The use of -omics tools for assessing biodeterioration of cultural heritage: A Review. *J. Cult. Herit.* **2020**, *45*, 351–361. [CrossRef]
9. Vilanova, C.; Porcar, M. Art-omics: Multi-omics meet archaeology and art conservation. *Microb. Biotechnol.* **2020**, *13*, 435–441. [CrossRef]
10. Sterflinger, K.; Pinar, G. Molecular-based techniques for the study of microbial communities in artworks. In *Microorganisms in the Deterioration and Preservation of Cultural Heritage*; Joseph, E., Ed.; Springer: Cham, Germany, 2021; pp. 59–77, ISBN 978-3-030-69410-4.
11. Pinar, G.; Sterflinger, K. Two decades using molecular techniques to study biodeterioration of cultural heritage: An amazing biotechnological development. In *Conserving Cultural Heritage*; Mosquera, M.J., Almoraima Gil, M.L., Eds.; CRC Press/Balkema, Taylor & Francis Group: London, UK; Leiden, The Netherlands, 2018; pp. 299–302, ISBN 9781315158648.
12. Zilberstein, G.; Zilberstein, R.; Zilberstein, S.; Maor, U.; Cohen-Ofri, I.; Shor, P.; Bitler, T.; Riestra, B.; Righetti, P.G. Proteomics and metabolomics composition of the ink of a letter in a fragment of a dead sea scroll from cave 11 (P1032-Fr0). *J. Proteom.* **2021**, *249*, 104370. [CrossRef]
13. Creydt, M.; Fischer, M. Mass spectrometry-based proteomics and metaproteomics analysis of ancient manuscripts. In *Exploring Written Artefacts, Objects, Methods, and Concepts*; Quenzer, J.B., Ed.; De Gruyter: Hamburg, Germany; pp. 183–212, ISBN 9783110753301.
14. Smith, O.; Gilbert, M.T.P. Ancient RNA. In *Paleogenomics. Population Genomics*; Lindqvist, C., Rajora, O.P., Eds.; Springer: Cham, Germany, 2019; pp. 53–74, ISBN 978-3-030-04753-5.
15. Johnson, C.; Patterson, A.; Idle, J.; Gonzalez, F. Xenobiotic metabolomics: Major impact on the metabolome. *Annu. Rev. Pharmacol. Toxicol.* **2011**, *52*, 37–56. [CrossRef]
16. Paul, A.; de Boves Harrington, P. Chemometric applications in metabolomic studies using chromatography-mass spectrometry. *TrAC Trends Anal. Chem.* **2021**, *135*, 116165. [CrossRef]
17. Patti, G.J.; Yanes, O.; Siuzdak, G. Metabolomics: The apogee of the omics trilogy. *Nat. Rev. Mol. Cell Biol.* **2012**, *13*, 263–269. [CrossRef]
18. Burgio, L. Pigments, dyes and inks: Their analysis on manuscripts, scrolls and papyri. *Archaeol. Anthropol. Sci.* **2021**, *13*, 194. [CrossRef]
19. Sharma, D.; Singh, M.R.; Dighe, B. Chromatographic study on traditional natural preservatives used for palm leaf manuscripts in india. *Restaur. Int. J. Preserv. Libr. Arch. Mater.* **2018**, *39*, 249–264. [CrossRef]

20. Anava, S.; Neuhof, M.; Gingold, H.; Sagy, O.; Munters, A.; Svensson, E.M.; Afshinnekoo, E.; Danko, D.; Foox, J.; Shor, P.; et al. Illuminating genetic mysteries of the dead sea scrolls. *Cell* **2020**, *181*, 1218–1231. [CrossRef] [PubMed]
21. Creydt, M.; Ludwig, L.; Köhl, M.; Fromm, J.; Fischer, M. Wood profiling by non-targeted high-resolution mass spectrometry: Part 1, Metabolite profiling in cedrela wood for the determination of the geographical origin. *J. Chromatogr. A* **2021**, *1641*, 461993. [CrossRef]
22. Creydt, M.; Lautner, S.; Fromm, J.; Fischer, M. Wood Profiling by non-targeted liquid chromatography high-resolution mass spectrometry: Part 2, Detection of the geographical origin of spruce wood (*Picea abies*) by determination of metabolite pattern. *J. Chromatogr. A* **2022**, *1663*, 462737. [CrossRef]
23. Migliore, L.; Thaller, M.C.; Vendittozzi, G.; Mejia, A.Y.; Mercuri, F.; Orlanducci, S.; Rubechini, A. Purple spot damage dynamics investigated by an integrated approach on a 1244 A.D. Parchment roll from the secret vatican archive. *Sci. Rep.* **2017**, *7*, 9521. [CrossRef]
24. Perini, N.; Mercuri, F.; Orlanducci, S.; Thaller, M.C.; Migliore, L. The integration of metagenomics and chemical physical techniques biodecoded the buried traces of the biodeteriogens of parchment purple spots. *Front. Microbiol.* **2020**, *11*, 598945. [CrossRef]
25. Raeisnia, N.; Arefian, E.; Amoozegar, M.A. Microbial community of an 11th century manuscript by both culture-dependent and -independent approaches. *Microbiology* **2022**, *91*, 313–323. [CrossRef]
26. D'Amato, A.; Zilberstein, G.; Zilberstein, S.; Compagnoni, B.L.; Righetti, P.G. Of mice and men: Traces of life in the death registries of the 1630 plague in milano. *J. Proteom.* **2018**, *180*, 128–137. [CrossRef] [PubMed]
27. Zilberstein, G.; Maor, U.; Baskin, E.; D'Amato, A.; Righetti, P. Unearthing Bulgakov's trace proteome from the master i margarita manuscript. *J. Proteom.* **2016**, *152*, 102–108. [CrossRef] [PubMed]
28. Zilberstein, G.; Maor, U.; Baskin, E.; Righetti, P.G. Maestro, marguerite, morphine: The last years in the life of Mikhail Bulgakov. *J. Proteom.* **2016**, *131*, 199–204. [CrossRef] [PubMed]
29. Zilberstein, G.; Zilberstein, S.; Maor, U.; Baskin, E.; D'Amato, A.; Righetti, P.G. De re metallica. Johannes Kepler and alchemy. *Talanta* **2019**, *204*, 82–88. [CrossRef]
30. Saravayskaya, Y.; Zilberstein, G.; Zilberstein, R.; Zilberstein, S.; Maor, U.; D'Amato, A.; Righetti, P.G. "1984": What Orwell could not predict. proteomic analysis of his scripts. *Electrophoresis* **2020**, *41*, 1931–1940. [CrossRef]
31. Zilberstein, G.; Zilberstein, R.; Zilberstein, S.; Fau, G.; D'Amato, A.; Righetti, P.G. Il n'y a Pas d'amour heureux pour Casanova: Chemical- and bio-analysis of his memoirs. *Electrophoresis* **2019**, *40*, 3050–3056. [CrossRef]
32. Zilberstein, G.; Zilberstein, S.; Righetti, P.G. Stalin's "black dog": A postmortem diagnosis. *Anal. Bioanal. Chem.* **2020**, *412*, 7701–7708. [CrossRef]
33. Oliveira, F.M.; Araujo, C.A.R.; Macario, K.D.; Cid, A.S. Radiocarbon analysis of the torah scrolls from the national museum of brazil collection. *Nucl. Instrum. Methods Phys. Res. Sect. B Beam Interact. Mater. Atoms* **2015**, *361*, 531–534. [CrossRef]
34. Jull, A.J.; Burr, G. Some interesting applications of radiocarbon dating to art and archaeology. *Archeometriai Muh.* **2014**, *11*, 139–148.
35. Giussani, B.; Monticelli, D.; Rampazzi, L. Role of laser ablation–inductively coupled plasma–mass spectrometry in cultural heritage research: A review. *Anal. Chim. Acta* **2009**, *635*, 6–21. [CrossRef]
36. Wagner, B.; Czajka, A. Non-invasive approximation of elemental composition of historic inks by LA-ICP-MS measurements of bathophenanthroline indicators. *Talanta* **2021**, *222*, 121520. [CrossRef] [PubMed]
37. Hu, C.; Jia, W. Multi-Omics Profiling: The way toward precision medicine in metabolic diseases. *J. Mol. Cell Biol.* **2021**, *13*, 576–593. [CrossRef] [PubMed]
38. Sandhu, C.; Qureshi, A.; Emili, A. Panomics for precision medicine. *Trends Mol. Med.* **2018**, *24*, 85–101. [CrossRef]
39. Yılmaz, D. Archaeology as an interdisciplinary science at the cross-roads of physical, chemical, biological, and social sciences: New perspectives and research. In *Transdisciplinarity*; Rezaei, N., Ed.; Springer International Publishing: Cham, Germany, 2022; pp. 435–455, ISBN 978-3-030-94651-7.
40. Roberts, J.; Robinson, P. Paleography and codicology. In *A Companion to the History of the Book*; Eliot, S., Rose, J., Eds.; Wiley: Chichester, UK, 2019; pp. 51–64, ISBN 9781119018193.
41. Rehbein, M.; Sahle, P.; Schassan, T.; Assmann, B. *Kodikologie nud Paläographie im Digitalen Zeitalter—Codicology and Palaeography in the Digital Age*; Kodikologie und Paläographie im digitalen Zeitalter; BoD—Books on Demand: Norderstedt, Germany, 2009; ISBN 9783837098426.
42. Online Etymology Dictionary, Definition: Paleo-. Available online: https://www.etymonline.com/word/paleo- (accessed on 29 May 2023).
43. Online Etymology Dictionary, Definition: Pan-. Available online: https://www.etymonline.com/word/pan- (accessed on 29 May 2023).
44. Creydt, M.; Fischer, M. Panomics—Fingerprinting approaches for food fraud detection. In *Encyclopedia of Food Safety*; Smithers, G., Ed.; Elsevier: Amsterdam, The Netherlands, 2023; *in press*, ISBN 978-0-08-100596-5.
45. Pinzari, F.; Montanari, M.; Michaelsen, A.; Pinar, G. Analytical protocols for the assessment of biological damage in historical documents. *Coalit. Newsl.* **2010**, *19*, 6–13.
46. Piñar, G.; Sterflinger, K.; Pinzari, F. Unmasking the measles-like parchment discoloration: Molecular and microanalytical approach. *Environ. Microbiol.* **2015**, *17*, 427–443. [CrossRef] [PubMed]

47. Karakasidou, K.; Nikolouli, K.; Amoutzias, G.D.; Pournou, A.; Manassis, C.; Tsiamis, G.; Mossialos, D. Microbial diversity in biodeteriorated greek historical documents dating back to the 19th and 20th century: A Case Study. *Microbiologyopen* **2018**, *7*, e00596. [CrossRef]
48. Piñar, G.; Tafer, H.; Schreiner, M.; Miklas, H.; Sterflinger, K. Decoding the biological information contained in two ancient slavonic parchment codices: An added historical value. *Environ. Microbiol.* **2020**, *22*, 3218–3233. [CrossRef]
49. Rabin, I.; Hahn, O. Detection of fakes: The merits and limits of non-invasive materials analysis. In *Fakes and Forgeries of Written Artefacts from Ancient Mesopotamia to Modern China*; Michel, C., Friedrich, M., Eds.; De Gruyter: Berlin. Germany; Boston, MA, USA, 2020; pp. 281–290, ISBN 9783110714333.
50. Paiva de Carvalho, H.; Sequeira, S.O.; Pinho, D.; Trovão, J.; da Costa, R.M.F.; Egas, C.; Macedo, M.F.; Portugal, A. Combining an innovative non-invasive sampling method and high-throughput sequencing to characterize fungal communities on a canvas painting. *Int. Biodeterior. Biodegrad.* **2019**, *145*, 104816. [CrossRef]
51. Schulz, A.; Lautner, S.; Fromm, J.; Fischer, M. Not stealing from the treasure chest (or just a bit): Analyses on plant derived writing supports and non-invasive DNA sampling. *PLoS ONE* **2018**, *13*, e0198513. [CrossRef]
52. Ding, X.; Lan, W.; Gu, J.-D. A review on sampling techniques and analytical methods for microbiota of cultural properties and historical architecture. *Appl. Sci.* **2020**, *10*, 8099. [CrossRef]
53. Multari, D.H.; Ravishankar, P.; Sullivan, G.J.; Power, R.K.; Lord, C.; Fraser, J.A.; Haynes, P.A. Development of a novel minimally invasive sampling and analysis technique using skin sampling tape strips for bioarchaeological proteomics. *J. Archaeol. Sci.* **2022**, *139*, 105548. [CrossRef]
54. Titubante, M.; Giannini, F.; Pasqualucci, A.; Romani, M.; Verona-Rinati, G.; Mazzuca, C.; Micheli, L. Towards a non-invasive approach for the characterization of arabic/christian manuscripts. *Microchem. J.* **2020**, *155*, 104684. [CrossRef]
55. Fiddyment, S.; Holsinger, B.; Ruzzier, C.; Devine, A.; Binois, A.; Albarella, U.; Fischer, R.; Nichols, E.; Curtis, A.; Cheese, E.; et al. Animal origin of 13th-century uterine vellum revealed using noninvasive peptide fingerprinting. *Proc. Natl. Acad. Sci. USA* **2015**, *112*, 15066–15071. [CrossRef] [PubMed]
56. Righetti, P.G.; Zilberstein, G.; Zilberstein, S. New baits for fishing in cultural heritage's Mare Magnum. *J. Proteom.* **2021**, *235*, 104113. [CrossRef] [PubMed]
57. Righetti, P.G.; Zilberstein, G.; D'Amato, A. What Sherlock sorely missed: The EVA technology for cultural heritage exploration. *Expert. Rev. Proteom.* **2019**, *16*, 533–542. [CrossRef] [PubMed]
58. Calvano, C.D.; Rigante, E.; Picca, R.A.; Cataldi, T.R.I.; Sabbatini, L. An easily transferable protocol for in-situ quasi-non-invasive analysis of protein binders in works of art. *Talanta* **2020**, *215*, 120882. [CrossRef]
59. Cicatiello, P.; Ntasi, G.; Rossi, M.; Marino, G.; Giardina, P.; Birolo, L. Minimally invasive and portable method for the identification of proteins in ancient paintings. *Anal. Chem.* **2018**, *90*, 10128–10133. [CrossRef]
60. Ntasi, G.; Kirby, D.P.; Stanzione, I.; Carpentieri, A.; Somma, P.; Cicatiello, P.; Marino, G.; Giardina, P.; Birolo, L. A Versatile and user-friendly approach for the analysis of proteins in ancient and historical objects. *J. Proteom.* **2021**, *231*, 104039. [CrossRef]
61. Kirby, D.P.; Manick, A.; Newman, R. Minimally invasive sampling of surface coatings for protein identification by peptide mass fingerprinting: A case study with photographs. *J. Am. Inst. Conserv.* **2020**, *59*, 235–245. [CrossRef]
62. Newton, J.; Ramage, G.; Gadegaard, N.; Zachs, W.; Rogers, S.; Barrett, M.P.; Carruthers, G.; Burgess, K. Minimally-destructive atmospheric ionisation mass spectrometry authenticates authorship of historical manuscripts. *Sci. Rep.* **2018**, *8*, 10944. [CrossRef]
63. McEwen, C.N.; McKay, R.G.; Larsen, B.S. Analysis of solids, liquids, and biological tissues using solids probe introduction at atmospheric pressure on commercial LC/MS instruments. *Anal. Chem.* **2005**, *77*, 7826–7831. [CrossRef] [PubMed]
64. Creydt, M.; Fischer, M. Food Metabolomics: Latest hardware—Developments for nontargeted food authenticity and food safety testing. *Electrophoresis* **2022**, *43*, 2334–2350. [CrossRef]
65. Hiraoka, K.; Ariyada, O.; Usmanov, D.T.; Chen, L.C.; Ninomiya, S.; Yoshimura, K.; Takeda, S.; Yu, Z.; Mandal, M.K.; Wada, H.; et al. Probe electrospray ionization (PESI) and its modified versions: Dipping PESI (DPESI), sheath-flow PESI (SfPESI) and adjustable SfPESI (Ad-SfPESI). *Mass Spectrom.* **2020**, *9*, A0092. [CrossRef] [PubMed]
66. Hiraoka, K.; Nishidate, K.; Mori, K.; Asakawa, D.; Suzuki, S. Development of probe electrospray using a solid needle. *Rapid Commun. Mass Spectrom.* **2007**, *21*, 3139–3144. [CrossRef] [PubMed]
67. Manfredi, M.; Robotti, E.; Bearman, G.; France, F.; Barberis, E.; Shor, P.; Marengo, E. Direct analysis in real time mass spectrometry for the nondestructive investigation of conservation treatments of cultural heritage. *J. Anal. Methods Chem.* **2016**, *2016*, 6853591. [CrossRef]
68. Adams, J. Analysis of printing and writing papers by using direct analysis in real time mass spectrometry. *Int. J. Mass Spectrom.* **2011**, *301*, 109–126. [CrossRef]
69. Jones, R.W.; McClelland, J.F. Analysis of writing inks on paper using direct analysis in real time mass spectrometry. *Forensic Sci. Int.* **2013**, *231*, 73–81. [CrossRef]
70. Heaton, K.; Solazzo, C.; Collins, M.J.; Thomas-Oates, J.; Bergström, E.T. Towards the application of desorption electrospray ionisation mass spectrometry (DESI-MS) to the Analysis of ancient proteins from artefacts. *J. Archaeol. Sci.* **2009**, *36*, 2145–2154. [CrossRef]
71. Sun, Q.; Luo, Y.; Wang, Y.; Zhang, Q.; Yang, X. Comparative analysis of aged documents by desorption electrospray ionization–mass spectrometry (DESI-MS) imaging. *J. Forensic Sci.* **2022**, *67*, 2062–2072. [CrossRef]

72. Sun, Q.; Luo, Y.; Sun, N.; Zhang, Q.; Wang, Y.; Yang, X. Technical note: Analysis of biological substances in ink fingerprint by desorption electrospray ionization mass spectrometry. *Forensic Sci. Int.* **2022**, *336*, 111321. [CrossRef]
73. Lee, G.; Cha, S. Depth-dependent chemical analysis of handwriting by nanospray desorption electrospray ionization mass spectrometry. *J. Am. Soc. Mass Spectrom.* **2021**, *32*, 315–321. [CrossRef] [PubMed]
74. Bouvier, C.; Van Nuffel, S.; Walter, P.; Brunelle, A. Time-of-flight secondary ion mass spectrometry imaging in cultural heritage: A focus on old paintings. *J. Mass Spectrom.* **2022**, *57*, e4803. [CrossRef] [PubMed]
75. Spoto, G. Secondary ion mass spectrometry in art and archaeology. *Thermochim. Acta* **2000**, *365*, 157–166. [CrossRef]
76. He, A.; Karpuzov, D.; Xu, S. Ink identification by time-of-flight secondary ion mass spectroscopy. *Surf. Interface Anal.* **2006**, *38*, 854–858. [CrossRef]
77. Attard-Montalto, N.; Ojeda, J.J.; Reynolds, A.; Ismail, M.; Bailey, M.; Doodkorte, L.; de Puit, M.; Jones, B.J. Determining the chronology of deposition of natural fingermarks and inks on paper using secondary ion mass spectrometry. *Analyst* **2014**, *139*, 4641–4653. [CrossRef]
78. Coumbaros, J.; Kirkbride, K.P.; Klass, G.; Skinner, W. Application of time of flight secondary ion mass spectrometry to the in situ analysis of ballpoint pen inks on paper. *Forensic Sci. Int.* **2009**, *193*, 42–46. [CrossRef]
79. Moore, K.L.; Barac, M.; Brajković, M.; Bailey, M.J.; Siketić, Z.; Bogdanović Radović, I. Determination of deposition order of toners, inkjet inks, and blue ballpoint pen combining mev-secondary ion mass spectrometry and particle induced X-ray emission. *Anal. Chem.* **2019**, *91*, 12997–13005. [CrossRef]
80. Goacher, R.E.; DiFonzo, L.G.; Lesko, K.C. Challenges determining the correct deposition order of different intersecting black inks by time-of-flight secondary ion mass spectrometry. *Anal. Chem.* **2017**, *89*, 759–766. [CrossRef]
81. Vilde, V.; Abel, M.-L.; Watts, J.F. A surface investigation of parchments using ToF-SIMS and PCA. *Surf. Interface Anal.* **2016**, *48*, 393–397. [CrossRef]
82. Wagner, B.; Syta, O.; Sawicki, M. A moderate microsampling in laser ablation inductively coupled plasma mass spectrometry analysis of cultural heritage objects: A review. In *Lasers in the Conservation of Artworks XI, Proceedings of Lacona XI*; Targowski, P., Ed.; NCU Press: Toruń, Poland, 2017; pp. 155–178.
83. Gonzalez, J.; Mao, X.L.; Roy, J.; Mao, S.S.; Russo, R.E. Comparison of 193, 213 and 266 nm laser ablation ICP-MS. *J. Anal. At. Spectrom.* **2002**, *17*, 1108–1113. [CrossRef]
84. Horn, I.; von Blanckenburg, F. Investigation on elemental and isotopic fractionation during 196 nm femtosecond laser ablation multiple collector inductively coupled plasma mass spectrometry. *Spectrochim. Acta Part B At. Spectrosc.* **2007**, *62*, 410–422. [CrossRef]
85. Burger, M.; Glaus, R.; Hubert, V.; van Willigen, S.; Wörle-Soares, M.; Convertini, F.; Lefranc, P.; Nielsen, E.; Günther, D. Novel sampling techniques for trace element quantification in ancient copper artifacts using laser ablation inductively coupled plasma mass spectrometry. *J. Archaeol. Sci.* **2017**, *82*, 62–71. [CrossRef]
86. Knaf, A.; Londero, P.; Nikkel, J.; Hark, R.; Bezur, A. Novel portable laser ablation micro-sampling in cultural Heritage. *Microsc. Microanal.* **2021**, *27*, 3014–3016. [CrossRef]
87. Wagner, B.; Bulska, E. On the use of laser ablation inductively coupled plasma mass spectrometry for the investigation of the written heritage. *J. Anal. At. Spectrom.* **2004**, *19*, 1325–1329. [CrossRef]
88. Glaus, R.; Koch, J.; Günther, D. Portable laser ablation sampling device for elemental fingerprinting of objects outside the laboratory with laser ablation inductively coupled plasma mass spectrometry. *Anal. Chem.* **2012**, *84*, 5358–5364. [CrossRef] [PubMed]
89. Glaus, R.; Dorta, L.; Zhang, Z.; Ma, Q.; Berke, H.; Günther, D. Isotope ratio determination of objects in the field by portable laser ablation sampling and subsequent multicollector ICPMS. *J. Anal. At. Spectrom.* **2013**, *28*, 801–809. [CrossRef]
90. Kradolfer, S.; Heutschi, K.; Koch, J.; Günther, D. Tracking mass removal of portable laser ablation sampling by its acoustic response. *Spectrochim. Acta Part B At. Spectrosc.* **2021**, *179*, 106118. [CrossRef]
91. Merkel, S.W.; D'Imporzano, P.; van Zuilen, K.; Kershaw, J.; Davies, G.R. "Non-invasive" portable laser ablation sampling for lead isotope analysis of archaeological silver: A comparison with bulk and in situ laser ablation techniques. *J. Anal. At. Spectrom.* **2022**, *37*, 148–156. [CrossRef]
92. Seman, S.; Dussubieux, L.; Cloquet, C.; Pryce, T.O. Strontium isotope analysis in ancient glass from south asia using portable laser ablation sampling. *Archaeometry* **2021**, *63*, 88–104. [CrossRef]
93. Numrich, M.; Schwall, C.; Lockhoff, N.; Nikolentzos, K.; Konstantinidi-Syvridi, E.; Cultraro, M.; Horejs, B.; Pernicka, E. Portable laser ablation sheds light on early bronze age gold treasures in the old world: New insights from troy, poliochni, and related finds. *J. Archaeol. Sci.* **2023**, *149*, 105694. [CrossRef]
94. Venter, A.; Nefliu, M.; Graham Cooks, R. Ambient desorption ionization mass spectrometry. *TrAC Trends Anal. Chem.* **2008**, *27*, 284–290. [CrossRef]
95. Lawal, R.O.; Donnarumma, F.; Murray, K.K. Deep-ultraviolet laser ablation electrospray ionization mass spectrometry. *J. Mass Spectrom.* **2019**, *54*, 281–287. [CrossRef] [PubMed]
96. Voß, H.; Moritz, M.; Pelczar, P.; Gagliani, N.; Huber, S.; Nippert, V.; Schlüter, H.; Hahn, J. Tissue sampling and homogenization with NIRL enables spatially resolved cell layer specific proteomic analysis of the murine intestine. *Int. J. Mol. Sci.* **2022**, *23*, 6132. [CrossRef] [PubMed]

97. Chen, Z.-C.; Chang, T.-L.; Li, C.-H.; Su, K.-W.; Liu, C.-C. Thermally stable and uniform DNA amplification with picosecond laser ablated graphene rapid thermal cycling device. *Biosens. Bioelectron.* **2019**, *146*, 111581. [CrossRef] [PubMed]
98. Moretti, P.; Iwanicka, M.; Melessanaki, K.; Dimitroulaki, E.; Kokkinaki, O.; Daugherty, M.; Sylwestrzak, M.; Pouli, P.; Targowski, P.; van den Berg, K.J.; et al. Laser cleaning of paintings: In situ optimization of operative parameters through non-invasive assessment by optical coherence tomography (OCT), Reflection FT-IR spectroscopy and laser induced fluorescence spectroscopy (LIF). *Herit. Sci.* **2019**, *7*, 44. [CrossRef]
99. Lan, T.; Lindqvist, C. Paleogenomics: Genome-scale analysis of ancient DNA and population and evolutionary genomic inferences. In *Population Genomics: Concepts, Approaches and Applications*; Rajora, O.P., Ed.; Springer: Cham, Germany, 2019; pp. 323–360, ISBN 978-3-030-04589-0.
100. Heintzman, P.; Soares, A.; Chang, D.; Shapiro, B. Paleogenomics. *Rev. Cell Biol. Mol. Med.* **2015**, *1*, 243–267. [CrossRef]
101. Dabney, J.; Meyer, M.; Pääbo, S. Ancient DNA damage. *Cold Spring Harb. Perspect. Biol.* **2013**, *5*, a012567. [CrossRef]
102. Orlando, L.; Allaby, R.; Skoglund, P.; Der Sarkissian, C.; Stockhammer, P.W.; Ávila-Arcos, M.C.; Fu, Q.; Krause, J.; Willerslev, E.; Stone, A.C.; et al. Ancient DNA analysis. *Nat. Rev. Methods Prim.* **2021**, *1*, 14. [CrossRef]
103. Andreeva, T.V.; Malyarchuk, A.B.; Soshkina, A.D.; Dudko, N.A.; Plotnikova, M.Y.; Rogaev, E.I. Methodologies for ancient DNA extraction from bones for genomic analysis: Approaches and guidelines. *Russ. J. Genet.* **2022**, *58*, 1017–1035. [CrossRef]
104. Renaud, G.; Schubert, M.; Sawyer, S.; Orlando, L. Authentication and assessment of contamination in ancient DNA. *Methods Mol. Biol.* **2019**, *1963*, 163–194.
105. Jónsson, H.; Ginolhac, A.; Schubert, M.; Johnson, P.L.F.; Orlando, L. MapDamage2.0: Fast approximate bayesian estimates of ancient DNA damage parameters. *Bioinformatics* **2013**, *29*, 1682–1684. [CrossRef]
106. Fulton, T.; Shapiro, B. Setting up an ancient DNA laboratory. In *Ancient DNA. Methods in Molecular Biology, Vol 1963*; Shapiro, B., Barlow, A., Heintzman, P., Hofreiter, M., Paijmans, J., Soares, A., Eds.; Humana Press: New York, NY, USA, 2019; pp. 1–13, ISBN 978-1-61779-516-9.
107. Matsvay, A.D.; Alborova, I.E.; Pimkina, E.V.; Markelov, M.L.; Khafizov, K.; Mustafin, K.K. Experimental approaches for ancient DNA extraction and sample preparation for next generation sequencing in ultra-clean conditions. *Conserv. Genet. Resour.* **2019**, *11*, 345–353. [CrossRef]
108. Peyrégne, S.; Peter, B.M. AuthentiCT: A model of ancient DNA damage to estimate the proportion of present-day DNA contamination. *Genome Biol.* **2020**, *21*, 246. [CrossRef] [PubMed]
109. Neukamm, J.; Peltzer, A.; Nieselt, K. DamageProfiler: Fast damage pattern calculation for ancient DNA. *Bioinformatics* **2021**, *37*, 3652–3653. [CrossRef]
110. Peyrégne, S.; Prüfer, K. Present-Day DNA Contamination in ancient DNA datasets. *BioEssays* **2020**, *42*, 2000081. [CrossRef]
111. Lech, T. Ancient DNA in historical parchments—Identifying a procedure for extraction and amplification of genetic material. *Genet. Mol. Res.* **2016**, *15*, gmr8661. [CrossRef] [PubMed]
112. Marota, I.; Basile, C.; Ubaldi, M.; Rollo, F. DNA decay rate in papyri and human remains from egyptian archaeological sites. *Am. J. Phys. Anthropol.* **2002**, *117*, 310–318. [CrossRef]
113. Barta, J.L.; Monroe, C.; Teisberg, J.E.; Winters, M.; Flanigan, K.; Kemp, B.M. One of the key characteristics of ancient DNA, low copy number, may be a product of its extraction. *J. Archaeol. Sci.* **2014**, *46*, 281–289. [CrossRef]
114. Rohland, N.; Hofreiter, M. Comparison and optimization of ancient DNA extraction. *Biotechniques* **2007**, *42*, 343–352. [CrossRef]
115. Gamba, C.; Hanghøj, K.; Gaunitz, C.; Alfarhan, A.H.; Alquraishi, S.A.; Al-Rasheid, K.A.S.; Bradley, D.G.; Orlando, L. Comparing the performance of three ancient DNA extraction methods for high-throughput sequencing. *Mol. Ecol. Resour.* **2016**, *16*, 459–469. [CrossRef]
116. Shapiro, B.; Hofreiter, M. *Ancient DNA: Methods and Protocols*; Methods in Molecular Biology; Humana Press Incorporated: Totowa, NJ, USA, 2012.
117. Briggs, A.W.; Stenzel, U.; Meyer, M.; Krause, J.; Kircher, M.; Pääbo, S. Removal of deaminated cytosines and detection of in vivo methylation in ancient DNA. *Nucleic Acids Res.* **2010**, *38*, e87. [CrossRef]
118. Pervez, M.T.; Hasnain, M.J.U.; Abbas, S.H.; Moustafa, M.F.; Aslam, N.; Shah, S.S.M. A comprehensive review of performance of next-generation sequencing platforms. *Biomed Res. Int.* **2022**, *2022*, 3457806. [CrossRef] [PubMed]
119. Elizabeth Pennisi, A $100 Genome? New DNA Sequencers Could Be a 'Game Changer' for Biology, Medicine. Available online: https://www.science.org/content/article/100-genome-new-dna-sequencers-could-be-game-changer-biology-medicine (accessed on 21 December 2022).
120. Schlecht, U.; Mok, J.; Dallett, C.; Berka, J. ConcatSeq: A method for increasing throughput of single molecule sequencing by concatenating short DNA fragments. *Sci. Rep.* **2017**, *7*, 5252. [CrossRef]
121. Pavlovic, J.; Cavalieri, D.; Mastromei, G.; Pangallo, D.; Perito, B.; Marvasi, M. MinION Technology for microbiome sequencing applications for the conservation of cultural heritage. *Microbiol. Res.* **2021**, *247*, 126772. [CrossRef] [PubMed]
122. Piñar, G.; Poyntner, C.; Lopandic, K.; Tafer, H.; Sterflinger, K. Rapid diagnosis of biological colonization in cultural artefacts using the MinION Nanopore sequencing technology. *Int. Biodeterior. Biodegrad.* **2020**, *148*, 104908. [CrossRef]
123. Knapp, M.; Hofreiter, M. Next generation sequencing of ancient DNA: Requirements, strategies and perspectives. *Genes* **2010**, *1*, 227–243. [CrossRef]
124. Meyer, M.; Kircher, M. Illumina sequencing library preparation for highly multiplexed target capture and sequencing. *Cold Spring Harb. Protoc.* **2010**, *2010*, pdb-prot5448. [CrossRef]

125. Kircher, M.; Sawyer, S.; Meyer, M. Double indexing overcomes inaccuracies in multiplex sequencing on the illumina platform. *Nucleic Acids Res.* **2012**, *40*, e3. [CrossRef] [PubMed]
126. Gansauge, M.-T.; Meyer, M. Single-stranded DNA library preparation for the sequencing of ancient or damaged DNA. *Nat. Protoc.* **2013**, *8*, 737–748. [CrossRef]
127. Gansauge, M.-T.; Gerber, T.; Glocke, I.; Korlevic, P.; Lippik, L.; Nagel, S.; Riehl, L.M.; Schmidt, A.; Meyer, M. Single-stranded DNA library preparation from highly degraded DNA using T4 DNA Ligase. *Nucleic Acids Res.* **2017**, *45*, e79. [CrossRef]
128. Gansauge, M.-T.; Aximu-Petri, A.; Nagel, S.; Meyer, M. Manual and automated preparation of single-stranded DNA libraries for the sequencing of DNA from ancient biological remains and other sources of highly degraded DNA. *Nat. Protoc.* **2020**, *15*, 2279–2300. [CrossRef] [PubMed]
129. Kapp, J.D.; Green, R.E.; Shapiro, B. A fast and efficient single-stranded genomic library preparation method optimized for ancient DNA. *J. Hered.* **2021**, *112*, 241–249. [CrossRef] [PubMed]
130. Carøe, C.; Gopalakrishnan, S.; Vinner, L.; Mak, S.S.T.; Sinding, M.H.S.; Samaniego, J.A.; Wales, N.; Sicheritz-Pontén, T.; Gilbert, M.T.P. Single-tube library preparation for degraded DNA. *Methods Ecol. Evol.* **2018**, *9*, 410–419. [CrossRef]
131. Suchan, T.; Kusliy, M.A.; Khan, N.; Chauvey, L.; Tonasso-Calvière, L.; Schiavinato, S.; Southon, J.; Keller, M.; Kitagawa, K.; Krause, J.; et al. Performance and automation of ancient DNA Capture with RNA HyRAD probes. *Mol. Ecol. Resour.* **2022**, *22*, 891–907. [CrossRef] [PubMed]
132. Vai, S.; Lari, M.; Caramelli, D. DNA sequencing in cultural heritage. In *Analytical Chemistry for Cultural Heritage*; Mazzeo, R., Ed.; Springer: Cham, Germany, 2017; pp. 329–346, ISBN 978-3-319-52804-5.
133. Teasdale, M.D.; Fiddyment, S.; Vnouček, J.; Mattiangeli, V.; Speller, C.; Binois, A.; Carver, M.; Dand, C.; Newfield, T.P.; Webb, C.C.; et al. The york gospels: A 1000-year biological palimpsest. *R. Soc. Open Sci.* **2021**, *4*, 170988. [CrossRef]
134. Teasdale, M.D.; van Doorn, N.L.; Fiddyment, S.; Webb, C.C.; O'Connor, T.; Hofreiter, M.; Collins, M.J.; Bradley, D.G. Paging through history: Parchment as a reservoir of ancient DNA for next generation sequencing. *Philos. Trans. R. Soc. Lond. B. Biol. Sci.* **2015**, *370*, 20130379. [CrossRef]
135. Shepherd, L.D.; Whitehead, P.; Whitehead, A. Genetic analysis identifies the missing parchment of New Zealand's founding document, the treaty of Waitangi. *PLoS ONE* **2019**, *14*, e0210528. [CrossRef]
136. Cappa, F.; Piñar, G.; Brenner, S.; Frühmann, B.; Wetter, W.; Schreiner, M.; Engel, P.; Miklas, H.; Sterflinger, K. The Kiev Folia: An interdisciplinary approach to unravelling the past of an ancient Slavonic manuscript. *Int. Biodeterior. Biodegrad.* **2022**, *167*, 105342. [CrossRef]
137. Wagner, S.; Lagane, F.; Seguin-Orlando, A.; Schubert, M.; Leroy, T.; Guichoux, E.; Chancerel, E.; Bech-Hebelstrup, I.; Bernard, V.; Billard, C.; et al. High-throughput DNA sequencing of ancient wood. *Mol. Ecol.* **2018**, *27*, 1138–1154. [CrossRef]
138. Lendvay, B.; Hartmann, M.; Brodbeck, S.; Nievergelt, D.; Reinig, F.; Zoller, S.; Parducci, L.; Gugerli, F.; Büntgen, U.; Sperisen, C. Improved recovery of ancient DNA from subfossil wood—Application to the world's oldest late glacial pine forest. *New Phytol.* **2018**, *217*, 1737–1748. [CrossRef]
139. Boesi, A. Paper plants in the Tibetan world: A Preliminary study. In *Tibetan Printing: Comparison, Continuities, and Change*; Diemberger, H., Ehrhard, K., Kornicki, P.F., Eds.; Brill: Leiden, The Netherlands; Boston, MA, USA, 2016; pp. 501–530, ISBN 978-90-04-31625-6.
140. Cappitelli, F.; Pasquariello, G.; Tarsitani, G.; Sorlini, C. Scripta manent? Assessing microbial risk to paper heritage. *Trends Microbiol.* **2010**, *18*, 538–542. [CrossRef]
141. Pinheiro, A.C.; Sequeira, S.O.; Macedo, M.F. Fungi in archives, libraries, and museums: A Review on paper conservation and human health. *Crit. Rev. Microbiol.* **2019**, *45*, 686–700. [CrossRef] [PubMed]
142. Okpalanozie, O.E.; Adebusoye, S.A.; Troiano, F.; Cattò, C.; Ilori, M.O.; Cappitelli, F. Assessment of indoor air environment of a Nigerian museum library and its biodeteriorated books using culture-dependent and –independent techniques. *Int. Biodeterior. Biodegrad.* **2018**, *132*, 139–149. [CrossRef]
143. Lech, T. Evaluation of a parchment document, the 13th century incorporation charter for the city of Krakow, Poland, for microbial hazards. *Appl. Environ. Microbiol.* **2016**, *82*, 2620–2631. [CrossRef]
144. Ruppert, K.M.; Kline, R.J.; Rahman, M.S. Past, present, and future perspectives of environmental DNA (eDNA) metabarcoding: A systematic review in methods, monitoring, and applications of global eDNA. *Glob. Ecol. Conserv.* **2019**, *17*, e00547. [CrossRef]
145. Pérez-Cobas, A.E.; Gomez-Valero, L.; Buchrieser, C. Metagenomic approaches in microbial ecology: An update on whole-genome and marker gene sequencing analyses. *Microb. Genom.* **2020**, *6*, mgen000409. [CrossRef]
146. Piñar, G.; Cappa, F.; Vetter, W.; Schreiner, M.; Miklas, H.; Sterflinger, K. Complementary strategies for deciphering the information contained in ancient parchment documentary materials. *Appl. Sci.* **2022**, *12*, 10479. [CrossRef]
147. Pinheiro, C.; Miller, A.Z.; Vaz, P.; Caldeira, A.T.; Casanova, C. Underneath the purple stain. *Heritage* **2022**, *5*, 4100–4113. [CrossRef]
148. Melo, D.; Sequeira, S.O.; Lopes, J.A.; Macedo, M.F. Stains versus colourants produced by fungi colonising paper cultural heritage: A review. *J. Cult. Herit.* **2019**, *35*, 161–182. [CrossRef]
149. Mazzoli, R.; Giuffrida, M.G.; Pessione, E. Back to the Past: "Find the guilty bug-microorganisms involved in the biodeterioration of archeological and historical artifacts. " *Appl. Microbiol. Biotechnol.* **2018**, *102*, 6393–6407. [CrossRef]
150. Oetari, A.; Susetyo-Salim, T.; Sjamsuridzal, W.; Suherman, E.A.; Monica, M.; Wongso, R.; Fitri, R.; Nurlaili, D.G.; Ayu, D.C.; Teja, T.P. Occurrence of fungi on deteriorated old dluwang manuscripts from Indonesia. *Int. Biodeterior. Biodegrad.* **2016**, *114*, 94–103. [CrossRef]

151. Szulc, J.; Otlewska, A.; Ruman, T.; Kubiak, K.; Karbowska-Berent, J.; Koziełec, T.; Gutarowska, B. Analysis of paper foxing by newly available omics techniques. *Int. Biodeterior. Biodegrad.* **2018**, *132*, 157–165. [CrossRef]
152. Kraková, L.; Šoltys, K.; Otlewska, A.; Pietrzak, K.; Purkrtová, S.; Savická, D.; Puškárová, A.; Bučková, M.; Szemes, T.; Budiš, J.; et al. Comparison of methods for identification of microbial communities in book collections: Culture-dependent (sequencing and MALDI-TOF MS) and culture-independent (Illumina MiSeq). *Int. Biodeterior. Biodegrad.* **2018**, *131*, 51–59. [CrossRef]
153. Taha, A.S.; Salem, M.Z.M.; Abo Elgat, W.A.A.; Ali, H.M.; Hatamleh, A.A.; Abdel-Salam, E.M. Assessment of the impact of different treatments on the technological and antifungal properties of papyrus (*Cyperus papyrus* L.) sheets. *Materials* **2019**, *12*, 620. [CrossRef]
154. Saada, H.; Othman, M.; Khaleil, M. Mold-deteriorated archaeological Egyptian papyri: Biodeteriogens, monitoring the deterioration, and treatment approach. *Archaeometry* **2022**, *65*, 335–353. [CrossRef]
155. Šoltys, K.; Planý, M.; Biocca, P.; Vianello, V.; Bučková, M.; Puškárová, A.; Sclocchi, M.C.; Colaizzi, P.; Bicchieri, M.; Pangallo, D.; et al. Lead soaps formation and biodiversity in a XVIII century wax seal coloured with minium. *Environ. Microbiol.* **2020**, *22*, 1517–1534. [CrossRef]
156. Szulc, J.; Jablonskaja, I.; Jabłońska, E.; Ruman, T.; Karbowska-Berent, J.; Gutarowska, B. Metabolomics and metagenomics characteristic of historic beeswax seals. *Int. Biodeterior. Biodegradation* **2020**, *152*, 105012. [CrossRef]
157. Pavlović, J.; Sclocchi, M.C.; Planý, M.; Ruggiero, D.; Puškárová, A.; Bučková, M.; Šoltys, K.; Colaizzi, P.; Riccardi, M.L.; Pangallo, D.; et al. The microbiome of candle beeswax drops on ancient manuscripts. *Int. Biodeterior. Biodegrad.* **2022**, *174*, 105482. [CrossRef]
158. Ostrom, P.H.; Schall, M.; Gandhi, H.; Shen, T.-L.; Hauschka, P.V.; Strahler, J.R.; Gage, D.A. New strategies for characterizing ancient proteins using matrix-assisted laser desorption ionization mass spectrometry. *Geochim. Cosmochim. Acta* **2000**, *64*, 1043–1050. [CrossRef]
159. Warinner, C.; Korzow Richter, K.; Collins, M.J. Paleoproteomics. *Chem. Rev.* **2022**, *122*, 13401–13446. [CrossRef]
160. Hendy, J. Ancient protein analysis in archaeology. *Sci. Adv.* **2021**, *7*, eabb9314. [CrossRef] [PubMed]
161. Demarchi, B. Mechanisms of degradation and survival. In *Amino Acids and Proteins in Fossil Biominerals*; Demarchi, B., Ed.; Wiley-Blackwell: Hoboken, NJ, USA, 2020; pp. 23–42, ISBN 9781119089537.
162. Ramsøe, A.; van Heekeren, V.; Ponce, P.; Fischer, R.; Barnes, I.; Speller, C.; Collins, M.J. DeamiDATE 1.0: Site-specific deamidation as a tool to assess authenticity of members of ancient proteomes. *J. Archaeol. Sci.* **2020**, *115*, 105080. [CrossRef]
163. Leo, G.; Bonaduce, I.; Andreotti, A.; Marino, G.; Pucci, P.; Colombini, M.P.; Birolo, L. Deamidation at asparagine and glutamine as a major modification upon deterioration/aging of proteinaceous binders in mural paintings. *Anal. Chem.* **2011**, *83*, 2056–2064. [CrossRef]
164. Schroeter, E.; Cleland, T. Glutamine deamidation: An indicator of antiquity, or preservational quality? *Rapid Commun. Mass Spectrom.* **2015**, *30*, 251–255. [CrossRef]
165. Ying, Y.; Li, H. Recent progress in the analysis of protein deamidation using mass spectrometry. *Methods* **2022**, *200*, 42–57. [CrossRef]
166. Nair, B.; Palomo, I.R.; Markussen, B.; Wiuf, C.; Fiddyment, S.; Collins, M.J. Parchment glutamine index (PQI): A novel method to estimate glutamine deamidation levels in parchment collagen obtained from low-quality MALDI-TOF data. *bioRxiv* **2022**. [CrossRef]
167. Toniolo, L.; D'Amato, A.; Saccenti, R.; Gulotta, D.; Righetti, P.G. The Silk Road, Marco Polo, a Bible and its proteome: A detective story. *J. Proteom.* **2012**, *75*, 3365–3373. [CrossRef] [PubMed]
168. Demarchi, B. Ancient protein sequences. In *Amino Acids and Proteins in Fossil Biominerals*; Demarchi, B., Ed.; Wiley-Blackwell: Hoboken, NJ, USA, 2020; pp. 113–126, ISBN 9781119089537.
169. Haugg, S.; Creydt, M.; Zierold, R.; Fischer, M.; Blick, R.H. Booster-microchannel plate (BMCP) detector for signal amplification in MALDI-TOF mass spectrometry for ions beyond m/z 50,000. *Phys. Chem. Chem. Phys.* **2023**, *25*, 7312–7322. [CrossRef]
170. Miller, R.M.; Smith, L.M. Overview and considerations in bottom-up proteomics. *Analyst* **2023**, *148*, 475–486. [CrossRef]
171. Pandeswari, P.B.; Sabareesh, V. Middle-down approach: A choice to sequence and characterize proteins/proteomes by mass spectrometry. *RSC Adv.* **2019**, *9*, 313–344. [CrossRef]
172. Soltwisch, J.; Kettling, H.; Vens-Cappell, S.; Niehaus, M.; Müthing, J.; Dreisewerd, K. Mass spectrometry imaging with laser-induced postionization. *Science* **2015**, *348*, 211. [CrossRef]
173. Potthoff, A.; Dreisewerd, K.; Soltwisch, J. Detailed characterization of the postionization efficiencies in MALDI-2 as a function of relevant input parameters. *J. Am. Soc. Mass Spectrom.* **2020**, *31*, 1844–1853. [CrossRef] [PubMed]
174. Shan, L.; Jones, B.R. Nano-LC: An updated review. *Biomed. Chromatogr.* **2022**, *36*, e5317. [CrossRef] [PubMed]
175. Kumar, S. Developments, advancements, and contributions of mass spectrometry in omics technologies. In *Advances in Protein Molecular and Structural Biology Methods*; Tripathi, T., Dubey, V., Eds.; Academic Press: London, UK, 2022; pp. 327–356, ISBN 978-0-323-90264-9.
176. Nys, G.; Nix, C.; Cobraiville, G.; Servais, A.-C.; Fillet, M. Enhancing protein discoverability by data independent acquisition assisted by ion mobility mass spectrometry. *Talanta* **2020**, *213*, 120812. [CrossRef]
177. Zarrouk, E.; Lenski, M.; Bruno, C.; Thibert, V.; Contreras, P.; Privat, K.; Ameline, A.; Fabresse, N. High-resolution mass spectrometry: Theoretical and technological aspects. *Toxicol. Anal. Clin.* **2022**, *34*, 3–18. [CrossRef]
178. Creydt, M.; Fischer, M. Food phenotyping: Recording and processing of non-targeted liquid chromatography mass spectrometry data for verifying food authenticity. *Molecules* **2020**, *25*, 3972. [CrossRef] [PubMed]

179. Ying, Y.-L.; Hu, Z.-L.; Zhang, S.; Qing, Y.; Fragasso, A.; Maglia, G.; Meller, A.; Bayley, H.; Dekker, C.; Long, Y.-T. Nanopore-based technologies beyond DNA sequencing. *Nat. Nanotechnol.* **2022**, *17*, 1136–1146. [CrossRef]
180. Kirby, D.P.; Buckley, M.; Promise, E.; Trauger, S.A.; Holdcraft, T.R. Identification of collagen-based materials in cultural heritage. *Analyst* **2013**, *138*, 4849–4858. [CrossRef]
181. Buckley, M.; Whitcher Kansa, S.; Howard, S.; Campbell, S.; Thomas-Oates, J.; Collins, M. Distinguishing between archaeological sheep and goat bones using a single collagen peptide. *J. Archaeol. Sci.* **2010**, *37*, 13–20. [CrossRef]
182. Buckley, M.; Fraser, S.; Herman, J.; Melton, N.D.; Mulville, J.; Pálsdóttir, A.H. Species identification of archaeological marine mammals using collagen fingerprinting. *J. Archaeol. Sci.* **2014**, *41*, 631–641. [CrossRef]
183. Hickinbotham, S.; Fiddyment, S.; Stinson, T.L.; Collins, M.J. How to get your goat: Automated identification of species from MALDI-ToF spectra. *Bioinformatics* **2020**, *36*, 3719–3725. [CrossRef]
184. Ruffini-Ronzani, N.; Nieus, J.-F.; Soncin, S.; Hickinbotham, S.; Dieu, M.; Bouhy, J.; Charles, C.; Ruzzier, C.; Falmagne, T.; Hermand, X.; et al. A biocodicological analysis of the medieval library and archive from orval abbey, Belgium. *R. Soc. Open Sci.* **2021**, *8*, 210210. [CrossRef]
185. Sommer, D.; Mühlen Axelsson, K.; Collins, M.; Fiddyment, S.; Bredal-Jørgensen, J.; Simonsen, K.; Lauridsen, C.; Larsen, R. Multiple microanalyses of a sample from the Vinland map. *Archaeometry* **2017**, *59*, 287–301. [CrossRef]
186. Calà, E.; Agostino, A.; Fenoglio, G.; Capra, V.; Porticelli, F.; Manzari, F.; Fiddyment, S.; Aceto, M. The Messale Rosselli: Scientific investigation on an outstanding 14th century illuminated manuscript from Avignon. *J. Archaeol. Sci. Rep.* **2019**, *23*, 721–730. [CrossRef]
187. Hofmann, C.; Rabitsch, S.; Vnoucek, J.; Fiddyment, S.; Quandt, A.; Collins, M.; Kanold, I.; Aceto, M.; Melo, M.; Calà, E.; et al. *The Vienna Genesis: Material Analysis and Conservation of a Late Antique Illuminated Manuscript on Purple Parchment*; Hofmann, C., Ed.; Böhlau: Wien, Austria, 2020, ISBN 978-320521058-0.
188. Fiddyment, S.; Goodison, N.J.; Brenner, E.; Signorello, S.; Price, K.; Collins, M.J. Girding the loins? Direct evidence of the use of a medieval parchment birthing girdle from biomolecular analysis. *R. Soc. Open Sci.* **2021**, *8*, 202055. [CrossRef]
189. Doherty, S.P.; Henderson, S.; Fiddyment, S.; Finch, J.; Collins, M.J. Scratching the surface: The use of sheepskin parchment to deter textual erasure in early modern legal deeds. *Herit. Sci.* **2021**, *9*, 29. [CrossRef] [PubMed]
190. van der Werf, I.D.; Calvano, C.D.; Germinario, G.; Cataldi, T.R.I.; Sabbatini, L. Chemical characterization of medieval illuminated parchment scrolls. *Microchem. J.* **2017**, *134*, 146–153. [CrossRef]
191. Han, B.; Niang, J.; Rao, H.; Lyu, N.; Oda, H.; Sakamoto, S.; Yang, Y.; Sablier, M. Paper fragments from the Tibetan Samye Monastery: Clues for an unusual sizing recipe implying wheat starch and milk in early Tibetan papermaking. *J. Archaeol. Sci. Reports* **2021**, *36*, 102793. [CrossRef]
192. Zilberstein, G.; Zilberstein, S.; Maor, U.; Righetti, P.G. Surface analysis of ancient parchments via the EVA film: The Aleppo Codex. *Anal. Biochem.* **2020**, *604*, 113824. [CrossRef] [PubMed]
193. Ashfaq, M.Y.; Da'na, D.A.; Al-Ghouti, M.A. Application of MALDI-TOF MS for identification of environmental bacteria: A review. *J. Environ. Manag.* **2022**, *305*, 114359. [CrossRef] [PubMed]
194. Avataneo, C.; Sablier, M. New Criteria for the characterization of traditional East Asian papers. *Environ. Sci. Pollut. Res. Int.* **2017**, *24*, 2166–2181. [CrossRef]
195. Han, B.; Vial, J.; Sakamoto, S.; Sablier, M. Identification of traditional East Asian handmade papers through the multivariate data analysis of pyrolysis-GC/MS Data. *Analyst* **2019**, *144*, 1230–1244. [CrossRef] [PubMed]
196. Corsaro, C.; Mallamace, D.; Łojewska, J.; Mallamace, F.; Pietronero, L.; Missori, M. Molecular degradation of ancient documents revealed by 1H HR-MAS NMR spectroscopy. *Sci. Rep.* **2013**, *3*, 2896. [CrossRef] [PubMed]
197. Levy, I.K.; Neme Tauil, R.; Valacco, M.P.; Moreno, S.; Siracusano, G.; Maier, M.S. Investigation of proteins in samples of a mid-18th century colonial mural painting by MALDI-TOF/MS and LC-ESI/MS (Orbitrap). *Microchem. J.* **2018**, *143*, 457–466. [CrossRef]
198. Lattuati-Derieux, A.; Bonnassies-Termes, S.; Lavédrine, B. Characterisation of compounds emitted during natural and artificial ageing of a book. Use of headspace-solid-phase microextraction/gas chromatography/mass spectrometry. *J. Cult. Herit.* **2006**, *7*, 123–133. [CrossRef]
199. Stringer, K.A.; McKay, R.T.; Karnovsky, A.; Quémerais, B.; Lacy, P. Metabolomics and its application to acute lung diseases. *Front. Immunol.* **2016**, *7*, 44. [CrossRef]
200. Wishart, D.S. NMR metabolomics: A look ahead. *J. Magn. Reson.* **2019**, *306*, 155–161. [CrossRef]
201. Letertre, M.P.M.; Dervilly, G.; Giraudeau, P. Combined nuclear magnetic resonance spectroscopy and mass spectrometry approaches for metabolomics. *Anal. Chem.* **2021**, *93*, 500–518. [CrossRef]
202. Blümich, B.; Singh, K. Desktop NMR and its applications from materials science to organic chemistry. *Angew. Chem. Int. Ed.* **2018**, *57*, 6996–7010. [CrossRef]
203. Rehorn, C.; Bluemich, B. Cultural heritage studies with mobile NMR. *Angew. Chem. Int. Ed.* **2018**, *57*, 7304–7312. [CrossRef]
204. Manso, M.; Carvalho, M.L. Application of spectroscopic techniques for the study of paper documents: A survey. *Spectrochim. Acta Part B At. Spectrosc.* **2009**, *64*, 482–490. [CrossRef]
205. Mirsaeidi, M.; Banoei, M.; Winston, B.; Schraufnagel, D. Metabolomics: Applications and promise in mycobacterial disease. *Ann. Am. Thorac. Soc.* **2015**, *12*, 1278–1287. [CrossRef] [PubMed]
206. Yoon, D.; Lee, M.; Kim, S.; Kim, K. Applications of NMR spectroscopy based metabolomics: A review. *J. Korean Magn. Reson. Soc.* **2013**, *17*, 1–10. [CrossRef]

207. Emwas, A.-H.; Roy, R.; McKay, R.T.; Tenori, L.; Saccenti, E.; Gowda, G.A.N.; Raftery, D.; Alahmari, F.; Jaremko, L.; Jaremko, M.; et al. NMR spectroscopy for metabolomics research. *Metabolites* **2019**, *9*, 123. [CrossRef] [PubMed]
208. Bogolitsyna, A.; Becker, M.; Dupont, A.-L.; Borgards, A.; Rosenau, T.; Potthast, A. Determination of carbohydrate- and lignin-derived components in complex effluents from cellulose processing by capillary electrophoresis with electrospray ionization-mass spectrometric detection. *J. Chromatogr. A* **2011**, *1218*, 8561–8566. [CrossRef] [PubMed]
209. Ding, L.; Yang, Q.; Liu, J.; Lee, Z. Evaluating volatile organic compounds from chinese traditional handmade paper by SPME-GC/MS. *Herit. Sci.* **2021**, *9*, 153. [CrossRef]
210. Mallamace, D.; Vasi, S.; Missori, M.; Mallamace, F.; Corsaro, C. NMR Investigation of degradation processes of ancient and modern paper at different hydration levels. *Front. Phys.* **2017**, *13*, 138202. [CrossRef]
211. Kaszonyi, A.; Izsák, L.; Králik, M.; Jablonsky, M. Accelerated and natural aging of cellulose-based paper: Py-GC/MS method. *Molecules* **2022**, *27*, 2855. [CrossRef]
212. Ortiz-Herrero, L.; Blanco, M.E.; García-Ruiz, C.; Bartolomé, L. Direct and indirect approaches based on paper analysis by Py-GC/MS for estimating the age of documents. *J. Anal. Appl. Pyrolysis* **2018**, *131*, 9–16. [CrossRef]
213. Lucejko, J.J.; Colombini, M.P.; Ribechini, E. Chemical alteration patterns of ancient Egyptian papyri studied by pyrolysis-GC/MS with in Situ Silylation. *J. Anal. Appl. Pyrolysis* **2020**, *152*, 104967. [CrossRef]
214. Sebestyén, Z.; Badea, E.; Carsote, C.; Czégény, Z.; Szabó, T.; Babinszki, B.; Bozi, J.; Jakab, E. Characterization of historical leather bookbindings by various thermal methods (TG/MS, Py-GC/MS, and micro-DSC) and FTIR-ATR spectroscopy. *J. Anal. Appl. Pyrolysis* **2022**, *162*, 105428. [CrossRef]
215. Sebestyén, Z.; Czégény, Z.; Badea, E.; Carsote, C.; Şendrea, C.; Barta-Rajnai, E.; Bozi, J.; Miu, L.; Jakab, E. Thermal characterization of new, artificially aged and historical leather and parchment. *J. Anal. Appl. Pyrolysis* **2015**, *115*, 419–427. [CrossRef]
216. Han, B.; Vilmont, L.-B.; Kim, H.-J.; Lavédrine, B.; Sakamoto, S.; Sablier, M. Characterization of Korean handmade papers collected in a hanji reference book. *Herit. Sci.* **2021**, *9*, 96. [CrossRef]
217. Degano, I.; La Nasa, J. Trends in high performance liquid chromatography for cultural heritage. *Top. Curr. Chem.* **2016**, *374*, 20. [CrossRef] [PubMed]
218. Calà, E.; Benzi, M.; Gosetti, F.; Zanin, A.; Gulmini, M.; Idone, A.; Serafini, I.; Ciccola, A.; Curini, R.; Whitworth, I.; et al. Towards the identification of the lichen species in historical orchil dyes by HPLC-MS/MS. *Microchem. J.* **2019**, *150*, 104140. [CrossRef]
219. Barberis, E.; Manfredi, M.; Zilberstein, G.; Zilberstein, S.; Righetti, P. fiat lux... How Alessandro Volta illuminated his scripts. *Comptes Rendus. Chim.* **2021**, *24*, 361–371. [CrossRef]
220. Sawoszczuk, T.; Syguła-Cholewińska, J.; del Hoyo-Meléndez, J.M. Application of HS-SPME-GC-MS method for the detection of active moulds on historical parchment. *Anal. Bioanal. Chem.* **2017**, *409*, 2297–2307. [CrossRef]
221. Zilberstein, G.; Zilberstein, S.; Rocco, R.M.; Righetti, P.G. Jack London and White Fang: A lost struggle. *Comptes Rendus. Chim.* **2022**, *25*, 115–123. [CrossRef]
222. Lazic, V.; Vadrucci, M.; Fantoni, R.; Chiari, M.; Mazzinghi, A.; Gorghinian, A. Applications of laser induced breakdown spectroscopy for cultural heritage: A comparison with XRF and PIXE techniques. *Spectrochim. Acta Part B At. Spectrosc.* **2018**, *149*, 1–14. [CrossRef]
223. Ishii, K. PIXE and its applications to elemental analysis. *Quantum Beam Sci.* **2019**, *3*, 12. [CrossRef]
224. Rawat, K.; Sharma, N.; Singh, V.K. X-ray fluorescence and comparison with other analytical methods (AAS, ICP-AES, LA-ICP-MS, IC, LIBS, SEM-EDS, and XRD). In *X-ray Fluorescence in Biological Sciences: Principles, Instrumentation, and Applications*; Singh, V.K.; Kawai, J., Tripathi, D.K., Eds.; Wiley: Hoboken, NJ, USA, 2022; pp. 1–20, ISBN 9781119645719.
225. Watteeuw, L.; van Bos, M.; Gersten, T.; Vandermeulen, B.; Hameeuw, H. An applied complementary use of macro X-ray fluorescence scanning and multi-light reflectance imaging to study medieval illuminated manuscripts. The Rijmbijbel of Jacob van Maerlant. *Microchem. J.* **2020**, *155*, 104582. [CrossRef]
226. Mazzinghi, A.; Ruberto, C.; Castelli, L.; Ricciardi, P.; Czelusniak, C.; Giuntini, L.; Mandò, P.A.; Manetti, M.; Palla, L.; Taccetti, F. The importance of being little: MA-XRF on manuscripts on a venetian island. *X-ray Spectrom.* **2021**, *50*, 272–278. [CrossRef]
227. Melo, M.J.; Otero, V.; Nabais, P.; Teixeira, N.; Pina, F.; Casanova, C.; Fragoso, S.; Sequeira, S.O. Iron-gall inks: A review of their degradation mechanisms and conservation treatments. *Herit. Sci.* **2022**, *10*, 145. [CrossRef]
228. Cappa, F.; Sterflinger, K. Non-invasive physico-chemical and biological analysis of parchment manuscripts—An overview. *Int. J. Preserv. Libr. Arch. Mater.* **2022**, *43*, 127–142. [CrossRef]
229. Brill, R.H.; Felker-Dennis, C.; Shirahata, H.; Joel, E.C. *Lead Isotope Analyses of Some Chinese and Central Asian Pigments*; Conservation of Ancient Sites on the Silk Road; The Getty Conservation Institute: Los Angeles, CA, USA, 1997; pp. 369–378.
230. Nord, A.G.; Billström, K. Isotopes in cultural heritage: Present and future possibilities. *Herit. Sci.* **2018**, *6*, 25. [CrossRef]
231. Dolgin, B.; Chen, Y.; Bulatov, V.; Schechter, I. Use of LIBS for rapid characterization of parchment. *Anal. Bioanal. Chem.* **2006**, *386*, 1535–1541. [CrossRef] [PubMed]

232. Manso, M.; Costa, M.; Carvalho, M.L. X-ray fluorescence spectrometry on paper characterization: A case study on XVIII and XIX century documents. *Spectrochim. Acta Part B At. Spectrosc.* **2008**, *63*, 1320–1323. [CrossRef]
233. Romano, F.P.; Puglia, E.; Caliri, C.; Pavone, D.P.; Alessandrelli, M.; Busacca, A.; Fatuzzo, C.G.; Fleischer, K.J.; Pernigotti, C.; Preisler, Z.; et al. Layout of ancient Greek papyri through lead-drawn ruling lines revealed by macro X-ray fluorescence imaging. *Sci. Rep.* **2023**, *13*, 6582. [CrossRef]

Disclaimer/Publisher's Note: The statements, opinions and data contained in all publications are solely those of the individual author(s) and contributor(s) and not of MDPI and/or the editor(s). MDPI and/or the editor(s) disclaim responsibility for any injury to people or property resulting from any ideas, methods, instructions or products referred to in the content.

Review

Fatty Acids and Their Metal Salts: A Review of Their Infrared Spectra in Light of Their Presence in Cultural Heritage

Anna Filopoulou, Sophia Vlachou and Stamatis C. Boyatzis *

Department of Conservation of Antiquities and Works of Art, University of West Attica, 12243 Egaleo, Greece; annafilopoulou.s@gmail.com (A.F.); ca15011@uniwa.gr (S.V.)
* Correspondence: sboyatzis@uniwa.gr; Tel.: +30-21-0538-5464

Citation: Filopoulou, A.; Vlachou, S.; Boyatzis, S.C. Fatty Acids and Their Metal Salts: A Review of Their Infrared Spectra in Light of Their Presence in Cultural Heritage. *Molecules* **2021**, *26*, 6005. https://doi.org/10.3390/molecules26196005

Academic Editors: Maria Luisa Astolfi, Maria Pia Sammartino and Emanuele Dell'Aglio

Received: 8 September 2021
Accepted: 28 September 2021
Published: 3 October 2021

Publisher's Note: MDPI stays neutral with regard to jurisdictional claims in published maps and institutional affiliations.

Copyright: © 2021 by the authors. Licensee MDPI, Basel, Switzerland. This article is an open access article distributed under the terms and conditions of the Creative Commons Attribution (CC BY) license (https://creativecommons.org/licenses/by/4.0/).

Abstract: In a cultural heritage context, fatty acids are usually found as breakdown products of lipid-containing organic remains in archaeological findings, binders in aged oil paintings, and additives in modern art-related materials. They may further interact with the ionic environment transforming into metal soaps, a process that has been recognized as a threat in aged paintings but has received less attention in archaeological objects. The investigation of the above related categories of materials with infrared spectroscopy can provide an overall picture of the organic components' identity and demonstrate their condition and prehistory. The capability of investigating and distinguishing fatty acids and their metal soaps through their rich infrared features, such as the acidic carbonyl, the carboxylate shifts, the variable splits of alkyl chain stretching, bending, twisting, wagging, and rocking vibrations, as well as the hydroxyl peak envelopes and acid dimer bands, allows for their direct detailed characterization. This paper reviews the infrared spectra of selected saturated fatty monoacids and diacids, and their corresponding sodium, calcium, and zinc salts and, supported by newly recorded data, highlights the significance of their spectroscopic features.

Keywords: fatty acids; metal soaps; oil binder; archaeological organic remains; infrared spectroscopy

1. Introduction

Lipids (glycerol triesters), existing as organic residues in archaeological findings (such as pottery and metal vessels) or as binders in oil paintings, are prone to hydrolytic damage producing free fatty acids (FA) [1–4]. Furthermore, FA in aged oil paintings may further react with abundant pigment metal ions in their molecular vicinity to form fatty acid metal salts (FAMS, metal soaps) [5–7]. Similar reactions may occur with calcium and other metal ions in archaeological samples, although they have been investigated in relatively fewer circumstances [8–10].

The analytical investigation of organic archaeological residues has received attention due to the resilience of organic molecules, such as lipids and their breakdown products, which helps researchers access information on dietary habits, rituals, and other practices developed by cultures of the past [11–15]. For instance, extended research has been done in ancient unglazed pottery since they offer a convenient protective medium, though allowing for the intrusion of organic compounds, preserves said compounds from environmental degradation for millennia [16–38]. Relatively fewer cases have been reported in wooden containers such as coffins [39] or metal alloys such as copper [9,40]. Most of the above cases have been studied through gas chromatography combined either with mass spectrometry (GC-MS) [18,20,28,33,34,38,41–43] or isotope ratio mass spectrometry (GC-IRMS) [30,36–38,41]. This technique allows for the accurate quantification of analytes. For instance, the oil content was analyzed in ancient amphorae found in the Late Bronze Age shipwreck of Uluburun [26] and the site of Amarna (Egypt) [44], or shipwrecks dating from the First Punic War [45]. In addition, the profiling of the adsorbed oil gradient within the clay fabric of potsherds has been investigated [44].

The acquisition of infrared spectra significantly assists analytical efforts, such as the ones mentioned above. Although detailed infrared analysis, such as the chemical mapping of paint cross-sections has been successfully conducted [46–48], the technique has generally been restricted to simple 'screening' or 'fingerprinting' in organic residue analysis. Nevertheless, its value has been recognized for guiding further analytical efforts [31,49]. Infrared spectroscopic analysis of lipid-containing samples is a fast technique that generally demands minimal sample workup, which based on their molecular bonding [24,50–52], provides rich information on their chemical components [50,53–55], such as glycerol esters, fatty acids, and metal salts. However, interpretation of complex samples can be tedious and problematic since frequent band overlaps impose difficulties for a secure assignment [56–62]. Thus, a necessity emerges for the accurate identification of fatty components in samples characterizing their existing condition through close examination of their infrared spectra.

In addition, advanced technological capabilities have led to the development of reflection-based techniques in combination with microscopy [7,46,48,63–68], which vastly expanded the depth of information offered by the method.

In the field of art, infrared spectroscopy has played a significant role in identifying and studying the binding media in oil paintings [7,69,70], and possibly more importantly, the severe deterioration caused by the formation of metal soaps [7,47,69,71–79]. In archaeology, infrared spectroscopy has also played a decisive role in analyzing organic archaeological residues [9,11,32,39,40,79]. Identifying fatty substances through their infrared spectra can elucidate chemical transformations such as the mineralization of acidic organic components in aged samples. Notably, organic molecular markers that connect the current condition of the materials with their original one can assess the prehistory of cultural heritage objects and their possible uses in antiquity [9,11,25,32,39,66,80,81].

This paper reviews the infrared spectra of selected fatty monoacids, diacids, and their sodium, calcium, and zinc salts, and points out their detailed features for detecting these compounds directly in their samples and further assessing their structural condition. The rich information gained through the close inspection of spectroscopic features can help researchers benefit in their investigations to a level beyond typical infrared screening.

2. Results and Discussion
2.1. Molecular Structure Phenomena in Fatty Acids

The intermolecular attraction of acidic carbonyls due to hydrogen bonding induces a dimerization phenomenon where FA associate in a handshake mode forming head-to-head molecular pairs (or 'dimers') [82–86]. This was first shown experimentally by Pauling, followed by normal mode vibrational analysis [87,88]. In combination with hydrophobic alkyl chain associations, the above results in the bilayer coordination of fatty acid molecules described in Figure 1a [89–91]. Extensively studied through infrared spectroscopy, hydrogen bonding has been recognized and studied in alcohols and carboxylic acids since the 1930s [86,92–96]. As a concept, it gained profound attention in the second half of the previous century [87,88,95,97–101], while advanced computational capabilities during the last two decades have further led to elucidating intermolecular phenomena [102,103].

Carboxylic acids carry a hydrogen bond donor (the OH group) and a hydrogen bond acceptor (C=O) within the same unit and are, therefore, ideal molecules for studying the phenomenon. First, as a profound manifestation of the hydrogen bonding attraction, the hydroxyl stretch is considerably downshifted, compared to their 'free' state, and characteristically broadened [93–95,101]. Secondly, the dimerization phenomenon induces additional dimer infrared bands due to the impeded stretching of associated hydroxyls and carbonyls and restricted bending of the C-O-H group (see below). Moreover, intermolecular associations between long alkyl chains of FA induce crystal packing (Figure 1), affecting their melting points [104,105] and infrared spectra.

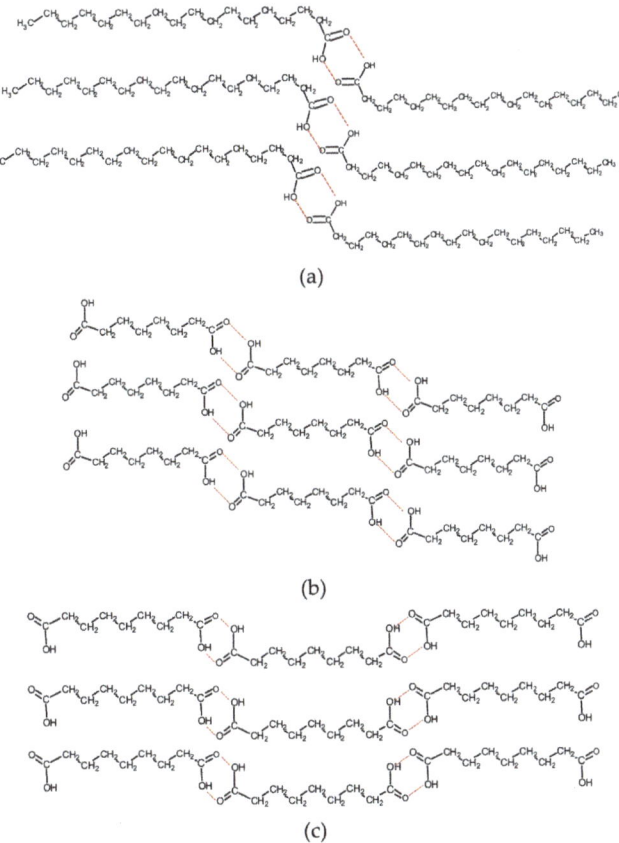

Figure 1. Molecular conformations of fatty acids. (**a**) Type B structure of stearic acid (adapted from [94]); (**b**) octanedioic acid (C8di) and (**c**) nonanedioic acid (C9di) (adapted from [96]).

2.2. Infrared Spectra of Saturated Monoacids and Their Metal Salts

2.2.1. Monoacids

The infrared spectra of saturated fatty acids (sFA) have been studied and their features have been previously reported [106–111]. In particular, studies for C8:0–C18:0 [112–116] may provide powerful identification tools. The published information, supported by newly recorded spectra, is presented and discussed below.

In sFA, the zig-zag hydrocarbon chains assume an overall linear direction, which generally allows for close chain approaching in the bulk phases, and finally, for diverse polymorphic structures, which subtly differ for odd- and even-numbered acids [108,109,115,116]. Fatty acids with chains shorter than that of C9:0 are liquid at room temperature, while C10:0, with a melting point of 31.3 °C, is marginally crystallizable [117]. The latter, along with the higher FA, show pronounced structural infrared features indicative of the crystalline state. Furthermore, even-numbered fatty acids, such as stearic (C18:0), crystallize in temperature-dependent polymorphs A (triclinic) and B, C, and E (monoclinic with orthorhombic sub-cells) [89,108,118]. Structure C is observed at temperatures higher than 35 °C. At room temperature, the metastable monoclinic structure E transforms to the dominating orthorhombic structure B (shown in Figure 1a), where head-to-head acid pairs in trans-planar (or fully extended) chain geometry are formed [89,103,118,119]. This allows the alignment and packing of hydrocarbon chains.

To better study FA infrared spectra, the spectra can be divided into seven regions, each generally having a particular vibrational character. Maxima and assignments are shown in Table 1.

Table 1. Main infrared peaks and assignments of saturated fatty carboxylic monoacids and diacids.

	Peak Maxima, Wavenumbers (cm^{-1}) [1]				
Monoacids	Octanedioic (Suberic) Acid	Nonanedioic (Azelaic) Acid	Assignment	Notes	
3600–2800 br	3037	3045	vOH	Typically, very broad with vague maximum.	
2960–2956 m-w	n.a.	n.a.	$v_{as}CH_3$	Variable, according to the number of carbons.	
2934–2919 s	2991, **2951**, 2941, 2911	2977, **2937**, 2915	$v_{as}CH_2$	Variable, according to the number of carbons; higher-frequency maxima correspond to a lower distance from the COOH group.	
2875–2872 w, s	n.a.	n.a.	v_sCH_3	Variable, according to the number of carbons.	
2851 m-s	2872, 2855	2875, 2858, 2847	v_sCH_2	Variable, according to the number of carbons.	
2670, 2565 br	2759, 2676, 2599, 2534	2774, 2698, 2621, 2553	vO-H•••O=C (dimer stretching band)	Structured with weak shoulders; more extended in diacids.	
1703 vs	1695	1694	vC=O, acidic		
1472, 1464, 1411 m	1470, 1410	1471, 1411	δCH_2 scissoring	Split into three components (two in diacids); the 1472 and 1464 cm^{-1} components are better resolved for monoacids C16-C24.	
1457–1450 w	n.a.	n.a.	$\delta_{as}CH_3$	Contribution is lower for higher carbon-number chains.	
1431 m-w	1426	1436	δ_{ip}COH	Relatively broad; often missed due to overlaps.	
1372 m-w			$\delta_s CH_3$ ('umbrella' vibration)		
1356–1347 m-w	1334	1360, 1346	τCH_2	Splitting in progressions for acids in their crystalline state (in room temperature, higher than C10).	
1318–1185 m-w	1360, 1346	1317, 1282, 1268, 1254, 1208, 1196	wCH_2	Splitting in extended progressions for acids in their crystalline state (in room temperature, higher than C10).	
1112–1075 m-w	n.o.	1105, 1098	vC-OH	Up-shifted for longer hydrocarbon chains. Weak, or not observed for diacids.	
943–935 m, br	932	920	δ_{oop}C-O-H•••O=C (bending dimer band)		
795–741, 725–710 w	796, 725	776, 726	ρCH_2	Stronger in long hydrocarbon chains; the 725 cm^{-1} peak is doubly split in crystalline monoacids.	
690	683	681	δ_{oop}C-O-H		

[1] in bold: the stronger peaks in the case of multiplets. v: stretching vibration; vs. symmetric stretching; v_{as} antisymmetric stretching; δ: bending vibration; δ_s symmetric bending; δ_{as} antisymmetric bending; δ_{ip}: in-plane bending; δ_{oop}: out-of-plane bending; ρ: rocking vibration; τ: twisting vibration; w: wagging vibration.

The Acidic Hydroxyl Stretching Region

The hydroxyl group is involved in the intense hydrogen bonding between molecules of the same type, which controls the structural conformations in the unit cell (Figure 1a). The O-H stretching region contains a typically broad, intense band, indicative of the acidic OH hydrogen bonding at 3600–2800 cm^{-1} with a vague maximum of around 3050–3000 cm^{-1} (Figure 2a).

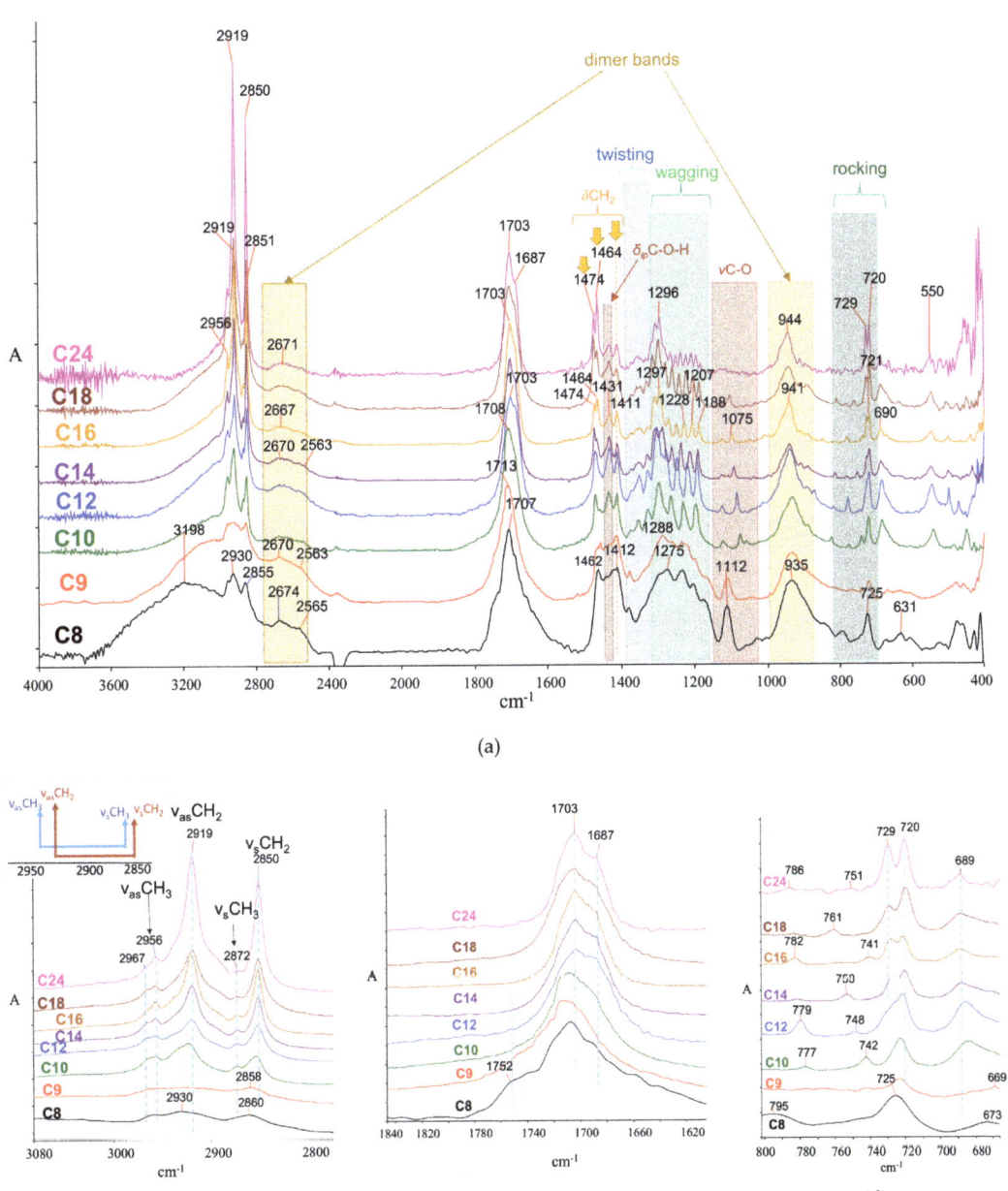

Figure 2. Infrared spectra of fatty acids. (**a**) Full spectra with shaded areas highlighting characteristic features; (**b**) C-H stretching; (**c**) carbonyl stretching; and (**d**) rocking vibrations. In all spectra: (i) C8, octanoic; (ii) C9, nonanoic; (iii) C10, decanoic; (iv) C12, dodecanoic; (v) C14, tetradecanoic; (vi) C16, hexadecanoic; (vii) C18, octadecanoic; (viii) C24, tetracosanoic. The negative peak at ~2380 cm^{-1} for C8 corresponds to carbon dioxide as a result of improper background subtraction.

An additional broad band at 2750–2500 cm^{-1} is assigned to hydroxyl-carbonyl interactions due to acid dimers [120]. Weak maxima are also observed in this region at ~2670 and 2560 cm^{-1}, often labeled as 'satellite bands,' and are somewhat unclear as they have been assigned to contributions from combination/summation of δOH and vC-O, Fermi resonance, as well as ionic resonance structures [57,121–123]. Nevertheless, they may work as supporting evidence for detecting FAs through infrared spectra of complex mixtures [57].

The C-H Stretching Region

The structural similarities between hydrocarbon chains in sFA and n-alkanes were recognized early and associated with their infrared spectra [107,109], although crucial differences were also pointed out [124]. The C-H stretch region contains the methyl (CH$_3$) and methylene (CH$_2$) vibrations, each of these sub-categorized into antisymmetric (*asym-*) and symmetric (*sym-*) vibrations [58,59,108,109,125], the former always observed at higher wavenumbers than the latter (Figure 2b). The maxima depend on the methyl or methylene group proximities to the carbonyl group (the closer to the carboxyl group, the higher the frequency) and the hydrocarbon chain length. Interestingly, the v_{as}CH$_2$ maxima for acids C3 and C4, with only α- and β- methylene groups (spectra not shown here), appear as high as 2990 and 2973 cm^{-1}, respectively [108,126]. Additional CH$_2$ groups that build longer chains contribute to the methylene stretching band with a maxima around 2932 cm^{-1} (as shown in the deconvolution analysis in Figure S3a–c). Further addition of CH$_2$ groups contributes to the sub-band at 2921–2919 cm^{-1} (Figure S3b,c). For C18:0, the 2919 cm^{-1} peak is the main contributor to the antisymmetric band.

A trend for v_{as}CH$_3$, v_{as}CH$_2$, v_sCH$_3$, and v_sCH$_2$ maxima as a function of the hydrocarbon chain length can be seen in Figure 3. The maxima decrease from short to mid-size (C12) chains, beyond which no significant change is marked (within the spectra acquisition resolution: 4 cm^{-1}). Moreover, the intensity ratios of methylene over methyl peaks are related to the number of CH$_2$ groups in the chain, thus reflecting the hydrocarbon chain length.

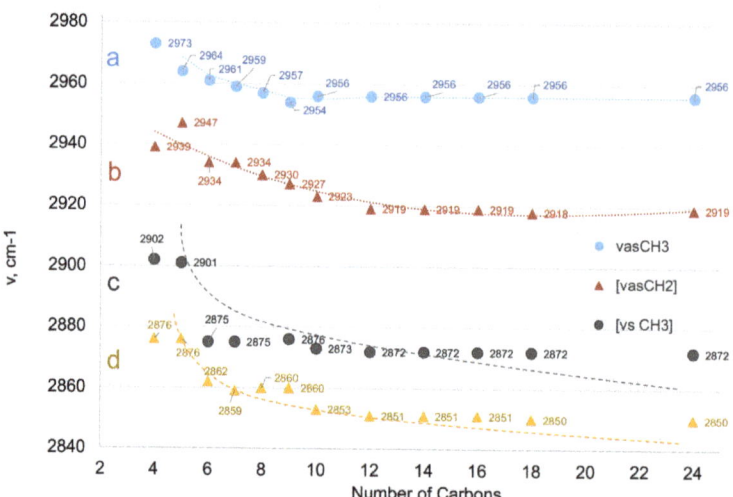

Figure 3. Graph illustrating the trends for (a) v_{as}CH$_3$, (b) v_{as}CH$_2$, (c) v_sCH$_3$, and (d) v_sCH$_2$ peak maxima with respect to the total number of carbons in sFA. Lower-chain acids, not discussed in the main text, are included to further emphasize the trend.

The Acidic Carbonyl Stretching Region

The carbonyl group absorption maxima depend on the conformation and packing of the hydrocarbon chains, which are more intense in long chains. The acidic carbonyl stretching vibration maximum (vC-O) is generally observed around 1710–1700 cm^{-1} (Figure 2a);

this maximum is considerably downshifted compared to that of isolated molecules (for instance, in the gas phase) [59]. Since the acidic carbonyl group participates in intermolecular hydrogen bonding with the hydroxyl group of a pairing molecule (Figure 1a), the acidic carbonyl downshifts and relatively broadens (Figure 2c) [58,59]. At room temperature, even-numbered acids from C10 and higher are solid and associate efficiently in their crystal; as a result, hydrogen bonding is stronger and results in moderately down-shifted peaks at ~1700 cm^{-1}. An additional significant feature is the shoulder at 1687 cm^{-1} observed in C14–C24 (Figure 2c), which can be assigned to a tighter association between dimers.

The CH$_2$ and CH$_3$ Bending Region

The methylene vibration is typically observed at 1465–1430 cm^{-1} for all sFA due to the symmetric in-plane bending (or 'scissoring') mode, as shown in Figure 2. A sharp, medium-strong peak around 1470 cm^{-1} is observed. For longer alkyl chain acids, the band is resolved in two sub-bands (~1473 and 1464 cm^{-1}) with a contribution of the α-CH$_2$ group at ~1411 cm^{-1}. On the other hand, the CH$_3$ symmetric bend (the 'umbrella' vibration) appears at around 1370–1350 cm^{-1}, while the antisymmetric appears at 1473 cm^{-1} [56,58,59,127,128]; the relative intensity of these bands generally diminishes with increasing hydrocarbon chain length. Figure S1 zooms in the 1570–1000 cm^{-1} region, showing bands more clearly for the crystallizable FAs.

The CH$_2$ Twisting and Wagging Region

This region is characteristic of crystallizable compounds and materials containing long saturated hydrocarbon chains; as a typical feature, band splitting in the form of band progressions appears [59] (shown in Figure 2a and, in more detail, in Figure S1). Early systematic studies that started in the 1950s, based on the coupled oscillator model of the zigzag-shaped hydrocarbon chain [107,124,129,130], offered a better understanding of this phenomenon. In the crystalline state, the twisting vibration of sFA is split throughout 1368–1329 cm^{-1} which better resolves upon chain increase; double-split CH$_2$ twisting peaks are observed in C14:0-C18:0, while four-split are observed in C24:0.

A similar phenomenon is even more pronounced for the methylene wagging vibrations, where extensive splitting in the form of progressions at 1320–1160 cm^{-1} are observed [107–110,131]. The multiplicity of progressions (which can be termed as w_1, w_2, etc.) increases with increasing hydrocarbon chain length (seen in detail in Supplementary Materials Figure S1), showing a multiplet of bands, the number w of which appears to follow the rule $w = \frac{n}{2} - 1$, where n is the total number of carbons in a fatty acid. Moreover, the distancing between consecutive w bands diminishes with increasing hydrocarbon chain, approx. 34 (for C10), 29 (C12), 24 (C16), 18 (C18), and 13 cm^{-1} (C24). The progressions may be stronger in well-crystallized samples, for instance, when the crystalline state is slowly formed upon the annealing of a fatty acid melt [108,109,132,133]; this is demonstrated in Figure S3 for stearic acid (C18:0), where a recrystallized C18:0 sample after a heating-annealing cycle is compared with the one that was recorded as purchased. Generally, on a practical level, when these bands are distinguishable, the carbon number of the fatty acid can be estimated.

Hydroxyl C-O Stretch and C-O-H Bending Region

The carbon-oxygen bond of the acidic hydroxyl stretches (vC-O) at around 1100 cm^{-1} and appears as a medium peak (Figure 2a); for liquid fatty acids, the band is relatively broad at 1100–1110 cm^{-1}, while for solid acids, it is split at 1124 cm^{-1} (weaker) and 1075–1122 cm^{-1} (stronger). The latter upshifts upon moving from C10:0 to C18:0, while in C24:0 this peak almost disappears (Figure S1).

The C-O-H in-plane bending vibration (δ_{ip} C-O-H) is observed as a medium, relatively broad peak at ~1431 cm^{-1} [59]. Although it is often partly overlapped in unknown, complex samples, its diagnostic value can be significant. Likewise, the out-of-plane bending vibration (δ_{oop} C-O-H), which typically appears as a weak peak at around 690 cm^{-1} and is also responsible for a medium-strong broad peak at 943–935 cm^{-1} [56,59] due to dimer formation, is also of diagnostic value (Figure 2a).

The CH2 Rocking Region

This band may span in the 780–700 cm^{-1} range [124]. In-phase CH_2 rocking vibration most prominently appears at 725–710 cm^{-1} (Figure 2d). In the crystalline state (and similarly with hydrocarbons), the band splits into a doublet with Δv 7–10 cm^{-1}, due to the anisotropic coupling of similar vibrations in adjacent CH_2 groups [109,134]. As shown in Figure 2d, splitting intensifies for long hydrocarbon chains (C14:0, C16:0, C18:0, and even more for C22:0).

Infrared Maxima and Crystallization

Most spectral characteristics discussed above depend on the state of fatty acids. Efficient crystal packing results in the downshifting of acidic carbonyl and the formation of CH_2 structural features (twisting, wagging progressions, and splitting of bending and rocking vibrations). On the other hand, non-crystallizable sFA, such as C8:0 and C9:0 (melting points 16.5 °C and 12.4 °C, respectively) show a relatively uniform broad band at 1380–1180 cm^{-1} with poorly resolved shoulders (Figure 2a); the lack of multiplicities reflects their non-crystalline nature at room temperature. A marginal case is capric acid (C10, melting point 34 °C), where crystallization at room temperature is not always efficient and although the wagging progressions are evident, the carbonyl peak appears upshifted (1708 cm^{-1}), and the rocking vibration is not split (Figure 2a,c).

The rocking band is significant and of diagnostic value regarding the chain length and effectiveness of crystallization. While the twisting and wagging progressions (discussed above) appear for any solid-state FA (i.e., for C10 and higher), the rocking band splits for chains longer than C14, as shown in Figure 2a,d. Furthermore, effective crystallization may affect the shape of the split band (i.e., even and deeper-split sub-bands in well-crystallized samples). The crystallization effect is shown in Figure S3 by comparing the CH_2 rocking band of C18:0 before and after the annealing-recrystallization cycle. In conditions favoring even more efficient packing (such as freezing in 10K [124]), the rocking vibration is even more split where coupling with the twisting vibration is evident as a progression ranging from 1065 to 720 cm^{-1}.

2.2.2. Fatty Monoacid Metal Salts

Fatty acids can be transformed into FAMS in the presence of metals [135,136] and reactive metallic compounds such as oxides, hydroxides, and salts [62,71,137–139]. In paintings, many of these metallic compounds exist in inorganic pigments and have been found to cause soap formation, particularly of zinc and lead, a phenomenon that has received considerable attention from researchers due to the detrimental effects in works of art [5,77,87,140–142]. Their presence in archaeological samples and other cultural heritage objects has also been documented [8,9]. In particular, calcium soaps are considered to be common products through FA adsorption to calcium carbonate in the geological environment and ceramics [27,143,144], since their formation is favored due to their very low solubilities in water (K_{sp} of the order of 10^{-17} mol^3/L^3). Other metal soaps can also be expected, such as copper, aluminum, cobalt, and iron [84,139,145]. This mineralization process, which can be followed by inspecting infrared spectra of samples, may prolong the preservation of the lipid fraction in the burial or other environments due to low solubility. Moreover, the above may lead to a severe underestimation of lipid yields during routine chromatographic analysis involving lipid extraction with organic solvents [27]. In the light of the above, chromatographic analysis methodologies measuring metal soaps as added components in lipid analysis of cultural heritage samples have recently been considered [146]. Finally, contamination by cleaning agents, including the sodium soaps (typically, stearates), may not be excluded. For all the above reasons, identifying FAMS in the sample prior to any chemical processing offers significant help in assessing the overall lipid load in a sample and further assessing its prehistory. To this end, the infrared spectra of calcium, zinc, and sodium fatty acid salts are presented, and their characteristic features are pointed out.

The infrared spectra of FAMS have been studied to some extent [78,87,133,138,147]. In support of the current discussion, selected sodium (representative of late-historical cleaning agents and modern additives in routinely used materials), calcium (typically expected in interaction products of FA with the calcareous environment) and zinc soaps (a usual case of metal soaps in oil paintings) were prepared in the laboratory (see Materials and Techniques). Their infrared spectra were recorded.

Although soap spectra generally share similar features, diagnostically significant differences can be observed regarding the carboxylate and some of the hydrocarbon chain vibrations. Moreover, in the case of the bivalent ions Ca and Zn, hydrated salts were produced, as evidenced by the intense, broad, crystalline water peaks (at ~3440 and ~1626 cm^{-1} for the calcium salt, and ~3565–3580 and 1620–1600 cm^{-1} for the zinc salt). Spectra are shown in Figure 3 (sodium), Figure 4 (calcium), and Figure 5 (zinc), while peaks with their assignments are listed in Table 2.

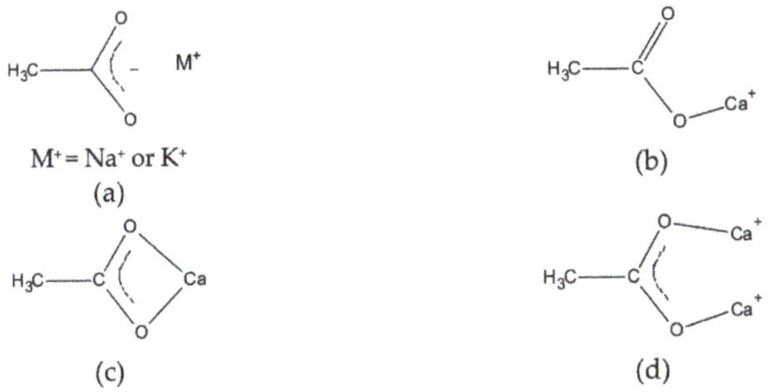

Figure 4. Chemical structures of metal carboxylates. (**a**) Ionic form of monovalent metal ions; (**b**) unidentate coordination; (**c**) chelating bidentate coordination; and (**d**) bridging bidentate coordination to calcium ions. Structures adapted from [148,149,152].

The Carboxylate Stretching

The carboxylate bands typically exist in pairs corresponding to the antisymmetric ($v_{as}COO^-$) and symmetric (v_sCOO^-) carboxylate vibrations at 1575–1530 and 1460–1400 cm^{-1}, respectively.

The coordination geometry plays a significant role in the band maxima [147–149]. Moreover, the separation between the antisymmetric and symmetric band, often called the Δ ('delta') value, depends on the coordination symmetry; the lower the symmetry, the higher the Δ. More specifically, carboxylates may assume various symmetry levels based on whether they exist in a pure ionic form (or a salt) or metal-coordinated forms such as the unidentate, the chelate bidentate, and the bridging bidentate [148–151] (Figure 4). In the lower-symmetry unidentate complexes (Figure 4b), the antisymmetric maximum is expected to appear close to that of the corresponding carboxylic acid. The more symmetric ionic and chelating bidentate structures (Figure 4a,c, respectively) appear at roughly similar frequencies, while the bridging bidentate is even lower [149]. The maxima are very sensitive to the metal type (i.e., ion mass, electronegativity, and effective charge [87]) but generally insensitive to alkyl chain length, as shown in Figure S4.

Table 2. Main infrared peaks and assignments of monoacid and diacid metal salts.

Peak Maxima, Wavenumbers (cm^{-1})			Assignment	Notes
Monoacid Metal Soaps	Octanedioate (Suberate) Metal Soaps	Nonanedioate (Azelate) Metal Soaps		
3600–3300	3590, 3521 Zn	3565, 3521 Zn	vO-H (cryst. water)	In Ca and higher Zn monoacid salts. Only in zinc diacid salts.
2956	n.a.	n.a.	v_{as}CH$_3$	
2940(sh), 2926–2916	2942, 2930, 2909 Na 2978, 2939, 2921, 2907 Ca 2939, 2927, 2905 Zn	2943, 2936, 2921, 2907 Na 2976, 2947, 2921, 2915 Ca 2939, 2927, 291 Zn	v_{as}CH$_2$	Downshifted (3–5 cm^{-1}) in monoacid salts, for >C16 (Ca, Na) and >C14 (Zn). Structured in diacid salts.
2874–2870	n.a.	n.a.	v_sCH$_3$	
2856–2850 Na salts 2851 Ca and Zn salts	2863 Na 2860, 2850 Ca 2867, 2851 Zn	2858, 2848 Na 2862, 2849 Ca 2867, 2850 Zn	v_sCH$_2$	Downshifted (~5 cm^{-1}) in monoacid Na salts, for >C9. Split into two components in diacid salts
1628 (br) Ca salt 1617–1619 Zn salts C12-C24	1607 Zn	1616 Zn	δH-O-H (cryst. water)	In Ca and higher Zn monoacid salts. Only in zinc diacid salts.
1560 Na salts 1579, 1542 Ca salts 1551, 1532 Zn salts, C8, C9 1540 (br) Zn salts C10-C24	1563 Na 1581, 1544 Ca 1551, 1537 Zn	1575 Na 1581, 1542 Ca 1556, 1534 Zn	v_{as}COO$^-$	Singlet for Na and higher Zn salts. Doublet for Ca and lower Zn salts.
1472–1459 Na salts 1473 Ca salts 1467–1464, Zn salts	1463, 14447, 1432, Na 1468, 1455, 1431, 1411 Ca 1467, 1460, 1450 Zn	1463, 1405 Na 1468, 1434, 1420 Ca 1468, 1456 Zn	δCH$_2$ scissoring	Single peak, a stable frequency for Ca and higher monoacid Zn salts. Four- or three-fold structuring in diacid salts.
1458 Zn salts	n.a.	n.a.	δ_{as}CH$_3$	Overlapped in Na salts; undetectable in Ca salts. Detectable as unresolved shoulder in Zn lower salts.
1423 Na salts 1433, 1411 Ca salts 1410, 1399 Zn salts C8, C9 1398 Zn salts C10-C24	1416 Na 1436, 1405 Ca 1412, 1399 Zn	1434 Na 1431, 1411 Ca 1407, 1398 Zn	v_sCOO$^-$	Singlet for Na and higher Zn salts. Doublet for Ca and lower Zn salts.
-	n.a.		δ_sCH$_3$ ('umbrella' vibration)	Undetectable in most monoacid salts
1400–1351 Na salts 1380–1287 Ca salts 1375–1260 Zn salts	1356 Na 1359, 1332 Ca 1366 Zn	1339 Na 1343, 1318 Ca 1353 Zn	τCH$_2$	Progressions of 4 or 5 sub-bands in monoacid salts. Singlet or doublet in diacid salts.
1341–1180 Na salts 1278–1190 Ca salts 1250–1045 Zn salts	1293, 1200 Na 1293, 1261, 1247, 1217, 1201 Ca 1303, 1207 Zn	1316, 1288, 1273, 1253, 1233, 1195 Na 1286, 1251, 1200 Ca 1285, 1243 Zn	wCH$_2$	Progressions following the n/2 pattern (n = total carbon atoms) in Na and Ca salts. Sub-bands in Zn salts.
1185–1100 Na 1114 Ca 1067–1105 Zn	1100	1100	Unassigned	Weak single band; up-shifting with longer carbon chains in monoacid Zn salts. Single band; not observed in Na and Zn C8di salts.
855–849			δCH$_3$ + vC-C.	Not observed in monoacid zinc and diacid soaps.
725–717, 698 721 Ca salts 747, 723 Zn salts			ρCH$_2$	Doublet for Na and Zn salts; single peak for Ca salts.

v: stretching vibration; v_s: symmetric stretching; v_{as}: antisymmetric stretching; δ: bending vibration; δ_s: symmetric bending; δ_{as}: antisymmetric bending; δ_{ip}: in-plane bending; δ_{oop}: out-of-plane bending; τ: twisting vibration; ρ: rocking vibration; w: wagging vibration.

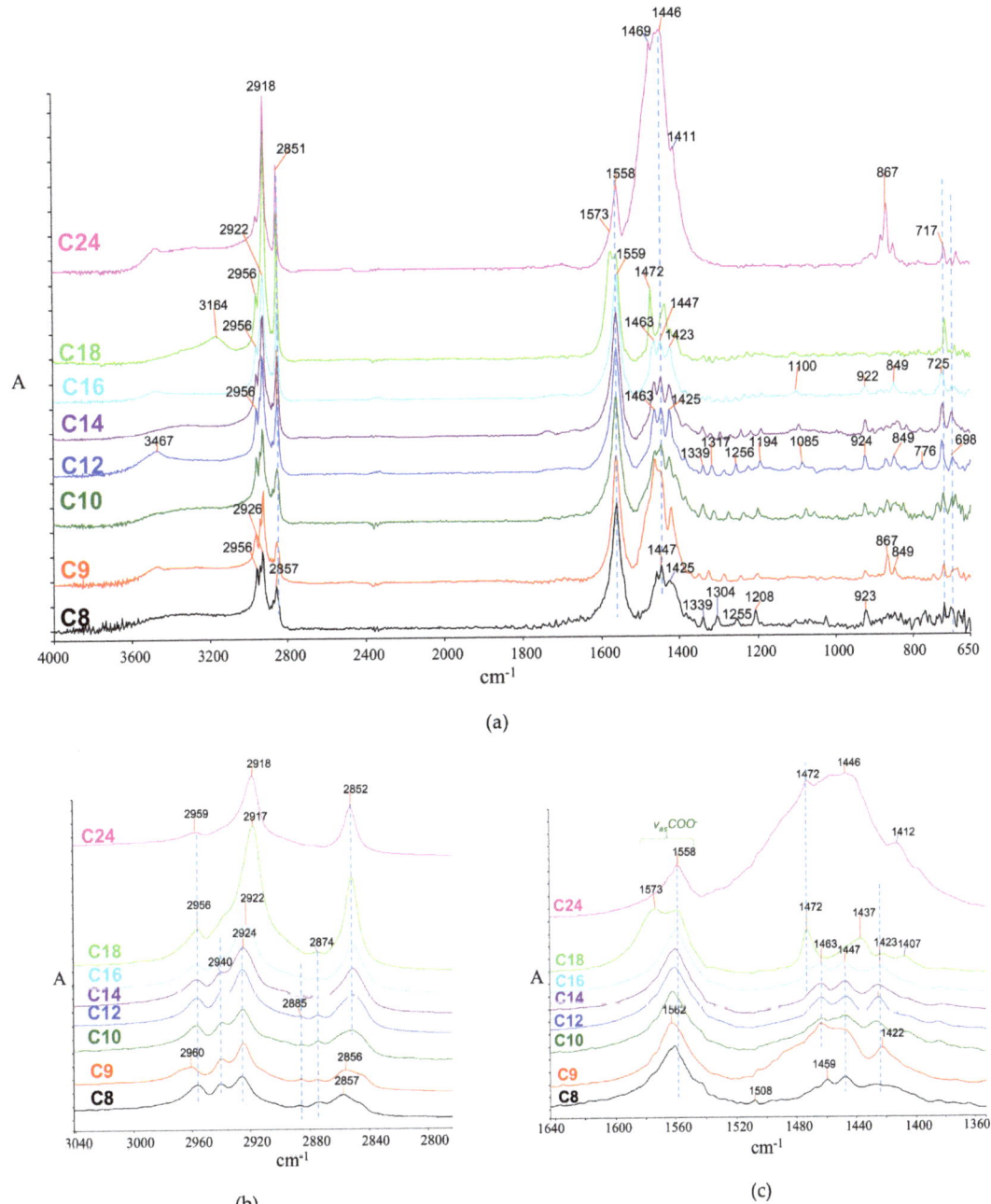

Figure 5. Infrared spectra of fatty acid sodium salts. (**a**) Full spectrum; (**b**) C-H stretching region; and (**c**) carboxylate region. In all spectra: C8, sodium octanoate; C9 sodium nonanoate; C10, sodium decanoate; C12, sodium dodecanoate; C14, sodium tetradecanoate; C16, sodium hexadecanoate; C18, sodium octadecanoate; C24, sodium tetracosanoate; C9 and C24 are contaminated with sodium carbonate (intense peaks at ~1450 and 867 cm^{-1}). The negative peaks at ~2380 cm^{-1} correspond to carbon dioxide as a result of improper background subtraction.

The sodium soaps can be correlated with the ionic structure (Figure 4a), showing carboxylate maxima at ~1560 cm^{-1} and 1425 cm^{-1} for $v_{as}COO^-$ and v_sCOO^-, respectively [149]. Specifically, for sodium stearate (C18Na), the antisymmetric band appears as a doublet at 1573 and 1559 cm^{-1} (Figure 5c), possibly due to two different sub-cell arrangements [153]. Interestingly, deviations of the above are observed in cases where sodium salts co-crystallize with their corresponding FA. A unified acid-soap carbonyl stretch is observed in above, at ~1740 cm^{-1}, attributed to a hydrogen bonding decrease combined with changes in acid-soap conformations [153].

For the calcium soaps, double-splitting is observed for all FAMS at 1580 and 1542 cm^{-1} ($v_{as}COO^-$) and 1435 and 1419 cm^{-1} (v_sCOO^-), indicating two different denticities in the coordination structure (Figure 6). Splitting of this type has been reported for precipitated calcium soap molecular units in bulk acquiring three-dimensional molecular geometry [84,154]. On the other hand, for two-dimensional calcium soap layers, such as those formed through the interaction of fatty acids with calcium-containing surfaces, the doublet becomes a singlet (1550 cm^{-1} for the antisymmetric band) [155]. Since the calcium soaps as reaction products of free fatty acids with calcium carbonate are generally expected in specific archaeological contexts [27,143,144], the above maxima are generally expected and have been in some instances observed [8,9].

The zinc salts (Figure 7) in most cases show single, relatively broadened carboxylate bands at 1540 ($v_{as}COO^-$) and 1398 cm^{-1} (v_sCOO^-). However, in the case of the shorter-chain fatty acid salts (C8$_2$Zn and C10$_2$Zn), splitting is observed at 1550 and 1532 cm^{-1} ($v_{as}COO^-$), and 1410 and 1399 cm^{-1} (v_sCOO^-) [72,73,155].

As a general remark, the splitting in both *asym-* and *sym-* carboxylate bands of the divalent metal salts into doublets has been previously observed and attributed to bidentate and unidentate salt types (Figure 4) [154,156]; the splitting can be as large as ~40 cm^{-1} for the calcium salts, while it is significantly smaller for the zinc salts. The trends in antisymmetric and symmetric carboxylate maxima for various sodium, calcium, and zinc soaps are shown in Figure S4. A significant feature of the carboxylate bands is the separation Δv between the antisymmetric and symmetric peaks, considered to depend on factors such as the type of metal, the ligand, and the coordination type and denticity [157,158], the packing of chains [158–161], as well as the molecular environment (solvent, or other materials in the molecular environment) [73,160]. As shown in Figure S4 (values drawn from spectra in Figure 4, Figure 5, and Figure 7), the Δv value in the lower-symmetry sodium soaps is generally expected to be ~137 cm^{-1}, while for the higher-symmetry calcium and zinc, this is expected to be higher at 142–144 cm^{-1} [158].

The calcium and zinc salts show similar Δv values at 143–144 cm^{-1} (with the exception of the C24 salt), suggesting higher-symmetry bidentate coordination. The differences in separation values between the sodium and the divalent salts are marginal as they are comparable with the wavenumber uncertainty limit (4 cm^{-1}, see Materials and Techniques). Our data shows a deviation for the C8Zn and C9Zn salts, showing Δv values at 131 cm^{-1}.

As a general remark, the carboxylate bands can be significantly broadened and shifted so that Δv decreases when soaps are formed in an amorphous molecular vicinity (as is the case of metal soaps formed in paintings with the glassy molecular environment of the binding medium) [6,73,160].

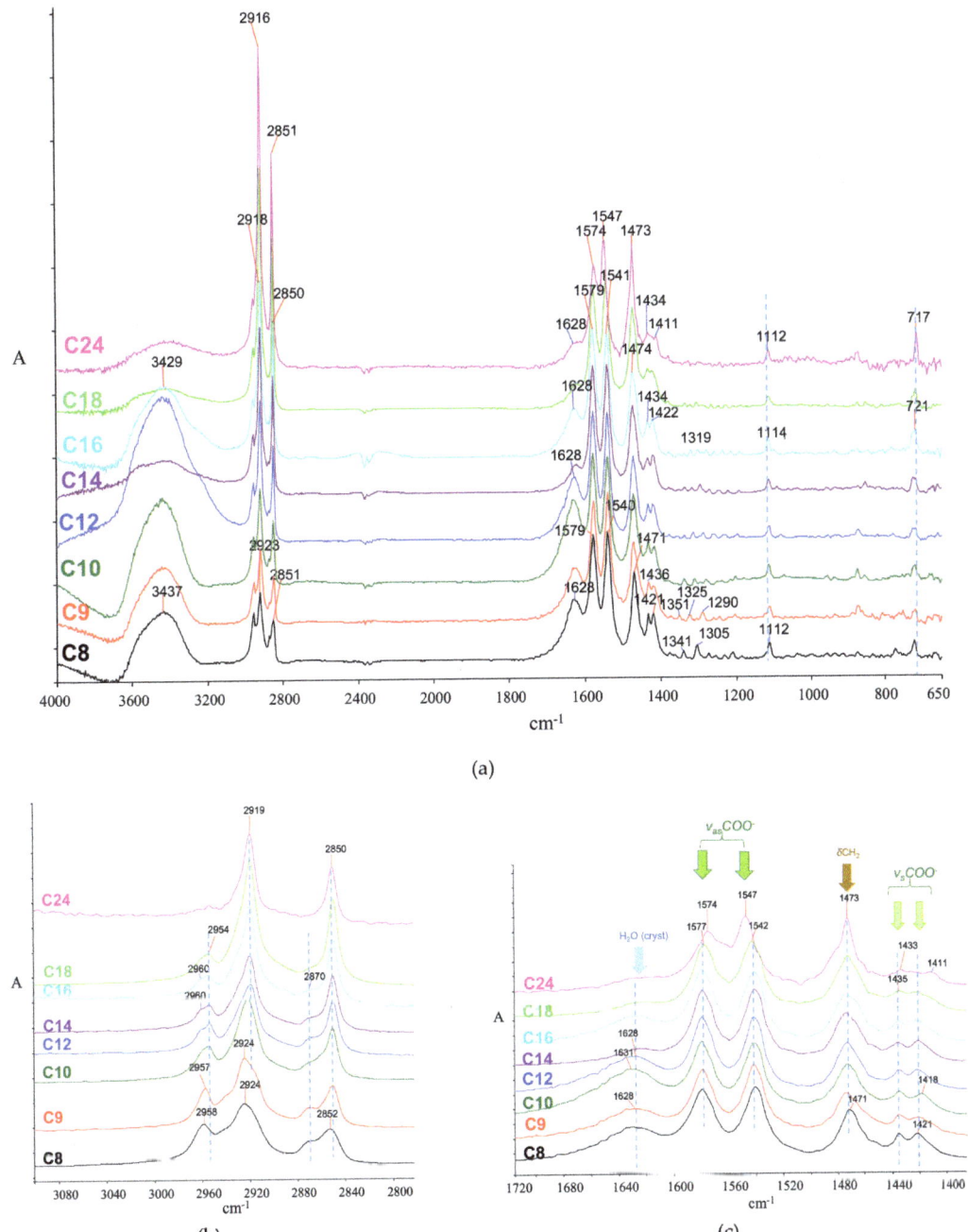

Figure 6. Infrared spectra of fatty acid calcium salts. (**a**) Full spectrum; (**b**) C-H stretching region; and (**c**) carboxylate region. In all spectra: C8, calcium octanoate; C9 calcium nonanoate; C10, calcium decanoate; C12, calcium dodecanoate; C14, calcium tetradecanoate; C16, calcium hexadecanoate; C18, calcium octadecanoate; C24, calcium tetracosanoate. Contaminated with small amounts of calcium carbonate (peaks at ~1435 and 874 cm^{-1}). The negative peaks at ~2380 cm^{-1} correspond to carbon dioxide as a result of improper background subtraction.

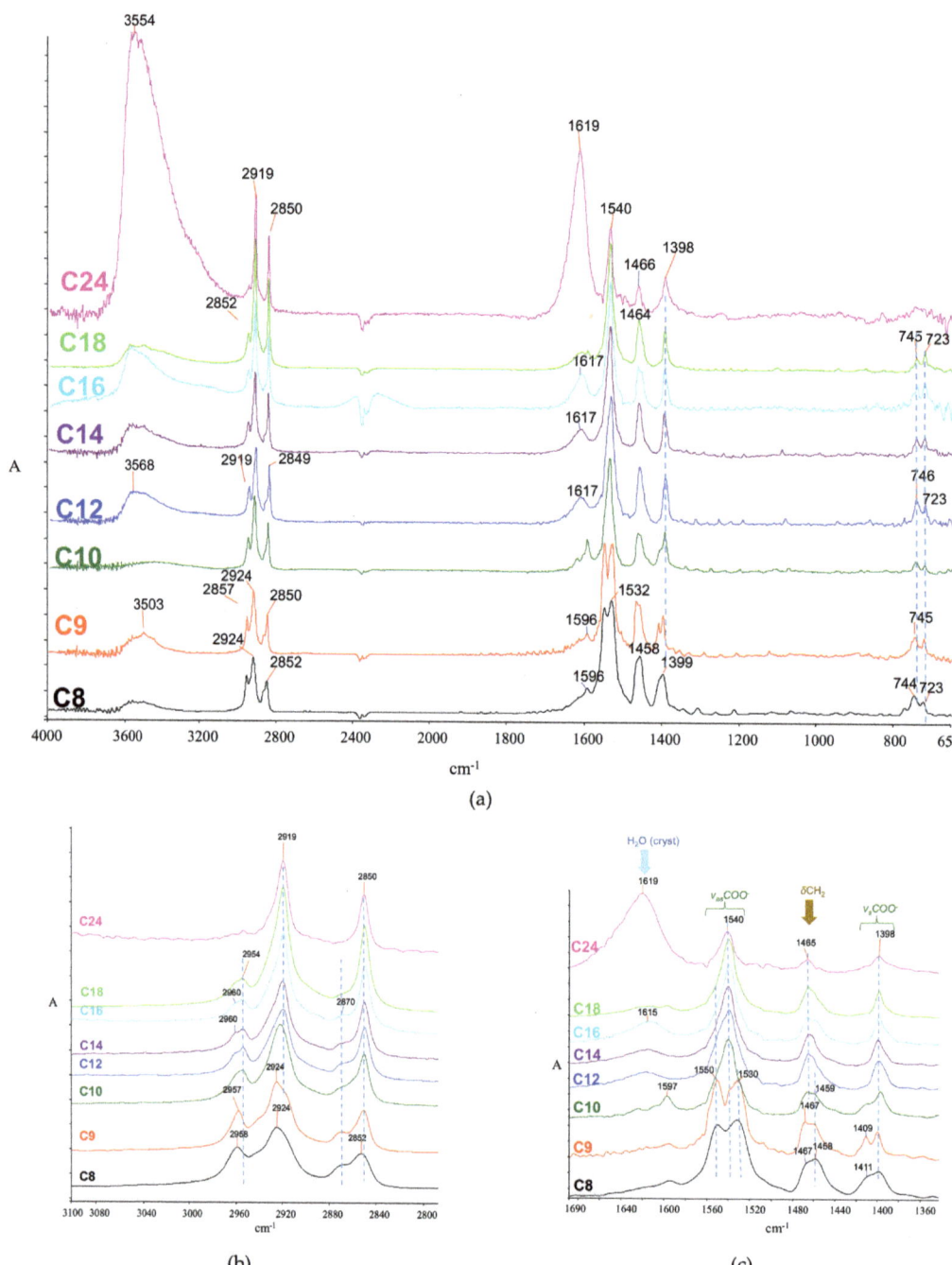

Figure 7. Infrared spectra of fatty acid zinc salts. (**a**) Full spectrum; (**b**) C-H stretching region; and (**c**) carboxylate region. In all spectra: C8, zinc octanoate; C9 zinc nonanoate; C10, zinc decanoate; C12, zinc dodecanoate; C14, zinc tetradecanoate; C16, zinc hexadecanoate; C18, zinc octadecanoate; C24, zinc tetracosanoate. The negative peaks at ~2380 cm^{-1} correspond to carbon dioxide as a result of improper background subtraction.

The C-H Vibrations

As a general remark, the CH$_2$ antisymmetric stretching band in FAMS shows fewer components and lower variability in their maxima than those in FAs. Spectra are shown in Figure 3b, Figure 4b, and Figure 5b, for the three metal salts, respectively. For the carbons closer to the carboxylate group, maxima appear at higher frequencies (2945–2935 cm^{-1}), while for the more distanced ones, these are downshifted at approximately 2920 cm^{-1}. Notably, for the sodium salts of fatty acids with chains shorter than C14, the peaks are resolved to ~2957, 2940, and 2920 cm^{-1}. The *sym-* CH$_2$ stretching typically appears at 2857–2850 cm^{-1}, and the CH$_3$ stretching appears at 2956 (*asym-*) and 2874 cm^{-1} (*sym-*).

The methylene bending vibrations were found at 1460–1470 cm^{-1} (Na salts), 1468–1473 cm^{-1} (Ca salts), and 1456–1467 cm^{-1} (Zn salts), respectively, while the methyl bending is either very weak or not observed at all. This region is shown in Figure 3c, Figure 4c, and Figure 5c. The twisting vibrations are observed at 1400–1260 cm^{-1} as progressions of four or five peaks, which are quite intense for Ca and Zn salts. The wagging vibrations are also observed as progressions, generally at 1270–1180 cm^{-1}, similar to those of the corresponding FA, although relatively weaker and at numbers that follow the n/1 pattern (where n is the number of total carbon atoms). The weak bands at 870–860 cm^{-1} are assigned to the coupling between δCH$_3$ and vC-C [84]. Finally, the splitting of rocking vibrations in the form of doublets at 745–723 cm^{-1} is observed only for zinc salts (Figure 7a). In most other cases, this is elusive, reflecting a misalignment of the hydrocarbon chains, possibly induced by the big difference between the ionic group and hydrocarbon chain polarities.

Based on the recorded spectra, the above results refer to the pure materials, and therefore 'ideal' situations in the absence of organic additives, inorganic materials, etc.; spectra of soaps present in 'real' samples may differ in their subtle features such as progressions and splitting. Metal soaps on surfaces may acquire liquid crystal character [138]; in the case of a divalent metal (calcium, zinc), the hydrocarbon chains arrange on both sides of the metal ion plane. When this arrangement is unperturbed (for instance, in pure salts), alkyl chains longer than C12 align well to show the progressions and splittings, as described above. However, the non-polar liquid crystal layer may occasionally accommodate foreign compounds of small MW (solvents, degradation products, etc.) that may disrupt the ordered chain alignment and, therefore, diminish the intensity of these features [138,145,161]. In these cases, the appearance of progressions and split peaks can be considered as a measure of phase purity. Nevertheless, more research is needed to investigate the crystallization phenomena of metal soaps in various environments.

2.3. Infrared Spectra of Selected Diacids and Their Metal Salts

2.3.1. Azelaic and Suberic Acids

During the last 50 years, research has shown that diacids, especially nonanedioic and octanedioic (azelaic, C9di, and suberic, C8di, respectively) acids, are standard oxidation products of polyunsaturated fatty acids [75,84,162]. They are commonly detected in organic remains of archaeological objects containing unsaturated oils [1,2,10,163–165], and in oil-based paintings [76,166–169]. Specific features in diacid spectra (shown in Figure 8, maxima and assignments in Table 1) can be helpful in their identification and distinction from monoacids.

The acidic carbonyl absorption for diacids is observed at ~1695 cm^{-1}, downshifted compared to monoacids due to more efficient hydrogen bonding between carboxyl groups. According to their structural characteristics, carboxyl groups are associated with intense intermolecular hydrogen bonding, resulting in monoclinic crystals. As seen in Figure 1b,c, the structures differ depending on odd- and even-numbered chains [91,170–173].

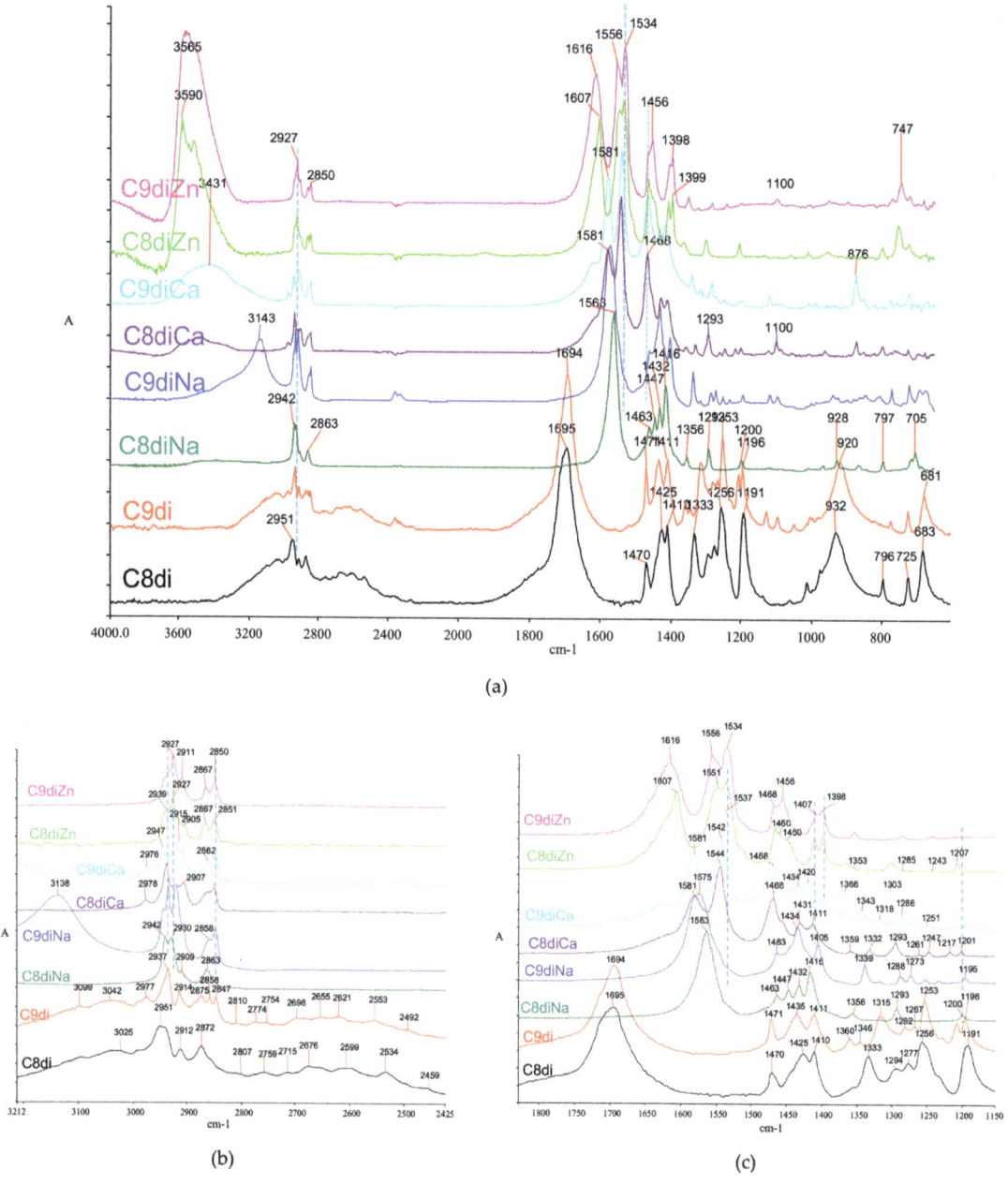

Figure 8. Infrared spectra of selected dicarboxylic acids and their sodium, calcium, and zinc salts. (**a**) Full spectrum; (**b**) C-H stretching region; and (**c**) 1800–1150 region. In all spectra: C8di, octanedioic acid (suberic acid); C9di, nonanedioic acid (azelaic acid); C8diNa, C8diCa, and C8diZn, sodium, calcium, and zinc octanedioates (suberates), respectively; C9diNa, C9diCa, and C9diZn, sodium, calcium, and zinc nonanedioates (azelates), respectively.

Similar to monoacids, the hydroxyl stretching is typically observed as a broad band, spanning 3200–2800 cm^{-1}. On the other hand, and in contrast to monoacids, the hydroxyl dimer band is structured, showing at least seven characteristic satellite bands spanning the 2810–2459 cm^{-1} region (Figure 8b).

Besides the expected absence of methyl bands, which can be seen as a diagnostic feature in spectra, the methylene stretching region is characteristically more complex than those of monoacids with split antisymmetric and symmetric bands. In the case of azelaic acid, this is more intense with both the $v_{as}CH_2$ and v_sCH_2 bands triply resolved at 2977–2914 cm^{-1} and 2875–2847 cm^{-1}, respectively (Figure 8b). The difference in the CH$_2$ stretching bands between the two acids can be attributed to a different crystal geometry allowing polymorphism in their monoclinic structures based on parallel and vertical orientations among the carboxyl group planes, resulting in different crystal packing between even- and odd-numbered diacids [90,91,115,172–179].

The methylene bending vibrations are observed as doubly split bands at ~1470 and 1410 cm^{-1} (Figure 8c). The twisting vibrations are triply split (1360, 1346, and 1315 cm^{-1}) in azelaic acid, while they show a uniform peak in suberic acid (1333 cm^{-1}). The wagging vibrations are triply split in both acids at 1290–1250 cm^{-1}. The rocking vibration appears as two medium-weak peaks at ~796 and 725 cm^{-1} for both acids with no evident splitting.

The hydroxyl bending (δC-O-H) appears as an obscure band at 1425 and 1435 cm^{-1}, for C8di and C9di, respectively [115] (Figure 8c). The acidic carbonyl C-O stretch is a very weak doublet at 1105 and 1098 cm^{-1} for azelaic, while it is not observed for suberic acid (Figure 8a). Finally, the out-of-plane C-O-H bending is shown at 681 cm^{-1} for both acids, while its dimer counterpart appears as a broad band at 932 (suberic) and 920 cm^{-1} (azelaic).

2.3.2. Diacid Metal Salts

In the case of diacid salts, most research has been done in the art paintings context. However, the literature for their systematic infrared spectra is limited [84,180], despite the fact that the corresponding fatty acids are widespread in oil paintings and specific archaeological samples, and the formation of their salts through ion exchange is more than expected. As a result of their difunctional character, diacid metal salts may associate with divalent metal ions with both their ionic ends; this way, metal-coordinated ionic networks can be formed with an ionomer character [6,75,145], which through their diminished mobility may act as a stabilizing means for the medium. This structural phenomenon has been proposed as a 'self-repair' mechanism by diacid salts which mitigates the detrimental effect caused by soap formation in paintings [84,145]. Although diacids have been investigated in lipid-containing archaeological samples [164], diacid salt formation has not been reported, and therefore, no data exist for similar phenomena in the various archaeological environments.

Similarly, with the previous cases, suberate (C8 dicarboxylate, or C8di), and azelate (C9 dicarboxylate, or C9di) salts with sodium, calcium, and zinc were also synthesized in the laboratory (see Materials and Methods). As seen in the spectra of Figure 8, the main infrared features of suberates and azelates (listed in Table 2) are similar to those of monoacids, although with subtle, but important differences. Specifically, the zinc salts appear in their hydrated form, with crystalline water observed at 3560–3520 cm^{-1} and 1607, and 1616 for C8di and C9di, respectively (Figure 8a). The carboxylate bands are intense and characteristic, appearing as an *asym-/sym-* pair (Figure 8c). For the sodium salts, a pair of single carboxylate peaks are observed at 1563–1575 (*asym-*) and 1416–1405 (*sym-*) (Figure 8c). On the other hand, each vibration is doubly split at ~1575–1544 (*asym-*) and 1431–1410 (*sym-*) for calcium and ~1551–1534 (*asym-*) and 1412–1400 (*sym-*) for zinc. By analogy to monoacid salts, the splitting can be attributed to differences in coordination geometry. The splitting value (Δv_{split}) is 40 cm^{-1} for the calcium salts, while it is significantly smaller, ~18 cm^{-1}, for the zinc salts.

Most methylene bands in diacid salts are characteristically structured, possibly as a result of interactions in the crystalline phase (Figure S1b,c). The antisymmetric CH$_2$ stretching band is split into four components for the calcium salts at 2978–2907 cm^{-1}, four components for sodium, and three for zinc salts at 2940–2905 cm^{-1} (see Figure 8b). The

CH$_2$ bending band (scissoring) is also structured at 1463–1405 cm^{-1}, with the suberate band being more extended than the azelate. The methylene twisting is observed as a single peak at ~1360 cm^{-1} for most salts, while wagging appears as a progression at 1318–1195 cm^{-1}, significantly more extended for the azelates. Finally, the CH$_2$ rocking is observed at 725–705 cm^{-1} for the sodium and calcium diacid salts and 747, 723 cm^{-1} for the zinc salts.

3. Materials and Methods

3.1. Materials

Fatty acids were purchased from Sigma-Aldrich (Kenilworth, NJ, USA): octanoic, or caprylic (CH$_3$(CH$_2$)$_6$COOH or C8:0, ≥99%), nonanoic, or pelargonic (CH$_3$(CH$_2$)$_7$COOH or C9:0, ≥99%), decanoic, or capric (CH$_3$(CH$_2$)$_8$COOH or C10:0, >98.0%), dodecanoic, or lauric (CH$_3$(CH$_2$)$_{10}$COOH or C12:0, ≥99%), tetradecanoic, or myristic (CH$_3$(CH$_2$)$_{12}$COOH or C14:0, ≥99%), hexadecanoic, or palmitic (CH$_3$(CH$_2$)$_{14}$COOH or C16:0, ≥99%), octadecanlic, or stearic (CH$_3$(CH$_2$)$_{16}$COOH or C18:0, ≥98.5%), tetracosanoic, or lignoceric (CH$_3$(CH$_2$)$_{22}$COOH or C24:0, ≥99%), octanedioic, or suberic (C$_6$H$_{12}$(COOH)$_2$ or C8di, 98%), and nonanedioic, or azelaic (C$_7$H$_{14}$(COOH)$_2$ or C9di, 98%).

Moreover, additional material and reagents were acquired as follows: sodium hydroxide (Sigma-Aldrich >88%), water (Honeywell, Charlotte, NC, USA, HPLC grade), calcium chloride (anhydrous, Sigma-Aldrich, ≥93%), zinc chloride (Sigma-Aldrich, ≥98%), ethanol (Merck, Kenilworth, NJ, USA, 99.5%), chloroform (Sigma-Aldrich, anhydrous, ≥99%), xylene (mixture of isomers, ≥98.5%), and acetone (Honeywell, ≥99.5%).

3.2. Synthesis of Fatty Acid Metal Salts

Sodium salts of fatty acids C8:0–C24:0 were prepared by adding 0.1 mmol of the corresponding acids to 1 mL of sodium hydroxide 0.1 M solutions in ethanol (Honeywell); for the diacid (C8di and C9di) sodium salts, two equivalents of the base was added. The solutions were initially warmed up to 80 °C on a heating plate, followed by sonication for 15 min. When precipitation was complete, the products were investigated with infrared spectroscopy, where the full conversion to the sodium salt was confirmed for all cases.

The calcium and zinc salts of fatty acids C8:0–C24:0 were prepared by substitution upon mixing aqueous solutions of the corresponding sodium salts (prepared in the previous step and pre-heated on a plate for full solubilization) with aqueous solutions of calcium chloride (Merck, Kenilworth, NJ, USA) and zinc chloride (Sigma-Aldrich, Kenilworth, NJ, USA). The mixtures were sonicated until full conversion was evident through infrared spectroscopy. Infrared spectra of obtained salts (shown in this paper) were in agreement with the literature [71,78,82–85].

3.3. Fourier Transform Infrared Spectroscopy

All samples were in powder form and analyzed according to the following procedure: each sample was mixed with KBr, pulverized in a pestle and mortar, and pressed in a 13 mm disc using a hydraulic press. Infrared spectra of the KBr discs were recorded in transmission mode using a Perkin Elmer Spectrum GX 1 FTIR spectrometer equipped with a DTGS detector at 4000–650 cm^{-1}, 32 scans, and 4 cm^{-1} resolution. Spectra of acids are shown normalized on the carbonyl peak. Spectra of salts are shown normalized on the highest carboxylate peak. Deconvolution of the C-H region (3000–2800 cm^{-1}) was done using the Peak Fitting application of GRAMS/AI (Thermo) software, using a mixed Gaussian (50%) and Lorentzian (50%) function, at low sensitivity. In all cases, the standard error was lower than 0.00647, while R^2 was better than 0.9987.

4. Conclusions

The investigation of fatty acids and their metal salts as degradation products of fatty substances in oil paintings and particular archaeological objects can be of high importance. This work explores the usability of infrared spectroscopy for characterizing these compounds in related complex samples. The spectra of some typical fatty monoacids, diacids, and their sodium, calcium, and zinc salts are investigated, and their features were examined in terms of their diagnostic usability.

Intermolecular interactions are fundamental for the acidic carbonyl maxima, which downshift from 1713 (for C9 monoacid) to 1703 cm^{-1} (C14–C24); these are further downshifted to 1694–1695 cm^{-1} in diacids. Crystallization and effective molecular associations in the higher members are key factors for this phenomenon, and in some cases, they are inhibited due to unfavorable molecular environments; acidic carbonyl maxima are relatively high, at ~1710 cm^{-1}. The same phenomenon also causes broadening of the dimer band at 945–920 cm^{-1}.

In salts, carboxylate absorptions exist in antisymmetric-symmetric pairs (at 1590–1530 and 1436–1398 cm^{-1}, respectively), often existing in doublets, with separations (Δv) depending on the metal. The distinctly higher frequency (181–1579 cm^{-1} of the calcium salts is helpful for their identification. In these maxima, chain lengths have little or no effect. On the other hand, inefficient crystallization may affect the band envelopes and the separation of both bands. The hydroxyl stretch spans the 3200–2800 cm^{-1} region of acids with a maximum of around 3000 cm^{-1}. Dimer formation is a typical feature in carboxylic acids, characteristically appearing as a broad maximum involving the stretching vibrations at 2800–2400 cm^{-1} (in diacids, it appears as a structured band) and as a relatively broad band involving the bending vibrations at ~940 cm^{-1}.

The various C-H vibrations significantly depend on monoacid chain lengths and intermolecular interactions among them. In crystallizable samples, splitting in stretching, bending, and rocking vibrations and the characteristic progressions in twisting and wagging for the longer-alkyl chain acids are often a diagnostic asset. Diacids also show splittings but no progressions. Progressions are also expected in metal monoacid soaps, although significantly weaker and more elusive, and therefore, their diagnostic capability is limited.

The rich spectroscopic features of fatty acids and their salts significantly benefit their diagnostic use. A close inspection of an infrared spectrum recorded from complex samples containing fatty acids, or their metal salts, can provide direct evidence for their condition.

Supplementary Materials: The following are available online, Figure S1: Infrared zoomed-in region (1570–1000 cm^{-1} of crystalline (at room temperature)) of even-numbered monoacids: decanoic (C10:0), dodecanoic (C12:0), tetradecanoic (C14:0), hexadecanoic (C16:0), octadecanoic (C18:0), and tetracosanoic (C24:0). Figure S2: Deconvolution and peak fitting of the vC-H band region of (a) octanoic (C8:0), (b) decanoic (C10:0), and (c) octadecanoic (C18:0) acids. Figure S3: Infrared spectrum of palmitic acid (C16:0) (a) as purchased from the vendor, and (b) after a heating (melting)-annealing cycle. Insets: (i) CH2 wagging region; (ii) CH2 rocking region.

Author Contributions: Conceptualization, S.C.B.; methodology, S.C.B.; writing—original draft preparation, S.C.B., supervision, A.F. and S V.; investigation, A.F. and S.V.; data curation, A.F. and S.V.; writing—review and editing, A.F. and S.V. All authors have read and agreed to the published version of the manuscript.

Funding: This research received no external funding.

Institutional Review Board Statement: Not applicable.

Informed Consent Statement: Not applicable.

Data Availability Statement: The data presented in this study are available on request from the corresponding author.

Conflicts of Interest: The authors declare no conflict of interest.

Abbreviations and Symbols

FA: fatty acids; FAMS: fatty acid metal salts (or metal soaps); FAE: fatty acid esters; FTIR: Fourier Transform Infrared Spectroscopy; C8:0–C24:0 saturated octanoic-tetracosanoic fatty acids; signifies total carbon number; C8di, C9di: octanedioic and nonanedioic acids; Δν: the difference between peak maxima between antisymmetric and symmetric carboxylate absorptions in metal soaps; sFA: saturated fatty acids.

References

1. Mills, J.S. *The Organic Chemistry of Museum Objects*, 2nd ed.; Butterworth-Heinemann: Oxford, UK, 2003.
2. Mills, J.S. The Gas Chromatographic Examination of Paint Media. Part, I. Fatty Acid Composition and Identification of Dried Oil Films. *Stud. Conserv.* **1966**, *11*, 92–107. [CrossRef]
3. Mills, J.S.; White, R. The gas-chromatographic examination of paint media. Part II. Some examples of medium identification in paintings by fatty acid analysis. *Stud. Conserv.* **1972**, *17*, 721–728. [CrossRef]
4. Nawar, W.W. Thermal degradation of lipids. *J. Agric. Food Chem.* **1969**, *17*, 18–21. [CrossRef]
5. Casadio, F.; Keune, K.; Noble, P.; Van Loon, A.; Hendriks, E.; Centeno, S.A.; Osmond, G. *Metal Soaps in Art*; Springer International Publishing: Cham, Switzerland, 2019.
6. Hermans, J.J. *Metal Soaps in Oil Paint: Structure, Mechanisms and Dynamics*; University of Amsterdam: Amsterdam, The Netherlands, 2017.
7. Keune, K.; Boon, J.J. Analytical Imaging Studies of Cross-Sections of Paintings Affected by Lead Soap Aggregate Formation. *Stud. Conserv.* **2007**, *52*, 161–176. [CrossRef]
8. Vichi, A.; Eliazyan, G.; Kazarian, S.G. Study of the Degradation and Conservation of Historical Leather Book Covers with Macro Attenuated Total Reflection-Fourier Transform Infrared Spectroscopic Imaging. *ACS Omega* **2018**, *3*, 7150–7157. [CrossRef]
9. Koupadi, K.; Boyatzis, S.C.; Roumpou, M. Organic remains in archaeological objects: Investigating their surviving profile in early Christian Egyptian vessels with Fou-rier Transform Infrared Spectroscopy and Gas Chromatog-raphy-Mass Spectrometry. *Heritage* **2021**. (submitted).
10. Regert, M.; Bland, H.A.; Dudd, S.N.; Bergen, P.V.; Evershed, R.P. Free and bound fatty acid oxidation products in archaeological ceramic vessels. *Proc. R. Soc. B Biol. Sci.* **1998**, *265*, 2027–2032. [CrossRef]
11. McGovern, P.E.; Hall, G.R. Charting a Future Course for Organic Residue Analysis in Archaeology. *J. Archaeol. Method Theory* **2016**, *23*, 592–622. [CrossRef]
12. Historic England. *Organic Residue Analysis and Archaeology*; Supporting Information; Historic England: London, UK, 2017; Volume 32.
13. Orna, M.V.; Lambert, J.B. New directions in archaeological chemistry. In *Archaeological Chemistry*; American Chemical Society: Washington, DC, USA, 1996; pp. 1–9.
14. Price, T.D.; Burton, J.H. Archaeological Chemistry. In *An Introduction to Archaeological Chemistry*; Springer: New York, NY, USA, 2011; pp. 1–24.
15. Lambert, J.B. *Traces of the Past: Unraveling the Secrets of Archaeology Through Chemistry*, 1st ed.; Perseus Books: Boston, MA, USA, 1998.
16. Evershed, R.P.; Payne, S.; Sherratt, A.G.; Copley, M.S.; Coolidge, J.; Urem-Kotsu, D.; Burton, M.M. Earliest date for milk use in the Near East and southeastern Europe linked to cattle herding. *Nature* **2008**, *455*, 528–531. [CrossRef]
17. Maniatis, Y.; Tsirtsoni, Z. Characterization of a black residue in a decorated Neolithic pot from Dikili Tash, Greece: An unexpected result. *Archaeometry* **2002**, *44*, 229–239. [CrossRef]
18. Spangenberg, J.E.; Jacomet, S.; Schibler, J. Chemical analyses of organic residues in archaeological pottery from Arbon Bleiche 3, Switzerland—Evidence for dairying in the late Neolithic. *J. Archaeol. Sci.* **2006**, *33*, 1–13. [CrossRef]
19. Maritan, L.; Nodari, L.; Mazzoli, C.; Milanob, A.; Russob, U. Influence of firing conditions on ceramic products: Experimental study on clay rich in organic matter. *Appl. Clay Sci.* **2006**, *31*, 1–15. [CrossRef]
20. Mitkidou, S.; Dimitrakoudi, E.; Urem-Kotsou, D.; Papadopoulou, D.; Kotsakis, K.; Stratis, J.A.; Stephanidou-Stephanatou, I. Organic residue analysis of Neolithic pottery from North Greece. *Microchim. Acta* **2008**, *160*, 493–498. [CrossRef]
21. Raven, A.M.; van Bergen, P.F.; Stott, A.W.; Dudd, S.N.; Evershed, R.P. Formation of long-chain ketones in archaeological pottery vessels by pyrolysis of acyl lipids. *J. Anal. Appl. Pyrolysis* **1997**, *40–41*, 267–285. [CrossRef]
22. Suryanarayan, A.; Cubas, M.; Craig, O.E.; Herone, C.P.; Shinde, V.S.; Singh, R.N.; O'Connell, T.C.; Petrie, C.A. Lipid residues in pottery from the Indus Civilisation in northwest India. *J. Archaeol. Sci.* **2021**, *125*, 105291. [CrossRef] [PubMed]
23. Lettieri, M. Infrared spectroscopic characterization of residues on archaeological pottery through different spectra acquisition modes. *Vib. Spectrosc.* **2015**, *76*, 48–54. [CrossRef]
24. Shillito, L.M.; Almond, M.J.; Wicks, K.; Marshall, L.-J.R.; Matthews, W. The use of FT-IR as a screening technique for organic residue analysis of archaeological samples. *Spectrochim. Acta Part A Mol. Biomol. Spectrosc.* **2009**, *72*, 120–125. [CrossRef]
25. McGovern, P.E.; Glusker, D.L.; Exner, L.J.; Voigt, M.M. Neolithic resinated wine. *Nature* **1996**, *381*, 480–481. [CrossRef]
26. Stern, B.; Heron, C.; Tellefsen, T.; Serpico, M. New investigations into the Uluburun resin cargo. *J. Archaeol. Sci.* **2008**, *35*, 2188–2203. [CrossRef]
27. Hammann, S.; Scurr, D.J.; Alexander, M.R.; Cramp, L.J.E. Mechanisms of lipid preservation in archaeological clay ceramics revealed by mass spectrometry imaging. *Proc. Natl. Acad. Sci. USA* **2020**, *117*, 14688–14693. [CrossRef]

28. Regert, M. Analytical strategies for discriminating archeological fatty substances from animal origin. *Mass Spectrom. Rev.* **2011**, *30*, 177–220. [CrossRef] [PubMed]
29. Mottram, H.R.; Evershed, R.P. Structure Analysis of Triacylglycerol Positional Isomers Using Atmospheric Pressure Chemical Ionisation Mass Spectrometry. *Tetrahedron Lett.* **1996**, *37*, 8593–8596. [CrossRef]
30. Mottram, H.; Dudd, S.; Lawrence, G.; Stott, A.W.; Evershed, R.P. New chromatographic, mass spectrometric and stable isotope approaches to the classification of degraded animal fats preserved in archaeological pottery. *J. Chromatogr. A* **1999**, *833*, 209–221. [CrossRef]
31. Heron, C.; Evershed, R.P. The Analysis of Organic Residues and the Study of Pottery Use The Analysis of Organic Residues and the Study of Pottery Use. *Archaeol. Method Theory* **1993**, *5*, 247–284.
32. McGovern, P.; Jalabadze, M.; Batiuk, S.; Callahan, M.P.; Smith, K.E.; Hall, G.R.; Kvavadze, E.; Maghradze, D.; Rusishvili, N.; Bouby, L.; et al. Early Neolithic wine of Georgia in the South Caucasus. *Proc. Natl. Acad. Sci. USA* **2017**, *114*, E10309–E10318. [CrossRef]
33. Izzo, F.C.; Zendri, E.; Bernardi, A.; Balliana, E.; Sgobbi, M. The study of pitch via gas chromatography–mass spectrometry and Fourier-transformed infrared spectroscopy: The case of the Roman amphoras from Monte Poro, Calabria (Italy). *J. Archaeol. Sci.* **2013**, *40*, 595–600. [CrossRef]
34. Regert, M.; Garnier, N.; Decavallas, O.; Cren-Olivé, C.; Rolando, C. Structural characterization of lipid constituents from natural substances preserved in archaeological environments. *Meas. Sci. Technol.* **2003**, *14*, 1620–1630. [CrossRef]
35. Craig, O.E.; Steele, V.J.; Fischer, A.; Hartz, S.; Andersen, S.H.; Donohoe, P.; Glykou, A.; Saul, H.; Jones, D.M.; Koch, E.; et al. Ancient lipids reveal continuity in culinary practices across the transition to agriculture in Northern Europe. *Proc. Natl. Acad. Sci. USA* **2011**, *108*, 17910–17915. [CrossRef]
36. Roffet-Salque, M.; Dunne, J.; Altoft, D.T.; Casanova, E.; Cramp, L.J.E.; Smyth, J.; Whelton, H.L.; Evershed, R.P. From the inside out: Upscaling organic residue analyses of archaeological ceramics. *J. Archaeol. Sci. Rep.* **2017**, *16*, 627–640. [CrossRef]
37. Gregg, M.W.; Slater, G.F. A new method for extraction, isolation and transesterification of free fatty acids from archaeological pottery. *Archaeometry* **2010**, *52*, 833–854. [CrossRef]
38. Šoberl, L.; Žibrat Gašparič, A.; Budja, M.; Evershed, R.P. Early herding practices revealed through organic residue analysis of pottery from the early Neolithic rock shelter of Mala Triglavca, Slovenia. *Doc. Praehist.* **2008**, *35*, 253–260. [CrossRef]
39. McGovern, P.E.; Glusker, D.L.; Moreau, R.A.; Nuñez, A.; Beck, C.W.; Simpson, E.; Butrym, E.D.; Exner, L.J.; Stout, E.C. A funerary feast fit for King Midas. *Nature* **1999**, *402*, 863–864. [CrossRef]
40. Boyatzis, S.C.; Kotzamani, D.; Phoca, A.; Karydi, G.M.Z.A.V.K. Characterization of Organic Remains Found in Copper Alloy Vessels of the Benaki Museum Collection with Fourier Transform Mid-Infrared Spectroscopy. In Proceedings of the 3rd ARCH_RNT, Kalamata, Greece, 3–5 October 2012; Zacharias, N., Ed.; University of the Peloponnese: Tripoli, Greece, 2014.
41. Evershed, R.P.; Dudd, S.N.; Copley, M.S.; Berstan, B.; Stott, A.W.; Mottram, H.; Buckley, S.A.; Crossman, Z. Chemistry of archaeological animal fats. *Acc. Chem. Res.* **2002**, *35*, 660–668. [CrossRef] [PubMed]
42. Harper, C.S.; Macdonald, F.V.; Braun, K.L. Lipid Residue Analysis of Archaeological Pottery: An Introductory Laboratory Experiment in Archaeological Chemistry. *J. Chem. Educ.* **2017**, *94*, 1309–1313. [CrossRef]
43. Heron, C.; Stacey, R. *Archaeology: Uses of Chromatography in*; Academic Press: Cambridge, MA, USA, 2000; pp. 2083–2089.
44. Stern, B.; Heron, C.; Serpico, M.; Bourriau, J. A comparison of methods for establishing fatty acid concentration gradients across potsherds: A case study using Late Bronze Age Canaanite amphorae. *Archaeometry* **2000**, *42*, 399–414. [CrossRef]
45. Pitonzo, R.; Armetta, F.; Saladino, M.L.; Oliveri, F.; Tusa, S.; Caponetti, E. Application of Gas Chromatography coupled with Mass Spectroscopy (GC/MS) to the analysis of archaeological ceramic amphorae belonging to the Carthaginian fleet that was defeated in the Egadi battle (241 B.C.). *Acta IMEKO* **2017**, *6*, 67–70. [CrossRef]
46. Sciutto, G.; Oliveri, P.; Prati, S.; Quaranta, M.; Lanteri, S.; Mazzeo, R. Analysis of paint cross-sections: A combined multivariate approach for the interpretation of µATR-FTIR hyperspectral data arrays. *Anal. Bioanal. Chem.* **2013**, *405*, 625–633. [CrossRef]
47. Mazzeo, R.; Prati, S.; Quaranta, M.; Joseph, E.; Kendix, E.; Galeotti, M. Attenuated total reflection micro FTIR characterisation of pigment-binder interaction in reconstructed paint films. *Anal. Bioanal. Chem.* **2008**, *392*, 65–76. [CrossRef]
48. Spring, M.; Ricci, C.; Peggie, D.A.; Kazarian, S.G. ATR-FTIR imaging for the analysis of organic materials in paint cross sections: Case studies on paint samples from the National Gallery, London. *Anal. Bioanal. Chem.* **2008**, *392*, 37–45. [CrossRef]
49. Bonaduce, I.; Ribechini, E.; Modugno, F.; Colombini, M.P. Analytical approaches based on gas chromatography mass spectrometry (GC/MS) to study organic materials in artworks and archaeological objects. *Top Curr. Chem.* **2016**, *374*, 1–37. [CrossRef] [PubMed]
50. Stuart, B.H. *Infrared Spectroscopy: Fundamentals and Applications*; John Wiley & Sons, Ltd.: Chichester, UK, 2004.
51. Griffiths, P.R.; Haseth, J.A. *Fourier Transform Infrared Spectrometry*, 2nd ed.; Wiley-Interscience: Hoboken, NJ, USA, 2007.
52. Griffiths, P.R. Introduction to Vibrational Spectroscopy. In *Handbook of Vibrational Spectroscopy*; John Wiley & Sons Inc.: Hoboken, NJ, USA, 2007. [CrossRef]
53. Derrick, M.R.; Stulik, D.; Landry, J.M. *Infrared Spectroscopy in Conservation Science*; Getty Conservation Institute: Los Angeles, CA, USA, 1999.
54. Stuart, B.H. *Analytical Techniques in Materials Conservation*; John Wiley & Sons: Chichester, UK, 2007.
55. Meilunas, R.J.; Bentsen, J.G.; Steinberg, A. Analysis of aged paint binders by ftir spectroscopy. *Stud. Conserv.* **1990**, *35*, 33–51. [CrossRef]

56. Coates, J. Interpretation of Infrared Spectra, A Practical Approach. In *Encyclopedia of Analytical Chemistry*; John Wiley & Sons Ltd.: Chichester, UK, 2006; pp. 10815–10837.
57. Bellamy, L.J. *The Infra-Red Spectra of Complex Molecules*; Springer: Dordrecht, The Netherlands, 1975.
58. Shurvell, H.F. Spectra—Structure Correlations in the Mid- and Far-Infrared. In *Handbook of Vibrational Spectroscopy*; Chalmers, J.M., Ed.; John Wiley & Sons, Ltd.: Chichester, UK, 2006.
59. Mayo, D.; Miller, F.; Hannah, R. *Course Notes on the Interpretation of Infrared and Raman Spectra*; John Wiley & Sons Inc.: Hoboken, NJ, USA, 2004.
60. Larkin, P. *Infrared and Raman Spectroscopy: Principles and Spectral Interpretation*; Elsevier: Amsterdam, The Netherlands, 2011.
61. Szymanski, H.A. *Interpreted Infrared Spectra_ Volume 1-Springer US (1964)*; Plenum Press: New York, NY, USA, 1981.
62. Silverwood, I.P.; Keyworth, C.W.; Brown, N.J.; Shaffer, M.S.P.; Williams, C.K.; Hellgardt, K.; Kelsall, G.H.; Kazarian, S.G. An attenuated total reflection fourier transform infrared (ATR FT-IR) spectroscopic study of gas adsorption on colloidal stearate-capped ZnO catalyst substrate. *Appl. Spectrosc.* **2014**, *68*, 88–94. [CrossRef] [PubMed]
63. Andrew Chan, K.L.; Kazarian, S.G. Attenuated total reflection Fourier-transform infrared (ATR-FTIR) imaging of tissues and live cells. *Chem. Soc. Rev.* **2016**, *45*, 1850–1864. [CrossRef]
64. Umemura, J. Reflection-Absorption Spectroscopy of Thin Films on Metallic Substrates. In *Handbook of Vibrational Spectroscopy*; Chalmers, J.M., Ed.; John Wiley & Sons Ltd.: Chichester, UK, 2006.
65. Claybourn, M. External Reflection Spectroscopy. In *Handbook of Vibrational Spectroscopy*; Chalmers, J.M., Ed.; John Wiley & Sons Ltd.: Chichester, UK, 2006.
66. Bitossi, G.; Giorgi, R.; Mauro, M.; Salvadori, B.; Dei, L. Spectroscopic Techniques in Cultural Heritage Conservation: A Survey. *Appl. Spectrosc. Rev.* **2005**, *40*, 187–228. [CrossRef]
67. Prati, S.; Sciutto, G.; Bonacini, I.; Mazzeo, R. New Frontiers in Application of FTIR Microscopy for Characterization of Cultural Heritage Materials. *Top. Curr. Chem.* **2016**, *374*, 1–32. [CrossRef] [PubMed]
68. Mazzeo, R.; Joseph, E.; Prati, S.; Millemaggi, A. Attenuated Total Reflection-Fourier transform infrared microspectroscopic mapping for the characterisation of paint cross-sections. *Anal. Chim. Acta* **2007**, *599*, 107–117. [CrossRef]
69. Van der Weerd, J.; van Loon, A.; Boon, J.J. FTIR Studies of the Effects of Pigments on the Aging of Oil. *Stud. Conserv.* **2005**, *50*, 3–22. [CrossRef]
70. Osmond, G.; Boon, J.J.; Puskar, L.; Drennan, J. Metal Stearate Distributions in Modern Artists' Oil Paints: Surface and Cross-Sectional Investigation of Reference Paint Films Using Conventional and Synchrotron Infrared Microspectroscopy. *Appl. Spectrosc.* **2012**, *66*, 1136–1144. [CrossRef]
71. Hermans, J.J.; Keune, K.; van Loon, A.; Stols-Witlox, M.J.N.; Corkery, R.W.; Iedema, P.D. The synthesis of new types of lead and zinc soaps: A source of information for the study of oil paint degradation. In Proceedings of the ICOM-CC 17th Triennial Conference, Melbourne, Australia, 15–19 September 2014; p. 1603.
72. Hermans, J.J.; Keune, K.; Van Loon, A.; Corkery, R.W.; Iedema, P.D. Ionomer-like structure in mature oil paint binding media. *RSC Adv.* **2016**, *6*, 93363–93369. [CrossRef]
73. Hermans, J.J.; Keune, K.; Van Loon, A.; Iedema, P.D. An infrared spectroscopic study of the nature of zinc carboxylates in oil paintings. *J. Anal. At. Spectrom.* **2015**, *30*, 1600–1608. [CrossRef]
74. Casadio, F.; Bellot-Gurlet, L.; Paris, C. Factors Affecting the Reactivity of Zinc Oxide with Different Drying Oils: A Vibrational Spectroscopy Study. In *Metal Soaps in Art*; Springer Publishing: Cham, Switzerland, 2019; pp. 153–170.
75. Banti, D.; La Nasa, J.; Tenorio, A.L.; Modugno, F.; Berg, K.J.V.A.; Lee, J.; Ormsby, B.; Burnstock, A.; Bonaduce, I. A molecular study of modern oil paintings: Investigating the role of dicarboxylic acids in the water sensitivity of modern oil paints. *RSC Adv.* **2018**, *8*, 6001–6012. [CrossRef]
76. Ma, X.; Beltran, V.; Ramer, G.; Georges, P.; Dilworthy, P.; Tyler, M.; Andrea, C.; Barbara, B. Revealing the Distribution of Metal Carboxylates in Oil Paint from the Micro- to Nanoscale. *Angew. Chem. Int. Ed.* **2019**, *58*, 11652–11656. [CrossRef]
77. Hermans, J.J.; Keune, K.; Van Loon, A.; Iedema, P.D. The crystallization of metal soaps and fatty acids in oil paint model systems. *Phys. Chem. Chem. Phys.* **2016**, *18*, 10896–10905. [CrossRef] [PubMed]
78. Robinet, L.; Corbeil, M.C. The characterization of metal soaps. *Stud. Conserv.* **2003**, *48*, 23–40. [CrossRef]
79. Oudemans, T.F.M.; Boon, J.J.; Botto, R.E. FTIR and solid-state 13C CP/MAS NMR spectroscopy of charred and non-charred solid organic residues preserved in roman iron age vessels from the Netherlands. *Archaeometry* **2007**, *49*, 571-294. [CrossRef]
80. McGovern, P.E. *Uncorking the Past. The Quest for Wine, Beer, and Other Alcoholic Beverages*; University of California Press: Berkeley, CA, USA, 2009.
81. Thickett, D.; Pretzel, B. Micro-spectroscopy: A powerful tool to understand deterioration. *e-Preserv. Sci.* **2010**, *7*, 158–164.
82. Plater, M.J.; De Silva, B.; Gelbrich, T.; Hursthouse, M.; Higgitt, C.; Saunders, D. The characterisation of lead fatty acid soaps in "protrusions" in aged traditional oil paint. *Polyhedron* **2003**, *22*, 3171–3179. [CrossRef]
83. Hermans, J.J.; Keune, K.; van Loon, A.; Corkery, R.W.; Iedema, P.D. The molecular structure of three types of long-chain zinc(II) alkanoates for the study of oil paint degradation. *Polyhedron* **2014**, *81*, 335–340. [CrossRef]
84. Otero, V.; Sanches, D.; Montagner, C.; Vilarigues, M.; Carlyle, L.; Lopes, J.A.; Melo, M.J. Supporting Information: Characterisation of metal carboxylates by Raman and infrared spectroscopy in works of art. *J. Raman. Spectrosc.* **2014**, *45*, 1197–1206. [CrossRef]
85. Gönen, M.; Öztürk, S.; Balköse, D.; Okur, S.; Ülkü, S. Preparation and characterization of calcium stearate powders and films prepared by precipitation and Langmuir-Blodgett techniques. *Ind. Eng. Chem. Res.* **2010**, *49*, 1732–1736. [CrossRef]

86. Pauling, L.; Brockway, L.O. The Structure of the Carboxyl Group: I. The Investigation of Formic Acid by the Diffraction of Electrons. *Proc. Natl. Acad. Sci. USA* **1934**, *20*, 336–340. [CrossRef]
87. Kishida, S.; Nakamoto, K. Normal coordinate analyses of hydrogen-bonded compounds. II. Dimeric formic acid and acetic acid. *J. Chem. Phys.* **1964**, *41*, 1558–1563. [CrossRef]
88. Witkowski, A. Infrared Spectra of the Hydrogen-Bonded Carboxylic Acids. *J. Chem. Phys.* **1967**, *47*, 3645–3648. [CrossRef]
89. Kaneko, F.; Tashiro, K.; Kobayashi, M. Polymorphic transformations during crystallization processes of fatty acids studied with FT-IR spectroscopy. *J. Cryst. Growth* **1999**, *198–199*, 1352–1359. [CrossRef]
90. Kaneko, F.; Ishikawa, E.; Kobayashi, M.; Suzuki, M. Structural study on polymorphism of long-chain dicarboxylic acids using oblique transmission method for micro FT-IR spectrometers. *Spectrochim. Acta Part A Mol. Biomol. Spectrosc.* **2004**, *60*, 9–18. [CrossRef]
91. Thalladi, V.R.; Nüsse, M.; Boese, R. The melting point alternation in α,ω-alkanedicarboxylic acids. *J. Am. Chem. Soc.* **2000**, *122*, 9227–9236. [CrossRef]
92. Badger, R.M.; Bauer, S.H. Spectroscopic studies of the hydrogen bond I. A photometric investigation of the association equilibrium in the vapor of acetic acid. *J. Chem. Phys.* **1937**, *5*, 605–608. [CrossRef]
93. Badger, R.M.; Bauer, S.H. Spectroscopic studies of the hydrogen bond. II. The shift of the O-H vibrational frequency in the formation of the hydrogen bond. *J. Chem. Phys.* **1937**, *5*, 839–851. [CrossRef]
94. Fox, J.J.; Martin, A.E. Investigations of infra-red spectra-absorption of some hydroxy compounds in the region of 3 μ. *Proc. R. Soc. Lond. Ser. A Math. Phys. Sci.* **1937**, *162*, 419–441. [CrossRef]
95. Buswell, A.M.; Rodebush, W.H.; Roy, M.F. Infrared Absorption Studies. V. Association in the Carboxylic Acids. *J. Am. Chem. Soc.* **1938**, *60*, 2239–2244. [CrossRef]
96. Davies, M.M.; Sutherland, G.B.B.M. Hydroxyl Frequency in Carboxylic Acids. *Nature* **1938**, *141*, 372–373. [CrossRef]
97. Cannon, C.G. The nature of hydrogen bonding. *Spectrochim. Acta* **1958**, *10*, 341–368. [CrossRef]
98. Hadži, D. Hydrogen Bonding. In Proceedings of the Symposium on Hydrogen Bonding, Ljubljana, Slovenia, 29 July–3 August 1957; Pergamon Press: Oxford, UK, 1959.
99. Nakamoto, K.; Margoshes, M.; Rundle, R. E Stretching Frequencies as a Function of Distances in Hydrogen Bonds. *J. Am. Chem. Soc.* **1955**, *77*, 6480–6486. [CrossRef]
100. Hadzi, D.; Sheppard, N. The infra-red absorption bands associated with the COOH and COOD groups in dimeric carboxylic acids. I. The region from 1500 to 500 cm^{-1}. *Proc. R. Soc. Lond. Ser. A Math. Phys. Sci.* **1953**, *216*, 247–266. [CrossRef]
101. Bratož, S.; Hadži, D.; Sheppard, N. The infra-red absorption bands associated with the COOH and COOD groups in dimeric carboxylic acid-II. The region from 3700 to 1500 cm^{-1}. *Spectrochim. Acta* **1956**, *8*, 249–261. [CrossRef]
102. Grabowski, S.J. *Hydrogen Bonding—New Insights*, 1st ed.; Springer: Dordrecht, The Netherlands, 2006.
103. Bezrodna, T. Temperature dynamics of dimer formation in behenic acid: FT-IR spectroscopic study. *J. Mol. Struct.* **2013**, *1040*, 112–116. [CrossRef]
104. Li, Y.-M.; Sun, S.-Q.; Zhou, Q.; Qin, Z.; Tao, J.-X.; Wang, J.; Fang, X. Identification of American ginseng from different regions using FT-IR and two-dimensional correlation IR spectroscopy. *Vib. Spectrosc.* **2004**, *36*, 227–232. [CrossRef]
105. Silva, L.F.; Andrade-Filho, T.; Freire, P.T.C. Polarized Raman and Infrared Spectroscopy and ab Initio Calculation of Palmitic and Stearic Acids in the Bm and C Forms. *J. Phys. Chem. A* **2004**, *121*, 4830–4842. [CrossRef]
106. Meiklejohn, R.A.; Meyer, R.J.; Aronovic, S.M.; Schuette, H.A.; Meloche, V.W. Characterization of Long-Chain Fatty Acids by Infrared Spectroscopy. *Anal. Chem.* **1957**, *29*, 329–334. [CrossRef]
107. De Ruig, W.G. *Infrared Spectra of Monoacid Triglycerides with Some Applications to Fat Analysis*; Center for Agricultural Publishing and Documentation: Wageningen, The Netherlands, 1971.
108. Corish, P.J.; Chapman, D. The infrared spectra of some monocarboxylic acids. *J. Chem. Soc.* **1957**, *18*, 1746. [CrossRef]
109. Chapman, D. Infrared spectroscopy of lipids. *J. Am. Oil Chem. Soc.* **1965**, *42*, 353–371. [CrossRef]
110. Sinclair, R.G.; McKay, A.F.; Jones, R.N. The Infrared Absorption Spectra of Saturated Fatty Acids and Esters. *J. Am. Chem. Soc.* **1952**, *74*, 2570–2575. [CrossRef]
111. Comí, M.; Fernández, M.; Santamaría, A.; Lligadas, G.; Ronda, J.C.; Galià, M.; Cadiz, V. Carboxylic Acid Ionic Modification of Castor-Oil-Based Polyurethanes Bearing Amine Groups: Chemically Tunable Physical Properties and Recyclability. *Macromol. Chem. Phys.* **2017**, *218*, 1700379. [CrossRef]
112. Cheng, Q.; Cao, Y.; Yang, L.; Zhang, P.-P.; Wang, K.; Wang, H.-J. Synthesis of titania microspheres with hierarchical structures and high photocatalytic activity by using nonanoic acid as the structure-directing agent. *Mater. Lett.* **2011**, *65*, 2833–2835. [CrossRef]
113. Ali, H.; Ghareeb, M.M.; Al-remawi, M.; Al-akayleh, F.T. New insight into single phase formation of capric acid/menthol eutectic mixtures by Fourier-transform infrared spectroscopy and differential scanning calorimetry. *Trop. J. Pharm. Res.* **2020**, *19*, 361–369. [CrossRef]
114. Roy, R.S. Spectroscopic studies of long chain fatty acids. Pyridine—Fatty acid—Carbon tetrachloride system. *Spectrochim. Acta* **1966**, *22*, 1877–1887. [CrossRef]
115. Suzuki, M.; Shimanouchi, T. Infrared and Raman spectra of adipic acid crystal. *J. Mol. Spectrosci.* **1969**, *29*, 415–425. [CrossRef]
116. Mikawa, Y.; Brasch, J.W.; Jakobsen, R.J. Polarized ir spectra of single crystals of propanoic acid. *J. Mol. Struct.* **1968**, *3*, 103–117. [CrossRef]
117. Lide, D.R. *Handbook of Chemistry and Physics*, 87th ed.; CRC: Boca Raton, FL, USA, 2006.

118. Zerbi, G.; Conti, G.; Minoni, G.; Pison, S.; Bigotto, A. Premelting phenomena in fatty acids: An infrared and Raman study. *J. Phys. Chem.* **1987**, *91*, 2386–2393. [CrossRef]
119. Conti, G.; Minoni, G.; Zerbi, G. E → C phase transition in fatty acids: A spectroscopic study. *J. Mol. Struct.* **1984**, *118*, 237–243. [CrossRef]
120. Flett, M.S.C. The characteristic infra-red frequencies of the carboxylic acid group. *J. Chem. Soc.* **1951**, *41*, 962. [CrossRef]
121. Ren, Z.; Ma, D.; Wang, Y.; Zhao, G. Molecular structure and hydrogen bonds in solid dimethylol propionic acid (DMPA). *Spectrochim. Acta Part A Mol. Biomol. Spectrosc.* **2003**, *59*, 2713–2722. [CrossRef]
122. Zhu, J.; Ren, Z.; Zhang, G.; Guo, X.; Ma, M. Comparative study of the H-bond and FTIR spectra between 2,2-hydroxymethyl propionic acid and 2,2-hydroxymethyl butanoic acid. *Spectrochim. Acta Part A Mol. Biomol. Spectrosc.* **2006**, *63*, 449–453. [CrossRef]
123. Pauling, L. *The Nature of the Chemical Bond and the Structure of Molecules and Crystals: An Introduction to Modern Structural Chemistry*, 3rd ed.; Cornell University Press: Ithaca, NY, USA, 1960.
124. Li, H.W.; Strauss, H.L.; Snyder, R.G. Differences in the IR methylene rocking bands between the crystalline fatty acids and n-alkanes: Frequencies, intensities, and correlation splitting. *J. Phys. Chem. A* **2004**, *108*, 6629–6642. [CrossRef]
125. Lewis, R.N.A.H.; McElhaney, R.N. Vibrational Spectroscopy of Lipids. In *Handbook of Vibrational Spectroscopy*; Griffiths, P.R., Ed.; John Wiley & Sons Ltd.: Chichester, UK, 2006.
126. Jones, R.N. The Effects of Chain Length on the Infrared Spectra of Fatty Acids and Methyl Esters. *Can. J. Chem.* **1962**, *40*, 321–333. [CrossRef]
127. Lambert, J.B.; Shurvell, H.F.; Cooks, R.G. *Introduction to Organic Spectroscopy*, 1st ed.; Macmillan: New York, NY, USA, 1987.
128. Smith, B.C. *Infrared Spectral Interpretation: A Systematic Approach*; CRC Press: Boca Raton, FL, USA, 1998.
129. Zbinden, R. *Infrared Spectroscopy of High Polymers*; Academic Press: New York, NY, USA; London, UK, 1964.
130. Snyder, R.G.; Schachtschneider, J.H. Vibrational analysis of the n-paraffins—I. Assignments of infrared bands in the spectra of C3H8 through n-C19H40. *Spectrochim. Acta* **1963**, *19*, 85–116. [CrossRef]
131. Jones, R.N.; McKay, A.F.; Sinclair, R.G. Band Progressions in the Infrared Spectra of Fatty Acids and Related Compounds. *J. Am. Chem. Soc.* **1952**, *74*, 2575–2578. [CrossRef]
132. Boyatzis, S.; Douvas, A.M.; Argyropoulos, V.; Siatou, A.; Vlachopoulou, M. Characterization of a water-dispersible metal protective coating with fourier transform infrared spectroscopy, modulated differential scanning calorimetry, and ellipsometry. *Appl. Spectrosc.* **2012**, *66*, 580–590. [CrossRef]
133. Kirby, E.M.; Evans-Vader, M.J.; Brown, M.A. Determination of the length of polymethylene chains in salts of saturated and unsaturated fatty acids by infrared spectroscopy. *J. Am. Oil Chem. Soc.* **1965**, *42*, 437–446. [CrossRef]
134. Chapman, D. The 720 cm−1 band in the infrared spectra of crystalline long-chain compounds. *J. Chem. Soc.* **1957**, 4489–4491. [CrossRef]
135. Ross, R.A.; Takacs, A. Heterogeneous Reactions of Aluminum and Copper Surfaces with Stearic Acid. *Ind. Eng. Chem. Prod. Res. Dev.* **1983**, *22*, 280–286. [CrossRef]
136. Ross, R.A.; Takacs, A.M. Surface reactions of ethyl stearate and stearic acid with zinc, manganese and their oxides. *Surf. Technol.* **1984**, *21*, 361–377. [CrossRef]
137. Dou, Q.; Ng, K.M. Synthesis of various metal stearates and the corresponding monodisperse metal oxide nanoparticles. *Powder Technol.* **2016**, *301*, 949–958. [CrossRef]
138. Corkery, R.W. *Artificial Biomineralisation and Metallic Soaps*; Australian National University: Canberra, Australia, 1998.
139. Nora, A.; Szczepanek, A.; Koenen, G. Metallic soaps. In *Ullmann's Encyclopedia of Industrial Chemistry*; Lauro, M.F., Ed.; Wiley–VCH: Weinheim, Germany, 2005; pp. 329–332.
140. Osmond, G. Zinc Soaps: An Overview of Zinc Oxide Reactivity and Consequences of Soap Formation in Oil-Based Paintings. In *Metal Soaps in Art*; Casadio, F., Noble, P., Hendricks, E., Eds.; Springer: Cham, Switzerland, 2019; pp. 25–46.
141. Noble, P. A Brief History of Metal Soaps in Paintings from a Conservation Perspective. In *Metal Soaps in Art*; Casadio, F., Noble, P., Hendricks, E., Eds.; Springer: Cham, Switzerland, 2019; pp. 1–22.
142. Raven, L.E.; Bisschoff, M.; Leeuwestein, M.; Geldof, M.; Hermans, J.J.; Stols-Witlox, M.; Keune, K. Delamination Due to Zinc Soap Formation in an Oil Painting by Piet Mondrian (1872–1944). In *Metal Soaps in Art*; Casadio, F., Noble, P., Hendricks, E., Eds.; Springer: Cham, Switzerland, 2019; pp. 343–358.
143. Thomas, M.M.; Clouse, J.A.; Longo, J.M. Adsorption of organic compounds on carbonate minerals. 1. Model compounds and their influence on mineral wettability. *Chem. Geol.* **1993**, *109*, 201–213. [CrossRef]
144. Frye, G.C.; Thomas, M.M. Adsorption of organic compounds on carbonate minerals. 2. Extraction of carboxylic acids from recent and ancient carbonates. *Chem. Geol.* **1993**, *109*, 215–226. [CrossRef]
145. Boon, J.J.; Hoogland, F.; Keune, K.; Parkin, H.M. Chemical processes in aged oil paints affecting metal soap migration and aggregation. In Proceedings of the AIC's 34th Annual Meeting, Providence, RI, USA, 16–19 June 2006; Volume 19, pp. 16–23.
146. La Nasa, J.; Modugno, F.; Aloisi, M.; Lluveras-Tenorio, A.; Bonaduce, I. Development of a GC/MS method for the qualitative and quantitative analysis of mixtures of free fatty acids and metal soaps in paint samples. *Anal. Chim. Acta* **2018**, *1001*, [CrossRef]
147. Hermans, J.; Helwig, K. The Identification of Multiple Crystalline Zinc Soap Structures Using Infrared Spectroscopy. *Appl. Spectrosc.* **2020**, *74*, 1505–1514. [CrossRef] [PubMed]
148. Nakamoto, K. *Infrared and Raman Spectra of Inorganic and Coordination Compounds, Part B, Applications in Coordination, Organometallic, and Bioinorganic Chemistry*, 6th ed.; John Wiley & Sons: Hoboken, NJ, USA, 2009.

149. Palacios-beas, E. Infrared spectroscopy of metal carboxylates: II. Analysis of Fe (III), Ni and Zn carboxylate solutions. *Hydrometallurgy* **2004**, *72*, 139–148. [CrossRef]
150. Deacon, G.B.; Phillips, R.J. Relationships between the carbon-oxygen stretching frequencies of carboxylato complexes and the type of carboxylate coordination. *Coord. Chem. Rev.* **1980**, *33*, 227–250. [CrossRef]
151. Zeleňák, V.; Vargová, Z.; Györyová, K. Correlation of infrared spectra of zinc(II) carboxylates with their structures. *Spectrochim. Acta Part A Mol. Biomol. Spectrosc.* **2007**, *66*, 262–272. [CrossRef]
152. Desseyn, H.O. Vibrational analysis of acid derivatives. In *The Chemistry of Acid Derivatives. The Chemistry of Functional Groups*; Patai, S., Ed.; John Wiley & Sons: Chichester, UK, 1992; Volume 2, Supplement B; pp. 271–304.
153. Lynch, M.L.; Pan, Y.; Laughlin, R.G. Spectroscopic and Thermal Characterization of 1:2 Sodium Soap/Fatty Acid Acid—Soap Crystals. *J. Phys. Chem.* **1996**, *100*, 357–361. [CrossRef]
154. Lu, Y.; Miller, J.D. Carboxyl stretching vibrations of spontaneously adsorbed and LB-transferred calcium carboxylates as determined by FTIR internal reflection spectroscopy. *J. Colloid Interface Sci.* **2002**, *256*, 41–52. [CrossRef]
155. Hermans, J.J.; Baij, L.; Koenis, M.; Keune, K.; Iedema, P.D. 2D-IR spectroscopy for oil paint conservation: Elucidating the water-sensitive structure of zinc carboxylate clusters in ionomers. *Sci. Adv.* **2019**, *5*, 1–10. [CrossRef] [PubMed]
156. Catalano, J.; Murphy, A.; Yao, Y.; Yap, G.P.A.; Zumbulyadis, N.; Centeno, S.; Dybowski, C. Coordination geometry of lead carboxylates—Spectroscopic and crystallographic evidence. *Dalton Trans.* **2015**, *44*, 2340–2347. [CrossRef]
157. Cotton, F.A. *Modern Coordination Chemistry*; Wiley-Interscience: New York, NY, USA, 1960.
158. Nickolov, Z.; Georgiev, G.; Stoilova, D.; Ivanov, I. Raman and IR study of cobalt acetate dihydrate. *J. Mol. Struct.* **1995**, *354*, 119–125. [CrossRef]
159. Martínez-Casado, F.J.; Rodríguez-Cheda, J.A.; Ramos-Riesco, M.; Redondo-Yélamos, M.I.; Cucinotta, F.; Fernández-Martínez, A. Physicochemistry of Pure Lead(II) Soaps: Crystal Structures, Solid and Liquid Mesophases, and Glass Phases—Crystallographic, Calorimetric, and Pair Distribution Function Analysis. In *Metal Soaps in Art*; Springer: Cham, Switzerland, 2019; pp. 227–239.
160. Hermans, J.J.; Keune, K.; Van Loon, A.; Iedema, P.D. Toward a Complete Molecular Model for the Formation of Metal Soaps in Oil Paints. In *Metal Soaps in Art*; Springer: Cham, Switzerland, 2019; pp. 47–67.
161. Corkery, R.W. Langmuir-Blodgett (L-B) multilayer films. *Langmuir* **1997**, *13*, 3591–3594. [CrossRef]
162. Passi, S.; Picardo, M.; De Luca, C.; Luca, C.D.; Nazzaro-Porro, M.; Rossi, L.; Rotilio, G. Saturated dicarboxylic acids as products of unsaturated fatty acid oxidation. *Biochim. Biophys. Acta Lipids Lipid Metab.* **1993**, *1168*, 190–198. [CrossRef]
163. Ioakimoglou, E.; Boyatzis, S.; Argitis, P.; Fostiridou, A.; Papapanagiotou, K.; Yannovits, N. Thin-Film Study on the Oxidation of Linseed Oil in the Presence of Selected Copper Pigments. *Chem. Mater.* **1999**, *11*, 2013–2022. [CrossRef]
164. Colombini, M.P.; Modugno, F.; Ribechini, E. Organic mass spectrometry in archaeology: Evidence for Brassicaceae seed oil in Egyptian ceramic lamps. *J. Mass Spectrom.* **2005**, *40*, 890–898. [CrossRef]
165. Ménager, M.; Azémard, C.; Vieillescazes, C. Study of Egyptian mummification balms by FT-IR spectroscopy and GC-MS. *Microchem. J.* **2014**, *114*, 32–41. [CrossRef]
166. Lee, J.; Bonaduce, I.; Modugno, F.; Nasa, J.L.; Ormsby, B.; Berg, K.J.v.d. Scientific investigation into the water sensitivity of twentieth century oil paints. *Microchem. J.* **2018**, *138*, 282–295. [CrossRef]
167. Modugno, F.; Di Gianvincenzo, F.; Degano, I.; Werf, I.D.v.d.; Bonaduce, I.; Berg, K.J.v.d. On the influence of relative humidity on the oxidation and hydrolysis of fresh and aged oil paints. *Sci. Rep.* **2019**, *9*, 5533. [CrossRef]
168. Colombini, M.P.; Modugno, F.; Giacomelli, M.; Francesconi, S. Characterisation of proteinaceous binders and drying oils in wall painting samples by gas chromatography-mass spectrometry. *J. Chromatogr. A* **1999**, *846*, 113–124. [CrossRef]
169. Bonaduce, I.; Carlyle, L.; Colombini, M.P.; Duce, C.; Ferrari, C.; Ribechini, E.; Selleri, P.; Tiné, M.R. New Insights into the Ageing of Linseed Oil Paint Binder: A Qualitative and Quantitative Analytical Study. *PLoS ONE* **2012**, *7*, e49333. [CrossRef]
170. Dunitz, J.D.; Robertson, J.M. The crystal and molecular structure of certain dicarboxylic acids. Part, I. Oxalic acid dihydrate. *J. Chem. Soc.* **1944**, 142–148. [CrossRef]
171. Morrison, J.D.; Robertson, J.M. The crystal and molecular structure of certain dicarboxylic acids. Part IV. β-Succinic acid. *J. Chem. Soc.* **1949**, 980–986. [CrossRef]
172. Morrison, J.D.; Robertson, J.M. The crystal and molecular structure of certain dicarboxylic acids. Part, V. Adipic acid. *J. Chem. Soc.* **1949**, 987–992. [CrossRef]
173. Morrison, J.D.; Robertson, J.M. The crystal and molecular structure of certain dicarboxylic acids. Part VI. Sebacic acid. *J. Chem. Soc.* **1949**, 993–1001. [CrossRef]
174. Susi, H. Infrared spectra of crystalline adipic acid and deuterated analogs. *Spectrochim Acta* **1956**, *15*, 1063–1071. [CrossRef]
175. Macgillavry, C.H.; Hoogschagen, G.; Sixma, F.L.J. The crystal structure of glutaric and pimelic acid. Alternation of properties in the series of dicarboxylic acids. *Recl. des Trav. Chim. des Pays-Bas* **2010**, *67*, 869–883. [CrossRef]
176. Kshnyakina, S.I.; Puchkovskaya, G.A. Spectroscopic investigations of crystals of the homologous series of dicarboxylic acids. *J. Appl. Spectrosc.* **1981**, *34*, 556–561. [CrossRef]
177. Nagakura, M.; Ogawa, Y.; Yoshitomi, K. Infrared Spectroscopic Determination of Alkyd Resin Components. *J. Jpn. Soc. Colour Mater.* **1968**, *41*, 542–553. [CrossRef]
178. Mishra, M.K.; Varughese, S.; Ramamurty, U.; Desiraju, G.R. Odd-Even effect in the elastic modulii of α,ω- alkanedicarboxylic acids. *J. Am. Chem. Soc.* **2013**, *135*, 8121–8124. [CrossRef] [PubMed]

179. Nattkemper, A.; Schleiden, T.; Migliavacca, J.M.; Melin, T. Monitoring Crystallization Kinetics of Azelaic Acid by in situ FTIR Spectroscopy in Three-Phase Systems. *Chem. Eng. Technol.* **2003**, *26*, 881–889. [CrossRef]
180. Li, M.; Zhang, J.; Huang, K.; Li, S.; Jiang, J.; Xia, J. Mixed calcium and zinc salts of dicarboxylic acids derived from rosin and dipentene: Preparation and thermal stabilization for PVC. *RSC Adv.* **2014**, *4*, 63576–63585. [CrossRef]

Review

Lipids in Archaeological Pottery: A Review on Their Sampling and Extraction Techniques

Anna Irto [1,*], Giuseppe Micalizzi [1,*], Clemente Bretti [1], Valentina Chiaia [1], Luigi Mondello [1,2,3] and Paola Cardiano [1]

1. Department of Chemical, Biological, Pharmaceutical and Environmental Sciences, University of Messina, 98168 Messina, Italy; clemente.bretti@unime.it (C.B.); valentina.chiaia@studenti.unime.it (V.C.); lmondello@unime.it (L.M.); paola.cardiano@unime.it (P.C.)
2. Chromaleont s.r.l., c/o Department of Chemical, Biological, Pharmaceutical and Environmental Sciences, University of Messina, 98168 Messina, Italy
3. Unit of Food Science and Nutrition, Department of Medicine, University Campus Bio-Medico of Rome, 00128 Rome, Italy
* Correspondence: anna.irto@unime.it (A.I.); giumicalizzi@unime.it (G.M.)

Abstract: Several studies have been performed so far for the effective recovery, detection and quantification of specific compounds and their degradation products in archaeological materials. According to the literature, lipid molecules are the most durable and widespread biomarkers in ancient pottery. Artificial ageing studies to simulate lipid alterations over time have been reported. In this review, specific lipid archaeological biomarkers and well-established sampling and extraction methodologies are discussed. Although suitable analytical techniques have unraveled archaeological questions, some issues remain open such as the need to introduce innovative and miniaturized protocols to avoid extractions with organic solvents, which are often laborious and non-environmentally friendly.

Keywords: lipids in pottery; archaeological biomarkers; ancient pottery; ageing study; sampling of lipids; lipid extraction; lipid derivatization

1. Introduction

Organic residues in archaeology refer to a wide variety of amorphous organic remains commonly associated with ceramic containers or tools, found in archaeological contexts. The extraction and analysis of such organic residues from pottery can provide answers to a variety of archaeological questions about diet, food storage and processing, rituals and medical practices, trade and the use of commodities, domestication of animals, etc., thus contributing to unveiling crucial hints about daily life of the ancient societies. In fact, it is well known that ceramics were used, not only for decorative purposes, but, significantly, for a wide variety of functions [1,2]. Broadly speaking, culinary pottery can be distinguished according to its use, i.e., storage containers, processing vessels (employed for grinding, crushing, mixing, marinating, boiling, roasting etc.) and tableware (for eating or serving) [3–5]. Organic residues can be found both in the inner and outer part of the pottery container as visible remains, surface deposits and encrustations. A surface residue on the vessel's outer walls may derive from soot deposited during cooking activities on the fire. In contrast, visible burnt residues adhering to the inner wall of a container can result from the charring of food [6–9]. More commonly, organic residues occur as invisible absorbed material within the porous unglazed vessel wall. Both visible and invisible organic residues can derive not only from the processing or storage of foodstuffs, but also from nonculinary practices, i.e., sealing or waterproofing purposes [8,10], to create coatings of the inner surface of ceramic [11]. However, containers could be used for multiple functions and/or reused or recycled for different purposes over time. From the aforementioned, it is clear that establishing specific functions and uses of pottery vessels is a truly daunting task.

Consequently, the most innovative technologies in analytical chemistry must be closely merged with adequate archaeological guidance in order to unravel the origin of organic residues and gain crucial hints on ancient human activities.

To date, lipids (i.e., fats, waxes and resins) represent one of the main chemical classes of substances investigated in archaeological pottery. This is due to their hydrophobicity that makes them less prone to loss by solubilization than other more soluble organic compounds (i.e., carbohydrates, proteins) [8,12], thus limiting their percolation and allowing their persistence in the original site. Conversely, more polar substances are more susceptible to decay, especially those containing nitrogen and phosphorous atoms [13]. Although the resistance to decay and the hydrophobic character of the lipids could make them excellent candidates as "archaeological biomarkers" [14–17], it should be highlighted that most of them are featured by reactive functional groups that fatally lead to their decay over time. In addition, the strict relationship between edible lipid substances and preservation should be kept in mind. The main characteristic of foodstuffs is their digestibility in the gastrointestinal tract. This means that lipid molecules are also likely susceptible to degradation by microorganisms in the burial environment [18]. Consequently, the lipids may undergo in situ chemical or microbiological degradations over time. This fact further complicates the interpretations on the origin of lipid matter [3].

It is well established, however, that some archaeological environments can retard the degradation of lipids, for instance, very dry climactic conditions [19–21] or acidic soils [4,22,23] may retard their decay. In addition, the entrapment of lipid molecules within a ceramic matrix preserves or, at least, retards their alteration [13,24]. The degree of preservation is highly dependent on the chemical and physical conditions (pH, temperature, biomass and humidity) of the burial environment. This means that preservation of the lipid matter in an archaeological context depends mainly on the presence of favorable conditions. The literature data indicate that pottery vessels possess suitable characteristics for absorbing organic material and preserving it during burial over millennia [25–27], whilst the contamination of organic residues from the burial soil occurs only rarely. The entrapment of lipids in organic or mineral matrices generally limits their loss by microbiological degradation. In fact, the access of exocellular enzymes produced by degrading microbes to lipid matter would be prevented, especially in highly dense or vitrified materials due to their low porosity and permeability [26]. Lipids are also well-preserved in carbonized organic residues on pottery [28], probably due to microencapsulation that inhibits microbial activities. The encapsulation of organic residues within clay surfaces may also limit the access of microorganisms, but the presence of water and other reactive species may cause some chemical degradation processes, such as hydrolysis or oxidation, leading to the formation of specific "archaeological biomarkers" [26]. For example, the partial hydrolysis of triacylglycerols (TAGs) leads to the formation of diacylglycerols (DAGs) and monoacylglycerols (MAGs), while the complete hydrolysis involves to the formation of free fatty acids (FFAs) [29]. A detailed overview of such hydrolysis products in ancient pottery is elucidated in the next paragraph.

2. Lipids and Archaeological Biomarkers

In the last decades, the analysis of organic residues coming from foodstuffs, balms and perfumes found in pottery has been mainly focused on the determination of lipids. Archaeological biomarkers are specific molecules, often detected at trace levels, providing useful information on the origin of organic residues and clues about the potential function of the ceramic container in which they have been found [30,31]. In this section, the distribution and composition of the main lipid constituents detected in pottery such as acylglycerols (TAGs, DAGs and MAGs) and FFAs, and minor lipid constituents such as sterols (STs), natural waxes and terpenoids, will be discussed. The chemical structures of the most common lipids reported in the literature are illustrated in Figure 1.

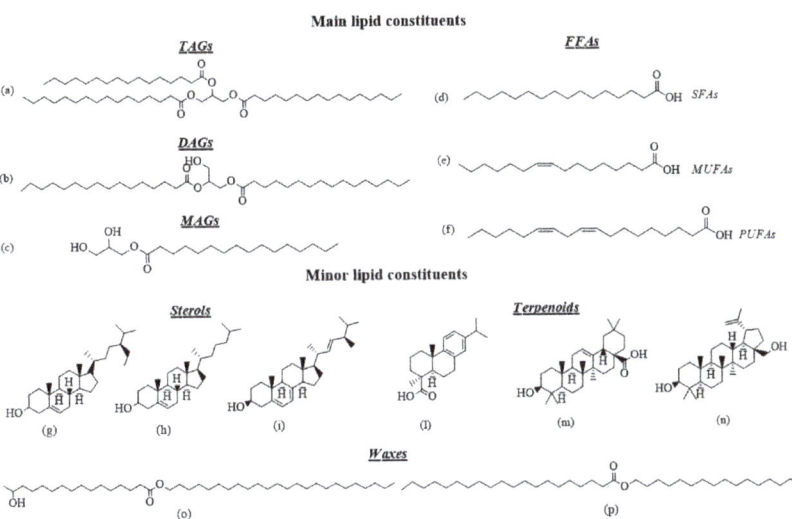

Figure 1. Examples of lipids detected in archaeological pottery samples: (**a**) tripalmitoylglycerol, (**b**) 1,3-dipalmitoyl-glycerol, (**c**) 1-palmitoyl-glycerol, (**d**) palmitic acid, (**e**) palmitoleic acid, (**f**) linoleic acid, (**g**) sitosterol, (**h**) cholesterol, (**i**) ergosterol, (**l**) dehydroabietic acid, (**m**) oleanolic acid, (**n**) betulin, (**o**) tetracosanyl 15-hydroxypalmitate, (**p**) hexadecyl eicosanoate.

2.1. Triacylglycerols, Diacylglycerols and Monoacylglycerols in Pottery

TAGs have sometimes been found in organic residues from pottery, since they usually readily decompose by means of chemical and microbial processes [13,32]. In the case of fats derived from ruminants (cattle, goats, sheep) and non-ruminant (pigs) animals, a narrow distribution of TAGs from C_{42}, and from traces of C_{44} to C_{54}–C_{56} carbon atoms has been detected, respectively. In dairy products from ruminant animals, a large distribution of TAGs from C_{40} to C_{54} was identified, whereas in marine and freshwater fish, they were not found [3,33].

As aforementioned, TAG decomposition reactions lead to the formation of DAGs, MAGs and FFAs. A low abundance of DAGs containing C_{32}, C_{34}, C_{36} long-chain acyl carbon atoms, together with significant concentrations of C_{16} and C_{18} MAGs, as well as C_{40}–C_{48} wax esters, have been detected in archaeological pottery [33]. MAGs and DAGs containing $C_{16:0}$ and $C_{18:0}$ acyl moieties are degradation products of TAGs possibly present in raw animal fats [34]. The presence of these specific DAGs and MAGs, together with high concentrations of $C_{18:0}$ and $C_{16:0}$ FFAs, is representative of the degradation of animal fats [19] due to the use and/or subsequent burial of pottery for many centuries, as observed by Evershed et al. [35] by performing in-laboratory decay studies on animal fats.

2.2. Free Fatty Acids as Archaeological Biomarkers

FFAs are the principal constituents of hydrolyzed fats and oils, the most encountered and investigated lipid types associated with archaeological pottery [36]. Despite many FFAs being identified in archaeological ceramic sherds [37,38], only some of them were detected in significant amounts, especially if the ceramic containers have been treated at high temperatures for cooking purposes or they have been subjected to burial, being exposed to chemical reactions (oxidation, hydrolysis, condensation) [1,39]. FFAs consist, in most cases, of an unbranched hydrocarbon chain, mainly containing an even number of carbon atoms, commonly from 12 to 24, and a terminal carboxyl group. FFAs can differ from each other, not only in the carbon chain length, but also in the number of double bonds along the carbon chain. In such a respect, FFAs can be classified as saturated FAs (SFAs), monounsaturated FAs (MUFAs) and polyunsaturated FAs (PUFAs) [40].

The distribution of SFAs is strongly related to the nature of the organic residue detected in ceramics [30]. The most abundant medium chain SFAs found in pottery samples contain an even carbon number, such as palmitic ($C_{16:0}$, hexadecanoic acid) and stearic ($C_{18:0}$, octadecanoic acid) acids. These compounds are ubiquitous, since they can be identified both in animal and vegetable products [41]. On the other hand, lauric ($C_{12:0}$, dodecanoic acid), arachidic ($C_{20:0}$, eicosanoic acid) and behenic ($C_{22:0}$, docosanoic acid) acids can be detected in significant amounts in coconut, palm and peanut oils, while myristic acid ($C_{14:0}$, tetradecanoic acid) can be found in plant seed oils and dairy products [42]. Short chain FAs containing an even carbon number, namely butyric ($C_{4:0}$, butanoic acid), caproic ($C_{6:0}$, hexanoic acid), caprylic ($C_{8:0}$, octanoic acid) and capric ($C_{10:0}$, decanoic acid) acids, were identified in pottery in which ruminant milk fats, palm or coconut oil were contained [41]. SFAs with an odd carbon number such as pentadecylic ($C_{15:0}$, pentadecanoic acid), margaric ($C_{17:0}$, heptadecanoic acid) and nonadecylic ($C_{19:0}$, nonadecanoic acid) acids were also revealed in ceramics [40,41]. Their origin is mainly linked to bacterial, milk and ruminant fats [43]. Short and medium chain SFAs with an odd carbon number, namely valeric ($C_{5:0}$, pentanoic acid), enanthic ($C_{7:0}$, heptanoic acid), pelargonic ($C_{9:0}$, nonanoic acid), undecylic ($C_{11:0}$, undecanoic acid) and tridecylic ($C_{13:0}$, tridecanoic acid) acids, were detected in archaeological ceramics used as containers for the flowering plants valerian, rancid oils, pelargonium and other vegetable oils and dairy products, respectively [41]. In C_6–C_{24} saturated fatty acids, as well as unsaturated FFAs, such as oleic ($C_{18:1\omega9}$, cis-9-octadecenoic acid) and linoleic acid ($C_{18:2\omega6}$, cis-9, cis-12-octadecadienoic acid) acids, a variable composition of C_{14}–C_{20} alcohols, C_{16} and C_{18} MAGs, and C_{23}–C_{29} n-alkanes was detected, together with small organic acids and monosaccharides deriving from glucose and glycerol, in pottery jars, vessels and amphorae possibly employed to store, contain and transport, at the same time or in different moments, vegetable oils or animal products with fermented alcoholic beverages (grape juice, wine) or sauces (Roman sapuum, mulsum or defrutum) [44–46]. Although tartaric and syringic acids were traditionally considered as wine biomarkers, the identification of wine in archaeological pottery remains controversial, since the aforementioned compounds can come from different sources [45,47–49]. Glutaric, fumaric, lactic malic, succinic and malonic acids, together with proper archaeological and historical support, could provide a more reliable interpretation of the data [46].

The Isotopic analysis of the $\delta^{13}C$ values of the main $C_{16:0}$ and $C_{18:0}$ FFAs [50], their difference ($\Delta^{13}C = \delta^{13}C_{18:0} - \delta^{13}C_{16:0}$) [3], as well as the proportions of selected SFAs [42], are particularly useful to obtain information about the origin of lipids determined in pottery [3,14,51–58]. Regert suggested that if the lipidic residue derives from non-ruminant animals, $C_{16:0}$ and $C_{18:0}$ FFAs are isotopically enriched in ^{13}C with respect to those found in ruminants and $\Delta^{13}C > -1‰$, whereas goat adipose fats are featured by $-3‰ < \Delta^{13}C < -1‰$ [3]. Whenever the source of the organic residue is a dairy product of ruminant animals, $C_{18:0}$ acid is depleted in ^{13}C with respect to adipose animal fats, and $\Delta^{13}C < 3.3‰$. In the case of marine organisms, $C_{16:0}$ and $C_{18:0}$ FFAs are isotopically enriched in ^{13}C with respect to those of terrestrial animals, although their values are not so different than domestic pig adipose fats. Freshwater fish resources are isotopically depleted in ^{13}C for both $C_{16:0}$ and $C_{18:0}$ acids with respect to marine fats [59]. Copley et al. observed organic residue displaying $\Delta^{13}C > -1‰$ which derive from ruminant fats, since the $C_{18:0}$ free fatty acid is depleted in ruminant tissues because of bacterial processing in the rumen [60]. Shoda and colleagues confirmed the validity of the $C_{16:0}$ and $C_{18:0}$ FFA's carbon isotopic variation criteria for the identification of lipids from ruminant animals. The same author also analyzed organic residues containing free fatty acids enriched in ^{13}C, similar to those determined by performing analogous measurements on modern marine fish and salmonids [61]. This origin was further confirmed by the presence of a high (>80%) relative amount of the 3S,7R,11R,15-phytanic acid (SRR), typical of aquatic organisms. In the case of organic residues found in pottery vessels employed for cooking, containing and/or mixing different type of food such as lipids derived from acorns and chestnuts, freshwater fish, wild boar, wild ruminants and salmonids, the author [61] applied a concentration-dependent mixing model [62] taking into account $\delta^{13}C_{16:0}$ and $\delta^{13}C_{18:0}$ values together with the %SRR [63]. Dunne et al., investigating lipid residues found in ceramic bottles employed for childhood nutrition and determined $\Delta^{13}C$ values between −3.4‰ and −3.7‰, which is attributable to the use of dairy ruminant products. Only in the case of one sample were the $\Delta^{13}C$ values in the range between the dairy and non-ruminant fats, suggesting a possible mixing in the pottery vessel of pig or probably human milk with dairy products [64]. In another paper, Dunne and co-workers confirmed that $C_{16:0}$ and $C_{18:0}$ FFAs $\delta^{13}C$ values are particularly useful to gain information on biosynthetic and dietary origin of fats detected in pottery from different ancient periods. The author observed lipidic residues deriving from non-ruminant animals featured by $\delta^{13}C_{16:0}$ and $\delta^{13}C_{18:0}$ values in the ranges from −11.0‰ to −28.1‰ and −11.0‰ to −26.9‰, respectively, and $0.1‰ < \Delta^{13}C < 7.4‰$. A ruminant adipose origin was attributed to samples displaying $\delta^{13}C_{16:0}$ and $\delta^{13}C_{18:0}$ values from −14.0‰ to −28.8‰ and $\Delta^{13}C \leq -0.9‰$. Dairy fats were identified with $-3.0‰ < \Delta^{13}C < 5.7‰$, mixed ruminant and non-ruminant fats were featured by $\Delta^{13}C$ values in the range from −0.1‰ to −0.5‰ and mixed dairy and adipose fat with $\delta^{13}C_{16:0}$ and $\delta^{13}C_{18:0}$ values from −22.4‰ to −27.3‰ and −25.5‰ to −30.0‰, respectively, and $\Delta^{13}C = -3.2‰$ [65]. Whelthon and co-workers asserted a comparison between the $\delta^{13}C$ of archaeological $C_{16:0}$ and $C_{18:0}$ FFAs and modern reference animal fats can be considered reliable for animal fats only. Conversely, free fatty acids coming from plant product processing can cause a depletion in ^{13}C and influence the $\delta^{13}C$ $C_{16:0}$ and $C_{18:0}$ values. The author also stated that FFAs featured by a ^{13}C enrichment or depletion for both the $C_{16:0}$ and $C_{18:0}$ could derive from the marine or freshwater commodity processing, respectively [18].

The investigation of the $C_{16:0}$ and $C_{18:0}$ FFA ratio is correlated to the study of the effect of environmental conditions on the FA degradation. Accordingly, some factors such as temperature, humidity and oxygen presence, are useful for the determination of degradation processes and their influence on the FFA concentration [42]. Notarstefano et al. suggested that whenever the $C_{18:0}$ content is much higher than the $C_{16:0}$ one ($0.2 < C_{16:0}/C_{18:0} < 0.6$), the organic residues detected in pottery may come from herbivore animals [66]. If the amount of $C_{18:0}$ is slightly higher than $C_{16:0}$ ($0.9 < C_{16:0}/C_{18:0} < 1.3$), the residues may have an animal origin. On the contrary, a $C_{18:0}$ level slightly lower than the $C_{16:0}$ amount ($1.2 < C_{16:0}/C_{18:0} < 2.0$), may indicate a vegetable residue that, in the presence of long chain alcohols, may also suggest the presence of waxes. Gregg and Slater indicated that when the $C_{16:0}/C_{18:0}$ value is between 1.0 and 2.0, the residues could contain decomposed animal fats, while if its value is higher than 3.0, they may come from vegetable oils [53]. Kimpe and co-workers stated that when the ratio of $C_{16:0}/C_{18:0}$ is 1.0 ± 0.1, it means that the pottery vessels were employed for cooking purposes [54]. However, this conclusion was based on the analysis of only two vessel samples, and consequently should be considered with caution. The strategy of using the $C_{16:0}/C_{18:0}$ ratio as an archeological biomarker to evaluate the residue origin in pottery has not been completely accepted by some authors [3,67], since significant amounts of $C_{16:0}$ and $C_{18:0}$ FAs could also be the result of the conversion of unsaturated FAs to SFAs. Sikorski considered the interpretation of data based only on the $C_{16:0}/C_{18:0}$ ratio doubtful, since they can occur in high concentration either in plant or in animal sources [67]. Whelton et al. suggested a cautious interpretation of the calculation of the $C_{16:0}/C_{18:0}$ ratio coming from archaeological fats with respect to modern ones, due to the different solubilities and volatilities of FFAs, which could interfere with the correct ratio value estimation [18]. Other authors suggested that further FFAs or different ratios could provide a more reliable and accurate identification strategy [3,55,57,68,69]. Regert observed that, if the concentration of $C_{16:0}$ is lower than $C_{18:0}$, and a small concentration of $C_{15:0}$ and $C_{17:0}$ FFAs is also determined, together with oleic acid and its isomers, the organic residue may come from ruminant fats, including dairy products [3]. On the contrary, if the $C_{16:0}$ content is higher than $C_{18:0}$ and long chain FFAs containing three double bonds are detected, the organic residue could be related to fish fats. Olsson and Isaksson [57,70] proposed a $C_{18:0}/C_{16:0}$ inverse ratio combined with long-chain FA presence. This condition could suggest the nonvegetable nature of organic residues. Based on this concept, a $C_{18:0}/C_{16:0}$ ratio lower than 0.48 could be indicative of fish residues [57], whereas a value of $C_{18:0}/C_{16:0}$ higher than 0.48 could reveal the occurrence of decomposed fat in terrestrial animals [71]. In addition, Marchbanks and Malainey [55,69] proposed detailed ratios and indices among FAs for determining the nature of organic residues in archaeological pottery. For example, the percentage ratio $(C_{12:0} + C_{14:0})/(C_{12:0} + C_{14:0} + C_{18:2\omega6} + C_{18:3\omega3})$ is less than 18% for vegetable oils, between 22% and 39% for fish fats, and higher than 47% for terrestrial animal fats. Analogously, Malainey [69] considered the FFA ratio $(C_{15:0} + C_{17:0})/(C_{12:0} + C_{14:0} + C_{16:0} + C_{18:0})$ to discriminate against monogastric and ruminant animal fats residues. Eerkens and co-workers [68] stated that when this ratio exceeds 0.04, the organic residue may derive from ruminant fats. Last but not least, a $C_{17:0}$ branched/$C_{18:0}$ ratio was suggested by some authors [42,72,73] as archaeological biomarkers in order to identify organic residues coming from monogastric animal and ruminant fats.

Naturally occurring unsaturated FAs are featured by one to six double bonds along the carbon chain, in most cases with a *cis* configuration [40]. MUFAs are mainly distributed in plant oils such as olive, sesame and sunflower oils, or in avocados, peanuts, almonds, pecans, walnuts and cashews, while PUFAs are found in plant-based foods, oils and fish (trout, salmon, herring) [30,74]. The most common MUFAs detected in archaeological ceramic samples are palmitoleic acid ($C_{16:1\omega7}$, *cis*-9-hexadecenoic acid) and oleic acid. In the case of adipose fats derived from ruminant animals, a mixture of isomers of oleic acid with a double bond at the C_9, C_{11}, C_{13}, C_{14}, C_{15} and C_{16} positions has been also detected. In porcine fats and dairy products from ruminant animals only a single isomer of the oleic acid has been found as an unsaturated FA [3]. In addition, *cis*-vaccenic acid ($C_{18:1\omega7}$, *cis*-11-octadecenoic acid) was identified in milk and ruminant fat residues and erucic acid ($C_{22:1\omega9}$, *cis*-13-docosenoic acid) in rapeseed and mustard oils. Elaidic acid ($C_{18:1\omega9}$, *trans*-9-octadecenoic acid) was also detected in archaeological pottery used as containers for hydrogenated fats. The most abundant PUFA identified in archaeological ceramics is the already mentioned linoleic acid, which can be found in vegetable oils residues [41]. With respect to fish oils, FFA fingerprints are quite complex since SFAs ($C_{14:0}$, $C_{16:0}$ and $C_{18:0}$), MUFAs ($C_{16:1\omega7}$, $C_{18:1\omega9}$) and PUFAs containing 18, 20 and 22 carbon atoms with a high degree of unsaturation (and up to six double bonds) have been reported [75]. Considering the high degree of unsaturation of PUFAs in fish oils, the probability that these molecules may degrade over time (by both chemical and biological degradation) is significant [13,76,77]. This means that a loss of unsaturated FAs over SFA compounds can be expected, hampering their possible detection in pottery [3,18]. In marine and freshwater fish, isomers of ω-(o-alkylphenyl) alkanoic acids (APAAs) (see Figure 2) with 16, 18 and 20 carbon atoms and positional isomers could be produced by degradation of tri-unsaturated FAs [3]. This was explained in terms of a multi-step alteration process, starting with the alkali isomerization of the acids, very likely promoted by pottery clays, followed by a 1,5-hydrogen shift with the formation of a conjugated triene system. Then, a *cis/trans* isomerization and an intramolecular Diels–Alder mechanism with aromatization may occur and produce conjugated cyclic products. Alternatively, a 1,7-hydrogen shift, an intramolecular Diels–Alder reaction and a final step of aromatization could lead to the formation of the mentioned products [22]. Vicinal dihydroxy acids are other oxidation products of unsaturated FAs. The position of the hydroxyl groups along the carbon chain indicates the original double bond position in the FA precursor. Such molecules, containing from 16 to 22 carbon atoms, have been detected in ancient pottery [78]. Furthermore, high concentration levels of isoprenoid fatty acids (IFAs), such as pristanic acid and phytanic acid, depicted in Figure 2, and low levels of 4,8,12-trimethyltridecanoic acid (4,8,12-TMTD), can be found in marine animals. On the other hand, they are not present in terrestrial animals [3,37]. The branched structures of IFA compounds are particularly resistant to degradation; therefore, they are listed as archaeological biomarkers for the identification of fish oils contained in pottery vessels [79,80]. In the case of adipose ruminant and non-ruminant animal fats, oxidation reactions could cause the formation of unsaturated short-chain dicarboxylic, hydroxy- and dihydroxy carboxylic acids, both in the free and esterified form [19]. The detection of (α,ω)-dicarboxylic acids ranging from C_5–C_7 to C_{12}–C_{13} has been reported in the literature [38]. The azelaic acid (nonanedioic acid, Figure 2) represents one of the most detected (α,ω)-dicarboxylic acids, indicating that FA precursors were featured by a double-bond at the C_9-position (i.e., oleic acid). Alkaline hydrolysis of FFAs could also produce ω-hydroxy even-numbered saturated carboxylic acids with 8–12 carbon atoms. Heating animal fats could induce the condensation of FFAs and the formation of odd-numbered monounsaturated ketones ranging from C_{29} (nonacosan-15-one, Figure 2) to C_{33} (tritriacontan-16-one)-C_{35} (pentatriacontan-18-one) [3].

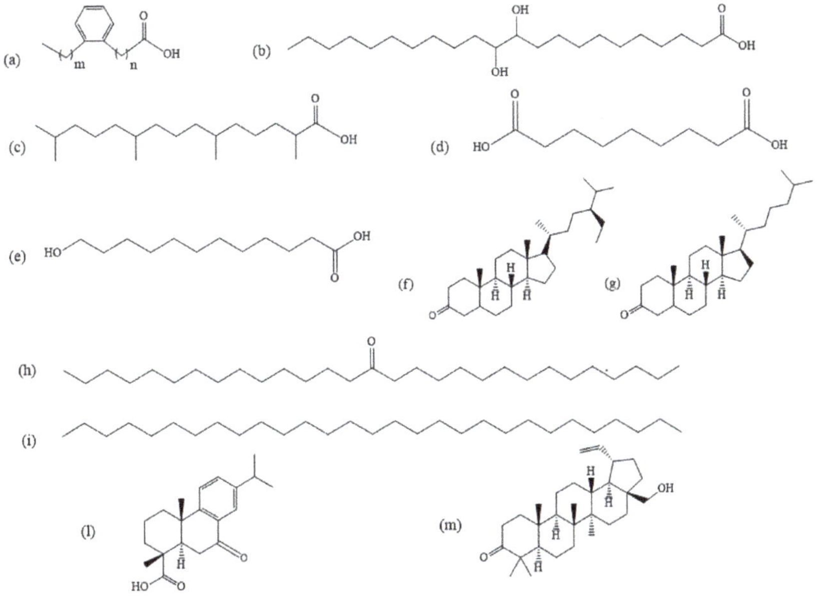

Figure 2. Examples of lipids degradation products detected in archaeological pottery samples: (**a**) ω-(o-alkylphenyl)alkanoic acid, (**b**) 11,12-dihydroxydocosanoic acid, (**c**) pristanic acid, (**d**) azelaic acid, (**e**) ω-hydroxydodecanoic acid, (**f**) sitostanone, (**g**) cholestanone, (**h**) nonacosan-15-one, (**i**) n-nonacosane, (**l**) 7-oxodehydroabietic acid, (**m**) betulone.

2.3. Minor Lipid Constituents in Archaeological Samples

Phospholipids (PLs) are structural components of biological cell membranes [8]. They consist of a phosphoric acid unit, often linked to a nitrogen-containing molecule and two FAs [81,82]. PLS are the constituents of the carcass fat of wild ruminants with higher concentrations with respect to the acylglycerols [83]. In soil, where pottery sherds are usually found after prolonged burial, PL's occurrence is transitory due to degradation processes that make them difficult to detect after many centuries, so that the only extractable compounds belonging to this class are hypothesized to derive from living biomass [83].

STs are precursors of some hormones and structural components of cell walls [30]. Their low concentrations in pottery, as well as the possible cross-contamination due to the handling of pottery during the excavation or post excavation phases, make the STs identification particularly questionable [18]. STs can be mainly grouped into phytosterols and zoosterols based on vegetable and animal origin, respectively [8]. Phytosterols, such as sitosterol, stigmasterol and campesterol, and their oxidation products such as sitostanone (Figure 2), sitostanol and campestanol, can be identified as organic residues coming from vegetable oils, cereal grains and nuts. For this reason, they are considered plant ST biomarkers in archaeological potsherds [42]. They were sometimes identified in visible carbonized residues on the inner surface of pottery vases or in the botanical remains of carbonized seeds [42,66]. The most abundant zoosterol is undoubtedly cholesterol, a component of the biological membranes of mammal cells and a precursor of estrogen and steroid hormones (progestogens, glucocorticoids, androgens, mineralocorticoids) [84]. Cholesterol either has an exogenous source, coming from animal food, and/or an endogenous origin, because it is produced in the animal liver [84]. Whelton et al. stated that its detection should be viewed with caution due to a possible contamination from human skin lipids [18]. Only if cholesterol is determined together with its hydroxy-, oxo-, epoxy, ketone oxidized derivatives, such as cholestanol and cholestanone (Figure 2), produced by the heating of animal fats in pottery vessels or the natural decay, can it be considered as an archaeological biomarker [33,85]. Saturated odd-numbered mid-chain ketones from C_{29} to C_{35}, such as nonacosan-15-one (Figure 2), triacontan-14-one, triacontan-15-one, hentriacontan-16-one, dotriacontan-15-one, dotriacontan-16-one, tritriacontan-16-one, tetratriacontan-17one and pentatriacontan-18-one, as well as C_{33} and C_{35} monounsaturated ketones, were also detected in pottery containers where animal fats were processed [3]. Ergosterol is a specific mycosterol, component of fungal cell membranes with the function of cholesterol in animals and precursor of ergocalciferol (vitamin D2). In lipid residues found in prehistoric pottery, it was considered a potential biomarker for alcohol fermentation in beer, bread or wine [42], but Isaksson et al. highlighted that ergosterol detection in pottery could also derive from modern contamination [71].

Other lipids identified in organic residues extracted from pottery could include natural waxes biosynthesized by insects, such as honey and beeswax, or by plants, that form hydrophobic coatings on their outer surface, as in the case of leaf or epicuticular waxes, and protective layers on the skin, hair and feathers of animals [3,30,74,82,86]. Wax composition is heterogeneous and varies with plant or animal type [87]. Waxes from leafy plants are characterized by long chains archaeological biomarkers, such as odd-numbered alkanes (C_{25}–C_{33}), even-numbered alcohols (C_{20}–C_{34}) and aldehydes (C_{24}–C_{28}), as well as C_{39}–C_{52} esters [88,89]. Some organic compounds are considered to be specific biomarkers of plant oils as a source of organic residues, as in the case of *Brassicaceae* seed oil, widely employed in ancient times, whose chemical fingerprints are gondoic acid ($C_{20:1\omega9}$, cis-11-eicosenoic acid) and the aforementioned erucic acid. Oxidation processes occurring on these compounds could produce vicinal dihydroxy acids and (α,ω)-dicarboxylic acids. Vicinal dihydroxy acids such as 11,12-dihydroxy arachidic (11,12-dihydroxyeicosanoic acid) and 13,14-dihydroxybehenic (13,14-dihydroxydocosanoic acid) acids, as well as the (α,ω)-undecanedioic and (α,ω)-tridecanedioic acids, are chemical fingerprints of *Brassicaceae* seed oils [29]. *n*-Nonacosane (Figure 2) and its oxygenated derivatives nonacosan-15-one and nonacosan-15-ol are indicators of the processing of cabbage, turnip, kale and broccoli vegetables [30,89]. In waxes, the formation of soluble salts and volatilization reactions caused by heating of the ceramic vessel, could produce loss of FAs and of *n*-alkanes, respectively [82]. Beeswax is featured by long-chain odd-numbered alkanes (C_{23}–C_{33}), even-numbered FAs and even-numbered C_{40}–C_{54} wax esters. Esters are more resistant to hydrolysis than TAGs so that wax can be considered less vulnerable to degradation processes and more likely detectable in archaeological samples with respect to TAGs [87]. Partial degradation products could occur over time on beeswax esters and produce long-chain even-numbered alcohols (C_{24}–C_{34}), FAs and n-alkanes.

Lipids detected in pottery could also originate from natural products not employed as foodstuffs, such as those coming from resins, tars, pitches and bitumen and wax from beeswax used for non-dietary purposes. These products were stored in vessels, used as sealants, for decoration, as adhesives for repair aims, as illuminants, ointments, cosmetics, balms and medicines. The main lipids detected in these matrices are terpenoids, whose structures are featured by isoprene (C_5H_8, 2-methylbutadiene) units that can be classified in monoterpenoids (C_{10}), sesquiterpenoids (C_{15}), diterpenoids (C_{20}) and triterpenoids (C_{30}) with 2, 3, 4 and 6 isoprene units, respectively. Terpenoids generally show a good preservation [90], even if their degradation products were detected in archaeological pottery samples. The loss of low molecular weight terpenoids, such as monoterpenoids and sesquiterpenoids, was observed due to their volatility [82]. Diterpenoids such as dehydroabietic and didehydroabietic acids, as well as their oxidation products, namely 7-oxo-dehydroabietic (Figure 2), 7-oxo-abietic and 15-hydroxydehydroabietic acids, are considered specific archaeological biomarkers of *Pinaceae* family resins [66,91,92]. Some of these compounds can provide information on the resin heating conditions, as for methyl dehydroabietate, which indicates that pine resin was heated at high temperatures and in the presence of wood [91]. Such molecules are often detected together with wine-related compounds such as tartaric acid, since resins were employed as sealing or waterproofing agents in pottery vessels, and for wines aromatization. In addition, pine resins were used in firing containers to improve the mechanical and thermal resistance of ceramic to heating [66]. Triterpenoids such as oleanonic and oleanolic acids, together with other compounds, are chemical fingerprints of storax resin [93]. Betulin, lupeol and their derivatives, such as lupenone, betulone (Figure 2) and betulinic acid, are archaeological biomarkers of birch bark [94]. In the last decade, a pentacyclic triterpene methyl ether named miliacin, commonly found in millet grains, was also identified in prehistoric pottery vessels [38,95].

3. Artificial Ageing Studies

In-laboratory alteration processes such as thermal decomposition, oxidation and hydrolysis of lipids have been investigated in order to simulate the natural degradation occurring in archaeological contexts. Such ageing studies can provide key elements to better interpret the origin of animal fats and plant oils that are partially or totally altered over time. The elucidation of the chemical and biochemical mechanisms responsible for the alteration of pristine molecules also allows for the unveiling of the life history of an organic residue. For example, long-chain ketones are formed via free radical-induced dehydration and decarboxylation mechanisms, which involve intensive heating of the carboxylic FAs up to over 300 °C [25]. As a result of such cooking activities, the degradation of unsaturated FAs over 270 °C with formation of APAAs was also reported. This means that their detection may provide clues on the original constituents fired in pottery.

FAs containing at least one double bond along the carbon chain are particularly sensitive to oxidation reactions. The oxidation process involves the inclusion of an oxygen atom in the carbon chain, the scission of the double bond and formation of lower molecular weight species [29]. As reported by Rottlander and Schlichtherle, the oxidation rate of FAs depends on the degree of unsaturation. This means that MUFAs are oxidated much more slowly than PUFAs [77]. (α,ω)-Dicarboxylic and ω-hydroxycarboxylic acids are often the main oxidation products, as reported by Colombini et al. [29]. The authors performed ageing experiments by heating gondoic and erucic acid standards at 120 °C for three weeks [29]. Gas chromatography–mass spectrometry (GC–MS) analyses revealed that oxidation products were strongly influenced by the double bond position along the carbon chain. Accordingly, (α,ω)-undecanedioic and (α,ω)-tridecanedioic acids were the most abundant oxidation compounds and their formation via radical oxidation mechanism was consistent with the gondoic and erucic structures, respectively. The GC–MS chromatograms, reported in Figure 3, also highlighted the presence of minor constituents with short- and medium-chains, indicating further reaction mechanisms such as migration of the radical adjacent to the carboxylic group [29]. Alternatively, Bondetti and co-workers focused their attention on the study of APAA species as degradation products of MUFAs and PUFAs [96]. Their formation was the result of double-bond rearrangements during protracted heating of lipids present in animal and plant tissues. The carbon chain length of APAAs allowed the organic residues coming from aquatic or terrestrial source to be distinguished. In fact, the presence of APAA with 20 and 22 carbon atoms could be directly related to the cooking of aquatic organisms such as freshwater and marine animals [97], since they derive from their long-chain FA precursors, eicosapentaenoic ($C_{20:5\omega3}$, eicosapentaenoic acid or EPA) and docosahexaenoic acid ($C_{22:6\omega3}$, docosahexaenoic acid or DHA), which are quite common in aquatic sources. In contrast, APAA-C_{16} and APAA-C_{18} species can be considered as archaeological biomarkers for revealing the terrestrial nature of the organic residue, belonging to both the animal and the plant kingdoms. In addition, the simulation of the degradation reactions allowed the authors to establish a further threshold value of 0.06 for the ratio APAA-C_{20}/APAA-C_{18}, in order to discriminate aquatic sources from terrestrial products. Moreover, Hammann et al. recently carried out artificial ageing of cholesterol in ancient clay potteries [84]. The authors demonstrated that cholesterol undergoes complete degradation at 100°C in the presence of a high content of FAs, consistent with those observed in animal fats (100:1 w/w, FA to cholesterol). However, the pro-oxidative behavior of FAs had a minimal effect on the cholesterol degradation when low concentration levels were registered (1:4 w/w, FA to cholesterol). In-laboratory heating experiments suggested that the clay contributed to the cholesterol degradation. In fact, different degradation products were observed during the experiments involving cholesterol heating with the clay material only. Since both FAs and the clay surface may contribute independently to degradation reactions, the absence of cholesterol in cooking lipid residues in ancient pottery is not surprising. Within such a context, the detection of cholesterol in clay pottery should be examined with caution [84].

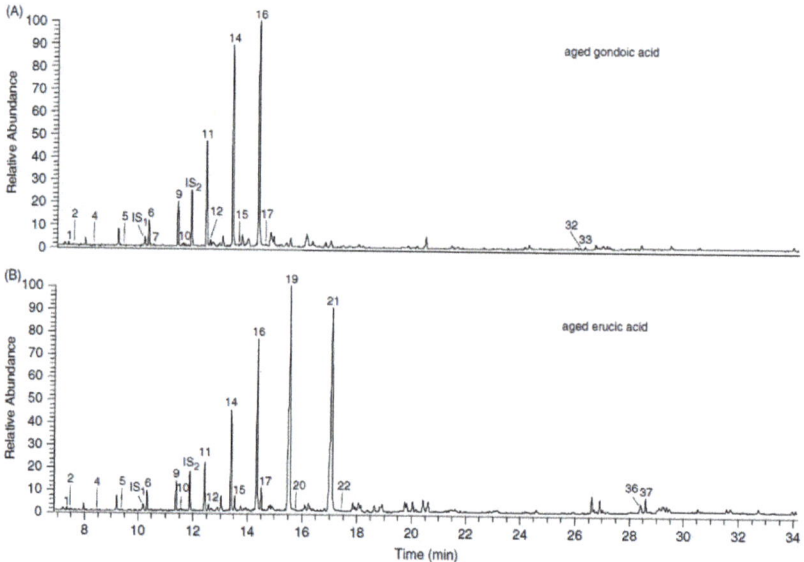

Figure 3. GC–MS chromatogram of aged gondoic (**A**) and erucic (**B**) acid standards. Reprinted with the permission from Ref. [91]. Copyright 2005 John Wiley and Sons, Ltd. Peak assignment is described as follows: (1) nonanoic acid, (2) (α,ω)-butanedioic acid, (3) decanoic acid, (4) (α,ω)-pentanedioic acid, (5) (α,ω)-hexanedioic acid, (6) (α,ω)-heptanedioic acid, (7) ω-hydroxyoctanoic acid, (8) dodecanoic acid, (9) (α,ω)-octanedioic acid, (10) ω-hydroxynonanoic acid, (11) (α,ω)-nonanedioic acid, (12) ω-hydroxydecanoic acid, (13) tetradecanoic acid, (14) (α,ω)-decanedioic acid, (15) ω-hydroxyundecanoic acid, (16) (α,ω)-undecanedioic acid, (17) ω-hydroxydodecanoic acid, (18) hexadecanoic acid, (19) (α,ω)-dodecanedioic acid, (20) ω-hydroxytridecanoic acid, (21) (α,ω)-tridecanedioic acid, (22) ω-hydroxytetradecanoic acid, (23) oleic acid, (24) octadecanoic acid, (25) (α,ω)-tetradecanedioic acid, (26) gondoic acid, (27) eicosanoic acid, (28) 9,10-dihydroxyoctadecanoic acid, (29) 9,10-dihydroxyoctadecanoic acid, (30) erucic acid, (31) docosanoic acid, (32) 11,12-dihydroxyeicosanoic acid, (33) 11,12-dihydroxyeicosanoic acid, (34) nervonic acid, (35) tetracosanoic acid, (36) 13,14-dihydroxydocosanoic acid and (37) 13,14-dihydroxydocosanoic acid. All compounds are intended as TMS derivatives.

The alteration of lipid substances over time is also partly mediated by bacterial action. In fact, it is well known that bacteria adapt readily where essential nutrients, including lipids, are available. Consequently, the effects of bacterial activity on lipids in archaeological organic residues have been discussed in the literature. Dudd et al. designed laboratory experiments in order to simulate the decay of absorbed lipid matter in ceramic vessels under oxic conditions [98]. In detail, two of the most popular foods in antiquity, milk and olive oil, were absorbed on sherds and incubated at 30 °C in a flask with mushroom compost (mushroom humix manure). The decay of lipid compounds was monitored at different times intervals. In-laboratory ageing experiments indicated that the degraded lipid profile of milk was indistinguishable from that of adipose fat. Extreme caution is therefore required in the assignation of such lipid matter. In addition, the bacterial-induced decay of lipids shows typical biomarkers such as branched-chain and odd carbon number FAs. However, such compounds naturally occur in milk fat due to the presence of bacteria in the rumen [99], thus the bacterial action on lipid degradation cannot be established with sufficient certainty [98]. On the other hand, the simpler lipid profile of olive oil allowed the authors to assess the contribution of bacteria to the decay of the original lipids, presumably resulting from the combination of microbiological and abiological hydrolysis [98]. In fact, the detection of branched chain FAs in olive oil indicated that bacterial organisms were actively responsible for the decay of the olive oil in potsherds, although to a lesser extent than other degradation mechanisms.

4. Sampling and Extraction Protocols of Lipids from Ancient Pottery

The capability of elucidating the lipid composition of materials used by the ancient societies can ensure that their practices remain part of our cultural heritage. Consequently, the chemical analysis of preserved lipid matter in archaeological contexts is not a simple and routine procedure, but it requires careful planning of the entire analytical workflow, from the sampling strategy to the interpretation of the analytical data. For a successful research study, a project design sampling strategy is the first and crucial aspect that must be emphasized. In fact, a significant criticism of the sampling methods has already been highlighted in the literature [8,18]. It seems quite clear that a single vessel or limited number of vessels cannot provide meaningful data, except for minor circumstances such as an archaeological sample coming from a "special depositional context" or showing "special typological characteristic" [18]. This means that a robust sampling strategy involving a large number of sherds (20–30) is necessary in order to statistically represent the time period, excavation site, burial conditions, object shape and pottery typologies. A large number of samples also enables the examination of a range of potential variables that may affect the quality of analytical data [8,18]. Another fundamental aspect involves the strict collaboration between the analyst and the archaeologist, essential to ensure a coherent strategy to the whole process. For example, the archaeological area can provide preliminary relevant information regarding the context and relationships within and between sites, as well as the absolute and relative chronological information, pottery typologies, materials etc. [18]. If such information is not included in the project design sampling strategy, the entire research may be of questionable quality. Appropriate handling and storage protocols are the key requirements to avoid the presence of contaminants. In general, the first source of contamination is linked to the organic matter in the burial environment. Animal and plant materials, as well as microbial synthesis from bacteria and fungi, represent all potential interfering agents that should be critically addressed. In order to exclude any possible lipid contamination, it is advisable to perform a comparison of the analytical data between the pottery excavated from the burial site and the surrounding soil. The effectiveness of such an approach was demonstrated by Heron et al. [100]. Contaminant agents can also come from handling of the sherds, both during excavation and post-excavation [18]. For instance, the human contact introduces lipid contaminants including cholesterol and squalene that can be mistaken with animal organic residues preserved in the vessels. Squalene is known to degrade over time; therefore, its detection is usually assigned to modern handling. On the other side, as aforementioned, the presence of cholesterol should be interpreted with caution due to its capability of surviving in organic residues of pottery, except when exposed to a high firing temperature [100]. Contaminants can also arise from materials used for the storage of samples such as phthalate plasticizers. In this case, paper bags, rather than plastic bags, solve the problem of plasticizer contamination, although these substances are easily identifiable and do not affect the analysis of lipids on pottery [8]. As routine practices, the monitoring of a blank sample is crucial for the detection of lipid contaminants. This measure allows not only for the examination of the purity of the chemicals (solvents and reagents), but also to evaluate the entire analytical protocol including the chromatographic instrumentation. Based on this assumption, the pottery analyses should be performed when a blank, without any lipid contaminants, is obtained. Of course, this approach can only be used to establish contaminants introduced in the laboratory, not those introduced through handling and storage protocols.

Organic residues in archaeological pottery represent very complex matrices, therefore it is necessary to optimize an adequate sample treatment. Generally, lipids are analyzed by using chromatographic techniques coupled to an MS detector [101] in order to reveal the correct identity of the lipids in a univocal manner. The extraction step is mandatory in order to purify and isolate the lipid compounds. As a rule, there is not a univocal protocol of sample preparation that is suitable and applicable for all archaeological matrices and compatible with all the analytical techniques. For this reason, numerous sample treatments have been described in the literature over the years for lipid analysis in pottery.

In term of lipid extraction, the literature data confirm that solvent-based extractions are the most common. Recently, Whelton et al. reported that chloroform:methanol (2:1 by volume) and dichloromethane:methanol (2:1 by volume) solvent mixtures have been used for the extraction of lipids from ceramics [18]. In these solid–liquid extraction techniques, widely utilized in foodomics and lipidomics research, lipid compounds are simultaneously extracted and isolated into a liquid layer. For example, Harper et al. utilized 5 mL of a 2:1 v/v dichloromethane:methanol solution to extract lipids from pottery [102]. Evershed et al. extracted lipids from powdered sherds (2 g) by using a chloroform:methanol (10 mL, 2:1 by volume) solvent mixture [103]. Both protocols demonstrated to be suitable in the extraction of FFAs, MAGs, DAGs and TAGs from archaeological samples. An alternative extraction strategy has recently been optimized by Tanasi et al. [104]: a solution of chloroform:methanol:water (1:1:0.9) was added to suspend the powder and extract "free" lipids. Compared to the waterless extraction procedures described above, such strategy requires that lipid compounds and interfering substances are partitioned into two layers: the upper methanolic phase containing virtually all of the non-lipid substances and lower chloroformic phase consisting of lipids [105]. From a methodological point of view, such a protocol requires the use of a centrifuge to allow the clear separation of the biphasic system. In order to improve the extraction of lipids, some authors also introduced sonication, as described by Regert et al. [3].

Alternative solvent-based extraction protocols have also been reported in the literature, especially whenever the recovery of polar lipids such as dicarboxylic acids and hydroxyl acids was incomplete, by using the well-established chloroform:methanol extraction. In fact, lipid molecules containing one or more polar functional groups, such as carboxyl (–COOH) and hydroxyl (–OH) portions, form strong intermolecular interactions (i.e., hydrogen bonds, dipole–dipole, ion–dipole and electrostatic interaction) with the polar surface of the ceramic matter. This means that their removal/extraction from ceramic is favored by more aggressive alkaline or acid reagents. The effectiveness of the alkaline treatment was demonstrated by Regert et al. [34]. In detail, a portion (1 g) of sherd, already subjected to a solvent-based extraction, was re-extracted with 10 mL of NaOH methanolic solution (0.5 M) at 70 °C for 90 min. Alkaline treatment confirmed the presence of (α,ω)-dicarboxylic acids (from C_7 to C_{12}) and α-hydroxy carboxylic acids (from C_8 to C_{12}) in the total lipid composition of organic residues that were not extractable by using a chloroform/methanol solvent mixture. The capability to extract the most polar lipids was also evaluated by using acidified methanolic solution (H_2SO_4–MeOH, 2% by volume) [51]. An alternative extraction technique is represented by the microwave-assisted extraction (MAE). One of the main advantages of this strategy is the reduction in the extraction time due to the different mechanism, by which heat is transferred to the solution. In fact, the use of microwaves allows the rapid heating of the solution, keeping the temperature gradient to a minimum. This actually increases the heating speed of the extraction mixture [106]. Although the MAE technique has great potential for heating materials, the correct selection of the extraction solvents is the fundamental requirement for a successful application. As a general principle, a solvent utilized in the conventional extraction procedures is not suitable in the MAE process if it is not able to absorb microwave energy. The use of MAE in the archaeological field has recently been reported by Blanco-Zubiaguirre et al. [24]. The extraction of TAGs was performed by mixing 1 g of the archaeological ceramic sample with 900 µL of the chloroform:hexane solution (3:2 by volume). The power of microwaves was fixed to 600 W for 25 min (temp. 80 °C). The authors also performed the extraction and simultaneous saponification of FAs by using a KOH ethanolic solution (10% w/v). In this case, the power of the microwave was 200 W for 60 min (temp. 80 °C). Other authors have performed the microwave-based extraction of lipids from ancient pottery [53,107], but the MAE approach is not widespread in archaeometry yet.

Last but not least, an innovative extraction technique based on the use of supercritical fluids has been recently explored by Devièse et al. [108]. As reported by the same authors, supercritical fluid extraction (SFE) has not yet been widely explored in the field of archaeological science, but the results of this pilot study are very promising. First of all, the SFE technique does not require the destruction of the sherd sample, since the lipid extraction can be performed without any further grinding step by pestle or mortar. This aspect is crucial for the reduction in the potential source of contamination that can substantially affect the analytical results. In addition, the methodology involves the use of supercritical carbon dioxide (CO_2), an excellent extraction solvent due to its relatively high density and consequently high solvation power. Further, it presents a low viscosity and high diffusion coefficient that allows fast extraction. From the environmental point of view, the supercritical CO_2 greatly reduces the use of organic solvents. In fact, only small portions of solvents are combined with the supercritical fluid in order to change its polarity. Devièse et al. [108] optimized the extraction of lipids from pottery by the SFE method using water and ethanol as cosolvents, which are less toxic than extraction solvents conventionally used in solid–liquid extraction. Regarding the extraction performance, the SFE allowed the authors to detect, in laboratory-made ceramic samples, higher quantities of lipids than those obtained by using a conventional solvent-based extraction (chloroform:methanol, 2:1 by volume) protocol. In addition, small amounts of unsaturated FAs were surprisingly detected in the SFE extracts, while they remained unrevealed using the solvent-based method. The SFE procedure thus showed a high extraction yield of lipids from ceramic containers, a fundamental requisite to establishing the correct functions of ceramic vessels. In light of the extraction procedures so far described, the solvent-based extractions are certainly the most widely used in the archaeological field, due to their reproducibility and reliability. However, these methodologies involve numerous sample preparation steps, sometimes more than necessary, that affect not only the duration of the entire analytical process, but also the quality of the analytical data in terms of extraction yield and contamination. With regard to the alternative and innovative strategies, both MAE and SFE techniques are still not particularly widespread for lipid analyses in archaeological pottery. As a general rule, the most innovative experiments take a long time to be accepted and used instead of the conventional procedures. Nevertheless, the effort required to migrate from a traditional extraction method towards an innovative one is counterbalanced by the equal or higher extraction yield that can be obtained in a shorter time. In addition, these alternative methodologies include practices that are in accordance with the miniaturization of chemicals consumption, preferential usage of low-toxicity reagents and a reduction in waste production.

Focusing on the total FA composition, widely elucidated by using GC techniques, a derivatization step is mandatory to convert the lipids into more volatile and less polar compounds [109]. In fact, if, on the one hand, high-molecular weight species such as FAs linked to glycerol (i.e., TAG), cholesterol or long-chain aliphatic alcohols (i.e., waxes) are not amenable for GC separation, on the other hand molecules containing functional groups such as –COOH or –OH (i.e., carboxylic acid, dicarboxylic acid, sterols, etc.) can form a hydrogen bond between compounds. This leads to poor volatility, insufficient thermal stability and low detectability due to the strong interaction between polar components and the stationary phase of the GC column [110]. Thus, the derivatization is a very useful procedure to modify the chemical structure of a compound and improve its chromatographic properties. In the archaeological field, different derivatization strategies are reported in the literature. The most common one is the saponification reaction followed by methylation or silylation, which is a classical method for the preparation of fatty acid methyl esters (FAMEs) from acylglycerols [111]. For example, Harper et al. performed the methylation of lipid residues by using sodium hydroxide (KOH) in a methanol solution (5 M) as basic catalyst [102]. The methylation reaction was carried out in an ultrasonic bath for 30 min. Based on the mechanism of transesterification (ester exchange), the acylglycerols were completely methylated into FAMEs. However, FFAs are not normally esterified by using a methanol solution in the presence of a basic catalyst [105]. In order to overcome this drawback, Romanus et al. investigated the FAME composition of archaeological potsherds by using a dual-stage derivatization approach [112]. In detail, the method involved the use of boron trifluoride (BF_3) in methanol (50% w/v) as an acid catalyst and KOH in methanol (1 M) as a basic catalyst in order to simultaneously methylate acylglycerols and FFAs. The addition of BF_3 was particularly advantageous, although methoxy artefacts can be produced by the addition of methanol across the double bond of FAs, especially in the presence of a very high concentration level of BF_3 (i.e., 50% w/v) [105]. For this reason, it is strongly suggested to use lower concentrations of BF_3 in methanol, in order to avoid the formation of the by-products that can alter the real FAMEs composition. In such a respect, some authors performed the lipid derivatization using a BF_3 methanolic solution 14% w/v at 70 °C for 1 h [6,79]. The silylation reaction involves the replacement of a labile hydrogen from acids or alcohols with a trimethylsilyl (TMS) group and the formation of derivative esters. Compared to the methylation approach, silylation allows the derivatization of multiple functional groups, such as carboxyl or hydroxyl, in one step. One of the most common silylation reagents used is N,O-bis-(trimethylsilyl)trifluoroacetamide (BSTFA) with the addition of 1% trimethylchlorosilane (TMCS) [29,34,91,103].

Many procedures have been described in the literature, in which lipids are extracted and methylated at the same time [4,51,113,114]. Such a strategy is the so-called "direct extraction-derivatization", which reduces the number of clean up-steps and is therefore particularly advantageous in terms of time and costs per analysis [109]. On the other side, the risk that non-lipid contaminants may interfere with subsequent GC analyses should be considered as well [105]. Such an approach was performed by Correa-Ascencio and Evershed [51]. The authors used an acidified methanolic solution (H_2SO_4-MeOH, 2% by volume) for extracting and simultaneously methylating lipid residues in powdered sherd samples. The developed strategy involved the heating of the reaction mixture at 70 °C for 1 h. Compared to the conventional solvent extraction ($CHCl_3$–MeOH, 2:1 v/v), the direct methanolic acid treatment enhanced the recovery of lipid residues, which is an ideal condition considering the low concentration levels of the lipid residues in pottery. In addition, the use of the acidified methanolic solution allowed the authors to reveal the presence of some polar lipid compounds such as dicarboxylic acids and long-chain ketones, which were unextracted using the $CHCl_3$–MeOH method. Both lipid extraction protocols were also evaluated by Reber [115]. The analytical workflows of the applied procedures are illustrated in Figure 4.

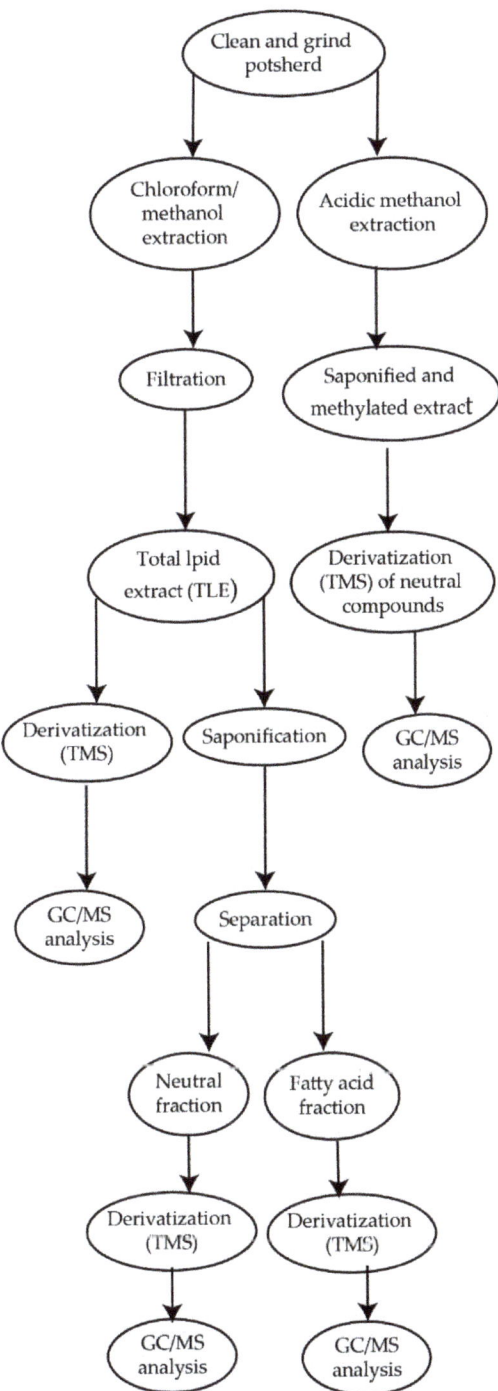

Figure 4. Analytical workflows of both conventional solvent extraction (chloroform/methanol) and direct extraction derivatization protocols. Reproduced with kind permission of MDPI [115].

The effectiveness of the direct extraction derivatization procedure was also demonstrated by Demirci et al. [114] and Papakosta et al. [116]. Besides the typical lipid distribution found in archaeological pottery (i.e., FAs, acylglycerols), the acidified methanol treatment allowed the extraction of APAAs, with carbon atoms ranging from 18 to 22, and isoprenoid FAs, including pristanic and phytanic acid. Some lipid-based archaeological studies and related analytical strategies, including extraction and derivatization approaches, are listed in Table 1.

Table 1. Examples of lipid-based archaeological studies and relative analytical approaches.

Pottery Samples and Archaeological Site	Lipid Biomarkers	Extraction	Derivatization	Analysis Method	Probable Origin	Ref.
n.35 from Zamostjen n.2 from Joton n.20 from Tianluoshan	APAAs	Solvent Extraction: MeOH/H_2SO_4 (4 h at 70 °C)	Direct extraction-derivatization	GC-MS GC-MSD	Aquatic	[96]
n.6 from Western Iberian Peninsula	ω-Hydroxy acids and cholesterol	Solvent Extraction: CH_2Cl_2/MeOH (2:1)	BSTFA + 1% TMCS *	GC-MS	Beeswax	[117]
n.14 from George Reeves, Mississippi Valley	Sterols, alkanols, alkanes and terpenoids	Solvent Extractions: $CHCl_3$/MeOH (2:1) and MeOH/H_2SO_4	BSTFA + 1% TMCS (70 °C for 1 h) NaOH in methanol (75 °C for 1 h)	GC-MS	Fish/shellfish and plants	[115]
n.20 from Pax Julia Civitas, Lusitania	FAs, acylglycerols and sterols	Solvent Extraction: $CHCl_3$/MeOH (2:1)	BSTFA + 1% TMCS (microwaveoven 700 W for 30 s)	GC-MS	Plant oil	[118]
n.172 from Northwest India	FAs	Solvent Extraction: MeOH/H_2SO_4 (4 h at 70 °C)	Direct extraction-derivatization	GC-MS GC-C-IRMS	Animal fat	[4]
n.12 from three sites: Jneneh, Sahab and Tell Abu al-Kharaz.	FAs, alkanols, MAGs, DAGs, sterols	Solvent Extraction: $CHCl_3$/MeOH (2:1)	BSTFA + 1% TMCS *	GC-MS	Plant oil and animal fat	[16]
958 potsherds from 14 different sites in Britain	$C_{16:0}$ and $C_{18:0}$	Solvent Extractions: $CHCl_3$/MeOH (2:1)	BSTFA + 1% TMCS (70 °C for 1 h) BF_3-methanol (14% w/v) (70 °C for 1 h)	GC-MS GC-C-IRMS	Ruminant adipose and dairy fats	[60]
n.63 from Samburu, Kenya	FAs	Solvent Extraction: MeOH/H_2SO_4 (1 h at 70 °C)	Direct extraction-derivatization	GC-MS GC-C-IRMS	Ruminant fats	[7]
n.15 from sites in Sardinia and Calabrian n.17 from Sicily	FAs, DAGs, TAGs and estolides	MAE extraction: KOH in ETOH (10% $w:v$) 200 W for 60 min	BSTFA + 1% TMCS (60 °C for 30 min)	GC-MS HPLC/ESI-Q-ToF	Cereal	[119]
n.101 from 13 different sites in Japan	FAs and isoprenoid FAs	Solvent Extraction: MeOH/H_2SO_4 (4 h at 70 °C)	Direct extraction-derivatization BF_3-methanol (14% w/v) (70 °C for 1 h)	GC-MS GC-C-IRMS	Aquatic oils and marine foods	[6]

Table 1. Cont.

Pottery Samples and Archaeological Site	Lipid Biomarkers	Extraction	Derivatization	Analysis Method	Probable Origin	Ref.
n. 5 from Sahab, Jordan	FAs	Solvent Extraction: $CHCl_3$/MeOH (2:1)	BSTFA + 1% TMCS *	GC-MS	Animal and ruminant fat	[17]
n. 12 from Chrysokamino	FAs	Solvent Extraction: CH_2Cl_2/Et_2O (1:1)	Diazomethane and KOH (25 °C for 24 h)	GC-MS	Plant oil	[120]
n. 10 from Qasr Ibrim, Egypt	TAGs, DAGs, MAGs, FAs, hydroxy FAs and (α, ω)-dicarboxylic acids	Solvent Extraction: $CHCl_3$/MeOH (2:1)	BSTFA + 1% TMCS (70 °C for 1 h) BF_3-methanol (14% w/v) (75 °C for 1 h)	GC-FID GC-MS GC-C-IRMS	Plant oil	[19]
n.6 from Florencen n.1 from the Pla d'Almatà site (Balaguer, Lleida, Spain)	FAs, MAGs and sterols	Solvent Extraction: $CHCl_3$/MeOH (2:1)	BSTFA + 1% TMCS (70 °C for 1 h)	GC-MS	Animal fats, ruminants and vegetable oil	[121]
n. 15 from two sites, one in East Asia and one in Europe (Poland)	FAs and APAAs	Solvent Extraction: MeOH/H_2SO_4 (4 h at 70 °C)	BSTFA + 1% TMCS (70 °C for 1 h)	GC-MS GC-C-IRMS	Plant oil	[95]
n.2 from Switzerland	FAs, hydroxy FAs, alkylresorcinols and (α,ω)-dicarboxylic	Solvent Extractions: CH_2Cl_2/MeOH (2:1) and MeOH/H_2SO_4	BSTFA + 1% TMCS (70 °C for 1 h)	GC-MS	Cereal grains	[122]
n.2 from old quarter of Lekeitio (Basque Country, northern Spain).	FAs, TAGs, (α,ω)-dicarboxylic acid and dihydroxy FAs	MAE extraction: (1) $CHCl_3$:Hex (3:2) 600 W for 25 min (2) KOH in ETOH (10%) 200 W for 60 min	BSTFA + 1% TMCS (60 °C for 30 min)	GC-MS HPLC-ESI-Q-ToF	Fish oil	[24]

* Temperature and duration conditions not defined.

5. Conclusions

The study of lipid residues in pottery is a key element to ensure that the practices of ancient societies remain part of our cultural heritage. The detailed composition of the lipid matter can unveil crucial hints about daily life, diet, food storage and processing, ritual and medical practices, etc. A large number of studies discussed in the first part of this review deal with specific archaeological lipid biomarkers. Within such a context, it appears that TAGs, DAGs, MAGs and FFAs represent the main classes of lipids detected in pottery. Nevertheless, the importance of other lipid classes such as (α,ω)-dicarboxylic acids, APAAs, beeswax, sterols, etc. has been highlighted.

As far as lipid alteration over time, mediated by thermal decomposition, oxidation and hydrolysis reactions is concerned, several authors have simulated the natural degradation of lipids occurring in archaeological contexts in the laboratory. Such artificial ageing studies are of significative importance since they allow the clarification of the degradation mechanisms responsible for the decay of lipid structures. From the discussion reported in the second part of the review, it can be concluded that careful planning of the entire analytical workflow is the pivotal step for a successful research study in the archaeological context. An adequate sampling strategy must be developed in order to ensure the quality of analytical data. First of all, a considerable number of samples guarantees the reliability of the study, especially from a statistical point of view, because it takes into account a range of potential variables affecting the quality of the data. In addition, special care should be given to the handling and storage protocols to avoid the presence of potential contaminant agents, both during excavation and post-excavation. In general, the main rules to be applied have been clarified in this review article.

Finally, several analytical approaches useful for the lipid characterization of ancient pottery have been discussed. From the literature data, organic solvent-based extractions resulted in well-established protocols. However, despite the fact that this type of approach guarantees reliability of the analytical data, no particular developments have been made in term of innovation or miniaturization for the reduction of chemicals. In the light of the studies reported so far, the extraction strategies need further optimization in order to be competitive with the well-established methodologies currently in use.

Author Contributions: Conceptualization, A.I., G.M., L.M. and P.C.; investigation, C.B. and V.C.; writing—original draft preparation, A.I., G.M. and C.B.; writing—review and editing, L.M. and P.C.; supervision, G.M., L.M. and P.C. All authors have read and agreed to the published version of the manuscript.

Funding: The research was conducted within the PON Research and Innovation 2014–2020 project funded by Italian Ministry of University and Research (MUR).

Institutional Review Board Statement: Not applicable.

Informed Consent Statement: Not applicable.

Conflicts of Interest: The authors declare no conflict of interest.

References

1. Dunne, J.; Chapman, A.; Blinkhorn, P.; Evershed, R.P. Fit for purpose? Organic residue analysis and vessel specialisation: The perfectly utilitarian medieval pottery assemblage from West Cotton, Raunds. *J. Archaeol. Sci.* **2020**, *120*, 105178. [CrossRef]
2. Lundy, J.; Drieu, L.; Meo, A.; Sacco, V.; Arcifa, L.; Pezzini, E.; Aniceti, V.; Fiorentino, G.; Alexander, M.; Orecchioni, P.; et al. New insights into early medieval Islamic cuisine: Organic residue analysis of pottery from rural and urban Sicily. *PLoS ONE* **2021**, *16*, e0252225. [CrossRef] [PubMed]
3. Regert, M. Analytical strategies for discriminating archeological fatty substances from animal origin. *Mass. Spectrom. Rev.* **2011**, *30*, 177–220. [CrossRef] [PubMed]
4. Suryanarayan, A.; Cubas, M.; Craig, O.E.; Heron, C.P.; Shinde, V.S.; Singh, R.N.; O'Connell, T.C.; Petrie, C.A. Lipid residues in pottery from the Indus Civilisation in northwest India. *J. Archaeol. Sci.* **2021**, *125*, 105291. [CrossRef] [PubMed]
5. Tanasi, D.; Greco, E.; Noor, R.E.; Feola, S.; Kumar, V.; Crispino, A.; Gelis, I. ^{1}H NMR, ^{1}H–^{1}H 2D TOCSY and GC-MS analyses for the identification of olive oil in Early Bronze Age pottery from Castelluccio (Noto, Italy). *Anal. Methods* **2018**, *10*, 2756–2763. [CrossRef]
6. Craig, O.E.; Saul, H.; Lucquin, A.; Nishida, Y.; Taché, K.; Clarke, L.; Thompson, A.; Altoft, D.T.; Uchiyama, J.; Ajimoto, M.; et al. Earliest evidence for the use of pottery. *Nature* **2013**, *496*, 351–354. [CrossRef]
7. Dunne, J.; Grillo, K.M.; Casanova, E.; Whelton, H.L.; Evershed, R.P. Pastoralist Foodways Recorded in Organic Residues from Pottery Vessels of Modern Communities in Samburu, Kenya. *J. Archaeol. Method Theory* **2019**, *26*, 619–642. [CrossRef]
8. Heron, C.; Evershed, R.P. The Analysis of Organic Residues and the Study of Pottery Use. *J. Archaeol. Method Theory* **1993**, *5*, 247–284.
9. Yoneda, M.; Kisida, K.; Gakuhari, T.; Omori, T.; Abe, Y. Interpretation of bulk nitrogen and carbon isotopes in archaeological foodcrusts on potsherds. *Rapid Commun. Mass. Spectrom.* **2019**, *33*, 1097–1106. [CrossRef]
10. Stern, B.; Heron, C.; Tellefsen, T.; Serpico, M. New investigations into the Uluburun resin cargo. *J. Archaeol. Sci.* **2008**, *35*, 2188–2203. [CrossRef]
11. Colombini, M.P.; Modugno, F.; Ribechini, E. Direct exposure electron ionization mass spectrometry and gas chromatography/mass spectrometry techniques to study organic coatings on archaeological amphorae. *J. Mass. Spectrom.* **2005**, *40*, 675–687. [CrossRef]
12. Tanasi, D.; Cucina, A.; Cunsolo, V.; Saletti, R.; Di Francesco, A.; Greco, E.; Foti, S. Paleoproteomic profiling of organic residues on prehistoric pottery from Malta. *Amino Acids* **2021**, *53*, 295–312. [CrossRef] [PubMed]

13. Evershed, R.P. Organic residue in archaeology: The archaeological biomarker revolution. *Archaeometry* **2008**, *50*, 895–924. [CrossRef]
14. Colonese, A.C.; Lucquin, A.; Guedes, E.P.; Thomas, R.; Best, J.; Fothergill, B.T.; Sykes, N.; Foster, A.; Miller, H.; Poole, K.; et al. The identification of poultry processing in archaeological ceramic vessels using in-situ isotope references for organic residue analysis. *J. Archaeol. Sci.* **2017**, *78*, 179–192. [CrossRef]
15. Evershed, R.P. Biomolecular archaeology and lipids. *World Archaeol.* **1993**, *25*, 74–93. [CrossRef] [PubMed]
16. Mayyas, A.S. Organic residues in ancient pottery sherds from sites in Jordan. *Mediterr. Archaeol. Archaeom.* **2018**, *18*, 61–75.
17. Mayyas, A.S.; Khrisat, B.R.; Hoffmann, T.; El Khalili, M.M. Fuel For Lamps: Organic Residues Preserved in Iron Age Lamps Excavated at the Site of Sahab in Jordan. *Archaeometry* **2017**, *59*, 934–948. [CrossRef]
18. Whelton, H.L.; Hammann, S.; Cramp, L.J.E.; Dunne, J.; Roffet-Salque, M.; Evershed, R.P. A call for caution in the analysis of lipids and other small biomolecules from archaeological contexts. *J. Archaeol. Sci.* **2021**, *132*, 105397. [CrossRef]
19. Copley, M.S.; Bland, H.A.; Rose, P.; Horton, M.; Evershed, R.P. Gas chromatographic, mass spectrometric and stable carbon isotopic investigations of organic residues of plant oils and animal fats employed as illuminants in archaeological lamps from Egypt. *Analyst* **2005**, *130*, 860–871. [CrossRef]
20. Steele, V.J. Organic residues in archaeology—The highs and lows of recent research. In *ACS Symposium Series*; Armitage, R.A., Burton, J.H., Eds.; ACS: New Orleans, LA, USA, 2013; Volume 1147, pp. 89–108.
21. Yang, Y.; Shevchenko, A.; Knaust, A.; Abuduresule, I.; Li, W.; Hu, X.; Wang, C.; Shevchenko, A. Proteomics evidence for kefir dairy in Early Bronze Age China. *J. Archaeol. Sci.* **2014**, *45*, 178–186. [CrossRef]
22. Evershed, R.P.; Copley, M.S.; Dickson, L.; Hansel, F.A. Experimental evidence for the processing of marine animal products and other commodities containing polyunsaturated fatty acids in pottery vessels. *Archaeometry* **2008**, *50*, 101–113. [CrossRef]
23. Evershed, R.P.; Vaughan, S.J.; Dudd, S.N.; Soles, J.S. Fuel for thought? Beeswax in lamps and conical cups from Late Minoan Crete. *Antiquity* **2015**, *71*, 979–985. [CrossRef]
24. Blanco-Zubiaguirre, L.; Ribechini, E.; Degano, I.; La Nasa, J.; Carrero, J.A.; Iñañez, J.; Olivares, M.; Castro, K. GC–MS and HPLC-ESI-QToF characterization of organic lipid residues from ceramic vessels used by Basque whalers from 16th to 17th centuries. *Microchem. J.* **2018**, *137*, 190–203. [CrossRef]
25. Evershed, R.P. Experimental approaches to the interpretation of absorbed organic residues in archaeological ceramics. *World Archaeol.* **2008**, *40*, 26–47. [CrossRef]
26. Jones, M.K.; Briggs, D.E.G.; Eglington, G.; Hagelberg, E.; Evershed, R.P.; Dudd, S.N.; Charters, S.; Mottram, H.; Stott, A.W.; Raven, A.; et al. Lipids as carriers of anthropogenic signals from prehistory. *Philos. Trans. R. Soc. Lond. B Biol. Sci.* **1999**, *354*, 19–31.
27. Krueger, M.; Wicenciak, U.; Kowarska, Z.; Niedzielski, P.; Kozak, L.; Jakubowski, K.; Proch, J.; Mleczek, M.; Waśkiewicz, A. First results of organic residue analysis on ceramic vessels (Jiyeh and Chhîm, Lebanon) by high perfomance liquid chromatography with tandem mass spectrometry. *Mediterr. Archaeol. Archaeom.* **2018**, *18*, 209–220.
28. Oudemans, T.F.M.; Boon, J.J. Molecular archaeology: Analysis of charred (food) remains from prehistoric pottery by pyrolysis—Gas chromatography/mass spectrometry. *J. Anal. Appl. Pyrolysis* **1991**, *20*, 197–227. [CrossRef]
29. Colombini, M.P.; Modugno, F.; Ribechini, E. Organic mass spectrometry in archaeology: Evidence for Brassicaceae seed oil in Egyptian ceramic lamps. *J Mass. Spectrom.* **2005**, *40*, 890–898. [CrossRef]
30. Dunne, J.; Evershed, R.P.; Heron, C.; Brettell, R.; Barclay, A.; Smyth, J.; Cramp, L. *Organic Residue Analysis and Archaeology Guidance for Good Practice*; Historic England Publishing: Liverpool, UK, 2018.
31. Hammann, S.; Scurr, D.J.; Alexander, M.R.; Cramp, L.J.E. Mechanisms of lipid preservation in archaeological clay ceramics revealed by mass spectrometry imaging. *Proc. Natl. Acad. Sci. USA* **2020**, *117*, 14688–14693. [CrossRef]
32. Dudd, S.N.; Evershed, R.P. Direct demonstration of milk as an element of archaeological economies. *Science* **1998**, *282*, 1478–1481. [CrossRef]
33. Hammann, S.; Cramp, L.J.E. Towards the detection of dietary cereal processing through absorbed lipid biomarkers in archaeological pottery. *J. Archaeol. Sci.* **2018**, *93*, 74–81. [CrossRef]
34. Regert, M.; Bland, H.A.; Dudd, S.N.; Bergen, P.F.V.; Evershed, R.P. Free and bound fatty acid oxidation products in archaeological ceramic vessels. *Proc. Royal Soc. B* **1998**, *265*, 2027–2032. [CrossRef]
35. Evershed, R.P.; Dudd, S.N.; Copley, M.S.; Berstan, R.; Stott, A.W.; Mottram, H.; Buckley, S.A.; Crossman, Z. Chemistry of archaeological animal fats. *Acc. Chem. Res.* **2002**, *35*, 660–668. [CrossRef]
36. Mahesar, S.A.; Sherazi, S.T.H.; Khaskheli, A.R.; Kandhro, A.A.; Uddin, S. Analytical approaches for the assessment of free fatty acids in oils and fats. *Anal. Methods* **2014**, *6*, 4956–4963. [CrossRef]
37. Casanova, E.; Knowles, T.; Bayliss, A.; Walton-Doyle, C.; Barclay, A.; Evershed, R. Compound-specific radiocarbon dating of lipid residues in pottery vessels: A new approach for detecting the exploitation of marine resources. *J. Archaeol. Sci.* **2022**, *137*, 105528. [CrossRef]
38. Han, B.; Sun, Z.; Chong, J.; Lyu, N.; Rao, H.; Yang, Y. Lipid residue analysis of ceramic vessels from the Liujiawa site of the Rui State (early Iron Age, north China). *J. Quat. Sci.* **2022**, *37*, 114–122. [CrossRef]
39. Skibo, J.M.; Malainey, M. *Residue Analysis*; Springer Science+Business Media: New York, NY, USA, 2013.
40. Gunstone, F.D.; Harwood, J.L.; Dijkstra, A.J. *The Lipid Handbook with CD-ROM*, 3rd ed.; CRC Press: Boca Raton, FL, USA, 2007.

41. Kałużna-Czaplińska, J.; Rosiak, A.; Kwapińska, M.; Kwapiński, W. Different Analytical Procedures for the Study of Organic Residues in Archeological Ceramic Samples with the Use of Gas Chromatography-mass Spectrometry. *Crit. Rev. Anal. Chem.* **2016**, *46*, 67–81. [CrossRef]
42. Rosiak, A.; Kałużna-Czaplińska, J.; Gątarek, P. Analytical Interpretation of Organic Residues from Ceramics As a Source of Knowledge About Our Ancestors. *Crit. Rev. Anal. Chem.* **2020**, *50*, 189–195. [CrossRef]
43. Keeney, M.; Katz, I.; Allison, M.J. On the probable origin of some milk fat acids in rumen microbial lipids. *J. Am. Oil Chem. Soc.* **1962**, *39*, 198–201. [CrossRef]
44. Blanco-Zubiaguirre, L.; Olivares, M.; Castro, K.; Carrero, J.A.; García-Benito, C.; García-Serrano, J.; Pérez-Pérez, J.; Pérez-Arantegui, J. Wine markers in archeological potteries: Detection by GC-MS at ultratrace levels. *Anal Bioanal Chem* **2019**, *411*, 6711–6722. [CrossRef]
45. Pecci, A.; Borgna, E.; Mileto, S.; Dalla Longa, E.; Bosi, G.; Florenzano, A.; Mercuri, A.M.; Corazza, S.; Marchesini, M.; Vidale, M. Wine consumption in Bronze Age Italy: Combining organic residue analysis, botanical data and ceramic variability. *J. Archaeol. Sci.* **2020**, *123*, 105256. [CrossRef]
46. Pecci, A.; Reynolds, P.; Mileto, S.; Vargas Girón, J.M.; Bernal-Casasola, D. Production and Transport of Goods in the Roman Period: Residue Analysis and Wine Derivatives in Late Republican Baetican Ovoid Amphorae. *Environ. Archaeol.* **2021**, 1–13. [CrossRef]
47. Amir, A.; Finkelstein, I.; Shalev, Y.; Uziel, J.; Chalaf, O.; Freud, L.; Neumann, R.; Gadot, Y. Residue analysis evidence for wine enriched with vanilla consumed in Jerusalem on the eve of the Babylonian destruction in 586 BCE. *PLoS ONE* **2022**, *17*, e0266085. [CrossRef] [PubMed]
48. Briggs, L.; Demesticha, S.; Katzev, S.; Wylde Swiny, H.; Craig, O.E.; Drieu, L. There's more to a vessel than meets the eye: Organic residue analysis of 'wine' containers from shipwrecks and settlements of ancient Cyprus (4th–1st century bce). *Archaeometry* **2022**, *64*, 779–797. [CrossRef]
49. Zhang, T.; Xu, S.; Li, Y.; Wen, R.; Yang, G. Orthogonal optimization of extraction and analysis for red wine residues in simulated and archaeological materials using LC/MS and HPLC methods. *Microchem. J.* **2018**, *142*, 175–180. [CrossRef]
50. Francés-Negro, M.; Iriarte, E.; Galindo-Pellicena, M.A.; Gerbault, P.; Carrancho, A.; Pérez-Romero, A.; Arsuaga, J.L.; Carretero, J.M.; Roffet-Salque, M. Neolithic to Bronze Age economy and animal management revealed using analyses lipid residues of pottery vessels and faunal remains at El Portalón de Cueva Mayor (Sierra de Atapuerca, Spain). *J. Archaeol. Sci.* **2021**, *131*, 105380. [CrossRef]
51. Correa-Ascencio, M.; Evershed, R.P. High throughput screening of organic residues in archaeological potsherds using direct acidified methanol extraction. *Anal. Methods* **2014**, *6*, 1330–1340. [CrossRef]
52. Evershed, R.P.; Mottram, H.R.; Dudd, S.N.; Charters, S.; Stott, A.W.; Lawrence, G.J.; Gibson, A.M.; Conner, A.; Blinkhorn, P.W.; Reeves, V. New Criteria for the Identification of Animal Fats Preserved in Archaeological Pottery. *Naturwissenschaften* **1997**, *84*, 402–406. [CrossRef]
53. Gregg, M.W.; Slater, G.F. A new method for extraction, isolation and transesterification of free fatty acids from archaeological pottery. *Archaeometry* **2010**, *52*, 833–854. [CrossRef]
54. Kimpe, K.; Jacobs, P.A.; Waelkens, M. Mass spectrometric methods prove the use of beeswax and ruminant fat in late Roman cooking pots. *J. Chromatogr. A* **2002**, *968*, 151–160. [CrossRef]
55. Marchbanks, M.L. *Lipid Analysis in Archaeology: An Initial Study of Ceramics and Subsistence at the George C. Davis Site*; University of Texas: Austin, TX, USA, 1989.
56. McClure, S.B.; Magill, C.; Podrug, E.; Moore, A.M.T.; Harper, T.K.; Culleton, B.J.; Kennett, D.J.; Freeman, K.H. Fatty acid specific δ13C values reveal earliest Mediterranean cheese production 7200 years ago. *PLoS ONE* **2018**, *13*, e0202807. [CrossRef] [PubMed]
57. Olsson, M.; Isaksson, S. Molecular and isotopic traces of cooking and consumption of fish at an Early Medieval manor site in eastern middle Sweden. *J. Archaeol. Sci.* **2008**, *35*, 773–780. [CrossRef]
58. Spangenberg, J.E.; Jacomet, S.; Schibler, J. Chemical analyses of organic residues in archaeological pottery from Arbon Bleiche 3, Switzerland—Evidence for dairying in the late Neolithic. *J. Archaeol. Sci.* **2006**, *33*, 1–13. [CrossRef]
59. Keute, J.; Isaksson, S.; Deviese, T.; Hein, A. Insights into ceramic use in prehistoric Northwest China obtained from residue analysis: A pilot study on the Andersson Collection at the Museum of Far Eastern Antiquities, Stockholm. *Bull. Mus. Far East. Antiq.* **2021**, *82*, 1–24.
60. Copley, M.S.; Berstan, R.; Harden, S.; Docherty, G.; Mukherjee, A.J.; Straker, V.; Payne, S.; Evershed, R. Direct Chemical Evidence for Widespread Dairying in Prehistoric Britain. *Proc. Natl. Acad. Sci. USA* **2003**, *100*, 1524–1529. [CrossRef] [PubMed]
61. Shoda, S.; Lucquin, A.; Yanshina, O.; Kuzmin, Y.; Shevkomud, I.; Medvedev, V.; Derevianko, E.; Lapshina, Z.; Craig, O.E.; Jordan, P. Late Glacial hunter-gatherer pottery in the Russian Far East: Indications of diversity in origins and use. *Quat. Sci. Rev.* **2020**, *229*, 106124. [CrossRef]
62. Fernandes, R.; Millard, A.R.; Brabec, M.; Nadeau, M.-J.; Grootes, P. Food Reconstruction Using Isotopic Transferred Signals (FRUITS): A Bayesian Model for Diet Reconstruction. *PLoS ONE* **2014**, *9*, e87436. [CrossRef]
63. Lucquin, A.; Colonese, A.C.; Farrell, T.F.G.; Craig, O.E. Utilising phytanic acid diastereomers for the characterisation of archaeological lipid residues in pottery samples. *Tetrahedron Lett.* **2016**, *57*, 703–707. [CrossRef]
64. Dunne, J.; Rebay-Salisbury, K.; Salisbury, R.B.; Frisch, A.; Walton-Doyle, C.; Evershed, R.P. Milk of ruminants in ceramic baby bottles from prehistoric child graves. *Nature* **2019**, *574*, 246–248. [CrossRef]

65. Dunne, J.; di Lernia, S.; Chłodnicki, M.; Kherbouche, F.; Evershed, R.P. Timing and pace of dairying inception and animal husbandry practices across Holocene North Africa. *Quat. Int.* **2018**, *471*, 147–159. [CrossRef]
66. Notarstefano, F. *Ceramica E Alimentazione: L'analisi Chimica Dei Residui Organici Nelle Ceramiche Applicata Ai Contesti Archeologici*; Edipuglia: Bari, Italy, 2012.
67. Sikorski, Z.E.; Staroszczyk, H. *Chemia Żywności*, 1st ed.; PWN: Warsaw, Poland, 2017; Volume 2.
68. Eerkens, J.W. GC–MS analysis and fatty acid ratios of archaeological potsherds from the western great basin of north america. *Archaeometry* **2005**, *47*, 83–102. [CrossRef]
69. Malainey, M.E. The Reconstruction and Testing of Subsistence and Settlement Strategies for the Plains, Parkland, and Southern Boreal Forest. Ph.D. Thesis, University of Manitoba, Winnipeg, MB, Canada, 1997.
70. Isaksson, S. Food and Rank in Early Medieval Time. Ph.D. Thesis, Comprehensive Summary, Archaeological Research Laboratory, Stockholm University, Stockholm, Sweden, 2000.
71. Isaksson, S.; Karlsson, C.; Eriksson, T. Ergosterol (5, 7, 22-ergostatrien-3β-ol) as a potential biomarker for alcohol fermentation in lipid residues from prehistoric pottery. *J. Archaeol. Sci.* **2010**, *37*, 3263–3268. [CrossRef]
72. Dudd, S.N.; Evershed, R.P.; Gibson, A.M. Evidence for Varying Patterns of Exploitation of Animal Products in Different Prehistoric Pottery Traditions Based on Lipids Preserved in Surface and Absorbed Residues. *J. Archaeol. Sci.* **1999**, *26*, 1473–1482. [CrossRef]
73. Hjulström, B.; Isaksson, S.; Karlsson, C. Prominent migration period building. *Acta Archaeol.* **2008**, *79*, 62–78. [CrossRef]
74. Dunne, J. *Organic Residue Analysis and Archaeology: Supporting Information*; Historic England Publishing: London, UK, 2017.
75. Rigano, F.; Oteri, M.; Micalizzi, G.; Mangraviti, D.; Dugo, P.; Mondello, L. Lipid profile of fish species by liquid chromatography coupled to mass spectrometry and a novel linear retention index database. *J. Sep. Sci.* **2020**, *43*, 1773–1780. [CrossRef]
76. Aillaud, S. Field and Laboratory Studies of Diagenetic Reactions Affecting Lipid Residues Absorbed in Unglazed Archaeological Pottery Vessels. Ph.D. Thesis, University of Bristol, Bristol, UK, 2002.
77. Rottlander, R.C.; Schlichtherle, H. Chemical analysis of fat residues in prehistoricvessels. *Naturwissenschaften* **1983**, *70*, 33–38.
78. Hansel, F.A.; Evershed, R.P. Formation of dihydroxy acids from Z-monounsaturated alkenoic acids and their use as biomarkers for the processing of marine commodities in archaeological pottery vessels. *Tetrahedron Lett.* **2009**, *50*, 5562–5564. [CrossRef]
79. Copley, M.S.; Hansel, F.A.; Sadr, K.; Evershed, R.P. Organic residue evidence for the processing of marine animal products in pottery vessels from the pre-colonial archaeological site of Kasteelberg D east, South Africa: Research article. *S. Afr. J. Sci* **2004**, *100*, 279–283.
80. Hansel, F.A.; Copley, M.S.; Madureira, L.A.S.; Evershed, R.P. Thermally produced ω-(o-alkylphenyl)alkanoic acids provide evidence for the processing of marine products in archaeological pottery vessels. *Tetrahedron Lett.* **2004**, *45*, 2999–3002. [CrossRef]
81. Budja, M. Neolithic pottery and the biomolecular archaeology of lipids. *Doc. Praehist.* **2014**, *41*, 196–224. [CrossRef]
82. Pollard, A.; Batt, C.M.; Stern, B.; Young, S.M.M. *Analytical Chemistry in Archaeology*; Cambridge University Press: Cambridge, UK, 2007; pp. 1–404.
83. Dudd, S.N. Molecular and Isotopic Characterisation of Animal Fats in Archaeological Pottery. Ph.D. Thesis, University of Bristol, Bristol, UK, 1999.
84. Hammann, S.; Cramp, L.J.E.; Whittle, M.; Evershed, R.P. Cholesterol degradation in archaeological pottery mediated by fired clay and fatty acid pro-oxidants. *Tetrahedron Lett.* **2018**, *59*, 4401–4404. [CrossRef]
85. Papakosta, V.; Oras, E.; Isaksson, S. Early pottery use across the Baltic—A comparative lipid residue study on Ertebølle and Narva ceramics from coastal hunter-gatherer sites in southern Scandinavia, northern Germany and Estonia. *J. Archaeol. Sci. Rep.* **2019**, *24*, 142–151. [CrossRef]
86. DeMan, J.M. *Principles of Food Chemistry*; Aspen Publishers, Inc.: Gaithersburg, MD, USA, 1999.
87. Zak, I.; Balcerzyk, A. *Chemia Medyczna*; Śląska Akademia Medyczna: Katowice, Poland, 2001.
88. Mukherjee, A.J.; Gibson, A.M.; Evershed, R.P. Trends in pig product processing at British Neolithic Grooved Ware sites traced through organic residues in potsherds. *J. Archaeol. Sci.* **2008**, *35*, 2059–2073. [CrossRef]
89. Smyth, J.; Evershed, R.P. The molecules of meals: New insight into Neolithic foodways. *Proc. R. Ir. Acad. C* **2015**, *115C*, 27–46. [CrossRef]
90. Eglinton, G.; Logan, G.A.; Ambler, R.P.; Boon, J.J.; Perizonius, W.R.K.; Eglinton, G.; Curry, G.B. Molecular preservation. *Philos. Trans. R. Soc. B Biol. Sci.* **1991**, *333*, 315–328.
91. Colombini, M.P.; Giachi, G.; Modugno, F.; Ribechini, E. Characterisation of organic residues in pottery vessels of the Roman age from Antinoe (Egypt). *Microchem. J.* **2005**, *79*, 83–90. [CrossRef]
92. Ribechini, E.; Modugno, F.; Colombini, M.P.; Evershed, R.P. Gas chromatographic and mass spectrometric investigations of organic residues from Roman glass unguentaria. *J. Chromatogr. A* **2008**, *1183*, 158–169. [CrossRef]
93. Modugno, F.; Ribechini, E.; Colombini, M.P. Aromatic resin characterisation by gas chromatography-mass spectrometry. Raw and archaeological materials. *J. Chromatogr. A* **2006**, *1134*, 298–304. [CrossRef]
94. Modugno, F.; Ribechini, E.; Colombini, M.P. Chemical study of triterpenoid resinous materials in archaeological findings by means of direct exposure electron ionisation mass spectrometry and gas chromatography/mass spectrometry. *Rapid Commun. Mass Spectrom.* **2006**, *20*, 1787–1800. [CrossRef]
95. Heron, C.; Shoda, S.; Breu Barcons, A.; Czebreszuk, J.; Eley, Y.; Gorton, M.; Kirleis, W.; Kneisel, J.; Lucquin, A.; Müller, J.; et al. First molecular and isotopic evidence of millet processing in prehistoric pottery vessels. *Sci. Rep.* **2016**, *6*, 38767. [CrossRef]

96. Bondetti, M.; Scott, E.; Courel, B.; Lucquin, A.; Shoda, S.; Lundy, J.; Labra-Odde, C.; Drieu, L.; Craig, O.E. Investigating the formation and diagnostic value of ω-(o-alkylphenyl)alkanoic acids in ancient pottery. *Archaeometry* **2020**, *63*, 594–608. [CrossRef] [PubMed]
97. Cramp, L.J.E.; Evershed, R.P. *Reconstructing Aquatic Resource Exploitation in Human Prehistory Using Lipid Biomarkers and Stable Isotopes*; Elsevier: Oxford, UK, 2014.
98. Dudd, S.N.; Regert, M.; Evershed, R.P. Assessing microbial lipid contributions during laboratory degradations of fats and oils and pure triacylglycerols absorbed in ceramic potsherds. *Org. Geochem.* **1998**, *29*, 1345–1354. [CrossRef]
99. Christie, W.W. The composition, structure and function of lipids in the tissues of ruminant animals. *Prog. Lipid Res.* **1978**, *17*, 111–205. [CrossRef]
100. Heron, C.; Evershed, R.P.; Goad, L.J. Effects of migration of soil lipids on organic residues associated with buried potsherds. *J. Archaeol. Sci.* **1991**, *18*, 641–659. [CrossRef]
101. Hua, P.-Y.; Manikandan, M.; Abdelhamid, H.N.; Wu, H.-F. Graphene nanoflakes as an efficient ionizing matrix for MALDI-MS based lipidomics of cancer cells and cancer stem cells. *J. Mater. Chem. B* **2014**, *2*, 7334–7343. [CrossRef]
102. Harper, C.S.; Macdonald, F.V.; Braun, K.L. Lipid Residue Analysis of Archaeological Pottery: An Introductory Laboratory Experiment in Archaeological Chemistry. *J. Chem. Educ.* **2017**, *94*, 1309–1313. [CrossRef]
103. Evershed, R.P.; Heron, C.; Goad, L.J. Analysis of organic residues of archaeological origin by high-temperature gas chromatography and gas chromatography-mass spectrometry. *Analyst* **1990**, *115*, 1339–1342. [CrossRef]
104. Tanasi, D.; Greco, E.; Pisciotta, F.; Hassam, S. Chemical characterization of organic residues on Late Roman amphorae from shipwrecks off the coast of Marsala (Trapani, Italy). *J. Archaeol. Sci. Rep.* **2021**, *40*, 103241. [CrossRef]
105. Christie, W.W.; Han, X. *Lipid Analysis. Isolation, Separation, Identification and Lipidomic Analysis*, 4th ed.; Oily Press: Middlesex, UK, 2010.
106. Eskilsson, C.S.; Björklund, E. Analytical-scale microwave-assisted extraction. *J. Chromatogr. A* **2000**, *902*, 227–250. [CrossRef]
107. Andreotti, A.; Bonaduce, I.; Colombini, M.P.; Gautier, G.; Modugno, F.; Ribechini, E. Combined GC/MS Analytical Procedure for the Characterization of Glycerolipid, Waxy, Resinous, and Proteinaceous Materials in a Unique Paint Microsample. *Anal. Chem.* **2006**, *78*, 4490–4500. [CrossRef]
108. Devièse, T.; Van Ham-Meert, A.; Hare, V.J.; Lundy, J.; Hommel, P.; Ivanovich Bazaliiskii, V.; Orton, J. Supercritical Fluids for Higher Extraction Yields of Lipids from Archeological Ceramics. *Anal. Chem.* **2018**, *90*, 2420–2424. [CrossRef] [PubMed]
109. Micalizzi, G.; Ragosta, E.; Farnetti, S.; Dugo, P.; Tranchida, P.Q.; Mondello, L.; Rigano, F. Rapid and miniaturized qualitative and quantitative gas chromatography profiling of human blood total fatty acids. *Anal. Bioanal. Chem.* **2020**, *412*, 2327–2337. [CrossRef] [PubMed]
110. Schummer, C.; Delhomme, O.; Appenzeller, B.M.; Wennig, R.; Millet, M. Comparison of MTBSTFA and BSTFA in derivatization reactions of polar compounds prior to GC/MS analysis. *Talanta* **2009**, *77*, 1473–1482. [CrossRef] [PubMed]
111. Ichihara, K.I.; Fukubayashi, Y. Preparation of fatty acid methyl esters for gas-liquid chromatography. *J. Lipid Res.* **2010**, *51*, 635–640. [CrossRef]
112. Romanus, K.; Poblome, J.; Verbeke, K.; Luypaerts, A.; Jacobs, P.; De Vos, D.; Waelkens, M. An evaluation of analytical and interpretative methodologies for the extraction and identification of lipids associated with pottery sherds from the site of Sagalassos, Turkey. *Archaeometry* **2007**, *49*, 729–747. [CrossRef]
113. Craig, O.E.; Steele, V.J.; Fischer, A.; Hartz, S.; Andersen, S.H.; Donohoe, P.; Glykou, A.; Saul, H.; Jones, D.M.; Koch, E.; et al. Ancient lipids reveal continuity in culinary practices across the transition to agriculture in Northern Europe. *Proc. Natl. Acad. Sci. USA* **2011**, *108*, 17910–17915. [CrossRef]
114. Demirci, Ö.; Lucquin, A.; Craig, O.E.; Raemaekers, D.C.M. First lipid residue analysis of Early Neolithic pottery from Swifterbant (The Netherlands, ca. 4300–4000 BC). *Archaeol. Anthropol. Sci.* **2020**, *12*, 105. [CrossRef]
115. Reber, E.A. Comparison of Neutral Compound Extraction from Archaeological Residues in Pottery Using Two Methodologies: A Preliminary Study. *Separations* **2021**, *8*, 6. [CrossRef]
116. Papakosta, V.; Smittenberg, R.H.; Gibbs, K.; Jordan, P.; Isaksson, S. Extraction and derivatization of absorbed lipid residues from very small and very old samples of ceramic potsherds for molecular analysis by gas chromatography–mass spectrometry (GC–MS) and single compound stable carbon isotope analysis by gas chromatography–combustion–isotope ratio mass spectrometry (GC–C–IRMS). *Microchem. J.* **2015**, *123*, 196–200.
117. Oliveira, C.; Araújo, A.; Ribeiro, A.; Delerue-Matos, C. Chromatographic analysis of honey ceramic artefacts. *Archaeol. Anthropol. Sci.* **2019**, *11*, 959–971. [CrossRef]
118. Manhita, A.; Martins, S.; Gomes da Silva, M.; Lopes, M.d.C.; Barrocas Dias, C. Transporting Olive Oil in Roman Times: Chromatographic Analysis of Dressel 20 Amphorae from Pax Julia Civitas, Lusitania. *Chromatographia* **2020**, *83*, 1055–1064. [CrossRef]
119. Lucejko, J.J.; La Nasa, J.; Porta, F.; Vanzetti, A.; Tanda, G.; Mangiaracina, C.F.; Corretti, A.; Colombini, M.P.; Ribechini, E. Long-lasting ergot lipids as new biomarkers for assessing the presence of cereals and cereal products in archaeological vessels. *Sci. Rep.* **2018**, *8*, 3935. [CrossRef] [PubMed]
120. Beeston, R.F.; Palatinus, J.; Beck, C.; Stout, E.C.; Appendix, M. Organic Residue Analysis of Pottery Sherds from Chrysokamino. In *The Chrysokamino Workshop and Its Territory (Hesperia Suppl. 36)*; American School: Princeton, NJ, USA, 2006; pp. 413–428.

121. Pecci, A.; Degl'Innocenti, E.; Giorgi, G.; Cau Ontiveros, M.Á.; Cantini, F.; Solanes Potrony, E.; Alós, C.; Miriello, D. Organic residue analysis of experimental, medieval, and post-medieval glazed ceramics. *Archaeol. Anthropol. Sci.* **2016**, *8*, 879–890. [CrossRef]
122. Colonese, A.C.; Hendy, J.; Lucquin, A.; Speller, C.F.; Collins, M.J.; Carrer, F.; Gubler, R.; Kühn, M.; Fischer, R.; Craig, O.E. New criteria for the molecular identification of cereal grains associated with archaeological artefacts. *Sci. Rep.* **2017**, *7*, 6633. [CrossRef]

MDPI
St. Alban-Anlage 66
4052 Basel
Switzerland
www.mdpi.com

Molecules Editorial Office
E-mail: molecules@mdpi.com
www.mdpi.com/journal/molecules

Disclaimer/Publisher's Note: The statements, opinions and data contained in all publications are solely those of the individual author(s) and contributor(s) and not of MDPI and/or the editor(s). MDPI and/or the editor(s) disclaim responsibility for any injury to people or property resulting from any ideas, methods, instructions or products referred to in the content.

www.ingramcontent.com/pod-product-compliance
Lightning Source LLC
LaVergne TN
LVHW070242100526
838202LV00015B/2167